Adaptation in Wireless Communications

Edited by
Mohamed Ibnkahla

ADAPTIVE SIGNAL PROCESSING
in WIRELESS COMMUNICATIONS

ADAPTATION and CROSS LAYER DESIGN
in WIRELESS NETWORKS

THE ELECTRICAL ENGINEERING
AND APPLIED SIGNAL PROCESSING SERIES
Edited by Alexander Poularikas

The Advanced Signal Processing Handbook: Theory and Implementation for Radar, Sonar, and Medical Imaging Real-Time Systems
Stergios Stergiopoulos

The Transform and Data Compression Handbook
K.R. Rao and P.C. Yip

Handbook of Multisensor Data Fusion
David Hall and James Llinas

Handbook of Neural Network Signal Processing
Yu Hen Hu and Jenq-Neng Hwang

Handbook of Antennas in Wireless Communications
Lal Chand Godara

Noise Reduction in Speech Applications
Gillian M. Davis

Signal Processing Noise
Vyacheslav P. Tuzlukov

Digital Signal Processing with Examples in MATLAB®
Samuel Stearns

Applications in Time-Frequency Signal Processing
Antonia Papandreou-Suppappola

The Digital Color Imaging Handbook
Gaurav Sharma

Pattern Recognition in Speech and Language Processing
Wu Chou and Biing-Hwang Juang

Propagation Handbook for Wireless Communication System Design
Robert K. Crane

Nonlinear Signal and Image Processing: Theory, Methods, and Applications
Kenneth E. Barner and Gonzalo R. Arce

Smart Antennas
Lal Chand Godara

Mobile Internet: Enabling Technologies and Services
Apostolis K. Salkintzis and Alexander Poularikas

Soft Computing with MATLAB®
Ali Zilouchian

Wireless Internet: Technologies and Applications
Apostolis K. Salkintzis and Alexander Poularikas

Signal and Image Processing in Navigational Systems
Vyacheslav P. Tuzlukov

Medical Image Analysis Methods
Lena Costaridou

MIMO System Technology for Wireless Communications
George Tsoulos

Signals and Systems Primer with MATLAB®
Alexander Poularikas

Adaptation in Wireless Communications - 2 volume set
Mohamed Ibnkahla

ADAPTIVE SIGNAL PROCESSING in WIRELESS COMMUNICATIONS

Edited by

Mohamed Ibnkahla

CRC Press
Taylor & Francis Group
Boca Raton London New York

CRC Press is an imprint of the
Taylor & Francis Group, an **informa** business

CRC Press
Taylor & Francis Group
6000 Broken Sound Parkway NW, Suite 300
Boca Raton, FL 33487-2742

© 2009 by Taylor & Francis Group, LLC
CRC Press is an imprint of Taylor & Francis Group, an Informa business

No claim to original U.S. Government works
Printed in the United States of America on acid-free paper
10 9 8 7 6 5 4 3 2 1

International Standard Book Number-13: 978-1-4200-4601-4 (Hardcover)

Library of Congress Cataloging-in-Publication Data

Adaptive signal processing in wireless communications / editor, Mohamed Ibnkahla.
    p. cm. -- (Electrical engineering and applied signal processing series)
    Includes bibliographical references and index.
    ISBN 978-1-4200-4601-4 (alk. paper)
    1. Adaptive signal processing. 2. Wireless communication systems. I. Ibnkahla, Mohamed. II. Title. III. Series.

TK5102.5.A296145 2008
621.382'2--dc22   2008025443

Visit the Taylor & Francis Web site at
http://www.taylorandfrancis.com

and the CRC Press Web site at
http://www.crcpress.com

Contents

Preface

Adaptive techniques play a key role in modern wireless communication systems. The concept of adaptation is emphasized in the *Adaptation in Wireless Communications Series* across all layers of the wireless protocol stack, ranging from the physical layer to the application layer.

This book is devoted to adaptation in the physical layer. It gives a tutorial survey of adaptive signal processing techniques used in wireless and mobile communication systems. The topics include adaptive channel modeling and identification, adaptive receiver design and equalization, adaptive modulation and coding, adaptive multiple-input-multiple-output (MIMO) systems, adaptive and opportunistic beam forming, and cooperative diversity. Moreover, the book addresses other important aspects of adaptation in wireless communications, such as software defined radio, reconfigurable devices, and cognitive radio. The book is supported by various new analytical, experimental, and simulation results and is illustrated by more than 160 figures, 20 tables, and 800 references.

I would like to thank all the contributing authors for their patience and excellent work. The process of editing started in June 2005. Each chapter has been blindly reviewed by at least two reviewers (more than 50% of the chapters received three reviews or more). I would like to thank the reviewers for their time and valuable contribution to the quality of the book.

Finally, a special thank you goes to my parents, my wife, my son, my daughter, and all my family. They all have been of great support for this project.

Mohamed Ibnkahla
Queen's University
Kingston, Ontario, Canada

Editor

Dr. Mohamed Ibnkahla earned an engineering degree in electronics in 1992, an M.Sc. degree in signal and image processing in 1992, a Ph.D. degree in signal processing in 1996, and the Habilitation à Diriger des Recherches degree in 1998, all from the National Polytechnic Institute of Toulouse (INPT), Toulouse, France.

Dr. Ibnkahla is currently an associate professor in the Department of Electrical and Computer Engineering, Queen's University, Kingston, Canada. He previously held an assistant professor position at INPT (1996–1999) and Queen's University (2000–2004).

Since 1996, Dr. Ibnkahla has been involved in several research programs, including the European Advanced Communications Technologies and Services (ACTS), and the Canadian Institute for Telecommunications Research (CITR). His current research is supported by industry and government agencies such as the Ontario Centers of Excellence (OCE), the Natural Sciences and Engineering Research Council of Canada (NSERC), the Ontario Ministry of Natural Resources, and the Ontario Ministry of Research and Innovation.

He is currently leading multidisciplinary projects designing, implementing and deploying wireless sensor networks for various applications in Canada. Among these applications are natural resources management, ecosystem and forest monitoring, species at risk tracking and protection, and precision agriculture.

Dr. Ibnkahla has published a significant number of journal papers, book chapters, technical reports, and conference papers in the areas of signal processing and wireless communications. He has supervised more than 40 graduate students and postdoctoral fellows. He has given tutorials in the area of signal processing and wireless communications in several conferences, including IEEE Global Communications Conference (GLOBECOM, 2007) and IEEE International Conference in Acoustics, Speech and Signal Processing (ICASSP, 2008).

Dr. Ibnkahla received the INPT Leopold Escande Medal for the year 1997, France, for his research contributions in signal processing; the Prime Minister's Research Excellence Award (PREA), Ontario, Canada in 2000, for his contributions in wireless mobile communications; and the Favorite Professor Award, Queen's University in 2004 for his excellence in teaching.

Contributors

Sofiène Affes
INRS-EMT
University of Quebec
Montreal, Quebec, Canada

Al-Mukhtar Al-Hinai
Department of Electrical and Computer
    Engineering
Queen's University
Kingston, Ontario, Canada

Mohamed-Slim Alouini
Department of Electrical and Computer
    Engineering
Texas A&M University (TAMU)-Qatar
Education City, Doha, Qatar

Moeness G. Amin
Center for Advanced Communications
College of Engineering
Villanova University
Villanova, Pennsylvania

Hüseyin Arslan
University of South Florida
Tampa, Florida

Bo-Yu Chang
National Tsing Hua University
Hsinchu, Taiwan

Alex B. Gershman
Darmstadt University of Technology
Darmstadt, Germany

Dennis L. Goeckel
Department of Electrical and Computer
    Engineering
University of Massachusetts
Amherst, Massachusetts

Philip K. F. Hölzenspies
University of Twente
Enschede, The Netherlands

Y.-W. Peter Hong
National Tsing Hua University
Hsinchu, Taiwan

Anders Høst-Madsen
Department of Electrical Engineering
University of Hawaii
Honolulu, Hawaii

Mohamed Ibnkahla
Department of Electrical and Computer
    Engineering
Queen's University
Kingston, Ontario, Canada

M. A. Khalighi
Institut Fresnel, UMR CNRS 6133
École Centrale Marseille
Marseille, France

Il-Min Kim
Department of Electrical and Computer
    Engineering
Queen's University
Kingston, Ontario, Canada

Young-Chai Ko
School of Electrical Engineering
Korea University
Seoul, Korea

André B. J. Kokkeler
University of Twente
Enschede, The Netherlands

Chun-Kuang Lin
National Tsing Hua University
Hsinchu, Taiwan

Richard K. Martin
Department of Electrical and Computer
    Engineering
Air Force Institute of Technology
Wright-Patterson AFB, Ohio

Haewoon Nam
Motorola, Inc.
Austin, Texas

Kyoung-Lae Noh
Department of Electrical and Computer
    Engineering
Texas A&M University
College Station, Texas

N. Prayongpun
GIPSA-Lab, UMR CNRS 5216
Département Images et Signal
ENSIEG, Domaine Universitaire
Saint Martin d'Hères, France

Khalid Qaraqe
Department of Electrical and Computer
    Engineering
Texas A&M University
College Station, Texas

K. Raoof
GIPSA-Lab, UMR CNRS 5216
Département Images et Signal
ENSIEG, Domaine Universitaire
Saint Martin d'Hères, France

Gerard K. Rauwerda
University of Twente
Enschede, The Netherlands

Erchin Serpedin
Department of Electrical and Computer
    Engineering
Texas A&M University
College Station, Texas

Besma Smida
Harvard University
Cambridge, Massachusetts

Gerard J. M. Smit
University of Twente
Enschede, The Netherlands

Vladimir Stanković
Department of Electronic and Electrical
    Engineering
University of Strathclyde
Glasgow, United Kingdom

Jitendra K. Tugnait
Department of Electrical and Computer
    Engineering
Auburn University
Auburn, Alabama

Shu-Hsien Wang
National Tsing Hua University
Hsinchu, Taiwan

Pascal T. Wolkotte
University of Twente
Enschede, The Netherlands

Yik-Chung Wu
Department of Electrical and Electronic
    Engineering
The University of Hong Kong
Hong Kong

Zixiang Xiong
Department of Electrical and Computer
    Engineering
Texas A&M University
College Station, Texas

Hong-Chuan Yang
Department of Electrical and Computer
    Engineering
University of Victoria
Greater Victoria, British Columbia, Canada

Zhihang Yi
Department of Electrical and Computer
    Engineering
Queen's University
Kingston, Ontario, Canada

Qiwei Zhang
University of Twente
Enschede, The Netherlands

1

Adaptation Techniques and Enabling Parameter Estimation Algorithms for Wireless Communications Systems

Hüseyin Arslan
University of South Florida

1.1   Introduction

Wireless communications systems have evolved substantially over the last two decades. The explosive growth of the wireless communication market is expected to continue in the future, as the demand for all types of wireless services is increasing. There is no doubt that the second generation of cellular wireless communications systems was a success. However, these systems were designed to provide good coverage for voice services so that a minimum required signal quality can be ensured over the coverage area. If the received signal quality is well above the minimum required level, the receivers do not exploit this. The speech quality does not improve much, as the quality is mostly dominated by the speech coder. On the other hand, if the signal quality is below the minimum required level, a call drop will be observed. Therefore, such a design requires the use of strong forward error correction (FEC) schemes, low-order modulations, and many other redundancies at the transmission and reception. In essence, the mobile receivers and transmitters are designed for the worst-case channel and received signal conditions. As a result, many users experience unnecessarily high signal quality from which they cannot benefit. While reliable communication is achieved, the system resources are not used efficiently.

New generations of wireless mobile radio systems aim to provide higher data rates and a wide variety of applications (like video, data, etc.) to mobile users while serving as many users as possible. However, this goal must be achieved under spectrum and power constraints. Given the high price of spectrum and its scarcity, the systems must provide higher system capacity and performance through better use of the available resources. Therefore, *adaptation techniques* have been becoming popular for optimizing mobile radio system transmission and reception at the physical layer as well as at the higher layers of the protocol stack.

Traditional system designs focus on allocating fixed resources to the user. Adaptive design methodologies typically identify the user's requirements and then allocate just enough resources, thus enabling more efficient utilization of system resources and consequently increasing capacity. Adaptive channel allocation and adaptive cell assignment algorithms have been studied since the early days of cellular systems. As the demand in wireless access for speech and data has increased, link and system adaptation algorithms have become more important.

For a given average transmit power, adaptation allows the users to experience better signal qualities. Adaptation reduces the average interference observed from other users, as they do not transmit extra power unnecessarily. As a result, the received signal quality will be improved over a large portion of the coverage area. These higher-quality signal levels can be exploited to provide increased data rates through rate adaptation. For a desired received signal quality, this might also translate into less transmit power, leading to improved power efficiency for longer battery life. On the other hand, for a desired minimum signal quality, this might lead to an increased coverage area or better frequency reuse. In addition, adaptive receiver designs allow the receiver to work with reduced signal quality values; i.e., a desired bit-error-rate (BER) or frame-error-rate (FER) performance can be achieved with a lower signal quality. Adaptive receivers can also enable reduced average computational complexities for the same quality of service, which again implies less power consumption. As can be seen, adaptation algorithms lead to improved performance, increased capacity, lower power consumption, increased radio coverage area, and eventually better overall wireless communications system design.

Many adaptation schemes require a form of measurement (or estimation) of various quantities (parameters) that might change over time. These estimates are then used to trigger or perform a multitude of functions, like the adaptation of the transmission and reception. For example, Doppler spread and delay spread estimations, signal-to-noise ratio (SNR) estimation, channel estimation, BER estimation, cyclic redundancy check (CRC) information, and received signal strength measurement are some of the commonly used measurements for adaptive algorithms. As the interest in the adaptation schemes increases, so does the research on improved (fast and accurate) parameter estimation techniques.

In this chapter, an overview of commonly used adaptation techniques and their applications for wireless mobile radio systems is given. Some of the commonly used parameters and their estimation using baseband signal processing techniques are explained in detail. Also, the current and future research issues regarding the improved parameter estimation and extensive use of adaptation techniques are discussed throughout the chapter. Note that there has been a significant amount of research on adaptation of wireless communications systems. This chapter is not intended to cover all these developments, but rather, it is intended to provide the readers an overview and conceptual understanding of adaptation techniques and related parameter estimation algorithms. More emphasis is given on signal processing perspectives of the adaptation of wireless communications systems.

1.2   Overview of Adaptation Schemes

In wireless mobile communications systems, information is transmitted through a radio channel. Unlike other guided media, the radio channel is highly dynamic. The transmitted signal reaches the receiver by undergoing many effects, corrupting the signal, and often placing limitations on the performance of the system.

Figure 1.1 illustrates a wireless communications system that includes some of the effects of the radio channel. The received signal strength varies depending on the distance relative to the transmitter, shadowing caused by large obstructions, and fading due to reflection, diffraction, and scattering. Mobility of the transmitter, receiver, or scattering objects causes the channel to change over time. Moreover, the interference conditions in the system change rapidly. Most important of all, the radio channel is highly random and the statistical characteristics of the channel are environment dependent. In addition to these changes, the traffic load, type of services, and mobile user characteristics and requirements might also vary in time. Adaptive techniques can be used to address all of these changing conditions.

The adaptation strategy can be different depending on the application and services. Constant BER constraint for a given fixed transmission bandwidth and constant throughput constraint are two of the most popular criteria for adaptation. In constant BER, a desired average or instantaneous BER is defined to satisfy the acceptable quality

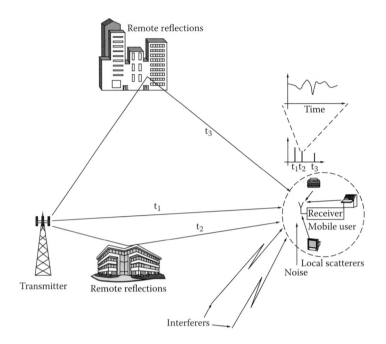

FIGURE 1.1 Illustration of some of the effects of a radio channel. Local scatterers cause fading; remote reflectors cause multipath and time dispersion, leading to ISI; mobility of the user or scatterers causes a time-varying channel; reuse of frequencies and adjacent carriers cause interference.

of service. Then the system is adapted to the varying channel and interference conditions so that the BER is maintained below the target value. In order to ensure this for all types of channel and interference conditions, the system changes power, modulation order, coding rate, spreading factor, etc. Note that this changes the throughput as the channel quality changes. On the other hand, for the constant throughput case, the adaptations are done to make sure that the effective throughput is constant, where the BER might change.

In general, it is possible to classify the adaptation algorithms as link and transmitter adaptation, adaptation of system resource allocation, and receiver adaptation. In the following sections, brief discussions of these adaptation techniques will be given.

1.2.1 Link and Transmitter Adaptation

A reliable link must ensure that the receiver is able to capture and reproduce the transmitted information bits. Therefore, the target link quality must be maintained all the time in spite of the changes in the channel and interference conditions. As mentioned earlier, one way to achieve this is to design the system for the worst-case scenario so that the target link quality can always be achieved.

If the transmitter sends more power for a specific user, the user benefits from it by having a better link quality, but the level of interference for the other users increases accordingly. On the other hand, if the user does not receive enough power, a reliable link cannot be established. In order to establish a reliable link while minimizing interference to other users, the transmitter should continuously control the transmitted power level. Power control is a simple form of adaptation that compensates for the variation of the received signal level due to path loss, shadowing, and sometimes fading. Numerous studies on power control schemes have been performed for various radio communications systems (see [1] and the references listed therein). In code division multiple-access (CDMA) systems, signals having widely different power levels at the receiver cause strong signals to swamp out weaker ones in a phenomenon known as the near–far effect. Power control mitigates the near–far problem by controlling the transmitted power.

It is possible to trade off power for bandwidth efficiency; i.e., a desired BER (or FER) can be achieved by increasing the power level or by reducing the bandwidth efficiency. One way of establishing a reliable link is to add redundancy to the information bits through FEC techniques. With no other changes, this would normally reduce the information rate (or bandwidth efficiency) of the communication. In the same way, high-quality links can be obtained by transmitting the signals with spectrally less efficient modulation schemes, like binary phase shift keying (BPSK) and quaternary PSK (QPSK). On the other hand, new-generation wireless systems aim for higher data rates made possible through spectrally efficient higher-order modulations. Therefore, a reliable link with higher information rates can be accomplished by continuously controlling the coding and modulation levels. Higher modulation orders with less powerful coding rates are assigned to users that experience good link qualities, so that the excess signal quality can be used to obtain higher data rates. Recent designs have exploited this with adaptive modulation techniques that change the order of the modulation [1, 2], as well as with adaptive coding schemes that change the coding rate [3, 4]. For example, the Enhanced General Packet Radio Service (EGPRS) standard introduces both Gaussian minimum shift keying (GMSK) and 8-PSK modulations with different coding rates through link adaptation and hybrid automatic repeat request (ARQ) [5]. The channel quality is estimated at the receiver, and the information is passed to the transmitter through appropriately defined messages. The transmitter adapts the coding and modulation based on this channel quality feedback. Similarly, variable spreading and coding techniques are present in third-generation CDMA-based systems [3], cdma2000 and wideband CDMA (WCDMA, or Universal Mobile Telecommunications System [UMTS]). Higher data rates can be achieved by changing the spreading factor and coding rate, depending on the perceived communication link qualities.

Adaptive antennas and adaptive beam-forming techniques have also been studied extensively to increase the capacity and to improve the performance of wireless communications systems [6]. The adaptive antenna systems shape the radiation pattern in such a way that the information is transmitted (for example, from a base station) directly to the mobile user in narrow beams. This reduces the probability of another user experiencing interference in the network, resulting in improved link quality, which can also be translated into increased network capacity. Although adaptive beam forming is an excellent way to utilize multiple-antenna systems to enhance the link quality, recently different flavors of the usage of multiantenna systems have gained significant interest. Space-time processing and multiple-input multiple-output (MIMO) antenna systems are some new developments that will allow further usage of multiple-antenna systems in wireless communications. Adaptive implementation of these technologies is important for successful and efficient integration of them into wireless communications systems.

1.2.2 Adaptive System Resource Allocation

In addition to physical link adaptation, system resources can also be allocated adaptively to reduce the interference and to improve the overall system quality. This includes adaptive power control, adaptive channel allocation, adaptive cell assignment, adaptive resource scheduling, adaptive spectrum management, congestion, handoff (mobility), admission, and load control strategies. Adaptive system resource allocation considers the current traffic load, as well as the channel and interference conditions. For example, the system could assign more resources to the mobiles that have better link quality to increase the throughput. Alternatively, the system could assign the resources to the user in such a way that the user experiences better quality for the current traffic condition.

Adaptive channel allocation and adaptive cell assignment in hierarchical cellular systems have been studied since the early days of cellular systems. Adaptive channel allocation increases the system capacity through efficient channel utilization and decreased probability of blocked calls [7]. Unlike fixed channel allocation, where the channels are assigned to the cells permanently and the assignment is done based on the worst-case scenario, in adaptive channel assignment, a common pool of channels is shared by many cells, and the channels are assigned with regard to the interference and traffic conditions.

Adaptive cell assignment can increase capacity without increasing the handoff rate. The cells can be assigned to the users depending on their mobility level. Fast-moving mobiles can be assigned to larger umbrella cells (to reduce the number of handoffs), while slow-moving mobiles are assigned to microcells (to increase capacity) [8].

Recently, research on increasing the average throughput of the system through water-filling-based resource allocation has gained significant interest [9–11]. The main idea is to allocate more resources to the users that experience better link quality, resulting in very efficient use of the available resources. The high-data-rate (HDR) system, which is based on a best-effort radio packet protocol, uses a water-filling-based approach in allocating system resources. Algorithms that deal with compromising the throughput to achieve fairness have also been studied [10, 11].

1.2.3 Receiver Adaptation

Digital wireless communication receiver performance is related to the required value of the signal-to-interference-plus-noise ratio (SINR) so that the BER (or FER) performance can be kept below a certain threshold for reliable communication. For a given complexity, if receiver A requires lower SINR than receiver B to satisfy the same error rate, receiver A is considered to perform better than receiver B.

Receiver adaptation techniques can increase the performance of the receiver, hence reducing the minimum required SINR. As mentioned before, this can be used to increase the coverage area for a fixed transmitted power, or it can be used to reduce the transmitted power requirement for a given coverage area. Moreover, receiver adaptation can reduce the average receiver complexity and the power drain from the battery for the same quality of service. In order to satisfy the desired BER performance, instead of running a computationally complex algorithm for all channel conditions, the receiver can choose the most appropriate algorithm given the system and channel conditions.

Advanced baseband signal processing techniques play a significant role in receiver adaptation. Baseband algorithms used for time and frequency synchronization, baseband filtering, channel estimation and tracking, demodulation and equalization, interference cancellation, soft information calculation, antenna selection and combining, decoding, etc., can be made adaptive depending on the channel and interference conditions.

Conventional receiver algorithms are designed for the worst-case channel and interferer conditions. For example, the channel estimation and tracking algorithms assume the worst-case mobile speed; the channel equalizers assume the worst-case channel dispersion; the interference cancellation algorithms assume that the interferer is always active and constant; and so on. Adaptive receiver design measures the current channel and interferer conditions and tunes the specific receiver function that is most appropriate for the current conditions. For example, a specific demodulation technique may work well in some channel conditions, but might not provide good performance in others. Hence, a receiver might include a variety of demodulators that are individually tuned to a set of channel classes. If the receiver could demodulate the data reliably with a simpler and less complex receiver algorithm under the given conditions, then it is desired to use that algorithm for demodulation.

1.3   Parameter Measurements

Many adaptation techniques require estimation of various quantities like channel selectivity, link quality, network load and congestion, etc. Here, we focus more on physical layer measurements from a digital signal processing perspective. As discussed earlier, link quality measures have many applications for various adaptation strategies. In addition, information on channel selectivity in time, frequency, and space is very useful for adaptation of wireless communications systems. In this section, these important parameters and their estimation techniques will be discussed.

1.3.1   Channel Selectivity Estimation

In wireless communications, the transmitted signal reaches the receiver through a number of different paths. Multipath propagation causes the signal to be spread in time, frequency, and angle. These spreads, which are related to the selectivity of the channel, have significant implications on the received signal. A channel is considered to be selective if it varies as a function of time, frequency, or space. The information on the variation of the channel in time, frequency, and space is very crucial in adaptation of wireless communications systems.

1.3.1.1   Time Selectivity Measure: Doppler Spread

Doppler shift is the frequency shift experienced by the radio signal when either the transmitter or receiver is in motion, and Doppler spread is a measure of the spectral

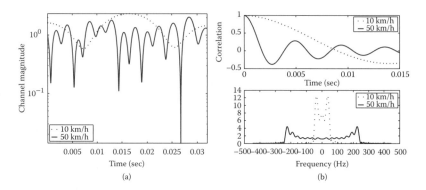

FIGURE 1.2 Illustration of the effect of mobile speed on time variation, time correlation, and Doppler spread of radio channel. (a) Channel time variation for different mobile speeds. (b) Time correlation of channel as a function of the time difference (separation in time) between the samples, and the corresponding Doppler spectrum in frequency.

broadening caused by the temporal rate of change of the mobile radio channel. Therefore, time-selective fading and Doppler spread are directly related. The coherence time of the channel can be used to characterize the time variation of the time-selective channel. It represents the statistical measure of the time window over which the two signal components have strong correlation, and it is inversely proportional to the Doppler spread. Figure 1.2 shows the effect of mobile speed on channel variation and channel correlation in time, as well as the corresponding Doppler spread values in frequency domain.

In an adaptive receiver, Doppler information can be used to improve performance or reduce complexity. For example, in channel estimation algorithms, whether using channel trackers or channel interpolators, instead of fixing the tracker or interpolation parameters for the worst-case Doppler spread value (as commonly done in practice), the parameters can be optimized adaptively based on Doppler spread information [12, 13]. Similarly, Doppler information could be used to control the receiver or transmitter adaptively for different mobile speeds, like variable coding and interleaving schemes [14]. Also, radio network control algorithms, such as handoff, cell assignment, and channel allocation in cellular systems, can utilize the Doppler information [8]. For example, as will be described later, in a hierarchical cell structure, the users are assigned to cells based on their speeds (mobility).

Doppler spread estimation has been studied for several applications in wireless mobile radio systems. Correlation and variation of channel estimates as well as correlation and variation of the signal envelope have been used for Doppler spread estimation [12]. One simple method for Doppler spread estimation is to use *differentials* of the complex channel estimates [15]. The differentials of the channel estimates are very noisy, which require low-pass filtering. The bandwidth of the low-pass filter is also a function of the Doppler estimate. Therefore, such approaches require adaptive receivers that continuously change the filter bandwidth depending on the previously obtained Doppler value. A Doppler estimation scheme based on the autocorrelation of complex channel estimates is described in [16]. Also, a maximum likelihood estimation-based approach, given the channel autocorrelation estimate, is utilized for Doppler spread estimation in

[17]. Channel autocorrelation is calculated using the channel estimates over the known field of the transmitted data.

Instead of using channel estimates, the received signal can also be used directly in estimating Doppler spread information. In [18], the Doppler frequency is extracted from the samples of the received signal envelope. Doppler information is calculated as a function of the squared deviation of the signal envelope. Similarly, in [19] the mobile speed is estimated as a function of the deviation of the averaged signal envelope in flat fading channels. For dispersive channels, pattern recognition, using the variation of pattern mean, can be used to quantify the deviation of signal envelope. In [20], the filtered received signal is used to calculate the channel autocorrelation values over each slot. Then, the autocorrelation estimate is used for identification of high- and low-speed mobiles. In [21], multiple antennas are exploited, where a linear relation between the switching rate of the antenna branches and Doppler frequency is given. Also, the level crossing rate of the average signal level has been used in estimating velocity [22, 23].

1.3.1.2 Frequency Selectivity Measure: Delay Spread

The multipath signals that reach the receiver have different delays as the paths that the signals travel through have different lengths. When the relative path delays are on the order of a symbol period or more, images of different transmitted symbols arrive at the same time, causing intersymbol interference (ISI). Delay spread is one of the most commonly used parameters that describes the time dispersiveness of the channel, and it is related to frequency selectivity of the channel. The frequency selectivity can be described in terms of coherence bandwidth, which is a measure of range of frequencies over which the two frequency components have a strong correlation. The coherence bandwidth is inversely proportional to the delay spread [24]. Figure 1.3 shows the effect of time dispersion on channel frequency variation and channel frequency correlation, as well as the corresponding power delay profiles.

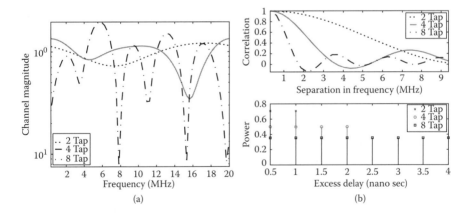

FIGURE 1.3   Illustration of the effect of time dispersion on channel frequency variation, channel frequency correlation, and delay spread. (a) Channel frequency variation for different delay spread values. (b) Channel frequency correlation as a function of separation in frequency and the corresponding power delay profiles.

Like time selectivity, the information about the frequency selectivity of the channel can be very useful for improving the performance of adaptive wireless radio systems. For example, in a time division multiple-access (TDMA)-based Global System for Mobile Communications (GSM), the number of channel taps needed for equalization might vary depending on channel dispersion. Instead of fixing the number of channel taps for the worst-case channel condition, we can change them adaptively [25], allowing simpler receivers with reduced battery consumption and improved performance. Similarly, in [26], a TDMA receiver with adaptive demodulator is proposed, using the measurement about the dispersiveness of the channel. Dispersion estimation can also be used for other parts of transmitters and receivers. For example, in frequency domain channel estimation using channel interpolators, instead of fixing the interpolation parameters for the worst expected channel dispersion, we can change the parameters adaptively depending on the dispersion information [27].

Although dispersion estimation can be very useful for many wireless communications systems, it is particularly crucial for orthogonal frequency division multiplexing (OFDM)-based wireless communications systems. OFDM, which is a multicarrier modulation technique, handles the ISI problem due to high-bit-rate communication by splitting the high-rate symbol stream into several lower-rate streams and transmitting them on different orthogonal carriers. The OFDM symbols with increased duration might still be affected by the previous OFDM symbols due to multipath dispersion. Cyclic prefix extension of the OFDM symbol avoids ISI from the previous OFDM symbols if the cyclic prefix length is greater than the maximum excess delay of the channel. Since the maximum excess delay depends on the radio environment, the cyclic prefix length needs to be designed for the worst-case channel condition. This makes the cyclic prefix a significant portion of the transmitted data, thereby reducing spectral efficiency. One way to increase spectral efficiency is to adapt the length of the cyclic prefix depending on the radio environment [28]. The adaptation requires estimation of maximum excess delay of the radio channel, which is also related to the frequency selectivity of the channel. In HiperLAN2, which is a wireless local area network (WLAN) standard, a cyclic prefix duration of 800 ns, which is sufficient to allow good performance for channels with delay spread up to 250 ns, is used. Optionally, a short cyclic prefix with 400 ns duration may be used for short-range indoor applications. Delay spread estimation allows adaptation of these various options to optimize the spectral efficiency. Other OFDM parameters that could be changed adaptively using the knowledge of the dispersion include OFDM symbol duration and OFDM subcarrier bandwidth.

Characterization of the frequency selectivity of the radio channel is studied in [29–31] using the level crossing rate (LCR) of the channel in frequency domain. Frequency domain LCR gives the average number of crossings per Hertz at which the measured amplitude crosses a threshold level. An analytical expression between LCR and the time domain parameters corresponding to a specific multipath power delay profile (PDP) is given. LCR is very sensitive to noise, which increases the number of level crossings and severely deteriorates the performance of the LCR measurement [31]. Filtering the channel frequency response reduces the noise effect, but finding the appropriate filter parameters is an issue. If the filter is not designed properly, one might end up smoothing the actual variation of frequency domain channel response. In [27], instantaneous root mean square (rms) delay spread, which provides information about local (small-scale) channel dispersion, is obtained by estimating the channel impulse response (CIR) in the time domain. The detected symbols in the frequency domain are used to regenerate the time domain signal through inverse fast Fourier transform (IFFT). This signal is then used to correlate the actual received signal to obtain CIR, which is then used for delay spread estimation. Since the detected symbols are random, they might not have good autocorrelation properties, which can be a problem, especially when the number of carriers is low. In addition, the use of detected symbols for correlating the received samples to obtain CIR provides poor results for low SNR values. In [28], the delay spread is also calculated from the instantaneous time domain CIR, wherein the CIR is obtained by taking IFFT of the frequency domain channel estimate. Channel frequency selectivity and delay spread information are calculated using the channel frequency correlation estimates in [24, 32]. An analytical expression between delay spread and coherence bandwidth is also given.

The level of time dispersion can be obtained by using known training sequences and a maximum likelihood-based algorithm. The channel can be modeled with different levels of dispersion. Using these various channel models, the corresponding channel estimates and the residual error can be calculated. From these residual error terms, a decision can be made about the level of dispersion. Note that when the channel is overmodeled, the residual error also becomes smaller. Hence, it is not necessarily true that the model that provides the smaller residual error is the most suitable one. The most appropriate model can be found by several information criteria algorithms, like Bayesian information criteria (BIC) or Akaike information criteria (AIC) [33].

1.3.1.3   Spatial Selectivity Measure: Angle Spread

Angle spread is a measure of how multipath signals are arriving (or departing) with respect to the mean arrival (departure) angle. Therefore, angle spread refers to the spread of angles of arrival (or departure) of the multipaths at the receiving (transmitting) antenna array [34]. Angle spread is related to the spatial selectivity of the channel, which is measured by coherence distance. Like coherence time and frequency, coherence distance provides the measure of the maximum spatial separation over which the signal amplitudes have strong correlation, and it is inversely proportional to angular spread, i.e., the larger the angle spread, the shorter the coherence distance. Figure 1.4 shows the effect of local scattering on angle of arrival. The local scattering in the vicinity of Receiver-2 results in larger angular spreads, as the received signals come from many different directions due to a richer local scattering environment. For a given receiver antenna spacing, this leads to less antenna correlations between the received antenna elements than in Receiver-1. Note that although the angular spread is described independent of the other channel selectivity values for the sake of simplicity, in reality the angle of arrival can be related to the path delay. The multipath components that arrive at the receiver earlier (with shorter delays) are expected to have similar angles of arrival (lower angle spread values).

Compared to time and frequency selectivity, spatial selectivity has not been studied widely in the past. However, recently there has been a significant amount of work in multiantenna systems. With the widespread application of multiantenna systems, it

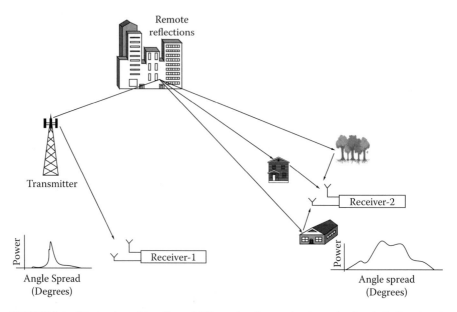

FIGURE 1.4   Illustration of the effect of different local scattering in angle of arrivals. Receiver-1 observes less angle spread than Receiver-2. Therefore, receiver antennas in Receiver-1 will have more correlations.

is expected that the need for understanding spatial selectivity and related parameter estimation techniques will gain momentum. Spatial selectivity will especially be useful when the requirement for placing antennas close to each other increases, as in the case of multiple antennas at the mobile units.

Spatial correlation between multiple-antenna elements is related to the spatial selectivity, antenna distance, mutual coupling between antenna elements, antenna patterns, etc. [35, 36]. Spatial correlation has significant effects on multiantenna systems. Full capacity and performance gains of multiantenna systems can only be achieved with low antenna correlation values. However, when this is not possible, maximum capacity can be achieved by employing efficient adaptation techniques. Adaptive power allocation is one way to exploit the knowledge of the spatial correlation to improve the performance of multiantenna systems [37]. Similarly, adaptive modulation and coding, which employs different modulation and coding schemes across multiantenna elements depending on the channel correlation, is possible [38, 39]. In MIMO systems, adaptive power allocation has been studied by using the knowledge of channel matrix estimate and eigenvalue analysis [40, 41].

1.3.2   Channel Quality Measurements

Channel quality estimation is by far the most important measurement that can be used in adaptive receivers and transmitters [3]. Different ways of measuring the quality of radio channel are possible, and many of these measurements are done in the physical layer using baseband signal processing techniques. In most of the adaptation algorithms,

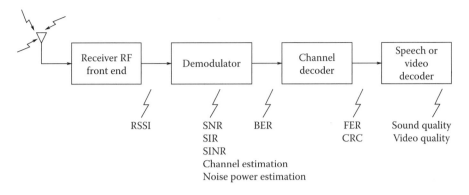

FIGURE 1.5   A simple wireless receiver that shows the estimation points of commonly used parameters.

the target quality measure is the FER or BER, as these are closely related to higher-level quality-of-service parameters like speech and video quality. However, reliable measurement of these qualities requires many measurements, and this causes delays in the adaptation as the process could be very long. Therefore, other types of channel quality measurements that are related to these might be preferred. When the received signal is impaired only by white Gaussian noise, analytical expressions can be found relating the BER to other measurements. For other impairment cases, like colored interferers, numerical calculations and computer simulations that relate these measurements to BER can be performed. Therefore, depending on the system, a channel quality is related to the BER. Then, for a target BER (or FER), a required signal quality threshold can be calculated to be used with the adaptation algorithm.

The measurements can be performed at various points of a receiver, depending on the complexity, reliability, and delay requirements. There are trade-offs in achieving these requirements at the same time. Figure 1.5 shows a simple example where some of these measurements can take place. In the following sections, these measurements will be discussed briefly.

1.3.2.1   Measures before Demodulation

Received signal strength (RSS) estimation provides a simple indication of the fading and path loss, and provides the information about how strong the signal is at the receiver front end. If the received signal strength is stronger than the threshold value, then the link is considered to be good. Measuring the signal strength of the available radio channels can be used as part of the scanning and intelligent roaming process in cellular systems. Also, other adaptation algorithms, like power control and handoff, can use this information. The RSS measurement is simply done by reading samples from a channel and averaging them [42]. Compared to other measurements, RSS estimation is simple and computationally less complex, as it does not require the processing and demodulation of the received samples. However, the received signal includes noise, interference, and other channel impairments. Therefore, receiving a good signal strength does not tell much about the channel and signal quality. Instead, it gives an indication of whether a strong signal is present in the channel of interest. For the measurement of RSS, the transmitter might send a pilot signal continuously, as in the WCDMA cellular system, or a link layer beacon can be transmitted at discrete time intervals, as in IEEE 802.11 WLANs.

Since the received signal power fluctuates rapidly due to fading, in order to obtain reliable estimates, the signal needs to be averaged over a time window to compensate for short-term fluctuations. The averaging window size depends on the system, application, variation of the channel, etc. For example, if multiple receiver antennas are involved at the receiver, the window can be shorter than that for a single-antenna receiver.

1.3.2.2 Measures during and after Demodulation

The signal-to-interference ratio (SIR), SNR, and SINR are the most common ways of measuring the channel quality during (or just after) the demodulation of the received signal. SIR (or SNR or SINR) provides information on how strong the desired signal is compared to the interferer (or noise or interference plus noise). Most wireless communications systems are interference limited; therefore, SIR and SINR are more commonly used. Compared to RSS, these measurements provide more accurate and reliable estimates at the expense of computational complexity and with additional delay.

There are many adaptation schemes where these measurements can be exploited. Link adaptation (adaptive modulation and coding, rate adaptation, etc.), adaptive channel assignment, power control, adaptive channel estimation, and adaptive demodulation are only a few of many applications.

SIR estimation can be employed by estimating signal power and interference power separately and then taking the ratio of these two. In many new-generation wireless communications systems, coherent detection, which requires estimation of channel parameters, is employed. These channel parameter estimates can also be used to calculate the signal power. The training (or pilot) sequences can be used to obtain the estimate of SIR. Instead of the training sequences, the data symbols can also be used for this purpose. For example, in [43], where SNR information is used as a channel quality indicator for rate adaptation, the cumulative Euclidean metric corresponding to the decoded trellis path is exploited for channel quality information. Another method for channel quality measurement is the use of the difference between the maximum likelihood decoder metrics for the best path and second-best path, as described in [44]. In a sense, in this technique, some sort of soft information is used for the channel quality indicator. However, this approach does not tell much about the strength of the interferer or the desired signal. There are several other ways of SNR measurement that are based on subspace projection techniques. These approaches can be found in [45] and in the references cited therein.

Often, in obtaining the estimates, the impairment (noise or interference) is assumed to be white and Gaussian distributed to simplify the estimation process. However, in wireless communications systems, the impairment might be caused by a strong interferer, which is colored. For example, in OFDM systems, where the channel bandwidth is wide and the interference is not constant over the whole band, it is very likely that some part of the spectrum is affected more by the interferer than the other parts. Figure 1.6 shows the OFDM frequency spectrum and two types of noise over this spectrum: colored and white. Hence, when the impairment is colored, estimates that take the color of the impairment into account might be needed [46].

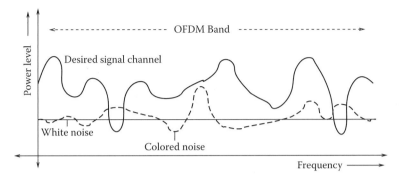

FIGURE 1.6   Representation of OFDM frequency channel response and noise spectrum. Spectrums for both white and colored noise are shown.

Note that since both the desired signal's channel and interferer conditions change rapidly, depending on the application, both short-term and long-term estimates are desirable. Long-term estimates provide information on long-term fading statistics due to shadowing and lognormal fading as well as average interference conditions. On the other hand, short-term estimates provide measurements of instantaneous channel and interference conditions. Applications like adaptive channel assignment and handoff prefer long-term statistics, whereas applications like adaptive demodulation, adaptive interference cancellation, etc., prefer short-term statistics.

For some applications, a direct measure of channel quality from channel estimates would be sufficient for adaptation. As mentioned above, channel estimates only provide information about the desired signal's power. It is a much more reliable estimate than RSS information, as it does not include the other impairments as part of the desired signal power. However, it is less reliable than SNR (or SINR) estimates, since it does not provide information about the noise or interference powers with respect to the desired signal's power.

Channel estimation for wireless communications systems has a very rich history. A significant amount of work has been done for various systems. In many systems, known information (like pilot symbols, pilot channels, pilot tones, training sequences, etc.) is transmitted along with the unknown data to help the channel estimation process. Blind channel estimation techniques that do not require known information transmission have also been studied extensively. For details on channel estimation for wireless communications systems, refer to [47, 48] and the references listed therein.

1.3.2.3 Measures after Channel Decoding

Channel quality measurements can also be based on postprocessing of the data (after demodulation and decoding). BER, symbol error rate (SER), FER, and CRC information are some of the examples of the measurements in this category. BER (or FER) is the ratio of the bits (or frames) that have errors relative to the total number of bits (or frames) received during the transmission. The CRC indicates the quality of a frame, which can be calculated using parity check bits through a known cyclic generator polynomial. FER can be obtained by averaging the CRC information over a number of frames. In order to calculate the BER, the receiver needs to know the actual transmitted bits, which is not possible in practice. Instead, BER can be calculated by comparing the bits before and after the decoder. Assuming that the decoder corrects the bit errors that appear before decoding, this difference can be related to BER. Note that the comparison makes sense only if the frame is error-free (good frame), which is obtained from the CRC information.

As mentioned earlier, although these estimates provide excellent link quality measures, reliable estimates of these parameters require observations over a large number of frames. Especially for low BER and FER measurements, extremely long transmission intervals will be needed. Therefore, for some applications these measures might not be appropriate. Note also that these measurements provide information about the actual operating condition of the receiver. For example, for a given RSS or SINR measure, two different receivers that have different performances will have different BER or FER measurements. Therefore, BER and FER measurements also provide information on the receiver capability as well as the link quality.

1.3.2.4   Measures after Speech or Video Decoding

The speech and video quality, the delays on data reception, and network congestion are some of the parameters that are related to the user's perception. Essentially, these are the ultimate quality measures that need to be used for adaptive algorithms. However, these parameters are not easy to measure, and in many cases, real-time measurement might not be possible. On the other hand, these measures are often related to the other measures mentioned above. For example, speech quality for a given speech coder can be related to FER of a specific system under certain assumptions [49]. However, as discussed in [49], some frame errors cause more audible damage than others. Therefore, it is still desired to find ways to measure the speech quality more reliably (and timely) and adapt the system parameters accordingly. Speech (or video) quality measures that take the human perception of the speech (or video) into account are highly desirable.

Perceptual speech quality measurements have been studied in the past. Both subjective and objective measurements are available [50]. Subjective measurements are obtained from a group of people who rate the quality of the speech after listening to the original and received speech. Then a mean opinion score (MOS) is obtained from their feedback. Although these measurements reflect the exact human perception that is desired for adaptation, they are not suitable for adaptation purposes because the measurements are not obtained in real time. On the other hand, the objective measurements can be implemented at the receiver in real time [51]. However, these measurements require a sample of the original speech at the receiver to compare the received voice with the original, undistorted voice. Therefore, they are also not applicable for many scenarios.

1.4   Applications of Adaptive Algorithms: Case Studies

1.4.1   Examples for Adaptive Receiver Algorithms

In this section, some representative examples for adaptive receiver algorithms will be discussed briefly. These algorithms can be employed in both base stations and mobile terminals, as well as in many other wireless receivers.

1.4.1.1 Channel Estimation with *A Priori* Information

Channel estimation is an integral part of standard adaptive receiver designs used in digital wireless communications systems. For conventional, coherent receivers, the effect of the channel on the transmitted signal must be estimated to recover the transmitted information. As long as the receiver estimates what the channel did to the transmitted signal, it can accurately recover the information sent.

The estimation of time-varying channel parameters is often based on an approximate underlying model of the radio channel. In fading environments, the coefficients of a channel model exhibit typical trends or quasi-periodic behavior in time, frequency, and space. The ability to track channel variation depends on how fast the channel changes in time, frequency, and space. As mentioned before, this is related to Doppler spread (time variation), delay spread (frequency variation), and angle spread (space variation). By utilizing *a priori* information about the channel variation, adaptive algorithms with larger memories can be designed without sacrificing tracking capability [15]. In contrast to the algorithms that do not exploit this information, adaptive algorithms provide a means of extrapolation of the channel coefficients in time, frequency, and space [13, 52]. For example, in [53], the step size of a simple least mean square (LMS) channel tracker is changed using the Doppler spread information. Similarly, the window size of a sliding window (moving average filtering)-based channel tracking algorithm can be adapted depending on Doppler spread and SNR information [54]. Wiener filtering, which is one of the most popular techniques for channel estimation using interpolation, is an excellent example in exploiting *a priori* information, as the optimal Wiener filter design requires knowledge of Doppler spread and noise power. In most conventional Wiener filtering designs, the worst-case expected Doppler spread values are used, degrading the performance of the algorithm for other Doppler spread values [55]. Recently, two-dimensional interpolation using Wiener filtering for OFDM-based wireless communications systems gained significant interest [28]. In this case, both Doppler spread and delay spread information, as well as noise variance estimates, can be used to optimize the channel tracker performance. Although we have mentioned a few examples, the usage of *a priori* information in channel estimation has been considered by many other authors. Further information can be found in [47, 48].

Figure 1.7 shows a simple coherent receiver structure with an adaptive channel tracker. The receiver includes a parameter measurement block that estimates the necessary parameters for the adaptation of the channel tracker. The necessary parameters can be estimated using the received signal and the output of the detector as described before. The detector requires the channel estimates that can be obtained from the channel tracker.

1.4.1.2 Adaptive Channel Length Truncation for Equalization

Time dispersion in wireless systems can cause ISI, which degrades the performance, often severely. Equalization is a technique used to counter the effects of ISI. In the Telecommunications Industry Association/Electronics Industry Association/Interim Standard 136 (TIA/EIA/IS-136, or simply IS-136) system, the channel can be assumed to be flat (nondispersive) with respect to the symbol duration most of the time. Equalization does

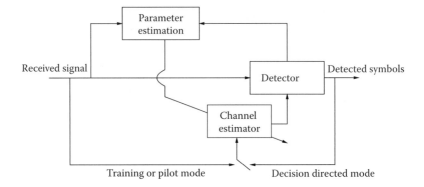

FIGURE 1.7   A simple adaptive channel estimation receiver.

not help much in nondispersive environments, and in fact hurts performance by trying to model dispersion that does not exist. However, in hilly terrain channel conditions, the channel is dispersive and requires equalization. Therefore, to design the receiver for the worst-case condition, equalization needs to be used for all the geographical conditions unnecessarily, resulting in a loss due to the mismatch of the implemented receiver to the fading scenario. An adaptive receiver, on the other hand, can have an algorithm that measures the dispersiveness of the channel and uses the appropriate demodulator based on the measurement [26]. This also results in conserving battery power.

   In another cellular communications system, GSM, the symbol duration is relatively short compared to that in IS-136. Also, the pulse shaping itself introduces intentional ISI, so that equalization is required even in nondispersive channels. However, the number of channel taps needed for equalization might vary depending on the dispersion (the geographical area). Instead of fixing the number of channel taps for the worst-case condition, the number can be made adaptive [25], allowing simpler receivers with reduced battery consumption and improved performance. Again, the point emphasized here is to avoid overmodeling the signal. Figure 1.8 shows a simple example of an adaptive receiver that measures the level of dispersion and adapts the equalizer number of taps accordingly.

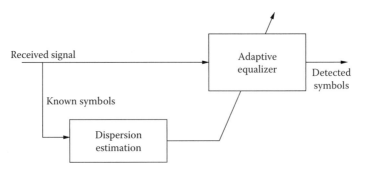

FIGURE 1.8   An adaptive receiver that uses the delay spread (time dispersion) estimate to adjust the equalizer.

1.4.1.3   Adaptive Interference Cancellation Receivers

The impairment sources in wireless mobile radio systems are numerous. Co-channel interference, which is caused by the reuse of carrier frequencies in nearby cells, is one of the major contributors. Another major interference source is adjacent channel interference, which is caused by the spectral overlap between adjacent channel users. Also, thermal noise and other impairment sources that are commonly modeled as additive white Gaussian noise (AWGN) degrade the performance of a receiver. The statistics of these disturbance sources are different. Conventional receivers commonly assume that the impairment at the receiver is white, which causes performance loss if the actual impairment is colored. By exploiting the statistics of the impairments, better receivers can be designed. For example, *interference whitening* is one such technique that partially suppresses the interference and optimizes the demodulator performance. However, at any given time, the kind of disturbance that is dominant at the receiver is not known before. In order to achieve the best possible performance in all situations, the receiver should estimate the possible disturbance source and adapt the receiver to the second-order statistics of the impairment. Such an adaptive receiver described in [56] improves the performance of the maximum likelihood–based receiver.

The interference can also be suppressed by employing interference cancellation techniques in the receivers. For example, joint demodulation (JD) of co-channel signals is a powerful technique for cancelling co-channel interference. In [57], it was shown that the capacity of the IS-136 system can be increased significantly by using a JD receiver. However, the JD receiver given in [57] works well only when there is a single dominant interferer, the mobile speed is low, and the channel is nondispersive. Otherwise, the conventional single-user demodulator (CD) works better than joint demodulation at the targeted operating SINR level. A simple and efficient solution to the above problem is an adaptive receiver that adapts the detector to the system conditions. Figure 1.9 illustrates

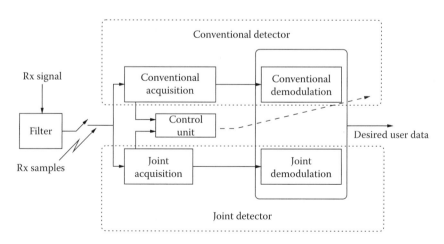

FIGURE 1.9   Example for adaptive interference cancellation receiver. A complex joint demodulation and a less complex single-user demodulation used adaptively based on the measured parameters.

the schematic of such an adaptive receiver. It contains the conventional detector, the joint detector, and a control unit to control the two detectors. For each slot, the control unit determines which of the two demodulators to use to recover the data symbols of the desired user. The control unit makes this decision on the basis of certain information obtained from conventional and joint acquisitions. The demodulator selected by the control unit outputs an estimate for the data symbols of the desired user. The details regarding the two demodulators can be found in [58].

The choice for the demodulator can be based on several criteria. Ideally, one would like to know the SNR, SIR, dominant interferer ratio ($I_1/I - I_1$, where I_1 is the dominant interferer and I is the total impairment, including the dominant interferer), and extent of ISI present in the system, among other parameters. Although these quantities are not generally available at the receiver, they can be estimated. For example, carrier and dominant interferer powers are estimated by averaging the corresponding channel tap strengths over multiple slots. The unmodeled impairment power is estimated from the accumulated Euclidean distance metric during the acquisition process (joint or conventional) over the training sequence of the desired signal.

1.4.1.4   Adaptive Soft Information Generation and Decoding

In digital wireless communications systems, forward error correction encoding is commonly used to provide a robust communication link. At the receiver, the decoder performance is optimized when the demodulator provides soft information for the encoded bits. The better soft information generation schemes require knowledge of the noise covariance, and often the noise covariance changes across the interleaving length. Therefore, a receiver should continuously measure the noise covariance and use these estimates for the improvement of soft bit values.

1.4.2   Examples for Link Adaptation and Adaptive Resource Allocation

In this section, some examples for adaptation of radio link and adaptive resource allocation will be discussed briefly. Examples in this area are numerous, and there has been a significant amount of research in this area.

1.4.2.1   Adaptive Power Control

Power control has a long and rich history in wireless communications systems [59–61]. Specifically, for CDMA-based cellular systems, adaptive power control has a significant role, as the performance and capacity of the CDMA systems are normally interference limited. Without power control, an interfering transmitter that is closer to the receiver than the desired signal's transmitter will cause a significant degradation, and this phenomenon is commonly referred to as the *near-far problem*. Power control handles this problem by adaptively controlling the user's power depending on the link quality and desired quality of service (QoS). As a result, the interference observed by other users due to this user will be less, which in turn reduces the average interference observed at the receivers. This results in a high-capacity system with improved battery life for the mobile terminals.

In voice-dominated cellular systems, the objective of the power control was mainly to maintain the minimal (target) link quality at a constant level for individual users. The data rates for all users are constant in this case, and each user experiences roughly the same quality of service. While this was appropriate for voice, recently, with the increased demand for multimedia services and high-speed data access, different objectives and cost functions to optimize the use of power resources have been developed. In mixed-traffic environments, the cost function for each service will be different, leading to different power allocation strategies [62]. Use of constant power along with variable coding, modulation, and spreading that adapts the data rate to the channel variations is one objective that some new-generation wireless systems have been adopting (rate control or rate adaptation) [43]. Also, water-filling types of power assignments, which assign more power to the users that have favorable channels, are being studied extensively [63].

In adaptive power control mechanisms, estimation of the link quality parameters is the key factor. Typical parameters used for adaptation include SIR, FER, and RSS. Doppler spread estimate can also be used to adjust the adaptation rate. Depending on the adaptation rate, power control can be classified as *fast power control* and *slow power control*. Fast power control compensates the changes in power level due to Rayleigh fading (small-scale fading), while slow power control is used for lognormal fading (shadowing) and path loss. The parameters that are used for them can also be different. For example, for fast power control, instantaneous SIR, SNR, SINR, and RSS can be more suitable than FER and BER, which might better suit slow power control. As mentioned in the parameter estimation section, parameter selection depends on the delay, complexity, and accuracy requirements. The estimation errors and delays, between measurements and adaptation of power, limit the efficient application of power control schemes. Therefore, more accurate and practical algorithms that estimate and predict the parameters to be used in adaptation are needed.

1.4.2.2 Adaptive Modulation and Channel Coding

Given the high price of spectrum and its scarcity, it is in the interest of operators to continue evolving their networks toward higher capacity and quality. Adaptive modulation and coding provide a framework to adjust modulation level and FEC coding rate depending on the link quality. Higher-order modulations (HOMs) allow more bits to be transmitted for a given symbol rate. On the other hand, HOM is less power efficient, requiring higher energy per bit for a given BER. Therefore, HOMs should be used only when the link quality is high, as they are less robust to channel impairments. Similarly, strong FEC and interleaving provide robustness against channel impairments at the expense of lower data rate and spectral efficiency, suggesting adaptation of coding rate based on the link quality. Figure 1.10 illustrates the capacity gain that can be achieved by employing adaptive modulation only. First, the BER performances of different modulations as a function of SNR are given in Figure 1.10(a). As can be seen, a desired BER can be achieved with low-order modulations for lower SNRs. Higher-order modulations need better link quality (higher SNRs) in order to obtain the same BER performance. Figure 1.10(b) shows the spectral efficiencies of different uncoded modulations, where an arbitrary packet size of 200 bits is used. Notice that the optimal spectral efficiency for

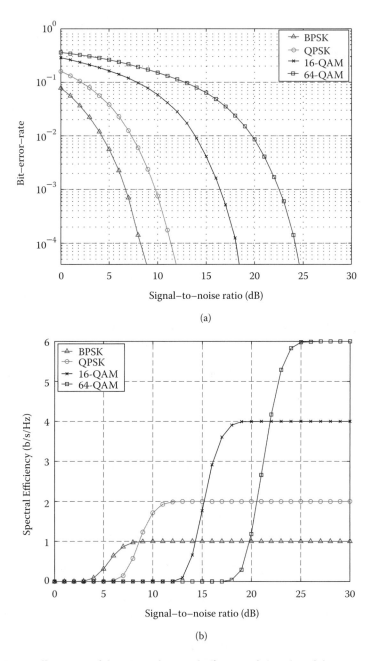

(a)

(b)

FIGURE 1.10 llustration of the BER and spectral efficiency of several modulation options. (a) BER plots of different modulations as a function of SNR. (b) Spectral efficiency of different modulation as a function of SNR.

TABLE 1.1   EGPRS Modulation and Coding
Schemes and Peak Data Rates

Scheme	Modulation	Maximum Rate per Slot (kb/s)	Code Rate
MCS-1	GMSK	8.8	0.53
MCS-2	GMSK	11.2	0.66
MCS-3	GMSK	14.8	0.80
MCS-4	GMSK	17.6	1.00
MCS-5	8-PSK	22.4	0.37
MCS-7	8-PSK	44.8	0.76
MCS-6	8-PSK	29.6	0.49
MCS-8	8-PSK	54.5	0.92
MCS-9	8-PSK	59.2	1.00

different SNR regions can be obtained through the use of different modulations depending on the SNR.

Link adaptation using adaptive coding and modulation is deployed in some of the new-generation wireless communications systems. For example, EGPRS, which is the evolution of the second-generation GSM, employs two different modulation options (GMSK and 8-PSK) along with different coding rates, resulting in nine different modulation/coding options, as shown in Table 1.1 [5, 43]. In addition, EGPRS introduces the use of a type II hybrid ARQ system, commonly known within the specification as incremental redundancy. In link adaptation, the link quality is measured regularly and the most appropriate modulation and coding scheme is assigned for the next transmission interval. On the other hand, in the incremental redundancy scheme, information is first sent with low coding power (high coding rate). This results in a high bit rate if decoding is successful with this rate. However, if decoding fails with such a high rate, additional coded bits (redundancy) should be sent so that the transmitted bits can be decoded successfully. However, sending extra coded bits incrementally reduces the resulting bit rate and introduces undesired extra delay. Therefore, the initial code rate and modulation for the incremental redundancy scheme should be based on measurements of the link quality, instead of starting with any arbitrary rate [5]. As a result, by combining incremental redundancy with adaptive initial code rate, lower delays with lower memory requirements, and high data rates can be achieved. The different initial code rates are obtained by puncturing a different number of bits from a common convolution code (rate 1/3). Incremental redundancy operation is enabled by puncturing a different set of bits each time a block is retransmitted, whereby the code rate is gradually decreased toward one-third for every new transmission of the block.

Recent studies introduce new modulation and coding options together with other capacity enhancement techniques to further increase the data rate and throughput of EGPRS [64, 65]. Higher-order modulations like 16-QAM and 64-QAM are being proposed along with some more coding options to optimize the performance.*

---

* 16-QAM and 64-QAM stand for 16-level and 64-level quadrature amplitude modulation, respectively.

Adaptive modulation and coding are also successfully employed for new-generation WLAN systems. HiperLAN2 and IEEE 802.11a, both of which use OFDM technology at the physical layer, allow four different modulation options (BPSK, QPSK, 16-QAM, and 64-QAM) with different coding rates. The coding rates are obtained with different puncturing patterns to a mother convolutional code, resulting in eight different modulation and coding options [66]. Similar to link adaptation in EGPRS, an appropriate modulation and coding scheme is used depending on the link quality. Therefore, a data rate ranging from 6 to 54 Mbit/s can be obtained by using various modes. BPSK, QPSK, and 16-QAM are used as mandatory modulation formats, whereas 64-QAM is applied as an optional mode.

Although only a couple of cases are given above, adaptive modulation and coding have attracted many new-generation wireless standards to consider them as options to increase the data rates, and there has been a significant amount of research in this area. Especially in conjunction with the advanced receiver algorithms that reduce the required SINR to lower values, the better link quality values can be exploited to increase the data rates further. Combining adaptive modulation with multiantenna transmitter and MIMO schemes based on the feedback-related channel estimates, channel quality, channel correlation, etc., is one of these interesting research areas. Based on the channel feedback information, the modulation type on multiantenna transmitters can be varied. Similarly, adapting the source coding with the channel coding or modulation is another interesting area of focus for link adaptation. For example, adaptive multirate (AMR) codec allows changing of the compression rate of speech depending on the link quality, as in GSM AMR. For weak link conditions, where heavy FEC is required, AMR has the ability to decrease the codec rate (more speech compression) to allocate more bits for FEC [49].

1.4.2.3   Adaptive Cell and Frequency Assignment

As mentioned before, radio spectrum is very expensive and limited. Efficient use of radio spectrum is very important to maximize the system capacity. The introduction of cellular technology was a major step toward efficient usage of finite spectrum through a concept called *frequency reuse*. The capacity of cellular systems is interference limited, dominated by co-channel interference (CCI) and adjacent channel interference (ACI). Early cellular systems aimed to avoid these major interference sources by designing systems for the worst-case interference conditions along with fixed channel allocation. This is often achieved by employing higher-frequency reuse and by allowing enough carrier spacing between adjacent channels. Both of these reduce the spectral efficiency. Later, more efficient spectrum usage strategies were developed that dynamically assign frequencies relative to current interference, propagation, and traffic conditions. In traditional cellular system designs, the allocation of frequency channels to cells is fixed, which means that each cell can use only a set of frequencies. Even if the other cells are not fully loaded, the cell that does not have any available frequency (fully loaded cell) cannot take advantage of it. In dynamic channel allocation, all the channels belong to a global pool and the channels are assigned according to a cost function that considers the

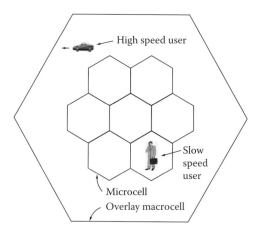

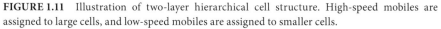

FIGURE 1.11 Illustration of two-layer hierarchical cell structure. High-speed mobiles are assigned to large cells, and low-speed mobiles are assigned to smaller cells.

CCI and ACI [67]. As a result, for nonuniform traffic conditions, the available channels can be used more efficiently.

Resource utilization has also evolved by employing a concept called hierarchical cellular structures (HCSs) [68]. The use of HCS has become a major component in third-generation mobile systems such as UMTS and IMT-2000. In an HCS, various cell sizes are deployed and small cell clusters are overlaid by larger cells. For example, Figure 1.11 shows a two-layer (e.g., microcell and macrocell) hierarchical system. Microcells increase capacity within a coverage area, but radio resource management becomes more difficult. The number of handoffs per cell is increased by an order of magnitude, and the time available to make a handoff is decreased. HCSs handle this by assigning cells to the mobiles depending on their speeds (Doppler spread estimate). For example, in the two-layer structure given in Figure 1.11, low-speed mobiles are assigned to microcells, whereas high-speed mobiles are assigned to macrocells. Hence, the macrocell-microcell overlay architecture provides a balance between maximizing the capacity per unit area and minimizing the number of handoffs [69]. As a result, the risk of call dropping is reduced, and there are other benefits, like lower handover delays, reduced switching load, and increased QoS. The HCS can be more than two layers (multilayer HCS). For example, picocellular layers can also be included in multilayer HCS. Similarly, communication satellite beams can overlay all the terrestrial layers at the highest hierarchical level.

Recently, dynamic allocation and multitiered design strategies are further generalized to take power control, cell handoff, traffic classes (like multimedia), and user priorities into account. Also, there are several studies toward combining link adaptation schemes with adaptive resource allocation. For example, adaptive modulation (and coding) can be combined with dynamic channel allocation. Similarly, adaptive modulation (and coding) can be combined with handover algorithms to introduce more intelligent handover strategies. All these developments require more sophisticated adaptation of the network, and they are based on many parameter measurements.

1.5   Future Research for Adaptation

Most of the techniques and issues that were described in the previous sections still need further research for the development of more efficient adaptation and parameter estimation algorithms. At the same time, there is a significant amount of effort in evolving current wireless communications systems to provide higher data rates, higher capacity, and better performance. New technologies are being introduced to accommodate these goals, like multicarrier wireless communications, MIMO, and ultrawideband (UWB). Adaptation techniques will be a significant factor in efficient and successful deployment of these technologies.

UWB is a promising technology for future data communications systems, high-accuracy (indoor) geolocation devices, sensor applications, etc. Any signal that occupies more than 500 MHz of bandwidth and meets the spectrum mask requirements enforced by spectrum regulation agencies is considered a UWB signal [70]. For example, in the United States, the Federal Communications Commission (FCC) has allocated 7.5 GHz of spectrum (between 3.1 and 10.6 GHz) for unlicensed use of UWB devices. One of the most popular UWB systems, which is based on impulse radio (IR), utilizes carrierless transmission with very low-power spectral density. IR-based UWB techniques are based on the transmission of nanosecond-level short pulses that generate extremely wide spectrums. This results in a covert noise-like signal in a radio channel. Note that within the transmission band of UWB, other technologies also coexist. For example, the OFDM-based WLAN technology at the 5 GHz U-NII band is a big concern for UWB signals, as it might create significant interference for UWB signals. In order to provide robustness against narrowband interference, adaptive implementation of UWB systems is very important. For this purpose, several strategies have been developed recently. Multiband UWB is one of the techniques proposed to reduce the effect of narrowband interference. In such techniques, the whole 7.5 GHz bandwidth is divided into several narrower bands that are still wider than 500 MHz. The information is transmitted in these bands depending on the narrowband interference situation. Several versions of multiband schemes are available, some of which can be found in [70]. Estimations of the existence and level of narrowband interference are interesting research topics that fall under the parameter estimation algorithms described before.

The wide bandwidth of UWB offers a capacity much higher than the current narrowband systems. Short-range data transmission rates of over 500 Mbps have been theoretically shown [71, 72]. However, these high data rates are only possible with excellent signal quality values and for short-range communications. High data rates can be traded off with longer ranges and for lower link quality values. Depending on the link quality and distance between transmitter and receiver, the rate can be changed through adaptation of the processing gain. UWB achieves processing gain due to pulse repetition, i.e., transmitting more than one pulse within a bit. For example, by transmitting 100 pulses per bit, a processing gain of 20 dB is obtained. Additional processing gain is obtained due to the low duty cycle, which is the ratio of the pulse repetition interval and the pulse width. Adaptation of processing gain is a research topic that needs to be explored for UWB systems. Similarly, multiple-access capability of UWB systems, which is primarily determined by the processing gain, needs to be explored. Adaptive multiaccess code design depending on delay spread of the channel is one of the interesting research areas. Also, adaptive multiuser detection techniques need to be studied for cancelling multiaccess interference. In summary, adaptation algorithms and related parameter estimation techniques for jointly optimizing the multiaccess capability, power consumption, data rate, and range of UWB systems are needed.

Adaptation of multicarrier systems has already gained some momentum. In multicarrier systems, the transmission bandwidth is much wider than the coherence bandwidth of the channel, resulting in frequency-selective fading channels. Therefore, different carriers experience different channel qualities. This leads to adaptation of each subcarrier individually. Adaptive bit/power loading can be used as an effective tool to get the highest capacity from a multicarrier system provided that the transmitter has the link quality information for each carrier. For example, adaptation of the modulation level for OFDM-based multicarrier systems has been studied recently [73]. The modulation level on different carriers can be changed depending on the link quality observed at the carriers. Transmitting different modulations on each carrier requires a large overhead for signaling. Therefore, approaches that group the neighboring carriers into subsets and use the same modulation in each subset are preferred. The signaling can also be avoided in time division duplexing (TDD) systems under certain assumptions. Unlike frequency division duplexing (FDD) systems, where the channels on the downlink and uplink are different, in TDD systems, using the assumption of the reciprocal and slowly varying channel, the transmitter and receiver can be assumed to experience the same channel response. Therefore, this might eliminate the need for signaling of the channel state information to the transmitter, if the channel estimates are used as the link quality measures. However, in this case, the receiver needs to know which modulation is used at the transmitter for each group. Blind modulation detection techniques can be used for this purpose [74]. Note that although the channels could be the same, the interferences observed in the transmitter and receiver are not necessarily the same. Therefore, the observed link qualities would be different at each end. Issues like these need to be studied further for successful implementation of adaptation techniques in multicarrier systems. Multicarrier CDMA (MC-CDMA), which combines OFDM modulation with CDMA-type multiple accessing, is a technology that is being pushed for fourth-generation cellular networks. The previous adaptation algorithms proposed for CDMA and OFDM technologies need to be revisited and optimized, and new adaptation algorithms need to be developed for MC-CDMA.

MIMO and multiantenna systems bring about a new dimension to wireless channels. The spatial dimension will be used in future communications systems for further improvement of the bandwidth and power efficiency. However, this dimension and the related parameter estimates need to be understood better. Research on parameter estimation for fast and accurate calculation of spatial selectivity, angular spread, antenna correlation, etc., is needed. Also, further research is required on the effect of mutual coupling between antenna elements, the effect of near-field scatterers on antenna patterns and antenna correlations, the exploitation of pattern selectivity when the spatial selectivity is not enough, and the generation of a desired pattern selectivity between antenna elements adaptively.

As described earlier, MIMO systems, which employ multiple-transmit and multiple-receive antennas, can provide huge capacity and improved performance gains by exploiting spatial selectivity of the channel. However, these gains, in reality, depend heavily on the statistical properties of the channel and the correlations between antenna elements. Among the factors that affect the antenna correlation are the characteristics of the scattering environment. Therefore, an optimal way of using multiple-antenna systems depends on the situation awareness. If the transmitter knows the instantaneous channel gains (the MIMO channel matrix), it can adapt the transmission to maximize the capacity of the MIMO system [40, 41]. Similarly, the instantaneous antenna correlation values can be exploited to adapt the transmission. In many cases, estimating the perfect instantaneous channel state and antenna correlation information and feeding this information back to the transmitter might not be possible. This is the case especially when the mobility is high. Instead, other parameter measures like partial (statistical) channel information, average channel selectivity, or angular spread would be useful for adapting the transmitter and receiver. Advanced signal processing techniques to calculate this partial channel and correlation information are needed.

Most of the previous adaptation techniques take place in physical and medium access control (MAC) layers. Future-generation wireless systems will also allow adaptive strategies at the higher networking layers. The higher layers will be more aware of the situation in lower layers. Cross-layer optimization algorithms and cost functions that involve many layers will be developed. This will also create the need for the development of new adaptation parameters and algorithms for the estimation of these parameters.

Current wireless communications systems are based on layered protocol design, and each layer is often designed and operated independently. For example, the channel variation is addressed by adapting the link in the physical or MAC layer using signal processing techniques as described before. The traffic load and delays are adapted by changing the routing tables, by adaptive channel and cell assignment techniques. The layered structure and adaptation of layers locally (and independently) simplify the network design. But the performance and capacity of the network is suboptimal, especially for addressing the requirements of future multimedia wireless services. The future applications will have different data rate, delay, power, and QoS requirements. Cross-layer adaptation could address these requirements by jointly optimizing multidimensional cost functions that involve all protocol layers [75–77]. As a result, networks with improved end-to-end performance subject to constraints in link quality and available network resources can be obtained, while being aware of application trade-offs.

1.6   Conclusion

Recently, the use of adaptation algorithms for better utilization of the available resources, like power and spectrum, has grown significantly. Several adaptation strategies to increase performance, data rate, capacity, and QoS of wireless communications systems have been introduced. Many of these adaptation techniques depend on accurate estimation of the various parameters. Therefore, further research on efficient parameter estimation techniques is still needed.

There is a significant amount of work needed for the evolution of the current wireless communications systems to accommodate the future demands. Ultrawideband, MIMO, and multicarrier wireless communications are some of the technologies that are being studied extensively. All of them have a common point: their capability to adapt the changing radio channel conditions. Adaptation of multicarrier communications and MIMO schemes has already gained some momentum. Adaptation algorithms for UWB still need to be explored. The flexibility of UWB makes it very attractive for employing successful adaptation schemes.

Acknowledgment

The author thanks T. Yücek for his help and the anonymous reviewers for their comments.

References

[1] S. Sampei. 1997. *Applications of digital wireless technologies to global wireless communications*. Englewood Cliffs, NJ: Prentice Hall.

[2] T. Ikeda, S. Sampei, and N. Morinaga. 2000. TDMA-based adaptive modulation with dynamic channel assignment for high-capacity communication systems. *IEEE Transactions on Communications* 49:404–12.

[3] S. Nanda, K. Balachandran, and S. Kumar. 2000. Adaptation techniques in wireless packet data services. *IEEE Transactions on Communications* 38:54–64.

[4] R. V. Nobelen, N. Seshadri, J. Whitehead, and S. Timiri. 1999. An adaptive radio link protocol with enhanced data rates for GSM evolution. *IEEE Communications Magazine* 54–63.

[5] A. Furuskar, S. Mazur, F. Muller, and H. Olofsson. 1999. EDGE: Enhanced data rates for GSM and TDMA/136 evolution. *IEEE Personal Communications Magazine* 6:56–66.

[6] S. Anderson, H. Dam, U. Forssen, J. Karlsson, F. Kronestedt, S. Mazur, and K. J. Molnar. 1999. Adaptive antennas for GSM and TDMA systems. *IEEE Personal Communications Magazine* 6:74–86.

[7] W. C. Lee. 1995. *Mobile cellular telecommunications*. New York: McGraw-Hill.

[8] G. Pollini. 1996. Trends in handover design. *IEEE Communications Magazine* 34:82–90.

[9] P. Bender, P. Black, M. Grob, R. Padavoni, N. Sindhushayana, and S. Viterbi. 2000. CDMA/HDR: A bandwidth efficient high speed wireless data service for nomadic users. *IEEE Communications Magazine* 38:70–77.

[10] A. Jalali, R. Padovani, and R. Pankaj. 2000. Data throughput of CDMA-HDR: A high efficiency-high data rate personal communication wireless system. In *Proceedings of the IEEE Vehicular Technology Conference*, Tokyo, vol. 3, pp. 1854–58.

[11] J. Holtzman. 2000. CDMA forward link waterfilling power control. In *Proceedings of the IEEE Vehicular Technology Conference*, Tokyo, vol. 3, pp. 1663–67.

[12] H. Arslan, L. Krasny, D. Koilpillai, and S. Channakeshu. 2000. Doppler spread estimation for wireless mobile radio systems. In *Proceedings of the IEEE WCNC Conference*, Chicago, vol. 3, pp. 1075–79.

[13] M. Sakamoto, J. Huoponen, and I. Niva. 2000. Adaptive channel estimation with velocity estimator for W-CDMA receiver. In *Proceedings of the IEEE Vehicular Technology Conference*, Tokyo, vol. 3, pp. 2024–28.

[14] D. Mottier and D. Castelain. 1999. A Doppler estimation for UMTS-FDD based on channel power statistics. In *Proceedings of the IEEE Vehicular Technology Conference*, Amsterdam, vol. 5, pp. 3052–56.

[15] L. Lindbom. 1992. Adaptive equalization for fading mobile radio channels. Licentiate thesis, Technology Department, Uppasala University, Uppasala, Sweden.

[16] M. Morelli, U. Mengali, and G. Vitetta. 1998. Further results in carrier frequency estimation for transmissions over flat fading channels. *IEEE Communications Letters* 2:327–30.

[17] L. Krasny, H. Arslan, D. Koilpillai, and S. Channakeshu. 2001. Doppler spread estimation in mobile radio systems. In *Proceedings of the IEEE WCNC Conference*, vol. 5, pp. 197–99.

[18] J. H. A. Sampath. 1993. Estimation of maximum Doppler frequency for handoff decisions. In *Proceedings of the IEEE Vehicular Technology Conference*, Secaucus, NJ, pp. 859–62.

[19] L. Wang, M. Silventoinen, and Z. Honkasalo. 1996. A new algorithm for estimating mobile speed at the TDMA-based cellular system. In *Proceedings of the IEEE Vehicular Technology Conference*, Atlanta, GA, vol. 2, pp. 1145–49.

[20] C. Xiao, K. Mann, and J. Olivier. 1999. Mobile speed estimation for TDMA-based hierarchical cellular systems. In *Proceedings of the IEEE Vehicular Technology Conference*, Amsterdam, vol. 2, pp. 2456–60.

[21] K. Kawabata, T. Nakamura, and E. Fukuda. 1994. Estimating velocity using diversity reception. In *Proceedings of the IEEE Vehicular Technology Conference*, Stockholm, vol. 1, pp. 371–74.

[22] W. Lee. 1998. *Mobile communications engineering*. New York: McGraw-Hill.

[23] M. Austin and G. Stüber. 1994. Eigen-based Doppler estimation for differentially coherent CPM. *IEEE Transactions on Vehicular Technology* 43:781–85.

[24] H. Arslan and T. Yücek. 2003. Delay spread estimation for wireless communication systems. In *Proceedings of the Eighth IEEE Symposium on Computers and Communications (ISCC 2003)*, Antalya, Turkey, pp. 282–87.

[25] J.-T. Chen, J. Liang, H.-S. Tsai, and Y.-K. Chen. 1998. Joint MLSE receiver with dynamic channel description. *Journal on Selected Areas in Communications* 16:1604–15.

[26] L. Husson and J.-C. Dany. 1999. A new method for reducing the power consumption of portable handsets in TDMA mobile systems: Conditional equalization. *IEEE Transactions on Vehicular Technology* 48:1936–45.

[27] H. Schober and F. Jondral. 2002. Delay spread estimation for OFDM based mobile communication systems. In *Proceedings of the European Wireless Conference*, Florence, Italy, pp. 625–28.

[28] F. Sanzi and J. Speidel. 2000. An adaptive two-dimensional channel estimator for wireless OFDM with application to mobile DVB-T. *IEEE Transactions on Broadcasting* 46:128–33.

[29] K. Witrisal, Y.-H. Kim, and R. Prasad. 1998. RMS delay spread estimation technique using non-coherent channel measurements. *IEE Electronics Letters* 34:1918–19.

[30] K. Witrisal, Y.-H. Kim, and R. Prasad. 2001. A new method to measure parameters of frequency selective radio channel using power measurements. *IEEE Transactions on Communications* 49:1788–1800.

[31] K. Witrisal and A. Bohdanowicz. 2000. Influence of noise on a novel RMS delay spread estimation method. In *Proceedings of the IEEE PIMRC Conference*, London, vol. 1, pp. 560–66.

[32] H. Arslan and T. Yücek. 2003. Estimation of frequency selectivity for OFDM based new generation wireless communication systems. In *Proceedings of the 2003 World Wireless Congress*, San Francisco.

[33] T. Söderström and P. Stoica. 1997. *Applications of digital wireless technologies to global wireless communications*. Englewood Cliffs, NJ: Prentice Hall.

[34] A. Paulraj and B. Ng. 1998. Space-time modems for wireless personal communications. *IEEE Personal Communications Magazine* 5:36–48.

[35] M. K. Özdemir, H. Arslan, and E. Arvas. 2003. Mutual coupling effect in multi-antenna wireless communication systems. In *IEEE Globecom Conference*, San Francisco, vol. 2, pp. 829–833.

[36] M. K. Özdemir, H. Arslan, and E. Arvas. 2004. Dynamics of spatial correlation and implications on MIMO systems. *IEEE Communications Magazine*.

[37] D. S. Shiu, G. J. Foschini, M. J. Gans, and J. M. Kahn. 2000. Fading correlation and its effects on the capacity of multielement antenna systems. *IEEE Transactions on Communications* 48:502–13.

[38] M. Ivrlac, T. Kurpjuhn, C. Brunner, and W. Utschick. 2001. Efficient use of fading correlations in MIMO systems. In *Proceedings of the IEEE Vehicular Technology Conference*, Atlantic City, NJ, vol. 4, pp. 2763–67.

[39] S. Catreux, V. Erceg, D. Gesbert, and J. Heath. 2002. Adaptive modulation and MIMO coding for broadband wireless data networks. *IEEE Communications Magazine* 40:108–15.

[40] E. Telatar. 1995. *Capacity of multiantenna Gaussian channels*. Technical report. AT&T Bell Laboratories.

[41] G. J. Foschini and M. J. Gans. 1998. On limits of wireless communications in a fading environment when using multiple antennas. *Wireless Personal Communications* 6:311–35.

[42] TIA/EIA. 2000. *TDMA third generation wireless: Digital traffic channel layer 1*. TIA/EIA 136-131-B.

[43] K. Balachandran, S. Kabada, and S. Nanda. 1998. Rate adaptation over mobile radio channels using channel quality information. In *Proceedings of the IEEE Globecom'98 Communication Theory Mini Conference Record*, pp. 46–52.

[44] J. Jacobsmeyer. 1996. Adaptive data rate modem. U.S. Patent 5541955.

[45] M. Türkboylari and G. L. Stüber. 1998. An efficient algorithm for estimating the signal-to-interference ratio in TDMA cellular systems. *IEEE Transactions on Communications* 46:728–31.

[46] H. Arslan and S. Reddy. 2003. Noise variance and SNR estimation for OFDM based wireless communication systems. In *Proceedings of the 3rd International Conference on Wireless and Optical Communications*, Banff, Alberta, Canada.

[47] G. B. H. Arslan. 2001. Channel estimation in narrowband wireless communication systems. *Wireless Communications and Mobile Computing (WCMC) Journal* 1:201–19.

[48] G. Bottomley and H. Arslan. 2003. Channel estimation for time-varying channels in wireless communication systems. In *The Wiley encyclopedia of telecommunications*. New York: Wiley.

[49] K. Homayounfar. 2003. Rate adaptive speech coding for universal multimedia access. *IEEE Signal Processing Magazine* 20:30–39.

[50] S. Wolf, C. Dvorak, R. Kubichek, and C. South. 1991. How will we rate telecommunications system performance? *IEEE Communications Magazine* 29:23–29.

[51] S. Voran. 1999. Objective estimation of perceived speech quality: Part II. *IEEE Transactions on Speech and Audio Processing* 7:383–90.

[52] H. Arslan, R. Ramesh, and A. Mostafa. 1999. Interpolation and channel tracking based receivers for coherent Mary-PSK modulations. In *Proceedings of the IEEE Vehicular Technology Conference*, Houston, TX, vol. 3, pp. 2194–99.

[53] W. Liu. 1994. Performance of joint data and channel estimation using tap variable step size LMS for multipath fast fading channel. In *Proceedings of the IEEE Globecom Conference*, San Francisco, vol. 2, pp. 973–78.

[54] M. Benthin and K.-D. Kammeyer. 1997. Influence of channel estimation on the performance of a coherent DS-CDMA system. *IEEE Transactions on Vehicular Technology* 46:262–68.

[55] P. Schramm. 1999. Differentially coherent demodulation for differential BPSK in spread spectrum systems. *IEEE Transactions on Vehicular Technology* 48:1650–56.

[56] D. Hui and K. Zangi. 2001. An adaptive maximum-likelihood receiver for colored noise and interference. In *Proceedings of the IEEE Vehicular Technology Conference*, Atlantic City, NJ, vol. 4, pp. 2257–61.

[57] A. Hafeez, K. Molnar, and G. Bottomley. 1999. Co-channel interference cancellation for D-AMPS handsets. In *Proceedings of the IEEE Vehicular Technology Conference*, Houston, TX, vol. 2, pp. 1026–30.

[58] A. Hafeez, H. Arslan, and K. Molnar. 2001. Adaptive joint detection of co-channel signals for ANSI-136 handsets. In *Proceedings of the IEEE PIMRC Conference*, San Diego, vol. 2, pp. E-105–10.

[59] J. Zander. 1992. Performance of optimum transmitter power control in cellular radio systems. *IEEE Transactions on Vehicular Technology* 41:57–62.

[60] Z. Rosberg and J. Zander. 1998. Toward a framework for power control in cellular systems. *Wireless Networks* 4:215–22.

[61] S. Ulukus and R. D. Yates. 1998. Stochastic power control for cellular radio systems. *IEEE Transactions on Communications* 46:784–98.

[62] S. V. Hanly and D. N. Tse. 1999. Power control and capacity of spread-spectrum wireless networks. *Automatica* 35:1987–2012.

[63] A. Goldsmith. 1997. The capacity of downlink fading channels with variable rate and power. *IEEE Transactions on Vehicular Technology* 46:569–80.

[64] H. Arslan, T. J.-F. Cheng, and K. Balachandran. 2001. Physical layer evolution for GSM/EDGE. In *Proceedings of the IEEE Globecom Conference*, San Antonio, TX, vol. 5, pp. 8–27.

[65] H. Arslan, T. J.-F. Cheng, and K. Balachandran. 2001. Evolution of EDGE to higher data rates using QAM. In *Proceedings of the IEEE Vehicular Technology Conference*, Atlantic City, NJ, vol. 4, pp. 2267–71.

[66] A. Doufexi, S. Armour, M. Butler, A. Nix, D. Bull, J. McGeehan, and P. Karlsson. 2002. A comparison of the HIPERLAN/2 and IEEE 802.11a wireless LAN standards. *IEEE Communications Magazine* 40:172–80.

[67] I. Katzela and M. Naghshineh. 1996. Channel assignment schemes for cellular mobile telecommunication systems: A comprehensive survey. *IEEE Personal Communications Magazine* 3:10–31.

[68] N. D. Tripathi, N. J. H. Reed, and H. F. V. Landingham. 1998. Handoff in cellular systems. *IEEE Personal Communications Magazine* 5:26–37.

[69] G. Pollini. 1996. Trends in handover design. *IEEE Communications Magazine* 3:82–90.

[70] G. Aiello and G. Rogerson. 2003. Ultrawideband wireless systems. *IEEE Microwave Magazine* 4(2):36–47.

[71] J. Forrester, G. Evan, D. Leeper, and S. Srinivasa. 2001. Ultra-wideband technology for short or medium-range wireless communications. *Intel Technology Journal* 1–7.

[72] P. Mannion. 2002. Ultrawideband radio set to redefine wireless signaling. *EE Times*, pp. 71–84.

[73] L. Hanzo, C. H. Wong, and M. S. Yee. 2002. *Adaptive wireless transceivers: Turbo-coded, turbo-equalized and space-time coded TDMA, CDMA, and OFDM systems*. New York: John Wiley & Sons.

[74] S. Reddy, T. Yucek, and H. Arslan. 2003. An efficient blind modulation detection algorithm for adaptive OFDM systems. In *Proceedings of the IEEE Vehicular Technology Conference*, Orlando, FL, vol. 6, pp. 8–27.

[75] A. Goldsmith and S. Wicker. 2002. Design challenges for energy-constrained ad hoc wireless networks. *IEEE Wireless Communications* 9(4):8–27.

[76] T. Rappaport, A. Annamalai, R. Buehrer, and W. Tranter. 2002. Wireless communications: Past events and a future perspective. *IEEE Communications Magazine* 40:148–61.

[77] Z. Haas. 2001. Design methodologies for adaptive and multimedia networks. *IEEE Communications Magazine* 39:106–7.

2

Adaptive Channel Estimation in Wireless Communications

Jitendra K. Tugnait
Auburn University

2.1 Introduction

Propagation of signals through wireless channels (indoors or outdoors) results in the transmitted signal arriving at the receiver through multiple paths. These paths arise due to reflection, refraction, or diffraction in the channel. Multipath propagation results in a received signal that is a superposition of several delayed and scaled copies of the transmitted signal giving rise to frequency-selective fading. Frequency-selective fading (defined as changes in the received signal level in time) is caused by destructive interference among multiple propagation paths. The environment around the transmitter and the receiver can change over time, particularly in a mobile setting, leading to variations

35

in the channel response with time. This gives rise to time-selective fading. Also, the channels may have a dominant path (direct path in line-of-sight channels) in addition to several secondary paths, or they may be characterized as having multiple "random" paths with no single dominant path.

Multipath propagation leads to intersymbol interference (ISI) at the receiver, which in turn may lead to high error rates in symbol detection. Equalizers are designed to compensate for these channel distortions. One may directly design an equalizer given the received signal, or one may first estimate the channel impulse response and then design an equalizer based on the estimated channel. After some processing (matched filtering, for instance), the continuous-time received signals are sampled at the baud (symbol) or higher (fractional) rate before processing them for channel estimation or equalization. It is therefore convenient to work with a baseband-equivalent discrete-time channel model. Depending upon the sampling rate, one has either a single-input single-output (SISO) (baud-rate sampling) or a single-input multiple-output (SIMO) (fractional sampling), complex discrete-time-equivalent baseband channel.

In this chapter, we present a review of various approaches to channel estimation for wireless mobile systems. Since approaches to channel estimation depend upon the underlying channel model, we also review various approaches to channel modeling. In section 2.2 we present the relevant channel models, including time-variant and time-invariant models. In section 2.3 various channel estimation methods suitable for block-by-block tracking are presented. In section 2.4 adaptive channel estimation approaches are reviewed. In section 2.5 we illustrate some of the reviewed approaches via two simulation examples to conclude the chapter.

Notation: Superscripts H, *, T, and † denote the complex conjugate transpose, complex conjugation, transpose, and Moore-Penrose pseudo-inverse operations, respectively. $\delta(\cdot)$ is the Kronecker delta function and $\mathbf{I}_N$ is the $N \times N$ identity matrix. The symbol $\otimes$ denotes the Kronecker product, and $tr(\mathbf{A})$ is the trace of a square matrix $\mathbf{A}$. The $(n,m)^{\text{th}}$ entry of a matrix $\mathbf{C}$ is denoted by $[\mathbf{C}]_{n,m}$.

2.2 Wireless Channel Models

2.2.1 Time-Variant (Doubly Selective) Channels

Consider a time-varying (e.g., mobile wireless) channel (linear system) with complex baseband, continuous-time, received signal $x(t)$ and transmitted complex baseband, continuous-time information signal $s(t)$ (with symbol interval T_s seconds) related by [42]

$$x(t) = \int_{-\infty}^{\infty} \tilde{h}(t;\tau)s(t-\tau)d\tau + w(t) , \qquad (2.1)$$

where $\tilde{h}(t;\tau)$ is the time-varying impulse response of the channel denoting the response of the channel at time t to a unit impulse input at time $t - \tau$, and $w(t)$ is the additive noise (typically white Gaussian). A delay Doppler spread function $H(f;\tau)$ is defined as the Fourier transform of $\tilde{h}(t;\tau)$ [2, 42]:

$$H(f;\tau) = \int_{-\infty}^{\infty} \tilde{h}(t;\tau) exp^{-j2\pi ft} \, dt \; . \tag{2.2}$$

If $|H(f;\tau)| \approx 0$ for $|\tau| > \tau_d$, then τ_d is called the (multipath) delay spread of the channel. If $|H(f;\tau)| \approx 0$ for $|f| > f_d$, then f_d is called the Doppler spread of the channel. Equation (2.1) is the most general form of a mobile channel discussed in this chapter.

In order to capture the complexity of the physical interactions characterizing the transmission through a real channel, $\tilde{h}(t;\tau)$ is typically modeled as a two-dimensional zero-mean random process. If $\tilde{h}(t;\tau)$ is wide-sense stationary in variable t, and $\tilde{h}(t;\tau_1)$ is uncorrelated with $\tilde{h}(t;\tau_2)$ for $\tau_1 \neq \tau_2$ and any t, one obtains the well-known wide-sense stationary uncorrelated scattering (WSSUS) channel [2, 42, section 14].

In this chapter we will confine our attention to deterministic modeling of $\tilde{h}(t;\tau)$, which may be thought of as capturing realizations of the underlying random process.

2.2.1.1 Tapped Delay Line Model

We now consider a discrete-time channel model. If a linear modulation scheme is used, the baseband transmitted signal can be represented as

$$s(t) = \sum_{k=-\infty}^{\infty} s(k) g_T(t - kT_s), \tag{2.3}$$

where $\{s(k)\}$ is the information sequence and $g_T(t)$ is the transmit (low-pass) filter (typically a root raised cosine filter). Therefore, the baseband signal incident at the receiver is given by

$$x(t) = \sum_{k=-\infty}^{\infty} s(k) \int_{-\infty}^{\infty} \tilde{h}(t;\alpha) g_T(t - kT_s - \alpha) d\alpha + w(t). \tag{2.4}$$

After filtering with a receive filter with impulse response $g_R(t)$, the received baseband signal is given by

$$y(t) = \sum_{k=-\infty}^{\infty} s(k) \int_{-\infty}^{\infty} \int_{-\infty}^{\infty} g_R(t-\beta) \tilde{h}(\beta;\alpha) g_T(\beta - kT_s - \alpha) d\alpha d\beta + v(t), \tag{2.5}$$

where

$$v(t) = \int g_R(\tau) w(t - \tau) d\tau.$$

If the continuous-time signal $y(t)$ is sampled once every T_s seconds, we obtain the discrete-time sequence

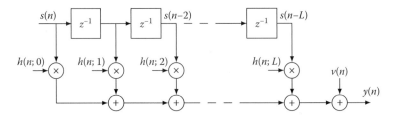

FIGURE 2.1 Tapped delay line model of frequency- and time-selective channel with finite impulse response. z^{-1} represents a unit (symbol duration) delay.

$$y(n) := y(t)|_{t=nT_s} = \sum_{k=-\infty}^{\infty} s(k)h(n;n-k)+v(n) = \sum_{l=-\infty}^{\infty} h(n;l)s(n-l)+v(n) , \quad (2.6)$$

where $h(n;l)$ is the (effective) channel response at time n to a unit input at time $n - l$ and

$$h(n;n-k) := \int_{-\infty}^{\infty} \int_{-\infty}^{\infty} g_R(nT_s - \beta)\tilde{h}(\beta;\alpha)g_T(\beta - kT_s - \alpha)d\alpha d\beta. \quad (2.7)$$

Note that the noise sequence $\{v(n)\}$ in (2.6) is no longer necessarily white; it can be whitened by further time-invariant linear filtering (see [42]). Henceforth, we assume that a whitening filter has been applied to $y(n)$, but with an abuse of notation, we will still use (2.6). For a causal system, $h(n;l) = 0$ for $l < 0$ ($\forall n$), and for a finite-length channel of maximum length $T_s L$, $h(n;l) = 0$ for $l > L$ ($\forall n$). In this case we modify (2.6) as (recall the noise whitening filter)

$$y(n) = \sum_{l=0}^{L} h(n;l)s(n-l)+v(n). \quad (2.8)$$

The model (2.6) represents a time- and frequency-selective linear channel. A tapped delay line structure for this model is shown in Figure 2.1. For a slowly (compared to the baud-rate) time-varying system, one often simplifies (2.8) to a time-invariant system as

$$y(n) = \sum_{l=0}^{L} h(l)s(n-l)+v(n) , \quad (2.9)$$

where $h(l) = h(0;l)$ is the time-invariant channel response to a unit input at time 0. The model (2.9) represents a frequency-selective linear channel with no time selectivity. It is the most commonly used model for receiver design.

Suppose that $h(n;l) = h(n)\delta(l,0)$ where $\delta(l,0)$ is the Kronecker delta located at 0, i.e., $\delta(l,0) = 1$ for $l = 0$ and $\delta(l,0) = 0$ for $l \neq 0$. Then we have the time-selective and frequency-non-selective channel whose output is given by

$$y(n) = h(n)s(n) + v(n). \tag{2.10}$$

Finally, a time-non-selective and frequency-non-selective channel is modeled as

$$y(n) = hs(n) + v(n), \tag{2.11}$$

where h is a random variable (or a constant).

2.2.1.2 Autoregressive (AR) Models

It is possible to accurately represent a WSSUS channel by a large-order AR model; see [34] and references therein. However, it is far more common to use a first-order AR model (AR (1)) given by [34, 35]

$$h(n;l) = \alpha_c h(n-1;l) + w_c(n), \tag{2.12}$$

where α_c is the AR coefficient, and the driving noise $w_c(n)$ is zero-mean complex Gaussian with variance σ_{wc}^2 and statistically independent of $h(n-1;l)$. Assume that $h(n;l)$ is also zero-mean complex Gaussian with variance σ_h^2. Then [35]

$$\alpha_c = \frac{1}{\sigma_h^2} E\{h(n;l)h^*(n-1;l)\}, \tag{2.13}$$

$$\sigma_{wc}^2 = \sigma_h^2 \left(1 - |\alpha_c|^2\right). \tag{2.14}$$

2.2.1.3 Basis Expansion Models

Recently, basis expansion models (BEMs) have been widely investigated to represent doubly selective channels in wireless applications [3, 13, 36, 46, 61], where the time-varying taps are expressed as superpositions of time-varying basis functions in modeling Doppler effects, weighted by time-invariant coefficients. Candidate basis functions include complex exponential (Fourier) functions [13, 36], polynomials [3], discrete prolate spheroidal sequences [61], etc. In contrast to AR models that describe temporal variation on a symbol-by-symbol update basis, a BEM depicts the evolution of the channel over a period (block) of time. Intuitively, the coefficients of the BEM approximation should evolve much more slowly in time than the channel, and hence are more convenient to track in a fast-fading environment.

Suppose that we include the effects of transmit and receive filters in the time-variant impulse response $h(t;\tau)$ in (2.1). Suppose that this channel has a delay spread τ_d and a Doppler spread f_d. Consider the kth block of data consisting of an observation window of T_B symbols where the baud-rate data samples in the block are indexed as $n = \bar{n}_k, \bar{n}_k + 1,$ $\cdots, \bar{n}_k + T_B - 1, \bar{n}_k := (k-1)T_B$. If $2f_d\tau_d < 1$ (underspread channel), the complex exponential basis expansion model (CE-BEM) representation of $h(n;l)$ in (2.6) is given by

$$h(n;l) = \sum_{q=1}^{Q} h_q(l)e^{j\omega_q n}, \quad n = \overline{n}_k, \overline{n}_k+1, \cdots, \overline{n}_k+T_B-1, \tag{2.15}$$

where one chooses ($l = 0, 1, \cdots, L$, and K is an integer)

$$T := KT_B, \quad K \geq 1, \tag{2.16}$$

$$Q \geq 2\lceil f_d T T_s \rceil + 1, \tag{2.17}$$

$$\omega_q := \frac{2\pi}{T}\left[q - (Q+1)/2\right], \quad q = 1, 2, \cdots, Q, \tag{2.18}$$

$$L := \lceil \tau_d / T_s \rceil. \tag{2.19}$$

The BEM coefficients $h_q(l)$ remain invariant during this block, but are allowed to change at the next block, and the Fourier basis functions $\{e^{j\omega_q n}\}$ ($q = 1, 2, \cdots, Q$) are common for each block. If the delay spread τ_d and the Doppler spread f_d of the channel (or at least their upper bounds) are known, one can infer the basis functions of the CE-BEM [36]. Treating the basis functions as known, estimation of a time-varying process is reduced to estimating the invariant coefficients over a block of length T_B symbols. Note that the BEM period is $T = KT_B$, whereas the block size is T_B symbols. If $K > 1$ (e.g., $K = 2$ or $K = 3$), then the Doppler spectrum is said to be oversampled [32] compared to the case $K = 1$, where the Doppler spectrum is said to be critically sampled. In [13, 36] only $K = 1$ (henceforth called CE-BEM) is considered, whereas [32] considers $K \geq 2$ (henceforth called oversampled CE-BEM).

CE-BEM has a finite impulse response (FIR) structure in both time and frequency domains [46]. This unique time-frequency duality makes it a widely used model depicting the temporal variations of wireless channels. For $K = 1$, the rectangular window of this truncated discrete Fourier transform (DFT)-based model introduces spectral leakage [43]. The energy at each individual frequency leaks to the full-frequency range, resulting in significant amplitude and phase distortion at the beginning and the end of the observation window [61]. To mitigate this leakage, the oversampled CE-BEM with $K = 2$ or 3 has been considered in [32].

Equation (2.15) applies to single-input single-output systems—one user and one receiver with symbol-rate sampling. It is easily modified to handle multiuser, multiple-transmit and -receive antennas, and higher-than-symbol-rate sampling—the basic representation remains essentially unchanged.

The representation $h(n;l)$ in (2.15) is a special case of a more general representation:

$$h(n;l) = \sum_{q=1}^{Q} h_q(l)\phi_q(n), \tag{2.20}$$

where $\{\phi_q(n)\}_{q=1}^{Q}$ are a set of orthogonal basis functions (over the time interval under consideration). Examples include wavelet-based expansions as in [37], polynomial bases as in [3], and other possibilities [40].

In discrete prolate spheroidal BEM (DPS-BEM), the ith DPS vector $\mathbf{u}_i := [u_i(0), \cdots, u_i(T_B - 1)]^T$ (called Slepian sequence in [61], which is a time-windowed [infinite] DPS sequence) is the ith eigenvector of a matrix $\mathbf{C}$ [48]: $\mathbf{C}\mathbf{U}_i = \lambda_i\mathbf{u}_i$, where

$$\left[\mathbf{C}\right]_{n,m} = \frac{\sin\left[2\pi\left(n-m\right)f_d T_s\right]}{\left[\pi\left(n-m\right)\right]}$$

is the $(n,m)^{th}$ entry of $\mathbf{C}$ and $\lambda_1 \geq \lambda_2 \geq \cdots \geq \lambda_{T_B}$ are the eigenvalues of $\mathbf{C}$. The Slepian sequences $\{u_q(n)\}$ are orthonormal over the finite time interval $[0,T_B - 1]$. The modeling error of the CE-BEM can result in a noticeable floor in BER curves [18]. The polynomial basis functions are neither time limited nor band limited, and their square bias varies heavily over the range of Doppler spread considered in [61]. DPS sequences are a good alternative as a basis set to approximate band-limited channels alleviating the spectral leakage of CE-BEM [61]. The (infinite) DPS sequences have their maximum energy concentration in an interval with length T while being band limited to $[-f_d T_s, f_d T_s]$, where $u_1(n)$ is the unique sequence that is band limited and most time concentrated, $u_2(n)$ is the next sequence having maximum energy concentration among the DPS sequences orthogonal to $u_1(n)$, and so on [48].

Figure 2.2 shows the channel modeling errors resulting from (critically sampled) CE-BEM, DPS-BEM, and oversampled CE-BEM ($K = 2$ or 3) when the underlying channel is a one-tap time-selective channel following Jakes' spectrum [22]. The results are

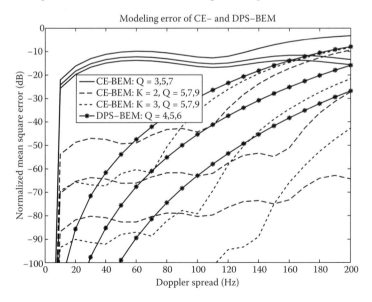

FIGURE 2.2 Channel modeling error for one-tap Jakes' channel, $T_B = 400$, $T_S = 25$ μs.

based on Monte Carlo averaging over one thousand runs with $T_B = 400$, $T_s = 25$ µs and varying Doppler spreads. (The results were obtained following the procedure in [61].) For a fixed value of Q, DPS-BEM provides the best fit, whereas CE-BEM (no oversampling) yields minor improvements with increasing Q. On the other hand, the basis functions in oversampled CE-BEM are not mutually orthogonal, leading to "analytical" difficulties. It is interesting to note that there exists a vast literature based on CE-BEM (no oversampling) where the large modeling errors are completely ignored and it is assumed that physical channel is accurately described by CE-BEM for both analysis and simulations; see, e.g., [1, 24–26, 31, 36, 45].

2.2.2   Time-Invariant Channels

After some processing (matched filtering, for instance), the continuous-time received signals are sampled at the baud (symbol) or higher (fractional) rate before being processed for channel estimation or equalization. It is therefore convenient to work with an equivalent baseband discrete-time white noise channel model [42, section 10.1]. For a baud-rate sampled system, the equivalent baseband channel model is given by

$$y(n) = \sum_{l=0}^{L} h(l)s(n-l) + v(n), \tag{2.21}$$

where $\{v(n)\}$ is a white Gaussian noise sequence with variance σ^2; $\{s(n)\}$ is the zero-mean, independent and identically distributed (i.i.d.), information (symbol) sequence, possibly complex, taking values from a finite set; $\{h(l)\}$ is an FIR linear filter (with possibly complex coefficients) that represents the equivalent channel, including the effects of the noise whitening filter; and $\{y(n)\}$ is the (possibly complex) equivalent baseband received signal. A tapped delay line structure for this model is shown in Figure 2.3.

The model (2.21) results in a single-input single-output (SISO) complex discrete-time baseband-equivalent channel model. The output sequence $\{y(n)\}$ in (2.21) is discrete-time stationary. When there is excess channel bandwidth [bandwidth $> \frac{1}{2} \times$ (baud rate)], baud-rate sampling is below the Nyquist rate, leading to aliasing and, depending upon the symbol timing phase, in certain cases, causing deep spectral notches in the sampled, aliased channel transfer function [15]. Linear equalizers designed on the basis of the

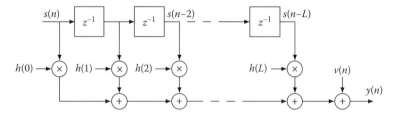

FIGURE 2.3   Tapped delay line model of the frequency-selective but time-non-selective baud-rate channel.

baud-rate sampled channel response are quite sensitive to symbol timing errors. Initially, in the trained case, fractional sampling was investigated to "robustify" the equalizer performance against timing errors. The model (2.21) does not apply to fractionally spaced samples, i.e., when the sampling interval is a fraction of the symbol duration. The fractionally sampled digital communications signal is a cyclostationary signal [9] that may be represented as a vector stationary sequence using a time series representation (TSR) [9, section 12.6]. Suppose that we sample at N times the baud rate with signal samples spaced T_s/N seconds apart where T_s is the symbol duration. Then a TSR for the sampled signal is given by

$$y_i(n) = \sum_{l=0}^{L} h_i(l)s(n-l) + v_i(n); \quad (i=1,2,\cdots,N), \qquad (2.22)$$

where now we have N samples every symbol period, indexed by i. Notice, however, that the information sequence $s(n)$ is still one sample per symbol. It is assumed that the signal incident at the receiver is first passed through a receive filter whose transfer function equals the square root of a raised cosine pulse, and that the receive filter is matched to the transmit filter. The noise sequence in (2.22) is the result of the fractional-rate sampling of a continuous-time filtered white Gaussian noise process. Therefore, the sampled noise sequence is white at the symbol rate, but correlated at the fractional rate. Stack N consecutive received samples in the nth symbol duration to form an N vector $y(n)$ satisfying

$$\mathbf{y}(n) = \sum_{l=0}^{L} \mathbf{h}(l)s(n-l) + \mathbf{v}(n), \qquad (2.23)$$

where $\mathbf{h}(n)$ is the vector impulse response of the SIMO-equivalent channel model given by

$$\mathbf{h}(n) = \begin{bmatrix} h_1(n) & h_2(n) & \cdots & h_N(n) \end{bmatrix}^T, \qquad (2.24)$$

and $\mathbf{y}(n)$ and $\mathbf{v}(n)$ are defined similarly. A block diagram of model (2.22) is shown in Figure 2.4.

2.3   Channel Estimation

We first consider three types of channel estimators within the framework of maximizing the likelihood function. (Unless otherwise noted, the underlying channel model is given by the time-invariant model (2.23).) In general, one of the most effective and popular parameter estimation algorithms is the maximum likelihood (ML) method. The class of maximum likelihood estimators are optimal asymptotically.

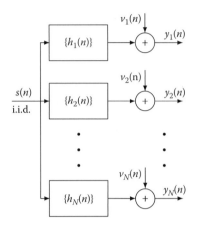

FIGURE 2.4 Block diagram of the fractionally sampled ($N \times$ baud-rate) frequency-selective but time-non-selective channel.

Let us consider the N vector channel model given in (2.23). Suppose that we have collected M samples of the observation $Y = [\mathbf{y}(T_B - 1), \cdots, \mathbf{y}^T(0)]^T$. We then have the following linear model:

$$
Y = \begin{pmatrix} s(T_B-1)\mathbf{I}_N & s(T_B-2)\mathbf{I}_N & \cdots & s(T_B-L-1)\mathbf{I}_N \\ \vdots & Block & Hankel & Matrix \\ s(0)\mathbf{I}_N & s(-1)\mathbf{I}_N & \cdots & s(-L)\mathbf{I}_N \end{pmatrix} \begin{pmatrix} \mathbf{h}(0) \\ \vdots \\ \mathbf{h}(L) \end{pmatrix} + \begin{pmatrix} \mathbf{v}(T_B-1) \\ \vdots \\ \mathbf{v}(0) \end{pmatrix}, \quad (2.25)
$$

$$
= \mathcal{T}(\mathbf{s})_{[T_B N]\times[N(L+1)]}\mathcal{H} + V
$$

where $\mathbf{I}_N$ is a $N \times N$ identity matrix, s and V are vectors consisting of samples of the input sequence $\{s(n)\}$ and noise $\{v(n)\}$, respectively, $\mathcal{H}$ is the vector of the channel parameters, and a block Hankel matrix has identical block entries on its block antidiagonals.

Let θ be the vector of unknown parameters that may include the channel parameters $\mathcal{H}$ and possibly the entire or part of the input vector s. Given the probability space that describes jointly the noise vector W and possibly the input data vector s, we can then obtain, in principle, the probability density function (pdf) of the observation Y. As a function of the unknown parameter θ, the pdf of the observation $f(Y|\theta)$ is referred to as the *likelihood function*. The maximum likelihood estimator is defined by the following optimization:

$$
\hat{\theta} = \arg\max_{\theta \in \Theta} f(Y|\theta), \quad (2.26)
$$

where Θ defines the domain of the optimization.

While the ML estimator is conceptually simple, and it usually has good performance when the sample size is sufficiently large, the implementation of the ML estimator is sometimes computationally intensive. Furthermore, the optimization of the likelihood function in (2.26) is often hampered by the existence of local maxima. Therefore, it is desirable that effective initialization techniques are used in conjunction with the ML estimation.

2.3.1 Training-Based Channel Estimation

The training-based channel estimation assumes the availability of the input vector s (as training symbols) and its corresponding observation vector Y. When the noise samples are zero mean, white Gaussian, i.e., v is a zero-mean Gaussian random vector with covariance $\sigma_v^2 \mathbf{I}_{T_{BN}}$, the ML estimator defined in (2.26), with $\theta = \mathcal{H}$, is given by

$$\hat{\mathcal{H}} = \arg\min_{\mathcal{H}} \left\| Y - \mathcal{T}(\mathbf{s})\mathcal{H} \right\|^2 = \mathcal{T}^{\dagger}(\mathbf{s})Y, \qquad (2.27)$$

where $\mathcal{T}^{\dagger}(\mathbf{s})$ is the Moore-Penrose pseudo-inverse of the $\mathcal{T}(\mathbf{s})$ defined in (2.25). This is also the classical linear least squares estimator, which can be implemented recursively, and it turns out to be the best (in terms of having minimum mean square error) among all unbiased estimators, and it is the most efficient in the sense that it achieves the Cramer-Rao lower bound. Various adaptive implementations can be found in [42].

2.3.1.1 Time-Variant Channels

In case of general time-varying channels represented by (2.8), a simple generalization of [4] (see also [36]) is to use a periodic Kronecker delta function sequence as training:

$$s(n) = \sum_{j} \delta(n - j\overline{P}). \qquad (2.28)$$

With (2.28) as input to model (2.8), one obtains

$$y(n) = \sum_{j} h(n; n - j\overline{P}) + v(n), \qquad (2.29)$$

so that if $\overline{P} > L$, we have for $0 \le i \le L$,

$$y(k\overline{P} + i) = \sum_{j} h(k\overline{P} + i; i) + v(k\overline{P} + i). \qquad (2.30)$$

Therefore, one may take the estimate of $h(k\overline{P}; i)$ as

$$\hat{h}(k\overline{P}; i) = y(k\overline{P} + i) = h(k\overline{P} + i; i) + v(k\overline{P} + i). \qquad (2.31)$$

For time samples between $k\overline{P}$ (k is an integer), linear interpolation may be used to obtain channel estimates.

If we use a CE-BEM representation (2.20), then one directly estimates the time-invariant parameters $h_q(l)$. From (2.8) and (2.20) we have

$$y(n) = \sum_{l=0}^{L} \sum_{q=1}^{Q} h_q(l)\phi_q(n)s(n-l) + v(n) \tag{2.32}$$

$$= \sum_{l=0}^{L} s(n-l) \underbrace{\left[\phi_q(n) \quad \cdots \quad \phi_q(n) \right]}_{E(n)} \underbrace{\begin{bmatrix} h_1(l) \\ \vdots \\ h_Q(l) \end{bmatrix}}_{H(l)} + v(n). \tag{2.33}$$

Collecting T_B samples of the observations $Y = [y(T_B - 1), \cdots, y(0)]^T$ we have the linear model

$$Y = \begin{pmatrix} s(T_B-1)E(T_B-1) & \cdots & s(T_B-L-1)E(T_B-1) \\ \vdots & & \vdots \\ I(0)E(0) & \cdots & I(-L)E(0) \end{pmatrix} \begin{pmatrix} H(0) \\ \vdots \\ H(L) \end{pmatrix}$$

$$+ \begin{pmatrix} v(T_B-1) \\ \vdots \\ v(0) \end{pmatrix}. \tag{2.34}$$

Now we have a model similar to (2.25) with a solution similar to (2.27).

2.3.2 Blind Channel Estimation

2.3.2.1 Combined Channel and Symbol Estimation

The simultaneous estimation of the input vector and the channel appears to be ill-posed; how is it possible that the channel and its input can be distinguished using only the observation? The key in blind channel estimation is the utilization of qualitative information about the channel and the input. To this end, we consider two different types of maximum likelihood techniques based on different models of the input sequence.

2.3.2.1.1 Stochastic Maximum Likelihood Estimation

While the input vector s is unknown, it may be modeled as a random vector with a known distribution. In such a case, the likelihood function of the unknown parameter $\theta = \mathcal{H}$ can be obtained by

$$f(Y|\mathcal{H}) = \int f(Y|s,\mathcal{H})f(s)ds, \tag{2.35}$$

where $f(\mathbf{s})$ is the marginal pdf of the input vector and $f(Y|\mathbf{s},\mathcal{H})$ is the likelihood function when the input is known. Assume, for example, that the input data symbol $s(k)$ takes, with equal probability, a finite number of values. Consequently, the input data vector s also takes values from the signal set $\{\mathbf{s}_1, \ldots, \mathbf{s}_K\}$. The likelihood function of the channel parameters is then given by

$$f(Y|\mathcal{H}) = \sum_{i=1}^{K} f(Y|\mathbf{s}_i,\mathcal{H})\text{Prob}(\mathbf{s}=\mathbf{s}_i) = C\sum_{i=1}^{K} \exp\left\{-\frac{\left\|Y-\mathcal{T}(\mathbf{s}_i)\mathcal{H}\right\|^2}{2\sigma^2}\right\}, \quad (2.36)$$

where C is a constant, $\|Y\|^2 := Y^H Y$, Y^H is the complex conjugate transpose of the complex vector Y, and the stochastic maximum likelihood estimator is given by

$$\hat{\mathcal{H}} = \arg\min_{\mathcal{H}} \sum_{i=1}^{K} \exp\left\{-\frac{\left\|Y-\mathcal{T}(\mathbf{s}_i)\mathcal{H}\right\|^2}{2\sigma^2}\right\}. \quad (2.37)$$

The maximization of the likelihood function defined in (2.35) is in general difficult because $f(Y|\theta)$ is nonconvex. The expectation-maximization (EM) algorithm can be applied to transform the complicated optimization to a sequence of quadratic optimizations. Kaleh and Vallet [23] first applied the EM algorithm to the equalization of communication channels with input sequence having the finite alphabet property. By using a *hidden Markov model* (HMM), they developed a batch (off-line) procedure that includes the so-called forward and backward recursions. Unfortunately, the complexity of this algorithm increases exponentially with the channel memory.

To relax the memory requirements and facilitate channel tracking, on-line sequential approaches have been proposed in [28] for input with finite alphabet properties under an HMM formulation. Given the appropriate regularity conditions and a good initialization guess, it can be shown that these algorithms converge to the true channel value.

2.3.2.1.2 *Deterministic Maximum Likelihood Estimation*

The deterministic ML approach assumes no statistical model for the input sequence $\{s(k)\}$. In other words, both the channel vector $\mathcal{H}$ and the input source vector s are parameters to be estimated. When the noise is zero-mean Gaussian with covariance $\sigma_v^2 \mathbf{I}_{T_B N}$, the ML estimates can be obtained by the nonlinear least squares optimization

$$\{\hat{\mathcal{H}},\hat{\mathbf{s}}\} = \arg\min \left\|Y-\mathcal{T}(\mathbf{s})\mathcal{H}\right\|^2. \quad (2.38)$$

The joint minimization of the likelihood function with respect to both the channel and the source parameter spaces is difficult. Fortunately, the observation vector Y is linear in both the channel and the input parameters individually. In particular, we have

$$Y = T(\mathbf{s})\mathcal{H} + V = F(\mathcal{H})\mathbf{s} + V, \tag{2.39}$$

where

$$F(\mathcal{H}) = \begin{pmatrix} \mathbf{h}(0) & \cdots & \mathbf{h}(L) & & \\ & \ddots & & \ddots & \\ & & \mathbf{h}(0) & \cdots & \mathbf{h}(L) \end{pmatrix} \tag{2.40}$$

is the the so-called filtering matrix. We therefore have a separable nonlinear least squares problem that can be solved sequentially:

$$\{\hat{\mathcal{H}}, \hat{\mathbf{s}}\} = \arg\min_{\mathbf{s}} \left\{ \min_{\mathcal{H}} \left\| Y - T(\mathbf{s})\mathcal{H} \right\|^2 \right\} \tag{2.41}$$

$$= \arg\min_{\mathcal{H}} \left\{ \min_{\mathbf{s}} \left\| Y - F(\mathcal{H})\mathbf{s} \right\|^2 \right\}. \tag{2.42}$$

If we are only interested in estimating the channel, the above minimization can be rewritten as

$$\hat{\mathcal{H}} = \arg\min_{H} \left\| \underbrace{(\mathbf{I} - F(\mathcal{H})\mathcal{F}^{\dagger}(\mathcal{H}))}_{P(\mathcal{H})} Y \right\|^2 = \arg\min_{\mathcal{H}} \left\| P(\mathcal{H})Y \right\|^2, \tag{2.43}$$

where $P(\mathcal{H})$ is a projection transform of Y into the orthogonal complement of the range space of $\mathcal{F}(\mathcal{H})$, or the noise subspace of the observation, and $\mathcal{F}^{\dagger}(\mathcal{H})$ denotes the pseudo-inverse of $\mathcal{F}(\mathcal{H})$. Discussions of algorithms of this type can be found in [50].

Similar to the HMM for the statistical maximum likelihood approach, the finite alphabet properties of the input sequence can also be incorporated into the deterministic maximum likelihood methods. These algorithms, first proposed by Seshadri [47] and Ghosh and Weber [12], iterate between estimates of the channel and the input. At iteration k, with an initial guess of the channel $\mathcal{H}^{(k)}$, the algorithm estimates the input sequence $\mathbf{s}^{(k)}$ and the channel $\mathcal{H}^{(k+1)}$ for the next iteration by

$$\mathbf{s}^{(k)} = \arg\min_{\mathbf{s} \in S} \left\| Y - \mathcal{F}(\mathcal{H}^{(k)})\mathbf{s} \right\|^2 \tag{2.44}$$

$$\mathcal{H}^{(k+1)} = \arg\min_{H} \left\| Y - T(\mathbf{s}^{(k)})\mathcal{H} \right\|^2, \tag{2.45}$$

where S is the (discrete) domain of $\mathbf{s}$. The optimization in (2.45) is a linear least squares problem, whereas the optimization in (2.44) can be achieved by using the Viterbi algorithm [42]. Seshadri [47] presented blind trellis search techniques. Reduced-state sequence estimation was proposed in [12]. Raheli et al. proposed a per-survivor processing technique in [44].

The convergence of such approaches is not guaranteed in general. Interesting examples have been provided in [5] where two different combinations of $\mathcal{H}$ and $\mathbf{s}$ lead to the same cost:

$$\left\| Y - \mathcal{T}(\mathbf{s})\mathcal{H} \right\|^2.$$

2.3.2.2   The Methods of Moments

Although the ML channel estimator discussed in section 2.3.2.1 usually provides better performance, the computation complexity and the existence of local optima are the two major difficulties. Therefore, simpler approaches have also been investigated.

2.3.2.2.1   *SISO Channel Estimation*

For baud-rate data, second-order statistics of the data do not carry enough information to allow estimation of the channel impulse response as a typical channel is nonminimum phase. On the other hand, higher-order statistics (in particular, fourth-order cumulants) of the baud-rate (or fractional-rate) data can be exploited to yield the channel estimates to within a scale factor.

Given the mathematical model (2.21), there are two broad classes of direct approaches to channel estimation, the distinguishing feature among them being the choice of the optimization criterion. All of the approaches involve (more or less) a least squares error measure. The error definition differs, however, as follows:

- *Fitting error*: Match the model-based higher-order (typically fourth-order) statistics to the estimated (data-based) statistics in a least squares sense to estimate the channel impulse response, as in [54] and [55], for example. This approach allows consideration of noisy observations. In general, it results in a nonlinear optimization problem. It requires availability of a good initial guess to prevent convergence to a local minimum. It yields estimates of the channel impulse response.

- *Equation error*: This is based on minimizing an "equation error" in some equation that is satisfied ideally. The approaches of [17] and [60] (among others) fall in this category. In general, this class of approaches results in a closed-form solution for the channel impulse response so that a global extremum is always guaranteed provided that the channel length (order) is known. These approaches may also provide good initial guesses for the nonlinear fitting error approaches. Quite a few of these approaches fail if the channel length is unknown.

Further details may be found in [14, 52, 56] and references therein.

2.3.2.2.2   *SIMO Channel Estimation*

Here we will concentrate upon second-order statistical methods, but first a few comments regarding indirect SIMO channel estimation. As noted in section 2.1, linear

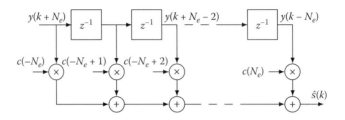

FIGURE 2.5   Structure of a baud-rate linear transversal equalizer.

equalizers designed on the basis of the baud-rate sampled received signal (see Figure 2.5 for a block diagram) are quite sensitive to symbol timing errors [15]. Therefore, fractionally spaced linear equalizers (typically with twice the baud-rate sampling: oversampling by a factor of two) are quite widely used to mitigate sensitivity to symbol timing errors. A fractionally spaced equalizer (FSE) in the linear transversal structure has the output

$$\hat{s}(k) = \sum_{n=-N_e}^{N_e} \left(\sum_{i=1}^{N} c_i(n) y_i(k-n) \right), \tag{2.46}$$

where $\{c_i(n)\}_{n=-N_e}^{n=N_e}$ are the $(2N_e + 1)$ tap weight coefficients of the ith subequalizer. Note that the FSE outputs data at the symbol rate. Similar to the SISO case, various criteria and cost functions exist to design the linear equalizers in both batch and recursive (adaptive) form.

Linear equalizers do not perform well when the underlying channels have deep spectral nulls in the passband. Several nonlinear equalizers have been developed to deal with such channels. Two effective approaches are:

- *Decision feedback equalizer* (DFE) is a nonlinear equalizer that employs previously detected symbols to eliminate the ISI due to the previously detected symbols on the current symbol to be detected. The use of the previously detected symbols makes the equalizer output a nonlinear function of the data. DFE can be symbol spaced or fractionally spaced.
- *Maximum likelihood sequence detector* estimates the information sequence to maximize the joint probability of the received sequence conditioned on the information sequence.

A detailed discussion may be found in [42].

Returning to the second-order statistical methods, for single-input multiple-output vector channels the autocorrelation function of the observation is sufficient for the identification of the channel impulse response up to an unknown constant [51, 53], provided that the various subchannels have no common zeros. This observation led to a number of techniques under both statistical and deterministic assumptions of the input sequence [50]. By exploiting the multichannel aspects of the channel, many of these techniques lead to a constrained quadratic optimization:

$$\hat{\mathcal{H}} = \arg\min_{\|\mathcal{H}\|=1} \mathcal{H}^H Q(Y)\mathcal{H}, \qquad (2.47)$$

where $Q(Y)$ is a positive definite matrix constructed from the observation. Asymptotically (as either the sample size increases to infinity or the noise variance approaches zero), these estimates converge to true channel parameters.

2.3.3 Semiblind Approaches

Semiblind approaches utilize a combination of training-based and blind approaches. Here we present a brief discussion about the idea and refer the reader to a recent survey [7] for details. The objective of semiblind channel estimation (and equalization) is to exploit the information used by blind methods as well as the information exploited by the training-based methods. Semiblind channel estimation assumes additional knowledge of the input sequence. Specifically, part of the input data vector is known. Both the statistical and deterministic maximum likelihood estimators remain the same except that the likelihood function needs to be modified to incorporate the knowledge of the input. However, semiblind channel estimation may offer significant performance improvement over either the blind or the training-based methods, as demonstrated in the evaluation of the Cramer-Rao lower bound in [7].

There are many generalizations of blind channel estimation techniques to incorporate known symbols. In [6], Cirpan and Tsatsanis extended the approach of Kaleh and Vallet by restricting the transition of the hidden Markov model. In [30], knowledge of the known symbol is used to avoid the local maxima in the maximization of the likelihood function. A popular approach is to combine the objective function used to derive the blind channel estimator with the least squares cost in the training-based channel estimation. For example, a weighted linear combination of the cost for blind channel estimator and that for the training-based estimator can be used [16, 29, 33].

2.3.4 Superimposed Training-Based Approaches

In the superimposed training (hidden pilots)-based approach, one takes

$$s(n) = b(n) + c(n), \qquad (2.48)$$

where $\{b(n)\}$ is the information sequence and $\{c(n)\}$ is a nonrandom periodic training (pilot) sequence. Exploitation of the periodicity of $\{c(n)\}$ allows identification of the channel without allocating any explicit time slots for training, unlike traditional training methods. There is no loss in data transmission rate. On the other hand, some useful power is wasted in superimposed training that could have otherwise been allocated to the information sequence. This lowers the effective signal-to-noise ratio (SNR) for the information sequence and affects the bit error rate (BER) at the receiver.

Superimposed training-based approaches have been discussed in [19, 20] and [38] for SISO systems. A block transmission method has been proposed in [10] and [11] where a data-dependent component is added to the superimposed training such that interference due to data (information sequence) is greatly reduced in channel estimation at the receiver. This method is applicable to time-invariant channels only, and it requires "data blocking" for block transmissions and insertion of a cyclic prefix in each data block. Its extension to a class of time-variant channels is given in [59]. The UTRA specification for third-generation (3G) systems [21] allows for a spread pilot (superimposed) sequence in the base station's common pilot channel, suitable for downlinks. Periodic superimposed training for channel estimation via first-order statistics for SISO systems has been discussed in [39, 41, 57, 58, 63]. In [8], performance bounds for training and superimposed training-based semiblind SISO channel estimation for time-varying flat fading channels have been discussed.

Suppose that the superimposed training sequence $c(n) = c(n + mP)$ $\forall m, n$ is a nonrandom periodic sequence with period P. Since $c(n)$ is P periodic, we have

$$c(n) = \sum_{m=0}^{P-1} c_m e^{j\alpha_m n}, \forall n, \quad \alpha_m := 2\pi m/P ,$$

(2.49)

where

$$c_m := \sum_{n=0}^{P-1} c(n) e^{-j\alpha_m n} / P.$$

Suppose that we use the Slepian (DPS) sequences $u_q(n)$ ($\equiv \phi_q(n)$) in the BEM. Then, by (2.20) and (2.49),

$$E\{\mathbf{y}(n)\} = \sum_{q=1}^{Q} \sum_{m=0}^{P-1} = \underbrace{\left[\sum_{l=0}^{L} c_m \mathbf{h}_q(l) e^{-j\alpha_m l} \right]}_{=:\mathbf{d}_{mq}} u_q(n) e^{j\alpha_m n}.$$

(2.50)

It follows that

$$\mathbf{y}(n) = \sum_{q=1}^{Q} \sum_{m=0}^{P-1} \mathbf{d}_{mq} u_q(n) e^{j\alpha_m n} + \mathbf{e}(n) ,$$

(2.51)

where $\{\mathbf{e}(n)\}$ is a zero-mean random sequence.

Define the cost function

$$J = \sum_{n=0}^{T-1} \left\| \mathbf{e}(n) \right\|^2.$$

(2.52)

Choose $\mathbf{d}_{mq}$'s to minimize J. We must have

$$\left.\frac{\partial J}{\partial \mathbf{d}_{mq}^*}\right|_{\mathbf{d}_{mq}=\hat{\mathbf{d}}_{mq}} = 0, \tag{2.53}$$

which leads to

$$\sum_{q'=1}^{Q}\sum_{m'=0}^{P-1}\hat{\mathbf{d}}_{m'q'}\left[\sum_{n=0}^{T-1}u_{q'}(n)u_q(n)e^{j(\alpha_{m'}-\alpha_m)n}\right] = \underbrace{\sum_{n=0}^{T-1}\mathbf{y}(n)u_q(n)e^{-j\alpha_m n}}_{=:\mathbf{g}_{mq}}. \tag{2.54}$$

Define

$$\mathbf{V} := \begin{bmatrix} 1 & 1 & \cdots & 1 \\ 1 & e^{-j\alpha_1} & \cdots & e^{-j\alpha_1 L} \\ \vdots & \vdots & \ddots & \vdots \\ 1 & e^{-j\alpha_{P-1}} & \cdots & e^{-j\alpha_{P-1}L} \end{bmatrix}, \tag{2.55}$$

$$\mathbf{D}_m := \left[\mathbf{d}_{m1}^T, \cdots, \mathbf{d}_{mQ}^T\right]^T, \quad \mathcal{D} := \left[\mathbf{D}_0^H, \cdots, \mathbf{D}_{P-1}^H\right]^H, \tag{2.56}$$

$$\mathbf{H}_l := \left[\mathbf{h}_1^T(l), \cdots, \ \mathbf{h}_Q^T(l)\right]^T, \quad \mathcal{H} := \left[\mathbf{H}_0^H, \cdots, \mathbf{H}_L^H\right]^H, \tag{2.57}$$

$$\mathcal{C} := \left(diag\{c_0, \cdots, c_{P-1}\}\mathbf{V}\right) \otimes \mathbf{I}_{NQ}. \tag{2.58}$$

By the definition of $\mathbf{d}_{mq}$ in (2.56), it then follows that

$$\mathcal{C}\mathcal{H} = \mathcal{D}. \tag{2.59}$$

It is shown in [57] that if $P \geq L + 1$, $rank(C) = NQ(L + 1)$. Hence, we can determine the $\mathbf{h}_q(l)$'s uniquely by using the estimates of $\mathbf{d}_{mq}$'s.

Define $\hat{\mathcal{D}}$ as in (2.56) with $\mathbf{d}_{mq}$ replaced with $\hat{\mathbf{d}}_{mq}$, and similarly define $\mathcal{G}$ as in (2.56) with $\mathbf{d}_{mq}$ replaced with $\mathbf{g}_{mq}$. Then (2.54) leads to

$$\left(\mathbf{\Psi} \otimes \mathbf{I}_N\right)\hat{\mathcal{D}} = \mathcal{G}, \tag{2.60}$$

where the entries of the $PQ \times PQ$ matrix $\mathbf{\Psi}$ are ($m, m' = 0, 1; \cdots, P-1, q, q' = 1, 2, \cdots, Q$)

$$\left[\Psi\right]_{mQ+q,m'Q+q'} = \sum_{n=0}^{T-1} u_{q'}(n)u_q(n)e^{j(\alpha_{m'}-\alpha_m)n}. \tag{2.61}$$

The estimate of $\mathcal{D}$ is given by

$$\hat{\mathcal{D}} = \left(\Psi^{-1}\otimes\mathbf{I}_N\right)\mathcal{G}. \tag{2.62}$$

By (2.59) and (2.78) we have the channel coefficient estimate

$$\hat{\mathcal{H}} = \mathcal{C}^\dagger\hat{\mathcal{D}} = \left(\mathcal{C}^H\mathcal{C}\right)^{-1}\mathcal{C}^H\left(\Psi^{-1}\otimes\mathbf{I}_N\right)\mathcal{G}. \tag{2.63}$$

The channel estimate is acquired by regenerating the DPS-BEM:

$$\hat{\mathbf{h}}(n;l) = \sum_{q=1}^{Q}\hat{\mathbf{h}}_q(l)u_q(n). \tag{2.64}$$

Remark 1: Using the fact that the (infinite) DPS sequences are band limited to the normalized frequency range $[-f_dT_s, f_dT_s]$, the time-limited DPS sequences, obtained by rectangular windowing over $0 \le n \le T-1$, approximately satisfy

$$\sum_{n=0}^{T-1} u_{q'}(n)u_q(n)e^{j(\alpha_{m'}-\alpha_m)n} \approx \delta(m'-m)\delta(q'-q), \tag{2.65}$$

if $f_dT_s \ll 1/P$ and T are multiples of P or if T is large, so that $\Psi \approx \mathbf{I}_{PQ}$. An estimate $\hat{\mathbf{d}}_{mq}$ of $\mathbf{d}_{mq}$, following (2.54) and (2.65), is given as

$$\hat{\mathbf{d}}_{mq} = \sum_{n=0}^{T-1} \mathbf{y}(n)u_q(n)e^{-j\alpha_m n}. \tag{2.66}$$

The estimation of the channel coefficients (2.63) is then given by

$$\hat{\mathcal{H}} = \left(\mathcal{C}^H\mathcal{C}\right)^{-1}\mathcal{C}^H\hat{\mathcal{D}}. \tag{2.67}$$

Remark 2: As noted in [57], if the mean of the noise $\mathbf{v}(n)$ is unknown, say

$$E\{\mathbf{v}(n)\} = \mathbf{m}, \tag{2.68}$$

one should omit the first row (corresponding to α_0) of V in (2.55) (denote the resulting matrix by $\tilde{V}$), and block $\mathbf{D}_0$ from $\mathcal{D}$ in (2.56) (denote the resulting matrix by $\tilde{\mathcal{D}}$). Define

$$\tilde{C} := \left(diag\{c_1 \cdots c_{P-1}\} \tilde{V} \right) \otimes \mathbf{I}_{NQ}. \tag{2.69}$$

Then we have $\tilde{C}\mathcal{H} = \tilde{\mathcal{D}}$ and

$$\hat{\mathcal{H}} = (\tilde{C}^H \tilde{C})^{-1} \tilde{C}^H \hat{\tilde{\mathcal{D}}}, \tag{2.70}$$

where $\hat{\tilde{\mathcal{D}}}$ follows from $\tilde{\mathcal{D}}$ by using estimate (2.66). For identifiability, we now need $P \geq L + 2$. All our subsequent results hold true if appropriate substitutions are used.

2.4 Adaptive Channel Estimation

2.4.1 Block-Adaptive Channel Estimation Using CE-BEM

Here we summarize the time-multiplexed training approach of [36]. In [36] each transmitted block of symbols $\{s(n)\}_{n=0}^{T_B-1}$ is segmented into $\overline{P}$ subblocks of time-multiplexed training and information symbols. Each subblock is of equal-length l_b symbols with l_d information symbols and l_t training symbols ($l_b = l_d + l_t$). If s denotes a column vector composed of $\{s(n)\}_{n=0}^{T_B-1}$, then s is arranged as

$$\mathbf{s} := \begin{bmatrix} \mathbf{b}_0^T & \mathbf{c}_0^T & \mathbf{b}_1^T & \mathbf{c}_1^T & \cdots & \mathbf{b}_{\overline{P}-1}^T & \mathbf{c}_{\overline{P}-1}^T \end{bmatrix}^T, \tag{2.71}$$

where $\mathbf{b}_p$ ($p = 1, 0, \cdots \overline{P} - 1$) is a column of l_d information symbols and $\mathbf{c}_p$ is a column of l_t training symbols. We clearly have TB $= \overline{P}l_b$. Given (2.69) and CE-BEM (2.15), [36] has shown that (2.77) is an optimum structure for $K = 1$ with $l_t = 2L + 1$, $\overline{P} \geq Q$, and

$$\mathbf{c}_p := \begin{bmatrix} \mathbf{0}_L^T & \gamma & \mathbf{0}_L^T \end{bmatrix}^T, \gamma > 0. \tag{2.72}$$

Thus, given a transmission block of size T_B, $(2L + 1)\overline{P}$ symbols have to be devoted to training and the remaining $T_B - (2L + 1)\overline{P}$ are available for information symbols.

Let $n_p := pl_b + l_d + L$ ($p = 0, 1, \ldots \overline{P} - 1$) denote the location of (nonzero) γ's in the optimum $\mathbf{c}_p$'s in the P subblocks. Then by design, received signal (assuming timing synchronization)

$$y(n_p + l) = \gamma h(n_p + l; l) + v(n_p + l) \tag{2.73}$$

for $l = 0, 1, \cdots, L$. Using (2.15) in these $y(n_p + l)$'s, one can uniquely solve for $h_q(l)$'s via a least squares approach. The channel estimates are given by the CE-BEM (2.15) using the estimated BEM coefficients.

2.4.2   Adaptive Channel Estimation via Subblock Tracking

Suppose that we collect the received signal over a time interval of $\overline{T}$ symbols. We wish to estimate the time-variant channel using a channel model and time-multiplexed training (such as that discussed in section 2.4.1 and [36]), and subsequently using the estimated channel, estimate the information symbols. For CE-BEM, if we choose $\overline{T}$ as the block size, then in general the Q value will be very high, requiring estimation of a large number of parameters, thereby degrading the channel estimation performance. If we divide $\overline{T}$ into blocks of size T_B, and then fit CE-BEM block by block, we need smaller Q; however, estimation of $h_q(l)$'s is now based on a shorter observation size of T_B symbols, which might also degrade channel estimation performance. Thus, one has to strike a balance between estimation variance and block size. Such considerations do not apply to the AR channel model fitting. In the sequel, we propose a novel subblock tracking approach to CE-BEM channel estimation where we update estimates of $h_q(l)$'s every subblock based on all of the past training symbols.

By exploiting the invariance of the coefficients of CE-BEM over each block, hence each of the $\overline{P}$ subblocks per block of length T_B symbols, we seek subblock-wise tracking of the BEM coefficients of the doubly selective channel. Consider two overlapping blocks that differ by just one subblock: blocks with $n = m, m + 1, \cdots, m + T_B - 1$, where $m = m_0$ for the past block and $m = m_0 + l_b$ for the current block. If the two blocks overlap so significantly, one would expect the BEM coefficients to vary only a little from the past block to the current overlapping block. Therefore, rather than estimate $h_q(l)$'s anew with every non-overlapping block, as in section 2.4.1 and [36], we propose to track the BEM coefficients subblock by subblock using a first-order AR model for their variations.

Stack the channel coefficients in (2.15) into vectors

$$\mathbf{h}_l := \begin{bmatrix} h_1(l) & h_2(l) & \cdots & h_Q(l) \end{bmatrix}^T, \tag{2.74}$$

$$\mathbf{h} := \begin{bmatrix} \mathbf{h}_0^T & \mathbf{h}_1^T & \cdots & \mathbf{h}_L^T \end{bmatrix}^T \tag{2.75}$$

of size Q and $M := Q(L + 1)$, respectively. The coefficient vector in (2.75) for the pth subblock ($p = 0, 1, \cdots$) will be denoted by $\mathbf{h}(p)$. We assume that the BEM coefficients over each subblock are Markovian: a simplified model is given by the first-order AR process, i.e.,

$$\mathbf{h}(p) = \alpha \mathbf{h}(p-1) + \mathbf{w}(p), \tag{2.76}$$

where α is the AR coefficient, and the driving noise vector $\mathbf{w}(p)$ is zero-mean complex Gaussian with variance $\sigma_w^2 \mathbf{I}_M$. If the channel is stationary and coefficients $h_q(l)$ are independent (as assumed in [36]), then by (2.76), $\sigma_w^2 = \sigma_h^2 (1 - |\alpha|^2)/Q$ with $\sigma_h^2 := E\{h(n;l)h^*(n;l)\}$. Since the coefficients evolve slowly, we have $\alpha \approx 1$ (but $\alpha < 1$ for tracking).

Under this formulation we do not have a strict definition of the block size T_B because, although we still use (2.15) for any n, we allow $h_q(l)$'s to change subblock by subblock based on the training symbols.

2.4.2.1 Subblock Tracking Using Kalman Filtering

Define $\varepsilon(n) := [e^{-j\omega_1 n} \; e^{-j\omega_2 n} \; \cdots \; e^{-j\omega_Q n}]^T$. If at time n the pth subblock is being received, by (2.63), (2.15)–(2.19), and (2.74) and (2.75), the received signal can be written as

$$y(n) = s^T(n)\left[\mathbf{I}_{L+1} \otimes \varepsilon(n)\right]^H \mathbf{h}(p) + v(n), \tag{2.77}$$

where $\mathbf{s}(n) := [s(n) \; s(n-1) \; \cdots \; s(n-L)]^T$. Treating (2.76) and (2.77) as the state and the measurement equations, respectively, Kalman filtering can be applied to track the coefficient vector $\mathbf{h}(p)$ for each subblock.

We will employ the time-multiplexed training scheme proposed in [36] (see section 2.4.1), where each subblock (of equal-length l_b symbols) consists of a data session (of length l_d symbols) and a succeeding training session (of length $l_b = 2L + 1$ symbols). Using (2.73), at time $n_p + l$ ($p = 0, 1, \cdots$ and $l = 0, 1, \cdots, L$)

$$y(n_p + l) = \gamma E^H(n_p + l)\mathbf{h}_l(p) + v(n_p + l). \tag{2.78}$$

We intend to use only training sessions for subblock-wise channel tracking. Defining

$$\mathbf{y}(p) := \left[y(n_p) \quad y(n_p + 1) \quad \cdots \quad y(n_p + L)\right]^T,$$

$$\mathbf{v}(p) := \left[v(n_p) \quad v(n_p + 1) \quad \cdots \quad v(n_p + L)\right]^T,$$

$$\Psi(p) := \begin{bmatrix} E(n_p) & & & \\ & E(n_p + 1) & & \\ & & \ddots & \\ & & & E(n_p + L) \end{bmatrix}^H,$$

by (2.78), we have

$$\mathbf{y}(p) = \gamma\Psi(p)\mathbf{h}(p) + \mathbf{v}(p). \tag{2.79}$$

Using this optimal training scheme, the measurement equation (2.77) is now simplified as (2.79). We have obtained a linear discrete-time system represented by (2.76) and (2.79). Kalman filtering is applied to track the channel BEM coefficients via the following steps [49]:

1. Initialization:

$$\hat{\mathbf{h}}(-1|-1)=\mathbf{0}_M, \mathbf{R}_h(-1|-1)=\sigma_w^2 \mathbf{I}_M;$$

2. Kalman recursion for $p = 0, 1, \cdots$

- Time update:

$$\hat{\mathbf{h}}(p|p-1)=\alpha \hat{\mathbf{h}}(p-1|p-1),$$

$$\mathbf{R}_h(p|p-1)=|\alpha|^2 \mathbf{R}_h(p-1|p-1)+\sigma_w^2 \mathbf{I}_M;$$

- Kalman gain:

$$\mathbf{R}_\eta(p)=|\gamma|^2 \Psi(p)\mathbf{R}_h(p|p-1)\Psi^H(p)+\sigma_v^2 \mathbf{I}_{L+1},$$

$$\mathbf{K}(p)=\mathbf{R}_h(p|p-1)\Psi^H(p)\mathbf{R}_\eta^{-1}(p);$$

- Measurement update:

$$\hat{\mathbf{h}}(p|p)=\hat{\mathbf{h}}(p|p-1)+\mathbf{K}(p)\left[\mathbf{y}(p)-\gamma\Psi(p)\hat{\mathbf{h}}(p|p-1)\right],$$

$$\mathbf{R}_h(p|p)=\left[\mathbf{I}-\gamma\mathbf{K}(p)\Psi(p)\right]\mathbf{R}_h(p|p-1),$$

where $\hat{\mathbf{h}}(p|m)$ is the estimate of $\mathbf{h}(p)$ given the observations $\{\mathbf{y}(0), \mathbf{y}(1), \cdots, \mathbf{y}(m)\}$, and $\mathbf{R}_h(p|m)$ is the error covariance matrix of $\hat{\mathbf{h}}(p|m)$, defined as

$$\mathbf{R}_h(p|m):=E\left\{\left[\hat{\mathbf{h}}(p|m)-\mathbf{h}(p)\right]\left[\hat{\mathbf{h}}(p|m)-\mathbf{h}(p)\right]^H\right\}.$$

Now we generate the channel for the entire p^{th} subblock by the estimate $\hat{\mathbf{h}}(p|p)$ via the CE-BEM (2.15) as

$$\hat{h}(n;l)=\varepsilon^H(n)\hat{\mathbf{h}}_l(p|p) \tag{2.80}$$

for $n = pl_b, pl_b + 1, \cdots, (p + 1)l_b - 1$. The definition of $\hat{\mathbf{h}}_l(p|p)$ is similar to (2.74).

2.4.2.2 Kalman Detector for Equalization

The channel estimate given by (2.80) is fed into an equalizer for symbol detection. Another Kalman filter, together with a quantizer, acts as the symbol detector at the receiver end. The state and the measurement equations are now given by

$$\mathbf{s}_d(n) = \Phi \mathbf{s}_d(n-1) + \Gamma \bar{s}(n) + \Gamma \tilde{s}(n), \tag{2.81}$$

$$y(n) = \tilde{\mathbf{h}}_d^T(n) \mathbf{s}_d(n) + v(n), \tag{2.82}$$

where

$$\mathbf{s}_d(n) := \begin{bmatrix} s(n) & s(n-1) & \cdots & s(n-d) \end{bmatrix}^T,$$

$$\bar{s}(n) := E\{s(n)\}, \ \tilde{s}(n) := s(n) - \bar{s}(n),$$

$$\Phi := \begin{bmatrix} 0_d^T & 0 \\ \mathbf{I}_d & 0_d \end{bmatrix}, \ \Gamma := \begin{bmatrix} 1 & 0_d^T \end{bmatrix}^T,$$

$$\tilde{\mathbf{h}}_d(n) := \begin{bmatrix} \hat{h}(n;0) & \hat{h}(n;1) & \cdots & \hat{h}(n;L) & 0_{d-L} \end{bmatrix}^T,$$

where integer $d \geq L$; it will also be the equalization delay. Assume data symbols are zero mean and white. If $s(n)$ is a data symbol, we have $\bar{s}(n) = 0$, $\tilde{s}(n) = s(n)$; if $s(n)$ is a training symbol, $\bar{s}(n) = s(n)$, $\tilde{s}(n) = 0$. Details of Kalman filtering of the system described by (2.81) and (2.82) can be found in [34].

2.4.3 Symbol-Adaptive Joint Channel Estimation and Data Detection

Representative approaches in this category are [27, 34] and references therein. A Gauss-Markov model for channel variations (typically an autoregressive model) is coupled with a state-space model for received data to form an augmented state-space model with nonlinear measurement equation. This results in a nonlinear state estimation problem. In [27] a finite-length minimum mean square error (MMSE) DFE is used during non-data-aided periods to generate hard decisions. Reference [34] presents a low-complexity turbo equalization receiver for coded signals where a nonlinear Kalman filtering–based adaptive equalizer is coupled with a soft-in soft-out decoder. These approaches work well so long as the channel does not fade too fast.

2.5   Simulation Examples

In this section we present two simulation examples to illustrate some of the approaches to channel estimation. In both examples a random time- and frequency-selective Rayleigh fading channel is considered. We take $L = 2$ (three taps) in (2.63), and $h(n;l)$ are zero-mean complex Gaussian with variance $\sigma_h^2 = 1/(L+1)$. For different l's, $h(n;l)$'s are mutually independent and satisfy Jakes' model [22]. To this end, we simulated each single tap following [62] (with a correction in the appendix of [61]).

We consider a communication system with carrier frequency of 2 GHz, data rate of 40 kBd (kilo-Bauds), therefore $T_s = 25$ μs, and a varying Doppler spread f_d in the range of 0 to 400 HZ, or the normalized Doppler spread $f_d T_s$ from 0 to 0.01 (corresponding to a maximum mobile velocity from 0 to 216 km/h). The additive noise was zero-mean complex white Gaussian. The (receiver) SNR refers to the average energy per symbol over one-sided noise spectral density. The time-multiplexed training scheme of [36] described in section 2.4.1 is adopted, where during data sessions the information sequence is modulated by binary phase-shift keying (BPSK) with unit power. The training session is described by (2.72) with $\gamma = \sqrt{2L+1}$ so that the average symbol power of training sessions is equal to that of data sessions.

We evaluate the performance of various approaches by considering the normalized channel mean square error (NCMSE) and the bit error rate (BER). The NCMSE is defined as

$$
\text{NCMSE} := \frac{\sum_{i=1}^{M_r} \sum_{n=0}^{T-1} \sum_{l=0}^{L} \left\| \hat{\mathbf{h}}^{(i)}(n;l) - \mathbf{h}^{(i)}(n;l) \right\|^2}{\sum_{i=1}^{M_r} \sum_{n=0}^{T-1} \sum_{l=0}^{L} \left\| \mathbf{h}^{(i)}(n;l) \right\|^2},
$$

where $\mathbf{h}^{(i)}(n;l)$ is the true channel and $\hat{\mathbf{h}}^{(i)}(n;l)$ is the estimated channel at the ith run, among total M_r runs.

2.5.1   Example 1

Here we consider block-adaptive channel estimation and restrict our attention to a single block. We compare superimposed training-based approaches with time-multiplexed training-based approaches. The superimposed training sequence was picked as a periodic repetition of a length 7 m-sequence (maximal length pseudo-random binary sequence) {1,−1,−1,1,1,1,−1}. In the simulations, the average transmitted power σ_c^2 in $c(n)$ was 0.3 of the power in $b(n)$, leading to a training-to-information power ratio TIR := $\sigma_c^2 / \sigma_b^2 = \beta/(1 − \beta) = 0.3$. We consider both critically sampled CE-BEM and DPS-BEM for channel modeling. For comparison, we consider a CE- or DPS-BEM-based periodically placed time-multiplexed training with zero padding, following the design of [36]. We took a training subblock of size $2L + 1 = 5$ symbols with the recommended structure

$$
\left\{ 0, 0, \sqrt{(2L+1)\left(\sigma_b^2 + \sigma_c^2\right)}, 0, 0 \right\},
$$

which follows an information data block of length 18 leading to a subblock of 23 symbols. This subblock was repeated over a record length of 418 symbols with a total of 16 subblocks. Thus, the training-to-information bit and power ratios are both 0.3 (the amplitude of the single nonzero training bit was picked to achieve this power ratio).

The results of our simulation averaged over five hundred runs are shown in Figures 2.6, 2.7, 2.8, and 2.9. For comparison, we plot the results of the CE- and DPS-BEM-based superimposed training schemes (denoted as SI in the figures), including the first-order

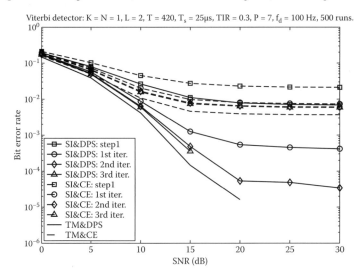

FIGURE 2.6 BER versus SNR for $f_d = 100$ Hz.

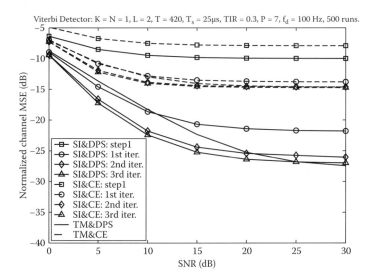

FIGURE 2.7 MSE versus SNR for $f_d = 100$ Hz.

Viterbi Detector: K = N = 1, L = 2, T = 420, T_s = 25μs, TIR = 0.3, P = 7, f_d = 200 Hz, 500 runs.

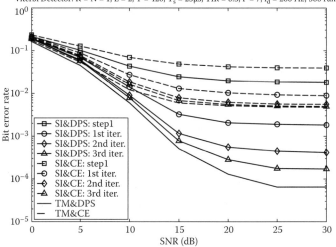

FIGURE 2.8 BER versus SNR for f_d = 200 Hz.

Viterbi Detector: K = N = 1, L = 2, T = 420, T_s = 25μs, TIR = 0.3, P = 7, f_d = 200 Hz, 500 runs.

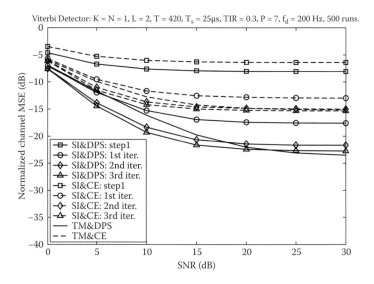

FIGURE 2.9 MSE versus SNR for f_d = 200 Hz.

statistics-based estimator (denoted as step 1 in the figures), and the DML approach after one, two, and three iterations (denoted as 1st iter., 2nd iter., and 3rd iter. in the figures), and TM training approaches (denoted as TM in the figures). From the four figures, we can see that after iterations, superimposed training-based estimation and detection performances improve a lot, because the information data that are viewed as interference by the first-order statistics-based estimator are now exploited to enhance the channel estimation for the next iteration. Therefore, the self-interference is effectively removed after iterations. The DML approach provides comparable error performance with TM training, but at a higher data transmission rate. The valuable bandwidth resources can thus be saved.

2.5.2 Example 2

Here we consider adaptive channel estimation at block-, subblock-, and symbol-adaptive levels using time-multiplexed training. Four estimation and tracking schemes are compared:

1. The block-adaptive channel estimation in [36] (see section 2.4.1). We consider two models corresponding to $T = 200$ and 400, respectively, so that $Q = 5$ and 9 by (2.17). In the figures, this scheme is denoted by BA.
2. The channel tracking scheme in [35] using the first-order AR model, where the time-varying channel is assumed to follow (2.12). Channel tracking is performed at training sessions only. For data sessions, the receiver updates the channel via $h(n;l) = \alpha_c h(n - 1;l)$. We assume that only the upper bound of the Doppler spread is known. Then by (2.13) and (2.14) and Jakes' model, $\alpha_c = J_0(2\pi f_d T_s) = 0.999$ for $f_d T_s = 0.01$, where $J_0(\cdot)$ denotes the zero-th Bessel function of the first kind. This scheme is denoted by $AR(1)$-KF in the figures.
3. We also compare the approach of joint channel estimation and data detection via extended Kalman filtering in [34]. For fairness, the Turbo equalization procedure in [34] is omitted. The AR parameter of the channel also follows (2.13) and (2.14), as suggested by [34]. This scheme is denoted by Joint-KF.
4. Our proposed subblock-wise tracking using CE-BEM, which is denoted by SUBBLOCK. We also take $T = 200$ and 400 for two different settings of CE-BEM, and $Q = 5$ and 9 correspondingly.

The BERs for the schemes of BA, $AR(1)$-KF, and SUBBLOCK are evaluated by employing the Kalman detector described in section 2.4.2.2 with delay $d = 5$, using the channel estimates obtained by each scheme. In each run, a symbol sequence of length 5,000 is generated and fed into a random doubly selective channel. The first two hundred symbols are discarded in evaluations. All the simulation results are based on five hundred runs.

In Figures 2.10 and 2.11, the performances of the four schemes over different Doppler spreads f_d are compared. We set SNR = 20 dB, $l_t = 2L + 1 = 5$, and $l_d = 15$ symbols, so that 25% of the transmitted symbols are dedicated to training. For the BA scheme, we use the oversampled CE-BEM with $T_B = T/2$ and $K = 2$ in order to suppress spectral leakage. For our SUBBLOCK scheme, we take $\alpha = 0.99$ for $T = 200$, and $\alpha = 0.995$ for $T = 400$. Since more unknown parameters (BEM coefficients) are involved in SUBBLOCK, and therefore result in higher estimation variance, SUBBLOCK is slightly inferior to $AR(1)$-KF for slow-fading channels. As f_d increases, SUBBLOCK gradually outperforms the other three schemes, since the time variations of the channel have been well captured in CE-BEM. For the BA scheme, more basis functions do not necessarily translate into better performance since estimation variance also increases, whereas this strategy can well improve the performance of SUBBLOCK since all past data are implicitly utilized in Kalman filtering–based subblock tracking. Figures 2.12 and 2.13 compare the

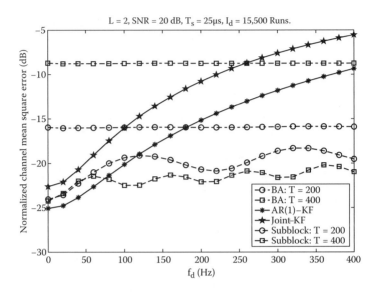

FIGURE 2.10   NCMSE versus f_d for SNR = 20 dB, l_d = 15, l_t = 5, l_c = 5, l_b = 5, $T_B = T/K$, and $K = 2$.

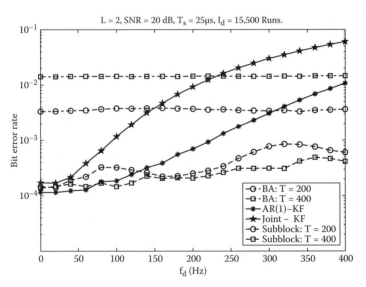

FIGURE 2.11   BER versus f_d for SNR = 20 dB, l_d = 15, l_t = 5, l_b = 20, $T_B = T/K$, and $K = 2$.

four schemes under different SNRs: results similar to those in Figures 2.10 and 2.11 are observed.

In Figures 2.14 and 2.15, longer data sessions are used to represent a more efficient transmission scenario, where we take l_d = 35 symbols (12.5% symbols are now devoted to training sessions). We now let α = 0.94 for T = 200, and α = 0.97 for T = 400 for

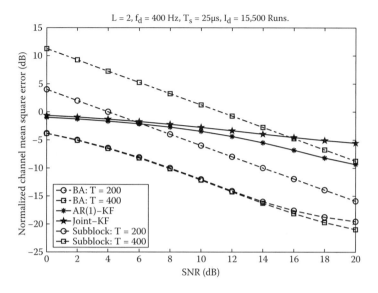

FIGURE 2.12 NCMSE versus SNR for $f_d T_s = 0.01$, $l_d = 15$, $l_t = 5$, $l_b = 20$, $T_B = T/K$, and $K = 2$.

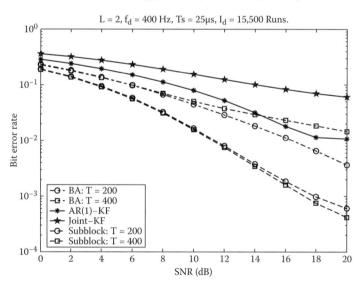

FIGURE 2.13 BER versus SNR for $f_d T_s = 0.01$, $l_d = 15$, $l_t = 5$, $l_b = 20$, $T_B = T/K$, and $K = 2$.

SUBBLOCK. For the BA scheme, an oversampled CE-BEM is not possible for $T = 200$, and we take $T_B = T$ for the two models. Comparing Figure 2.14 with Figure 2.12, we observe that the spectral leakage has worsened the channel estimation in BA (note the floors of the two BA curves in Figure 2.14), and thus the curves for symbol detection. Our SUBBLOCK maintains a satisfactory performance, while the performances of the schemes of $AR(1)$-KF and Joint-KF have severely deteriorated.

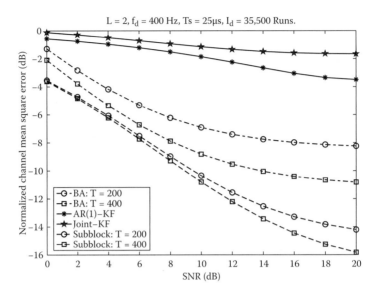

FIGURE 2.14 NCMSE versus SNR for $f_d T_s = 0.01$, $l_d = 35$, $l_c = 5$, $l_b = 40$, $T_B = T/K$, and $K = 1$.

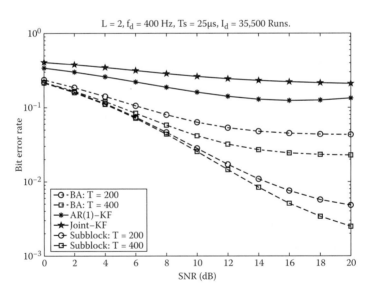

FIGURE 2.15 BER versus SNR for $f_d T_s = 0.01$, $l_d = 35$, $l_t = 5$, $l_b = 40$, $T_B = T/K$, and $K = 1$.

2.6 Conclusions

A review of various approaches to adaptive channel estimation for wireless mobile systems was presented. Emphasis was on linear baseband-equivalent models with a tapped delay line structure, and both time-invariant and time-variant (doubly-selective) models were discussed. Emphasis was on basis expansion modeling for time-variant channels where the basis functions are related to the physical parameters of the channel (such as Doppler and delay spreads). Channel modeling was followed by a discussion of various approaches to channel estimation, including training-based approaches, blind approaches, semiblind approaches and superimposed training-based approaches, and various approaches to channel adaptation. In the training-based approach a sequence known to the receiver is transmitted in the acquisition mode. In blind approaches no such sequence is available (or used), and the channel is estimated based solely on the noisy received signal exploiting the statistical and other properties of the information sequence. Semiblind approaches utilize a combination of training-based and blind approaches. In the superimposed training-based approaches a periodic (nonrandom) training sequence is arithmetically added (superimposed) at a low power to the information sequence at the transmitter before modulation and transmission. Channel adaptaion can be at the block level suitable for block transmissions, or the symbol-by-symbol level suitable for serial transmissions. Some of the approaches were illustrated via simulations.

Acknowledgment

This work was prepared in part under the support of the U.S. NSF under grant ECS-0424145.

References

[1] M.-A. R. Baissas and A. M. Sayeed. 2002. Pilot-based estimation of time-varying multipath channels for coherent CDMA receivers. *IEEE Trans. Signal Processing* 50:2037–49.

[2] P. A. Bello. 1963. Characterization of randomly time-variant channels. *IEEE Trans. Commun. Syst.* 11:360–93.

[3] D. K. Borah and B. Hart. 1999. Receiver structures for time-varying frequency-selective fading channels. *IEEE J. Selected Areas Commun.* 17:1863–75.

[4] J. K. Cavers. 1995. Pilot symbol assisted modulation and differential detection in fading and delay spread. *IEEE Trans. Commun.* 43:2206–12.

[5] K. M. Chugg. 1998. Blind acquisition characteristics of PSP-based sequence detectors. *IEEE J. Selected Areas Commun.* 16:1518–29.

[6] H. A. Cirpan and M. K. Tsatsanis. 1998. Stochastic maximum likelihood methods for semi-blind channel estimation. *IEEE Signal Processing Lett.* 5:21–24.

[7] E. de Carvalho and D. T. M. Slock. 2001. Semi-blind methods for FIR multichannel estimation. In *Signal processing advances in wireless and mobile communications*, ed. G. B. Giannakis, Y. Hua, P. Stoica, and L. Tong, chap. 7. Vol. I. Upper Saddle River, NJ: Prentice Hall.

[8] M. Dong, L. Tong, and B. M. Sadler. 2003. Optimal insertion of pilot symbols for transmissions over time-varying flat fading channels. In *Proceedings of the 2003 IEEE Workshop on Signal Processing Advances in Wireless Communications*, Rome, pp. 472–76.

[9] W. A. Gardner. 1989. *Introduction to random processes: With applications to signals and systems.* 2nd ed. New York: McGraw-Hill.

[10] M. Ghogho, D. McLernon, E. Alamdea-Hernandez, and A. Swami. 2005. Channel estimation and symbol detection for block transmission using data-dependent superimposed training. *IEEE Signal Processing Lett.* 12:226–29.

[11] M. Ghogho, D. McLernon, E. Alamdea-Hernandez, and A. Swami. 2005. SISO and MIMO channel estimation and symbol detection using data-dependent superimposed training. In *Proceedings of the 2005 IEEE ICASSP*, Philadelphia, pp. III-461–64.

[12] M. Ghosh and C. L. Weber. 1992. Maximum-likelihood blind equalization. *Optical Eng.* 31:1224–28.

[13] G. B. Giannakis and C. Tepedelenlioğlu. 1998. Basis expansion models and diversity techniques for blind identification and equalization of time-varying channels. *Proc. IEEE* 86:1969–86.

[14] G. B. Giannakis, Y. Hua, P. Stoica, and L. Tong, eds. 2001. *Signal processing advances in wireless and mobile communications: Trends in channel estimation and equalization,* ed. G. B. Giannakis, Y. Hua, P. Stoica, and L. Tong. Vol. 1. Upper Saddle River, NJ: Prentice Hall.

[15] R. D. Gitlin and S. B. Weinstein. 1981. Fractionally-spaced equalization: An improved digital transversal equalizer. *Bell Syst. Tech. J.* 60:275–96.

[16] A Gorokhov and P. Loubaton. 1997. Semi-blind second order identification of convolutive channels. In *Proceedings of the IEEE International Conference on Acoustics, Speech, Signal Processing*, Munich, Germany, pp. 3905–8.

[17] D. Hatzinakos and C. L. Nikias. 1991. Blind equalization using a tricepstrum based algorithm. *IEEE Trans. Commun.* 39:669–81.

[18] S. He and J. K. Tugnait. 2006. On bias-variance trade-off in superimposed training-based doubly selective channel estimation. In *Proceedings of the 2006 Conference on Information Systems and Sciences*, Princeton, NJ, pp. 1308–1313.

[19] P. Hoeher and F. Tufvesson. 1999. Channel estimation with superimposed pilot sequence. In *Proceedings of the IEEE GLOBECOM Conference*, Rio de Janeiro, pp. 2162–66.

[20] T. P. Holden and K. Feher. 1990. A spread spectrum based system technique for synchronization of digital mobile communication systems. *IEEE Trans. Broadcasting* 36:185–94.

[21] H. Holma and A. Toskala, eds. 2002. *WCDMA for UMTS: Radio access for third generation mobile communications.* 2nd ed. New York: Wiley.

[22] W. C. Jakes. 1974. *Microwave mobile communications.* New York: Wiley.

[23] G. K. Kaleh and R. Vallet. 1994. Joint parameter estimation and symbol detection for linear or nonlinear unknown dispersive channels. *IEEE Trans. Commun.* 42:2406–13.

[24] A. P. Kannu and P. Schniter. 2005. MSE-optimal training for linear time-varying channels. In *Proceedings of ICASSP*, vol. 3, pp. 789–92.

[25] A. P. Kannu and P. Schniter. 2005. Capacity analysis of MMSE pilot-aided transmission for doubly selective channels. In *Proceedings of the IEEE 6th Workshop on SPAWC*, pp. 801–5.

[26] A. P. Kannu and P. Schniter. 2006. Minimum mean-squared error pilot-aided transmission for MIMO doubly selective channels. In *Proceedings of the 2006 Conference on Information Systems and Sciences*, Princeton University, NJ.

[27] C. Komninakis, C. Fragouli, A. H. Sayed, and R. D. Wesel. 2002. Multi-input multi-output fading channel tracking and equalization using Kalman estimation. *IEEE Trans. Signal Processing* 50:1065–76.

[28] V. Krishnamurthy and J. B. Moore. 1993. On-line estimation of hidden Markov model parameters based on Kullback-Leibler information measure. *IEEE Trans. Signal Processing* 41:2557–73.

[29] S. Lasaulce, P. Loubaton, and E. Moulines. 2003. A semi-blind channel estimation technique based on second-order blind method for CDMA systems. *IEEE Trans. Signal Processing* 51:1894–904.

[30] J. Laurila, K. Kopsa, and E. Bonek. 1999. Semi-blind signal estimation for smart antennas using subspace tracking. In *Proceedings of the 1999 IEEE Workshop on Signal Processing: Advances in Wireless Communications*, Annapolis, MD, pp. 271–274.

[31] G. Leus, S. Zhou, and G. B. Giannakis. 2003. Orthogonal multiple access over time- and frequency-selective channels. *IEEE Trans. Inf. Theory* 49:1942–50.

[32] G. Leus. 2004. On the estimation of rapidly time-varying channels. In *Proceedings of the European Signal Processing Conference*, Vienna, pp. 2227–30.

[33] G. Leus. 2005. Semi-blind channel estimation for rapidly time-varying channels. In *Proceedings of the 2005 IEEE International Conference on Acoustics, Speech, Signal Processing*, Philadelphia, vol. III, pp. 773–76.

[34] X. Li and T. F. Wong. 2007. Turbo equalization with nonlinear Kalman filtering for time-varying frequency-selective fading channels. *IEEE Trans. Wireless Commun.* 6:691–700.

[35] Z. Liu, X. Ma, and G. B. Giannakis. 2002. Space-time coding and (Kalman) filtering for time-selective fading channels. *IEEE Trans. Commun.* 50:183–86.

[36] X. Ma, G. B. Giannakis, and S. Ohno. 2003. Optimal training for block transmissions over doubly selective wireless fading channels. *IEEE Trans. Signal Processing* 51:1351–66.

[37] M. Martone. 2000. Wavelet-based separating kernels for array processing of cellular DS/CDMA signals in fast fading. *IEEE Trans. Commun.* 48:979–95.

[38] F. Mazzenga. 2000. Channel estimation and equalization for M-QAM transmission with a hidden pilot sequence. *IEEE Trans. Broadcasting* 46:170–76.

[39] X. Meng and J. K. Tugnait. 2004. Superimposed training-based doubly-selective channel estimation using exponential and polynomial bases models. In *Proceedings of the 2004 Conference on Information Sciences and Systems*, Princeton University, NJ, pp. 621–626.

[40] M. Niedzwiecki. 2000. *Identification of time-varying processes*. New York: Wiley.

[41] A. G. Orozco-Lugo, M. M. Lara, and D. C. McLernon. 2004. Channel estimation using implicit training. *IEEE Trans. Signal Processing* 52:240–54.

[42] J. G. Proakis. 2001. *Digital communications*. 4th ed. New York: McGraw-Hill.

[43] J. G. Proakis and D. G. Manolaks. 1996. *Digital signal processing*. 3rd ed. Englewood Cliffs, NJ: Prentice-Hall.

[44] R. Raheli, A. Polydoros, and C. K. Tzou. 1995. Per-survivor processing: A general approach to MLSE in uncertain environments. *IEEE Trans. Commun.* 43:354–64.

[45] A. M. Sayeed, A. Sendonaris, and B. Aazhang. 1998. Multiuser detection in fast-fading multipath environment. *IEEE J. Selected Areas Commun.* 16:1691–701.

[46] A. M. Sayeed and B. Aazhang. 1999. Joint multipath-Doppler diversity in mobile wireless communications. *IEEE Trans. Commun.* 47:123–32.

[47] N. Seshadri. 1994. Joint data and channel estimation using blind trellis search techniques. *IEEE Trans. Commun.* 42:1000–11.

[48] D. Slepian. 1978. Prolate spheroidal wave functions, Fourier analysis, and uncertainty. V. The discrete case. *Bell Syst. Tech. J.* 57:1371–430.

[49] M. D. Srinath, P. K. Rajasekaran, and R. Viswanathen. 1996. *Introduction to statistical signal processing with applications.* Upper Saddle River, NJ: Prentice-Hall.

[50] L. Tong and S. Perreau. 1998. Multichannel blind channel estimation: From subspace to maximum likelihood methods. *Proc. IEEE* 86:1951–68.

[51] L. Tong, G. Xu, and T. Kailath. 1994. A new approach to blind identification and equalization of multipath channels. *IEEE Trans. Inform. Theory* 40:340–49.

[52] J. K. Tugnait. 1987. Identification of linear stochastic systems via second- and fourth-order cumulant matching. *IEEE Trans. Inform. Theory* 33:393–407.

[53] J. K. Tugnait. 1995. On blind identifiability of multipath channels using fractional sampling and second-order cyclostationary statistics. *IEEE Trans. Inform. Theory* 41:308–11.

[54] J. K. Tugnait. 1995. Blind estimation and equalization of digital communication FIR channels using cumulant matching. *IEEE Trans. Commun.* 43:1240–45.

[55] J. K. Tugnait. 1996. Blind equalization and estimation of FIR communications channels using fractional sampling. *IEEE Trans. Commun.* 44:324–36.

[56] J. K. Tugnait. 2001. Channel estimation and equalization using higher-order statistics. In *Signal processing advances in wireless and mobile communications: Trends in channel estimation and equalization,* ed. G. B. Giannakis, Y. Hua, P. Stoica, and L. Tong, chap. 1, pp. 1–39. Vol. 1. Upper Saddle River, NJ: Prentice Hall.

[57] J. K. Tugnait and W. Luo. 2003. On channel estimation using superimposed training and first-order statistics. In *Proceedings of the 2003 IEEE International Conference on Acoustics, Speech, Signal Processing,* Hong Kong, vol. 4, pp. 624–627.

[58] J. K. Tugnait and W. Luo. 2003. On channel estimation using superimposed training and first-order statistics. *IEEE Commun. Lett.* 8:413–15.

[59] J. K. Tugnait and S. He. 2006. Doubly-selective channel estimation using data-dependent superimposed training and exponential bases models. In *Proceedings of the 2006 Conference on Information Systems and Sciences,* Princeton University, NJ.

[60] J. Vidal and J. A. R. Fonollosa. 1996. Adaptive blind equalization using weighted cumulant slices. *Int. J. Adaptive Control Signal Processing* 10:213–38.

[61] T. Zemen and C. F. Mecklenbräuker. 2005. Time-variant channel estimation using discrete prolate spheroidal sequences. *IEEE Trans. Signal Processing* 53:3597–607.

[62] Y. R. Zheng and C. Xiao. 2003. Simulation models with correct statistical properties for Rayleigh fading channels. *IEEE Trans. Commun.* 51:920–28.

[63] G. T. Zhou, M. Viberg, and T. McKelvey. 2003. A first-order statistical method for channel estimation. *IEEE Signal Processing Lett.* 10:57–60.

3

Adaptive Coded Modulation for Transmission over Fading Channels

Dennis L. Goeckel
University of Massachusetts

3.1 Introduction

The main goal of modern communications systems is to make universal high-speed access to information a reality. Under nearly any realizable scenario, the end user in such a system is free of tethers, thus making robust high-speed data transfer across the wireless channel a topic of extreme importance. However, the wireless channel presents a number of challenges. In particular, the wireless signal experiences [55]:

1. Path loss: The signal attenuates with distance from the transmitter.
2. Shadowing: Large objects between the transmitter and receiver can obstruct the radio signal.
3. Multipath fading: Reflections from objects in the environment can add constructively or destructively at the receiver.

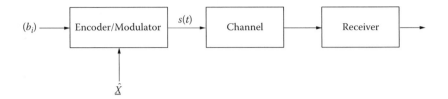

FIGURE 3.1 General adaptation framework: The transmitter sets parameters based on $\hat{X}$, which contains information about the channel state, while forming the transmitted signal $s(t)$ from the information bit sequence (b_i). Note that $\hat{X}$ can take many forms: path loss/shadowing estimates [47, 68, 70], number of errors corrected in previous packets [54, 55], explicit multipath fading estimates [31, 32], etc.

Since all three of these effects change with time or position, they cannot be known at the time that the system is designed. Thus, if no form of adaptation is performed while the system is operating, the system must be designed to in some sense deal with the worst case, which can be very expensive in terms of system resources. For example, if all users in a cellular system have to assume worst-case path loss and shadowing (e.g., behind a large building at the very edge of the cell) regardless of their location, they will transmit at maximum power, thus maximizing battery usage and the interference to other users. Therefore, wireless system adaptation has been a topic of critical interest in recent years.

Wireless link adaptation can be defined as shown in Figure 3.1: any altering of the parameters at the transmitter based on information about the current link state. Note that *link state* will be defined very generally; in particular, in addition to wireless channel conditions (path loss, shadowing, multipath fading), user data requirements will be included in the definition. Methods of adaptation can be classified by the type of adaptation performed (power, code rate, modulation, etc.) and the timescale of that adaptation.

The timescale of the adaptation depends on the link state phenomena for which measurements are provided to the adaptation algorithm. User data rate requirements, path loss, and shadowing change at a timescale that is long relative to the symbol rate. This makes measurements of such phenomena relatively robust [26, 64], particularly at high signal-to-noise ratios (SNRs), and has resulted in widespread penetration into current and pending systems of slow adaptations, such as power control [45, 66, 68] and data rate adaptation through variable spreading, code rate, or code aggregation [46]. Thus, throughout this chapter, it will be largely assumed that the measurements of the path loss and shadowing are accurate and known at both the transmitter and receiver, thus yielding a wireless system with a known given average received signal power but experiencing variable multipath fading.

Multipath fading is caused by the arrival at the receiver of many signal reflections, the superposition of which causes the instantaneous received signal power to vary widely [57, chapter 4] as described in section 3.2. Since significant nulls can occur, this signal fading is one of the most difficult problems to deal with in wireless communications systems. In particular, when the received power drops too low, a burst of bit errors can occur, and such bursts tend to dominate the error probability—even if the occurrence

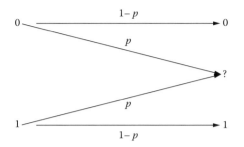

FIGURE 3.2 The binary erasure channel [16, p. 188] to represent a discrete-valued fading channel with channel state information available at the receiver.

of such system power drops is relatively unlikely. This results in a significantly higher required average received SNR for a given level of performance relative to systems operating over additive white Gaussian noise (AWGN) channels [53, p. 820]. However, unlike the effects of path loss and shadowing, which vary slowly and thus limit the ability of the system designer to average their effects over time, multipath fading varies relatively rapidly with time, position, and frequency. Thus, in many scenarios, well-designed systems achieve diversity, allowing them to average effectively over the effects of the multipath fading and thus significantly reduce its impact [53, p. 821]—even if there is no channel knowledge at the transmitter. For many systems, such nonadaptive solutions come at the cost of system latency or complexity, which motivates the consideration of transmission schemes that employ measurements of the multipath fading values.

Since this chapter will largely focus on the design of techniques for adaptation based on measurements of the multipath fading, it is important to understand the applicability of such adaptation. Thus, the gains from having knowledge of the channel at the transmitter will be a key topic discussed, and as motivated in the previous paragraph, it is often a question of the system complexity and latency allowable. To make this more concrete, consider the simple information theoretic example drawn from [16, p. 188], which is shown in Figure 3.2. First, to see how this represents a fading channel, consider a binary transmission system for which there are essentially no errors when the signal is transmitted over an AWGN channel (i.e., no fading); a simple example is coherently detected binary phase-shift keying (BPSK) with a relatively high SNR [53, p. 820]. Now assume that a BPSK system is operating over a discrete-valued fading channel described as follows. First, the state of the channel is independent for separate channel uses, which implies that a deep interleaver [53, p. 467] is employed. For a given channel use:

1. With probability p, the transmitted signal is multiplied by zero (hence disappears).
2. With probability $1 - p$, the transmitted signal is multiplied by $\alpha = 1/(1 - p)$ (hence amplified, which keeps the average received SNR identical to the AWGN case).

Assuming that channel state information (CSI), which throughout this paper will be the value of the multiplicative factor (in this case 0 or α), is available at the receiver, this yields the model shown in Figure 3.2. Consider signaling over the channel shown in

Figure 3.2 with and without channel state information at the transmitter. With CSI at the transmitter, transmission is halted when $\alpha = 0$ and a single bit is transmitted whenever $\alpha = 1/(1 - p)$. Thus, with a very simple receiver identical to that for the BPSK system operating over the AWGN channel, the system reliably transmits $1 - p$ information bits per channel use. Next, consider the case where there does not exist channel state information at the transmitter. Using information theoretic results [16, p. 188], the capacity of the channel without transmitter CSI is still $1 - p$ information bits per channel use, but now it requires very long code words and the typical sequence decoding employed for the achievability statement of Shannon's capacity [59]. This simple example captures the key idea to adaptive signaling in response to the multipath fading in many cases—it will often not make sense from a Shannon capacity, but it can greatly simplify system design in practical systems [30].

Thus, adaptation in response to transmitter knowledge of the multipath fading has the promise of greatly simplifying the system design or, for a fixed system complexity, has the promise of greatly improving system performance (such as average data rate) [31, 32]. However, it is the very property that makes adaptation fruitful that also complicates its implementation; in particular, adaptation can be exploited because of the time-varying nature of user needs, path loss, shadowing, and multipath fading. But this time-varying nature makes that adaptation difficult; in particular, although changes in user needs and path loss generally happen over a long enough timescale that they can be reliably estimated, the time-varying nature of the shadowing [66] and the multipath fading [24, 27] make channel measurements outdated by the time they are ready to be used. In other words, the channel has changed since the measurements were performed, and thus the utility of such measurements in representing the current state of the channel can be questioned. This will be particularly exacerbated, of course, in systems that seek to adapt to the multipath fading [48].

The consideration of the design of signaling schemes that employ inherently outdated or noisy measurements is best done by carefully considering the channel characteristics *conditioned* on the measurements available. Naturally, if the support of the probability density function of the conditional channel given the measurements is very narrow, indicating that the system is fairly certain of the channel value, one can design coded modulation structures and rules for adapting those structures based on the assumption that the channel is fully known [31, 32] and suffer only mild degradations. However, such schemes can be very sensitive, even if the probability density function only shows a little spread around the estimated value [24, 27]. In such cases, not only must the rules of adaptation consider such spread, but the spread often will affect the types of coding and modulation structures that are effective, as demonstrated in section 3.3.3.

This chapter is organized as follows. In section 3.2, the system model that will be used throughout this work is presented. Section 3.3 provides a detailed derivation of the key issues in adaptive signaling using the simplest case of a system where there is only a single antenna employed at each the transmitter and receiver. Section 3.4 discusses recent extensions of these results to systems with multiple antennas at the transmitter and receiver, and section 3.5 presents conclusions and avenues for future work in the field.

3.2 Adaptive System Model

3.2.1 Model for a Wireless Link

The transmitted signal in a wireless communications system is affected by three factors: path loss, shadowing, and multipath fading. In complex baseband notation [53], the signal $r(t)$ that is received when the signal $s(t)$ is transmitted can be written as

$$r(t) = L(t)X(t)s(t) + n(t), \tag{3.1}$$

where $L(t)$ is a real-valued random process that represents the combined effect of the path loss and shadowing, $X(t)$ is a complex random process representing the effect of the multipath fading, and $n(t)$ is a stationary complex Gaussian random process with (two-sided) power spectral density $S_N(f) = N_0/2$, representing additive noise. In equation (3.1), the fading has been assumed to be frequency-non-selective [53, p. 816]; this is appropriate for a narrowband single-carrier system or a single subcarrier of a wideband orthogonal frequency division multiplexing (OFDM) system [6, 71]. Extensions of the concepts presented in this chapter to frequency-selective channels are conceptually straightforward, although such channels offer inherent natural diversity with little system latency, and hence often reduce the gain available through adaptive signaling.

Understanding the characteristics of the processes $L(t)$ and $X(t)$ in equation (3.1) is crucial in determining methods of adaptation based on such. The random process $L(t)$ is caused by path loss, which is determined by the distance the receiver is from the transmitter, and shadowing, which is determined by the existence of large objects between the transmitter and receiver. For a stationary user, the path loss and shadowing are generally modeled as constant, despite the fact that it could be argued that the movement of large objects can affect the shadowing. For a user in motion, the shadowing will be the more variable of the two effects, and the distance over which it is highly correlated can be roughly modeled as 100 m in a macrocellular suburban environment [35]. For a user at walking speed (say, 2 m/s), this implies that the shadowing correlation time is on the order of 50 s; for a user in a vehicle (say, 88 km/h), this implies that the shadowing correlation time is on the order of 4 s. This suggests that it is quite plausible to make estimates of the path loss and shadowing and to employ such in wireless communications systems. In fact, this is very often done in current and next-generation cellular system implementations [48]. Also, since $L(t)$ varies at a relatively long timescale, it will be assumed throughout the remainder of this chapter that it is measured accurately and known at the transmitter and receiver.

In contrast, consider the random process $X(t)$, which represents the multipath fading. In a wireless environment, the signal $s(t)$ is reflected to the receiver from many objects. Because the propagation distance is different for each of these reflections, the reflected signals will arrive at slightly different times at the receiver. For a narrowband system, which has a relatively long symbol interval, there will not be appreciable intersymbol interference (ISI) [53, p. 817]. However, because of the large carrier frequencies typically employed in modern wireless communications systems, even a small difference in arrival times for two paths can result in a large phase difference between those paths. For example, a path-length difference of only 1 ft results in the signal being delayed by 1 ns, which causes a full 2π rotation in phase when the carrier frequency is 1 GHz. Hence, the phase of any given arriving path is generally modeled as uniformly distributed. Since the process $X(t)$ is caused by the sum of very many roughly independent paths, projected onto each of the in-phase (real) and quadrature (imaginary) components, the central limit theorem [52, p. 214] motivates its modeling as a complex Gaussian random process [3].

By considering the genesis of the multipath fading as described above, it is easy to observe that the phase of a given path will change greatly for each movement of the reflecting object, the receiver, or the transmitter by one wavelength. Hence, even with only walking speed mobility (say, 2 m/s), a system with a 1 GHz carrier will yield a process $X(t)$ that changes independently six times per second (or, as commonly stated, with a 6 Hz Doppler frequency) [51, 56], which makes adaptation challenging, since feedback of the channel characteristics provided to the transmitter at some delay must accurately model the current channel fading for adaptation to be effective. Note that this problem will be exacerbated at higher mobilities and higher carrier frequencies.

Mathematically, $X(t) = X_R(t) + jX_I(t)$ will be assumed to be a zero-mean stationary Gaussian random process with an autocorrelation function of the real part $X_R(t)$ (or imaginary part $X_I(t)$) defined as

$$R_X(\tau) = E\left[X_R(t)X_R(t+\tau)\right] = E\left[X_I X_I(t+\tau)\right],$$

and the real part $X_R(t)$ and imaginary part $X_I(t)$ will be assumed to be independent of one another. The zero-mean assumption implies that a line-of-sight path is not present, which corresponds to the most pessimistic case—Rayleigh fading. Throughout this chapter, the popular Jakes model [41] will generally be adopted, which is characterized by $R_X(\tau) = J_0(2\tau f_d \tau)$, where $J_0(\cdot)$ is the zero-order Bessel function of the first kind and f_d is the Doppler frequency, which is defined as the number of wavelengths of motion of an object per second. It will be assumed that although the Doppler frequency might be large, it will not approach the symbol rate, and thus the channel $X(t)$ can be assumed to be constant over the support of a single signaling pulse $p(t)$, which is termed the *slowly fading* assumption in most digital communications texts [53, p. 816]. We hasten to emphasize, however, that the use of the word *slow* in this context is with reference to the symbol interval—not the amount of time between channel estimation and signal transmission, where even such "slow" multipath fading can have a significant effect.

3.2.2   Adaptation in Response to Path Loss/Shadowing

There are many forms of adaptation currently employed in response to path loss/shadowing in wireless communications systems. In fact, even the base station selection process, where a mobile generally decides to associate with the base station from which it sees the largest average received signal strength, can be viewed as a form of adaptation.

Such adaptations will be called "slow" adaptations throughout this work, and they will be characterized by schemes that adapt the transmitter at an interval on the order of (at least) many (hundreds of) symbols. A good tutorial on slow adaptations, particularly in current standards, is provided in [48].

First, consider wireless system adaptations that adapt depending on user needs. In particular, one of the key features of third-generation cellular systems is supporting users with high data rates. This is often done by simply allocating more of the time/bandwidth/code space to the users. For example, in Enhanced Data Rates for GSM Evolution (EDGE) systems, which are built on a time-division multiple-access (TDMA) framework, users with high-data-rate needs are allocated more time slots. In the code-division multiple-access (CDMA)-based IS-95 Revision B, high-data-rate users are allocated multiple spreading codes, which is termed code aggregation [48].

Next, consider adaptations based on the current channel conditions for a given user. In fixed-rate systems, such as first- and second-generation cellular telephone systems, where the rate of the vocoder is generally fixed, the key is to adapt the system such that acceptable performance is maintained at this fixed rate. The transmission technology and channel assumptions fix a minimum average received SNR γ_0 required for acceptable operation—the goal of adaptation is to maintain γ_0, which can be done by adapting the transmitted power in response to measurements of the path loss and shadowing. Methods of performing such adaptation include channel inversion, where the transmitted power is set proportional to the channel loss, and truncated channel inversion [e.g., 17, 66]. Truncated channel inversion is defined by a threshold L_0, which breaks the policy into two cases:

1. $L(t_0) \geq L_0$: The transmitted power is set to $\gamma_0/L(t_0)$, which results in an average received SNR of γ_0.
2. $L(t_0) < L_0$: The transmitted power is set to zero, which results in an outage.

Using this policy, the required average received SNR (and no more) is obtained whenever possible, but excessive power is not wasted by inverting the channel when there are large losses on the channel.

Outside of current standards, the setting of the rate (coding, modulation, spreading factor) of the system to the current average received SNR has clearly emerged as a critical topic. In particular, turbo codes [4] and low-density parity-check (LDPC) codes [23, 46] are approaching channel capacities on a variety of channels. Thus, assuming enough receiver complexity for the decoding of such codes and enough latency to allow perfect interleaving of the coded symbols, the rate of nearly error-free systems should approach the Shannon capacity [59] of the independent and identically distributed (IID) discrete-time Rayleigh fading channel, defined by:

$$Y_i = \alpha_i X_i + n_i,$$

where Y_i is the received sequence, α_i is the IID sequence of Rayleigh channel fading values, X_i is the transmitted sequence, and n_i is the noise sequence. The Shannon capacity for

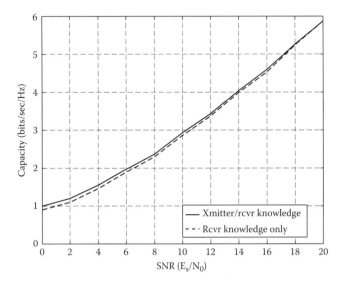

FIGURE 3.3 The Shannon capacity of an independent and identically distributed Rayleigh fading channel, assuming (1) perfect CSI available only at the receiver, and (2) perfect CSI available at both the transmitter and receiver [28]. Note that the gain in Shannon capacity resulting from having perfect CSI available only at the transmitter is only slight, as discussed in section 3.3.1.

such a channel without CSI at the transmitter is shown as the lower curve in Figure 3.3, where the SNR on the horizontal axis is the average received SNR (i.e., the SNR after the path loss and lognormal shadowing are considered). The only requirement for the highly efficient operation of such a system is the knowledge of this average received SNR at the transmitter so that the rate of the transmitter can be set appropriately, and this can be obtained by feedback of the path loss and shadowing. We emphasize that approaching the curves in Figure 3.3 still requires high decoding complexity and significant latency, which motivate whether adaptation with the additional knowledge of the values of the multipath fading can improve on the performance in Figure 3.3 in terms of performance versus system complexity.

3.2.3   Analytic Model for Fine-Scale Adaptation

The main portion of this chapter will be dedicated to the design and analysis of adaptive systems that use explicit measurements of the multipath fading to perform system adaptation. This is a topic that was considered in the 1970s [11, 37, 38] and then became popular again in the early 1990s [e.g., 1, 6, 14, 29, 30, 65].

A block diagram of the typically employed system is shown in Figure 3.4. Given the model shown in Figure 3.4 and channel model given in section 3.2.1, the key to designing adaptive signaling systems is considering signaling for the *conditional* channel for the symbol of interest (call it s_k) given the outdated measurement $\hat{\underline{X}} = (\hat{X}(t - \tau_1), \hat{X}(t - \tau_2),$ ..., $\hat{X}(t - \tau_N))^T$. It will be assumed that a measurement $\hat{X}(t - \tau_i)$ is equal to the true value $X(t - \tau_i)$ plus additive Gaussian noise of variance

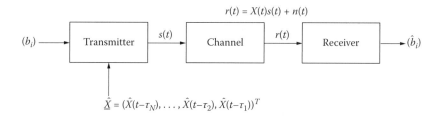

$$r(t) = X(t)s(t) + n(t)$$

$$\hat{\underline{X}} = (\hat{X}(t{-}\tau_N), \ldots, \hat{X}(t{-}\tau_2), \hat{X}(t{-}\tau_1))^T$$

FIGURE 3.4   A block diagram of the system, where (b_i) is the sequence of information bits to be transmitted across the channel, $s(t) = \sum_{k=-\infty}^{\infty} s_k p(t - kT_s)$ is the transmitted signal, $r(t)$ is the received signal, $n(t)$ is additive white Gaussian noise, $\hat{\underline{X}}$ is the vector of outdated channel measurements, and $(\hat{b}_i)$ is the sequence of information bit estimates output from the receiver.

$$\sigma_\varepsilon^2 = \frac{1}{\dfrac{E_p}{N_0}}$$

in each of the in-phase and quadrature components. For example, such would be the case in an adaptive system employing pilot symbol–assisted modulation (PSAM) [12] with a pilot symbol energy of E_p [7, 24].

Note that the model in Figure 3.4 captures the critical issue of delay in the feedback path from the receiver to the transmitter, since the most recent estimate is assumed to have been made τ_1 seconds ago. In other words, the "outdated" nature of the estimates takes into account this key implementation issue in adaptive communications systems.

Denoting Y as the magnitude of the fading that multiplies s_k in the matched filter output for the kth symbol and using the fact that linear functionals of a Gaussian random process are jointly Gaussian, Y is Rician when conditioned on the vector $\hat{\underline{X}}$, with probability density function [25]

$$p_{Y|\hat{\underline{X}}}\left(y \,|\, \underline{x}\right) = \frac{y}{\sigma^2} e^{-\frac{y^2+s^2}{2\sigma^2}} I_0\left(\frac{ys}{\sigma^2}\right), \quad y \geq 0, \tag{3.2}$$

where $I_0(\cdot)$ is the zero-order modified Bessel function. Using the assumption that $X(t)$ can be assumed constant over the support of $p(t)$ and normalizing the fading such that $E[(X_R(kTs))^2] = E[X_I(kT_s))^2] = 1$ (note that this simplification will make the average received energy twice that provided by simply the path loss and shadowing, which will be accounted for below), the noncentrality parameter in equation (3.2) is given by

$$s^2 = \left(\underline{\rho}^T\left(\Sigma_{\underline{X}} + \sigma_\varepsilon^2 I_N\right)^{-1} \underline{x}_R\right)^2 + \left(\underline{\rho}^T\left(\Sigma_{\underline{X}} + \sigma_\varepsilon^2 I_N\right)^{-1} \underline{x}_I\right)^2, \tag{3.3}$$

where I_N is an N by N identity matrix. The $(m, n)^{\text{th}}$ element of $\Sigma_{\underline{x}}$, the N by N autocorrelation matrix of the in-phase component of $\hat{\underline{X}}$ when the channel estimates are noiseless, is given by $R_x(\tau_{N-m+1} - \tau_{N-n+1})$, and the correlation vector of the in-phase component of $\hat{\underline{X}}$ with the in-phase component of the fading of interest is given by $\underline{\rho}$, where $\rho_i = R_x(\tau_{N-i+1})$. The parameter σ^2 in equation (3.2) is the mean square error of a minimum mean square error (MMSE) estimator [72, p. 54] of the in-phase (or quadrature) fading of interest, and is given by

$$\sigma^2 = 1 - \underline{\rho}^T \left(\Sigma_{\underline{x}} + \sigma_\epsilon^2 I_N \right)^{-1} \underline{\rho} . \qquad (3.4)$$

Understanding the Rician density in equation (3.2) and the expression for the Rician noncentrality parameter s in equation (3.3) is key to designing effective adaptive coded modulation schemes. In particular, for $s = 0$, the Rician probability density function in equation (3.2) is equivalent to a Rayleigh probability density function, indicating that coded modulation structures designed for Rayleigh fading channels are pertinent for application when s is small; likewise, as $s \to \infty$, the (properly normalized) Rician density function approaches a delta function, thus indicating that the effective channel approaches an AWGN channel. Since coded modulation schemes for Rayleigh fading channels differ greatly from AWGN schemes, the interpretation of equation (3.3) is used extensively in the design of structures, as demonstrated in section 3.3.3.

There is one limitation to directly employing the result in equation (3.2). In particular, it presumes that the autocorrelation function $R_X(\tau)$ of the random process $X(t)$ is known at the transmitter; however, this autocorrelation function can vary greatly in wireless systems [51, p. 88–89]. Thus, it must generally be estimated [18, 19], either implicitly or explicitly, or uncertainties in it must be worked into system design [24, 27]. To address both possibilities in one framework, the autocorrelation function will be assumed to lie in some uncertainty class $\mathcal{R}$, which matches the approach taken in [27] directly. If it can be accurately estimated through techniques as described in [18, 19], this class can be shrunk accordingly (in the limit to a single autocorrelation function).

Thus, given the model for the system measurements and the measurements of the autocorrelation function, one should be able to ascertain (1) how much predictor error σ^2 will generally be in the system, and (2) what is the uncertainty class $\mathcal{R}$ over which some sort of robustness will be maintained. Understanding both of these for a given system configuration will be the key to understanding the design of the coded modulation. In particular, the former will allow the choice of the coded modulation structure, while the latter will allow one to design on that structure.

3.3　Adaptivity in Single-Input Single-Output Systems

3.3.1　Information Theoretic Bounds

Before considering the derivation of practical signaling schemes that adapt to the multipath fading, it is instructive to consider the improvement in Shannon capacity that is available when CSI is made available to the transmitter. Assume that the sequence of zero-mean complex Gaussian channel fading coefficients affecting the transmitted symbols forms an IID sequence; in other words, an IID Rayleigh fading channel is assumed.

If the criterion is to maximize the average data rate under an average power constraint, Goldsmith [30] has demonstrated that the information theoretic capacity when perfect CSI is available at both the transmitter and receiver is achieved with variable-power Gaussian codebooks, where the power depends on the current value of the channel fading. A comparison of the Shannon capacity when perfect CSI is available at both the transmitter and receiver with the capacity when CSI is available only at the receiver is shown in Figure 3.3. Note that, somewhat surprisingly, the gain in channel capacity is only slight. However, as with the example of the erasure channel in section 3.1, it should be remembered that this assumes very long code words and large decoding complexities. In particular, the minimum distance of the codes becomes very large, and since IID Rayleigh fading is assumed, the diversity achieved by a given code is very large—even for the case of CSI only at the receiver.

The analysis in the previous paragraph and Figure 3.3 applies to the Shannon capacity [59], which is generally appropriate if the average rate of a system is being considered. Recently, however, there has been significant interest in whether a system can transmit a fixed amount of information within a given time constraint—such approaches lead to measures such as the outage capacity [22] or the delay-limited capacity [36]. Such analyses are generally still done under the assumption of infinite-length code words, which are required for the random coding arguments invoked, but now under the assumption that a given code word will only see some small number of fades. Thus, in essence, this yields a view at system operation in a diversity-limited context. The metric is based on the probability that the system experiences a set of fading values for which it can communicate at the desired rate. In contrast to Figure 3.3, such analyses [8, 49] have demonstrated the significant gains possible when knowledge of the channel fading values is provided to the transmitter in addition to the receiver. This has motivated work in the design of practical adaptation schemes that focus on outage probability [43]. These results lead to the preliminary conclusion that the gain from having estimates of the channel fading at the transmitter is highly reliant on the decoder complexity and system latency allowed.

3.3.2 Design for Uncoded Systems

In this section, adaptive uncoded modulation will be designed. Unlike coded schemes, where the design is complicated by memory in the trellis and questions about the proper structure, uncoded schemes present a simple framework to demonstrate many of the key issues.

As described in section 3.2.3, there is a key issue of robustness to uncertainties in the autocorrelation function $R_X(\tau)$, which can be captured by designing for an uncertainty class $\mathcal{R}$ that shrinks to a single point when the autocorrelation is known or can be accurately estimated. If the class $\mathcal{R}$ is a single point as is often considered for prediction-based methods [18, 19, 44, 45], the application of equation (3.2) is identical regardless of the number of outdated estimates N employed. This is observed by noting that, for a vector of random variables drawn as samples from a stationary Gaussian random process, (1) the marginal probability density function is independent of the sampling time, and (2) the conditional probability density function for any one of the variables is

Gaussian when conditioned on the others, with variance given by σ^2 in equation (3.4). Thus, given the predicted value and σ^2, the choice of the signal set is independent of N. Hence, when the autocorrelation function $R_X(\tau)$ is known exactly, the design for $N = 1$ with the appropriate σ^2 is sufficient to characterize performance of a given scheme.

When $\mathcal{R}$ is made larger to capture uncertainties in the knowledge of $R_X(\tau)$, the design becomes greatly complicated [26]. Thus, robust design with general $\mathcal{R}$ with only a single outdated estimate ($N = 1$) will be considered; however, we hasten to note that, per the previous paragraph, designing adaptive coded modulation for known $R_X(\tau)$ and *any* N is a simplification of this case.

Since the case $N = 1$ will be considered, the quantity $\rho = R_X(\tau_1)$, which, since $X_R(t)$ and $X_I(t)$ are normalized to have unit energy, is the correlation coefficient of the in-phase (or quadrature) components of the multipath fading process between the time of channel estimation and the time of data transmission, will be important. Assuming that the estimates are noiseless ($\sigma_{\tilde{e}}^2 = 0$) implies that $\sigma^2 = 1 - \rho^2$, and it is observed that the mean square prediction error increases rapidly with decreasing correlation between the estimate and the current value, as expected. Throughout much of this chapter, systems will be designed for a given ρ, which captures the amount of information in the channel estimate about the current fading value. A conversion to mean square predictor error, if desired, can be obtained by the transformation $\rho = \sqrt{1 - \sigma^2}$.

Designing robustly using a single outdate estimate requires performance to be guaranteed for all $\rho \in [\rho_{min}, 1]$, where ρ_{min} is the minimum value of $R_X(\tau_1)$.

3.3.2.1 Design Rules

The design rules for uncoded systems have been well established by a number of authors [7, 27, 31]. The signal sets considered in this section will be 0-QAM (quadrature amplitude modulation) (no data transmitted), 2-QAM, 4-QAM, 16-QAM, and 64-QAM with two-dimensional Gray mapping, although the extension to any set of signal sets is immediate.

Following [27], let P_b be the target bit error probability for the system, which operates at the average received SNR E_s/N_0, where E_s is the average received energy per QAM symbol. For now, it will be assumed that the average energy E_s is not varied over time; energy adaptation will be discussed in detail below. Specification of the adaptive transmitter requires finding $\tilde{M}(h)$, $\forall h$, where $\tilde{M}(h)$ is the number of signals in the QAM signal set employed when $|X(kT_s - \tau_1)| = h$. If $\tilde{M}(h)$ is chosen such that P_b is maintained for each h,

$$\tilde{M}(h) = \max\left\{ M : \sup_{\rho_{min} \leq \rho \leq 1} \tilde{P}_M\left(\tfrac{E_s}{N_0}, h, \rho\right) \leq P_b \right\}, \tag{3.5}$$

where $\tilde{P}_M(E_s/N_0, h, \rho)$ is defined as the bit error probability of the M-QAM signal set at average received SNR E_s/N_0 when $R_X(\tau_1) = \rho$ and $|\hat{X}(kT_s - \tau_1)| = h$. Assume that maximum likelihood symbol detection, given the current channel fading amplitude, is employed on the samples of the matched filter output at the receiver. A tight approximation to the bit error rate of M-QAM modulations is given by [31]

$$P_M\left(y^2\,\frac{E_s}{N_0}\right) \approx 0.2\exp\left(-\frac{3}{4(M-1)}\,\frac{E_s}{N_0}\,y^2\right), \tag{3.6}$$

which will be employed for all M for much of the design work for uncoded systems. If errors in channel estimation *at the receiver* are considered, the right side of equation (3.6) will increase, of course, but it will often fit into the same functional form [7], which is convenient, since the same optimization will apply. Using equation (3.6) yields

$$\tilde{P}_M\left(\frac{E_s}{N_0},h,\rho\right) = E\left[P_M\left(Y^2\,\frac{E_s}{N_0}\right)\Big|\hat{X}(kT_s-\tau_1)\Big|=h\right]$$

$$\approx \begin{cases} \dfrac{0.2\exp\left[-\dfrac{h^2\rho^2}{2(1-\rho^2)}\left(1-\dfrac{1}{1+\dfrac{3}{2}\dfrac{E_s}{N_0}\dfrac{(1-\rho^2)}{(M-1)}}\right)\right]}{1+\dfrac{3}{2}\dfrac{E_s}{N_0}\dfrac{(1-\rho^2)}{(M-1)}} & \rho<1 \\[6ex] 0.2\exp\left(-\dfrac{3}{4}\dfrac{E_s}{N_0}\dfrac{h^2}{(M-1)}\right) & \rho=1 \end{cases} \tag{3.7}$$

where the second line is obtained by substituting equation (3.2) and equation (3.6) into the first line and evaluating the expectation over Y using [34, 6.614.3].

From equation (3.5), equation (3.7) must be evaluated at its supremum on $\rho \in [\rho_{\min}, 1]$. Since the right side of equation (3.7) is a continuous function on this closed interval, it achieves its maximum on this interval at a point that will be denoted ρ^*. The following solution is found by standard calculus techniques. Let

$$\tilde{\rho} = \begin{cases} 0 & h \geq \sqrt{2} \\[2ex] \sqrt{\left(1+\dfrac{2(M-1)}{3}\dfrac{N_0}{E_s}\right)\dfrac{(2-h^2)}{2}} & 0 \leq h \leq \sqrt{2} \end{cases}.$$

The worst-case autocorrelation is then given by

$$\rho^* = \begin{cases} \rho_{\min} & \tilde{\rho} \leq \rho_{\min} \\ \tilde{\rho} & \rho_{\min} < \tilde{\rho} < 1 \\ 1 & 1 \leq \tilde{\rho} \end{cases}. \tag{3.8}$$

The signal set is specified using equations (3.7) and (3.8) in

$$\tilde{M}(h) = \max\left\{ M : \tilde{P}_M\left(\frac{E_s}{N_0}, h, \rho^*\right) \leq Pb \right\}.$$

Note that $\tilde{M}(h)$ is nondecreasing in h. Thus, the adaptive scheme can be specified by the values h_m, $m = 2, 4, 16, 64$, where h_m is defined as the threshold such that for $h \geq h_m$, m-QAM can be employed.

The discrete nature of the set of rates for any finite collection of signal sets hurts the performance of the system; in particular, for all h such that $h_m < h < h_{m+1}$, the estimate is better than that required to use m-QAM but not good enough to use $(m + 1)$-QAM. Energy adaptation provides a means to solve this problem [31]. Rather than employing the method of [31], an alternate method, which is analogous to truncated channel inversion and the power pruning of [20], is described here. The advantage of this method is that, with very little loss of optimality, it is easily extended to coded modulation structures, where the overall optimization problem of [31] is not easily framed when channel prediction is not perfect [27]. Once a signal set has been chosen, the system is essentially a fixed-rate system; thus, the goal changes from maximizing average rate to attempting to allow communication at this fixed rate with the least amount of power. Thus, after the signal set is chosen, equations (3.7) and (3.8) are used to decide the minimum energy required to maintain P_b given the channel estimate h, and this energy is employed rather than the average energy. Any excess energy is put into a "bank" on which successive symbols can draw.

3.3.2.2   Numerical Results

As discussed in section 3.3.1, systems with a significant amount of decoding complexity and allowable latency only have the potential for a small amount of improvement when CSI is provided to the transmitter. As might be expected, uncoded systems, which have the least decoder complexity and essentially no latency, benefit the most when transmitter CSI is available. In particular, uncoded systems operating over frequency-non-selective Rayleigh fading channels perform very poorly, because they do not achieve diversity. Because of this, coherently decoded quadrature phase-shift keying (QPSK) with only receiver CSI requires an SNR of 34 dB to achieve a bit error rate of 10^{-4} on a frequency-non-selective Rayleigh fading channel [53, p. 829], whereas the same technique requires an SNR of less than 10 dB to achieve the same bit error rate on an AWGN channel. The reason for this discrepancy is that the QPSK system operating over the Rayleigh fading channel is extremely susceptible to deep signal fades. Although the occurrence of such is relatively uncommon, the error rate during a bad fade can be orders of magnitude above that occurring when the average received SNR is observed, and thus these bad fades dominate the error rate.

In adaptive signaling, CSI is available at the transmitter. Arguably, the greatest utility of such information is that signaling can be avoided when bad fades are present. In particular, with perfect transmitter CSI [31], average rates in excess of 2 bits per symbol are possible at bit error rates of 10^{-5} for average received SNRs under 20 dB. Thus, there is a significant gain in system performance when transmitter CSI is available in uncoded systems.

However, as pointed out in [24, 27], the assumption of perfect CSI is dangerous when channel estimates are outdated or noisy, as would be the case with realistic delay in the feedback path from the receiver to the transmitter. In particular, the conditional density function given in equation (3.2) becomes Rician (rather than a delta function), and hence the conditional channel acts like a fading channel. For example, for the example described in section 3.3.2.1, adaptive signaling assuming perfect channel estimation can miss its target bit error rate by two orders of magnitude—even for the relatively high correlation coefficients of $\rho = 0.96$. In this case, bad predictions, which are relatively uncommon, lead to instantaneous error rates that are orders of magnitude above the target and thus dominate system performance. Using the design method of equation (3.2) reveals that there are still significant gains in adaptive signaling versus nonadaptive signaling when transmitter CSI is not perfect—even when the correlation coefficient drops as low as $\rho = 0.96$. We conclude from this section that adaptive signaling is particularly effective for simple, low-latency systems such as adaptive uncoded QAM systems [24, 27].

3.3.3   Coded Modulation Structures

As discussed in section 3.2.3, the conditional channel given an outdated measurement can vary from almost Rayleigh to almost AWGN—depending on both the channel estimate and the mean square prediction error σ^2. For small σ^2, the conditional channel is nearly always Rician with a large noncentrality component [53, p. 811] (hence approaching AWGN), whereas for large σ^2, the channel often approaches Rayleigh. It is well known that coded modulation structures optimized for AWGN channels [e.g., 63] are not well matched to Rayleigh channels [58]. Thus, the characterization of the mean square prediction error in a given system determines the types of coded modulation structures to be employed.

For systems where the mean square prediction error is anticipated to be nearly zero, coded modulation structures designed for the AWGN channel can be employed without interleaving [31]. For systems with a moderate amount of channel prediction error, structures designed for a Rayleigh fading channel can have aspects of structures designed for an AWGN channel embedded in them [27]. Finally, for adaptive systems where the mean square prediction error is expected to be large, adaptive bit-interleaved coded modulation (BICM) [50] is preferable. These three types of structures are presented below.

3.3.3.1   Coding Structures with (Nearly) Perfect Prediction

If the current channel fading $X(t)$ is known accurately at the transmitter (i.e., σ^2, the prediction error of an MMSE predictor, is small), the effective channel given the outdated estimate is roughly AWGN. Thus, coding structures designed for AWGN channels [63] should be employed [32]. In particular, a base trellis-coded modulation scheme [63] tuned to the average received SNR can be selected, and then uncoded bits can be added or deleted based on the channel estimate. This structure is shown in Figure 3.5. As

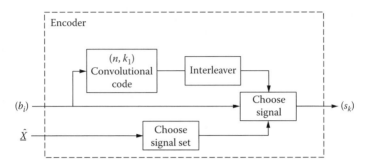

FIGURE 3.5 The adaptive trellis coding diagram—to be employed when performing adaptive signaling with low to moderate prediction errors. When the prediction error is low, the interleaver can be removed.

described in the legend to Figure 3.5, the interleaver can be removed in this case, since the Euclidean distance between two possible paths at the receiver can be precisely controlled. This structure allows symbol-by-symbol adaptation (unlike changing the rate of a convolutional encoder) and, because parallel branches are generally effective when communicating over AWGN channels, is an efficient coding structure over a wide range of instantaneous rates of the system.

3.3.3.2 Coding Structures with Moderate Prediction Error Statistics

When there is a moderate amount of predictor error power (i.e., moderate values of σ^2), the use of uncoded bits can be detrimental, since there will be a high number of channels that are not strongly Rician per equation (3.2), and it is well known that the use of uncoded bits on fading channels is problematic, per section 3.3.2.2. However, note that the channels become more Rician *as the predicted value increases.* This is fortuitous, because it allows the retention of the adaptive coded modulation structure shown in Figure 3.5, except that the base convolutional code is chosen to have no parallel branches [27]. Thus, when the estimate is small (and the channel nearly Rayleigh), a code appropriate for such a fading channel is employed [58]. When the channel estimate is large (and thus the channel strongly Rician), the structure in Figure 3.5 adds uncoded bits, which are appropriate in such a situation.

3.3.3.3 Coding Structures with Large Prediction Error Statistics

When there is a significant amount of prediction error (i.e., large values of σ^2), the use of parallel branches (i.e., uncoded bits) is not possible under almost any channel measurement, since from equation (3.2) it can be seen that the channel will be nearly Rayleigh with very high probability. Thus, symbol-by-symbol rate adaptation is desirable, but parallel branches are not allowable. A structure that allows such was presented in [50] and is shown in Figure 3.6. Note that the instantaneous rate is adapted, but all bits are coded, and thus the scheme retains diversity (in this case, against bad predictions) equal to the minimum Hamming distance of the convolutional code.

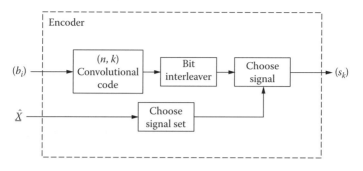

FIGURE 3.6 The adaptive bit-interleaved coded modulation scheme—to be employed when performing adaptive signaling with relatively frequent large prediction errors.

3.3.4 Designing with a Given Coded Modulation Structure

3.3.4.1 Design Rules

Unlike uncoded systems, which do not possess memory, and thus allow simple symbol-by-symbol adaptation, as demonstrated in section 3.3.2.1, the memory in coded modulation schemes complicates design. The main techniques that have been developed for coded modulation systems are described in [27] and [32]. Since the techniques in the later work of [27] include those in [32], the techniques of [27] will be briefly described. In particular, it is important to protect both the coded and uncoded information bits. This is done by maintaining the intersubset and intrasubset differences, which, roughly stated (see [27] for details), is the pairwise error probability between the two signal points in different and the same subsets [63], respectively. When the prediction error power is small, the intrasubset and intersubset differences for the coded modulation structures described by Figure 3.5 can be maintained by simply preserving the *received* Euclidean distance between adjacent points in the signal set [32]. When the prediction error power is moderate, the intersubset and intrasubset differences must be maintained separately [27]. For the adaptive BICM structure of Figure 3.6, there are only intersubset differences to be maintained [50].

3.3.4.2 Performance Results

As demonstrated in sections 3.3.1 and 3.3.2.2, the gains when CSI is available at the transmitter should decrease as system decoding complexity and latency are increased. Thus, it is anticipated that the gains described in this section will be smaller than those shown in section 3.3.2.2, and this is indeed observed. In particular, nonadaptive systems employing coded modulation and interleaving over frequency-non-selective Rayleigh fading channels have an enormous potential for gain, as evidenced by the vast difference between the performance of nonadaptive uncoded systems [53, p. 829] and the channel capacity shown in Figure 3.3. In contrast, adaptive signaling schemes with perfect prediction are signaling for a channel that is conditionally AWGN, which implies that the gains between uncoded systems [31] and channel capacity (see Figure 3.3) for systems with CSI at the transmitter are not so vast; in fact, they are similar to those attainable for AWGN channels [32].

The performance loss when perfect channel predictions are not available at the transmitter can be mitigated by employing the techniques of sections 3.3.3.2 and 3.3.3.3. Doing such with eight-state trellis codes provides gains over nonadaptive schemes that are on the order of 25 to 75% in data rate [27, 50], which, as expected, do not match the exorbitant gains seen in the uncoded case. One would expect the gains to decrease even further for more complicated codes and, in the complexity/latency limit, almost disappear as prescribed by Figure 3.3.

3.4   Adaptivity in Multiantenna Systems

Wireless systems employing multiple antennas at the transmitter or receiver have demonstrated both the theoretical [21, 62] and practical [21] ability to greatly increase system capacities far beyond those previously imagined. In particular, the disparate fading values between different pairs of antennas in multiple-input multiple-output (MIMO) systems lead to a large increase in the number of degrees of freedom of the system, and capacities can even dwarf those attainable for the AWGN channel at the same average transmitted SNR—even if channel state information is not available at the transmitter [21, 62].

Throughout this section, a multiple-antenna system will be referred to as an (M, N) system if it employs M transmit and N receive antennas. The system model generally employed for a narrowband MIMO system is given by

$$\underline{Y} = H\underline{X} + \underline{Z}, \tag{3.9}$$

where $\underline{X}$ is an $M \times 1$ vector whose j^{th} component represents the signal transmitted by the jth antenna. Similarly, the received signal and received noise are represented by $N \times 1$ complex vectors, $\underline{Y}$ and $\underline{Z}$, respectively. Generally, it is assumed that the entries of the $N \times M$ matrix H, whose entry (i, j) represents the fading from transmitter j to receiver i, are identically distributed zero-mean jointly complex Gaussian random variables. For many of the early results, the entries of H were considered to be independent [21, 62], although the impact of correlation of the entries has been widely considered in recent years [13, 60, 67].

First, a review of recent information theoretic results for systems that employ some form of channel information at the transmitter is considered. Although this topic is relatively new, single-user MIMO information theory has progressed very rapidly, and interesting results in multiuser MIMO information theory are starting to appear. Next, work concerned with the adaptation of practical structures built on equation (3.9) with knowledge of the values of H at the transmitter is considered.

3.4.1   Information Theoretic Considerations

An excellent recent tutorial of information theoretic considerations for MIMO systems, including adaptation-based ones on transmitter knowledge of the channel, can be found in [33].

3.4.1.1 MIMO Single-User Systems

It was established in early work on MIMO systems [62] that the Shannon capacity of an (N, N) MIMO system operating over a block-fading channel with CSI available at the transmitter and receiver is obtained by decomposing the channel into its eigenmodes, and then performing water filling [16, p. 349] on the eigenmodes based on the corresponding eigenvalues, where an eigenvalue indicates the SNR of the corresponding eigenmode. The system takes the convenient form of a single codebook designed for the AWGN channel followed by a beamformer that is adapted to each block [5].

As noted throughout this work, the assumption of perfect channel state information at the transmitter is problematic on wireless communication channels due to their time-varying nature. This is particularly true in the case of MIMO channels, since there are far more numerous coefficients to estimate than in the single-input single-output case. There has been a recent set of papers [40, 42, 61, 64] that consider the Shannon capacity of MIMO systems when there is mean and covariance feedback. The results in [40, 42, 61, 64] demonstrate the trade-offs in Shannon capacity associated with having channel state information available at the transmitter; in particular, they generalize the results in [62] and reveal when Shannon capacity can be obtained by beamforming—only a scalar codebook followed by a beamformer [33] rather than vector coding. In all cases, the Shannon capacity with knowledge of the channel at the transmitter only grows linearly with the number of antennas—only the leading constant and the simplicity of the system are possibly improved [2, 21], thus echoing the result of section 3.3 with respect to gains in ergodic capacity when CSI is available at the transmitter.

In [5], information theoretic measures based on the notion of outage are considered when there is perfect CSI available at both the transmitter and receiver. Recall that such measures (see section 3.3.1) attempt to capture the notion of system latency by limiting the number of fading blocks (and hence diversity) that a given code word experiences. In this case, a large number of antennas allows spatial diversity to be exploited, and in the limit, the effective channel can be made to look AWGN by employing a beamforming approach. Hence, an error control code designed for an AWGN channel concatenated with a beamformer is the optimal approach [5].

3.4.1.2 Multiuser MIMO Systems

The information theory of multiuser MIMO systems has only recently been explored. The most striking result in this context is the key gains that CSI at the transmitter [9, 10, 69] can provide, and this has generated a lot of interest in "dirty paper coding" methods [15]. In particular, unlike single-user MIMO systems, where generally only capacity-multiplying factors are improved or system complexity is decreased, CSI at the transmitter can allow the number of degrees of freedom to be increased [10, 33]. The facilitation of the penetration of such results into practical multiuser systems, where obtaining CSI at the transmitter can be complicated (and, at best, noisy and outdated), is an area that promises to be fruitful for research in the future.

3.4.2 Adaptive Coded Modulation for MIMO Systems

As demonstrated in section 3.4.1, there has been significant recent work establishing information theoretic bounds and methods for achieving those bounds for MIMO systems. It is now incumbent upon the communication theory community to translate those gains into practice. In particular, it will be important to consider the nature of obtaining channel state information at the transmitter. If the CSI provided to the transmitter is reliable and the channel is relatively constant over a long block, the path will be clear—standard scalar coding followed by beamforming matched to the current CSI. Code rate (and power) adaptation will only need to be done on a block-by-block basis. However, if the CSI is noisy or outdated, it will be interesting to consider whether the results of [40, 42, 61, 64] apply; that is, it will be interesting to consider the robustness of beamforming when practical coded modulation schemes are employed.

Recent work on adaptive coded modulation for multiple-antenna systems has followed the basic tenets above. In particular, there has been some consideration about how to perform adaptation in practical systems in the MIMO environment. In [2], consideration is given to a scheme that tracks the eigenspace of the system so that water-filling-type schemes can be employed—the parallel channel idea of [62]. Recent papers [e.g., 39] have considered different forms of adaptation—antenna selection, beamforming, and space-time coding. Because such papers generally rely on (nearly) perfect prediction assumptions, there has been little consideration of the effects of imperfect channel state information in these works. Recent work [73] has considered the impact of imperfect channel state information. In particular, robustness under imperfect channel state information is obtained by adding an Alamouti scheme over an inner beamformer (which would be ideal with perfect channel state information), and the optimality of such is shown for a system needing to transmit two information bits across the channel. These papers represent the start of research in a very important area that will bring the information theoretic gains of section 3.4.1 to application.

3.5 Conclusions

Adaptive signaling, where the transmitted signal in a wireless system is adjusted based on channel state information available to the transmitter, has been an area of significant research interest for over a decade. For single-user systems, the gain from having channel state information at the transmitter is generally a reduction in complexity and latency. Conversely, the gain of adaptive signaling is generally a function of the system complexity—large gains for very simple uncoded systems and almost no gain in the system Shannon capacity. However, for systems characterized by outage capacity or those employing multiple antennas, channel state information at the transmitter can have a significant role, particularly in systems with multiple users. This indicates the importance of the design of practical adaptive coded modulation for multiple-input multiple-output systems. As in single-antenna systems, variability of the channel coefficients between the time of channel estimation and the time of data transmission is a key concern that will need to be addressed.

References

[1] S. Alamouti and S. Kallel. 1994. Adaptive trellis-coded multiple-phase-shift keying for Rayleigh fading channels. *IEEE Transactions on Communications* 42:2305–14.

[2] B. Bannister and J. Zeidler. 2003. Feedback assisted transmission subspace tracking for MIMO systems. *Journal on Selected Areas in Communications*, 21:452–463.

[3] P. Bello. 1963. Characterization of randomly time-variant linear channels. *IEEE Transactions on Communications Systems* 11:360–93.

[4] C. Berrou, A. Glavieux, and P. Thitimajshima. 1993. Near Shannon limit error-correcting coding and decoding: Turbo-codes. In *Proceedings of the International Conference on Communications*, 1064–1070.

[5] E. Biglieri, G. Caire, and G. Taricco. 2001. Limiting performance of block-fading channels with multiple antennas. *IEEE Transactions on Information Theory* 47:1273–89.

[6] J. Bingham. 1990. Multicarrier modulation for data transmission: An idea whose time has come. *IEEE Communications Magazine* 28(5):5–14.

[7] X. Cai and G. Giannakis. 2003. Adaptive modulation with adaptive pilot symbol assisted estimation and prediction of rapidly fading channels. In *Proceedings of the Conference on Information Sciences and Systems*.

[8] G. Caire, G. Taricco, and E. Biglieri. 1999. Optimal power control for the fading channel. *IEEE Transactions on Information Theory* 45:1468–89.

[9] G. Caire and S. Shamai. 2000. On achievable rates in a multi-antenna broadcast downlink. In *Proceedings of the 38th Annual Allerton Conference on Communication, Control, and Computing*, pp. 1188–93.

[10] G. Caire and S. Shamai. On the achievable throughput of a multi-antenna Gaussian broadcast channel. *IEEE Transactions on Information Theory* 49(7):1691-1706.

[11] J. Cavers. 1972. Variable-rate transmission for Rayleigh fading channels. *IEEE Transactions on Communications* 20:15–22.

[12] J. Cavers. 1991. An analysis of pilot symbol assisted modulation for Rayleigh fading channels. *IEEE Transactions on Vehicular Technology* 40:686–93.

[13] C. Chuah, D. Tse, J. Kahn, and R. Valenzuela. 2002. Capacity scaling in MIMO wireless systems under correlated fading. *IEEE Transactions on Information Theory* 48(3):637–51.

[14] L. Cimini, Jr. 1995. Performance studies for high-speed indoor wireless communications. *Wireless Personal Communications* 2:67–85.

[15] M. Costa. 1983. Writing on dirty paper. *IEEE Transactions on Information Theory* 29(3):439-441.

[16] T. Cover and J. Thomas. 1991. *Elements of information theory*. New York: Wiley.

[17] L. Ding and J. S. Lehnert. 2000. Performance analysis of an uplink power control using truncated channel inversion for data traffic in a cellular CDMA system. In *Proceedings of the IEEE Vehicular Technology Conference*, 1673–1677.

[18] A. Duel-Hallen, S. Hu, and H. Hallen. 2000. Long-range prediction of fading signals: Enabling adaptive transmission for mobile radio channels. *IEEE Signal Processing Magazine* 17:62–75.

[19] T. Eyceoz, A. Duel-Hallen, and H. Hallen. 1998. Deterministic channel modeling and long range prediction of fast fading mobile radio channels. *IEEE Communication Letters* 2:254–56.

[20] E. Feig. 1990. Practical aspects of DFT-based frequency division multiplexing for data transmission. *IEEE Transactions on Communications* 38:929–32.

[21] G. Foschini. 1996. Layered space-time architecture for wireless communication in a fading environment when using multi-element antennas. *Bell Labs Technical Journal* 41–59.

[22] G. Foschini and M. Gans. 1998. On limits of wireless communications in a fading environment when using multiple antennas. *Wireless Personal Communications* 6:311–35.

[23] R. Gallager. 1963. *Low density parity-check codes.* Cambridge, MA: MIT Press.

[24] D. Goeckel. 1997. Robust adaptive coded modulation for time-varying channels with delayed feedback. In *Proceedings of the Thirty-Fifth Annual Allerton Conference on Communication, Control, and Computing,* pp. 370–79.

[25] D. Goeckel. 1998. Adaptive coding for fading channels using outdated fading estimates. In *Proceedings of the IEEE Vehicular Technology Conference,* pp. 1925–29.

[26] D. Goeckel. 1998. Strongly robust adaptive signaling for time-varying channels. In *Proceedings of the 1998 International Conference on Communications,* pp. 454–58.

[27] D. Goeckel. 1999. Adaptive coding for time-varying channels using outdated fading estimates. *IEEE Transactions on Communications* 47:844–55.

[28] A. Goldsmith, L. Greenstein, and G. Foschini. 1994. Error statistics of real-time power measurements in cellular channels with multipath and shadowing. *IEEE Transactions on Vehicular Technology* 43:439–46.

[29] A. Goldsmith. 1994. Variable-rate coded MQAM for fading channels. In *Proceedings of the IEEE Global Communications Conference: Communication Theory Miniconference,* pp. 186–90.

[30] A. Goldsmith. 1995. Capacity and dynamic resource allocation in broadcast fading channels. In *Proceedings of the Allerton Conference on Communications, Control, and Computing,* pp. 915–24.

[31] A. Goldsmith and S. Chua. 1997. Variable-rate variable-power MQAM for fading channels. *IEEE Transactions on Communications* 45:1218–30.

[32] A. Goldsmith and S. Chua. 1998. Adaptive coded modulation for fading channels. *IEEE Transactions on Communications* 46:595–602.

[33] A. Goldsmith, S. Jafar, N. Jindal, and S. Vishwanath. 2003. Capacity limits of MIMO channels. *IEEE Journal on Selected Areas in Communications* 21(5):684–702.

[34] I. Gradshteyn and I. Rhyzhik. 1980. *Table of integrals, series, and products.* New York: Academic Press.

[35] M. Gudmundson. 1991. Correlation model for shadow fading in mobile radio systems. *Electronics Letters* 27:2145–46.

[36] S. Hanly and D. Tse. 1998. Multi-access fading channels. Part II. Delay-limited capacities. *IEEE Transactions on Information Theory* 44:2816–31.

[37] J. Hayes. 1968. Adaptive feedback communications. *IEEE Transactions on Communications Technology* 16:29–34.

[38] V. Hentinen. 1974. Error performance for adaptive transmission on fading channels. *IEEE Transactions on Communications* 22:1331–37.

[39] S. Hu and A. Duel-Hallen. 2001. Combined adaptive modulation and transmit diversity using long range prediction for flat mobile radio channels. In *Proceedings of the Global Communications Conference,* 1257–1262.

[40] S. Jafar and A. Goldsmith. Transmitter optimization and optimality of beamforming for multiple antenna systems with imperfect feedback. *IEEE Transactions on Wireless Communications.*

[41] W. Jakes, Jr. 1974. *Microwave mobile communications.* New York: Wiley.

[42] E. Jorswieck and H. Boche. 2004. Channel capacity and capacity-range of beamforming in MIMO wireless systems under correlated fading with covariance feedback. *IEEE Transactions on Wireless Communications* 3(5): 1543-1553.

[43] K. Kamath and D. Goeckel. 2004. Adaptive modulation schemes for minimum outage probability in wireless systems. *IEEE Transactions on Communications* 52(10): 1632–1635.

[44] V. Lau and M. Macleod. 1998. Variable rate adaptive trellis coded QAM for high bandwidth efficiency applications in Rayleigh fading channels. In *Proceedings of the Vehicular Technology Conference,* pp. 348–52.

[45] V. Lau and M. Macleod. 2001. Variable-rate trellis coded QAM for flat-fading channels. *IEEE Transactions on Communications* 49:1550–60.

[46] D. MacKay. 1999. Good error-correcting codes based on very sparse matrices. *IEEE Transactions on Information Theory* 45:399–431.

[47] A. Monk and L. Milstein. 1995. Open-loop power control error in a land mobile satellite system. *Journal on Selected Areas in Communications* 13:205–12.

[48] S. Nanda, K. Balachandran, and S. Kumar. 2000. Adaptation techniques in wireless packet data services. *IEEE Communications Magazine 38(1):* 54–64.

[49] R. Negi and J. Cioffi. 2002. Delay-constrained capacity with causal feedback. *IEEE Transactions on Information Theory* 48:2478–94.

[50] P. Örmeci, X. Liu, D. Goeckel, and R. D. Wesel. 2001. Adaptive bit-interleaved coded modulation. *IEEE Transactions on Communications* 49:1572–81.

[51] K. Pahlavan and A. Levesque. 1995. *Wireless information networks.* New York: John Wiley & Sons.

[52] A. Papoulis. 1991. *Probability, random variables, and stochastic processes.* 3rd ed. New York: McGraw-Hill.

[53] J. Proakis. 2001. *Digital communications.* 4th ed. New York: McGraw-Hill.

[54] M. Pursley and C. Wilkins. 1999. Adaptive-rate coding for frequency-hop communications over Rayleigh fading channels. *Journal on Selected Areas in Communications* 17:1224–32.

[55] M. Pursley and J. Shea. 2000. Channel quality estimation with channel error counters for adaptive signaling in wireless communications. In *Proceedings of the International Symposium on Information Theory,* 109.

[56] T. Rappaport and C. McGillen. 1989. UHF fading in factories. *Journal on Selected Areas in Communications* 7:40–48.

[57] T. Rappaport. 1996. *Wireless communications.* New York: Prentice Hall.

[58] C. Schlegel and D. Costello, Jr. 1989. Bandwidth efficient coding for fading channels: Code construction and performance analysis. *Journal on Selected Areas in Communications* 7:1356–68.

[59] C. Shannon. 1948. A mathematical theory of communications. *Bell Systems Technical Journal* 27:379–423, 623–56.

[60] D. Shiu, G. Foschini, M. Gans, and J. Kahn. 2000. Fading correlation and its effect on the capacity of multielement antenna systems. *IEEE Transactions on Communications* 48:502–13.

[61] S. Simon and A. Moustakas. 2003. Optimizing MIMO antenna systems with channel covariance feedback. *Journal on Selected Areas in Communications 21(3):406-417.*

[62] I. Telatar. 1999. Capacity of multi-antenna Gaussian channels. *European Transactions on Telecommunications* 10:586–95.

[63] G. Ungerboeck. 1982. Channel coding with multilevel/phase signals. *IEEE Transactions on Information Theory* 28:55–67.

[64] E. Visotsky and U. Madhow. 2001. Space-time transmit precoding with imperfect feedback. *IEEE Transactions on Information Theory* 47:2632–39.

[65] B. Vucetic. 1991. An adaptive coding scheme for time-varying channels. *IEEE Transactions on Communications* 39:653–63.

[66] S. Wei and D. Goeckel. 2002. Error statistics for average power measurements in wireless communication systems. *IEEE Transactions on Communications* 50:1535–46.

[67] S. Wei, D. Goeckel, and R. Janaswamy. 2005. On the asymptotic capacity of MIMO systems with antenna arrays of fixed length. *IEEE Transactions on Wireless Communications 4(4):1608-1621.*

[68] R. Yates. 1995. A framework for uplink power control in cellular radio systems. *Journal on Selected Areas in Communications* 13:1341–47.

[69] W. Yu and J. Cioffi. 2001. Trellis precoding for the broadcast channel. In *Proceedings of the Global Communications Conference*, pp. 1344–48.

[70] J. Zander. 1992. Performance of optimum transmitter power control in cellular radio systems. *IEEE Transactions on Vehicular Technology* 41:57–62.

[71] R. van Nee and R. Prasad. 2000. *OFDM for wireless multimedia communications.* Norwood, MA: Artech House.

[72] H. Van Trees. 1968. *Detection, estimation, and modulation theory.* Vol. I. New York: Wiley.

[73] S. Zhou and G. Giannakis. 2002. Optimal transmitter eigen-beamforming and space-time block coding based on channel mean feedback. *IEEE Transactions on Signal Processing 50(10):2599-2613.*

4

MIMO Systems: *Principles, Iterative Techniques, and Advanced Polarization*

K. Raoof
Domaine Universitaire

M. A. Khalighi
École Centrale Marseille

N. Prayongpun
Domaine Universitaire

95

4.1 Introduction

This chapter considers the principles of multiple-input multiple-output (MIMO) wireless communication systems as well as some recent accomplishments concerning their implementation. By employing multiple antennas at both transmitter and receiver, very high data rates can be achieved under the condition of deployment in a rich-scattering propagation medium. This interesting property of MIMO systems suggests their use in the future high-rate and high-quality wireless communication systems. Several concepts in MIMO systems are reviewed in this chapter. We first consider MIMO channel models and recall the basic principles of MIMO structures and channel modeling. We next study the MIMO channel capacity and present the early developments in these systems concerning the information theory aspect. Iterative signal detection is considered next; it considers iterative techniques for space-time decoding. As the capacity is inversely proportional to the spatial channel correlation, MIMO antennas should be sufficiently separated, usually by several wavelengths. In order to minimize antennas' deployment, we present advanced polarization diversity techniques for MIMO systems and explain how they can help to reduce the spatial correlation in order to achieve high transmission rates. We end the chapter by considering the application of MIMO systems in local area networks, as well as their potential in enhancing range, localization, and power efficiency of sensor networks.

4.2 MIMO Systems and Channel Models

In this section, we present briefly a general MIMO communication structure. The definition of MIMO channel is then described with some related characteristics of wireless communication channels. Some recent MIMO channel models are also reported.

4.2.1 MIMO Communication Systems

Using multiple antennas [1] at both transmitter and receiver permits the increasing of the data rate by creating multiple spatial channels. Multiple receiving antennas can also be used to combat fading without expanding the bandwidth of the transmitted signal. In particular, with M_T transmitting and M_R receiving antennas, it is possible to achieve an M-time capacity of a single transmitting and single receiving antenna configuration where $M = \min\{M_T, M_R\}$. Figure 4.1 demonstrates a general system employing multiple transmitting and multiple receiving antennas to increase the data rate. A sequence of

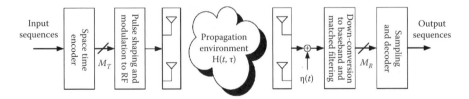

FIGURE 4.1   General configuration of a MIMO communication system.

input symbols is encoded by a space-time encoding function into an $M_T \times 1$ discrete-time complex baseband sequence $x[n]$ (n is a discrete-time index). The $x[n]$ sequence is subsequently transformed by a pulse-shaping filter into an $M_T \times 1$ continuous-time complex baseband sequence $x(t)$, and then the baseband signal is modulated with a transmission carrier. The transmission channel $\mathbf{H}$ superposes the transmitted signal due to the distortions of environment. At the reception side, under the assumption of synchronous sampling, the received signal $y(t)$ with additive noise is downconverted to baseband and sampled to produce a discrete-time signal sequence. Finally, the estimated symbols are decoded by the space-time decoding block.

If the channel is time invariant, the equivalent received signal at the receiving antenna with M_T elements at the transmitter (Tx) and M_R elements at the receiver (Rx) can be written as

$$y(t) = \sum_{l=0}^{L-1} \mathbf{H}_l x(t - \tau_l) + \eta(t). \tag{4.1}$$

If the channel is time variant, the overall MIMO relation can be formulated as

$$y(t) = \int_\tau \mathbf{H}(t, \tau) x(t - \tau) + \eta(t), \tag{4.2}$$

where $x(t)$ and $y(t)$ represent the transmitted and received signals,

$$x(t) = [x_1(t), x_2(t), \ldots, x_{M_T}(t)]^T, \tag{4.3}$$

$$y(t) = [y_1(t), y_2(t), \ldots, y_{M_R}(t)]^T, \tag{4.4}$$

and L denotes the number of resolvable multipaths, τ is the propagation delay, $\mathbf{H}(t, \tau)$ is the $M_R \times M_T$ time-variant channel matrix, $\mathbf{H}_l$ is the $M_R \times M_T$ channel matrix of resolvable path l, and finally, $\eta(t)$ is an additive noise.

4.2.2 MIMO Channel Stationarity Definition

In wireless communication, a stochastic time-variant linear channel usually employs wide-sense stationarity uncorrelated scattering (WSSUS) for stationary property [2, 3]. This WSSUS channel expresses uncorrelated attenuation in both time-delay and Doppler-shift domains. The quasi-WSSUS channel [2] is usually applied to real radio systems. It has the properties of a WSSUS channel for a limited bandwidth and for a limited time or within a limited environment. This kind of assumption is exceptionally useful in communication systems. For example, ergodic MIMO channel performance can be given by averaging the channel performances over many independent channel realizations considering that they have the same statistics.

In many practical communication systems, the WSSUS property is not truly satisfied because of path loss, shadowing, and varying propagation conditions such as delay-Doppler dispersion by band or time limitations at both Tx and Rx [4]. These effects cause the channel to be nonstationary, i.e., non-WSSUS.

4.2.3 Classification of MIMO Channel Models

Numerous MIMO channel models have already been proposed. They can be classified into physical and analytical models, as shown in Figure 4.2. On the one hand, the physical channel models focus on the characteristics of the environment and the electromagnetic wave propagation between the Tx and Rx, and consider the antenna configurations at both ends. On the other hand, the analytical models do not provide site-specific descriptions, as they do not take into account the wave propagation characteristics. The model impulse response is mathematically generated and related to the statistical properties of the propagation environment. However, due to its simplicity, an analytical channel model is very useful for producing a MIMO channel matrix for different kinds of communication systems.

In the literature, the physical models are decomposed into deterministic models [5–8] and geometry-based stochastic models [9–13]. Deterministic models, such as ray tracing and recording impulse response models, start by creating an artificial environment, and the channel response is then calculated for simulation purposes. However, these methods have high computational complexity. Geometry-based stochastic channel models (GSCMs), on the other hand, calculate the channel response by taking into account the characteristic of wave propagation, both site-specific Tx-Rx environments, and the scattering mechanisms. All parameters are statistically set to closely match the measured channel observation. By this approach, the channel response can be rapidly computed even for a multibounce scattering mechanism.

Analytical channel models can be further classified into correlation-based models, statistical cluster models, and propagation-based models. Correlation-based models

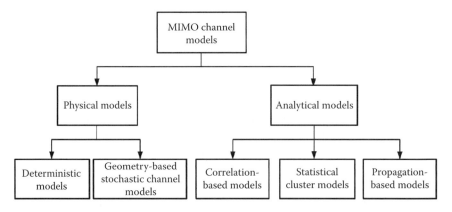

FIGURE 4.2 MIMO channel and propagation models.

contain the Tx-Rx channel correlation matrices. For example, the independent identically distributed (IID) model is proposed in the case of a rich-scattering environment with no spatial correlation [1, 14, 15]. The Kronecker model [16–18] assumes that the channel correlation is a product of the correlations at the Tx and Rx sides. Statistical cluster models determine physical parameters in a random manner without referring to the geometry of a physical medium. For example, the Saleh-Valenzuela model [19] uses two exponentially decaying amplitudes varying in time and distance of the clusters, while increasing delay time with the assumptions that the direction of departure (DOD) and the angle of arrival (AOA) are independent and identically distributed. The other models are propagation based, such as keyhole channel models [20], finite scattered model [21], maximum entropy model [22], virtual channel representation [23], etc.

In addition to the discussed models, there are several organizations that have proposed different MIMO channel models, such as COST 207, COST 231, COST 259 [24], COST 273 [25], 3GPP [26], and IEEE 802.16a, e.

4.3   MIMO Channel Capacity

At the end of the 1990s, pioneering works in Bell Laboratories showed for the first time that the use of multiple antennas at both sides of the transmission link can result in tremendous channel capacities, provided that the propagation medium is rich scattering [1, 14, 27–29]. This increase in capacity is obtained without any need for extra bandwidth or extra transmission power. Multipath propagation, previously regarded as an impediment to reliable communication, was shown to be exploitable for increasing the data throughput. In this section, we explain briefly how exploiting the spatial dimension can lead to an increase in the system spectral efficiency. Only single-user applications are considered. For more discussions and details, the reader is referred to [15] and the references therein.

4.3.1   Capacity of a Fading Channel

Let us first recall the definition of the capacity for a fading channel. For a time-varying channel, the capacity C becomes a random variable whose instantaneous value depends on the channel realization [30]. In such a case, the Shannon capacity of the channel may even be zero. Indeed, if we choose a transmission rate for communication, there may be a nonzero probability that the channel realization is incapable of supporting it. The two most used definitions for the channel capacity are *ergodic* capacity and *outage* capacity. The ergodic capacity, C_{erg}, which is the expected value of C, is suitable for fast varying channels. The outage capacity, C_{out}, is usually used when considering packet-based transmission systems where the block-fading model properly describes the channel. If the preassumed channel capacity is too optimistic, i.e., larger than the instantaneous capacity, a channel outage may occur. The outage capacity, or more correctly, the capacity versus outage, as seen from its name, is the channel capacity conditioned to an outage probability P_{out}. Obviously, there is a trade-off between the expected data throughput and the outage probability.

In the expressions that we will provide in this section, C will denote the instantaneous channel capacity.

4.3.2 MIMO Capacity

We present the MIMO channel capacity for different cases of channel state information (CSI) at the transmitter (Tx) and receiver (Rx).

4.3.2.1 General Assumptions

The global scheme of the transmission link is shown in Figure 4.3. We denote by M_T and M_R the number of antennas at Tx and Rx, respectively. The communication channel includes the effect of transmit/receive antennas and the propagation medium. We neglect the effect of the antenna patterns and assume the *far-field* conditions; that is, dominant reflectors are assumed to be sufficiently far from the Tx and Rx. We consider

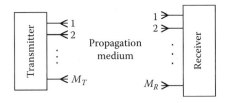

FIGURE 4.3   Global scheme of a MIMO communication structure.

the simple case of a frequency-non-selective (flat) fading channel that is true for narrowband communications. This assumption is mostly valid in indoor applications [31, 32]. In the equivalent baseband representation, each subchannel is characterized by a complex, circularly symmetric random variable that is assumed to be normalized in power. The entire MIMO channel is described by a channel matrix H of dimension $(M_R \times M_T)$. For instance, the entry H_{ij} of H, characterizes the subchannel between the ith receive and the jth transmit antenna. Finally, we assume that the total transmit power at each sample time (corresponding to each channel use) is constrained to P_T.

4.3.2.2 CSI Known to Rx but Unknown to Tx

This is the classical case that is usually considered in the literature. The channel is assumed to be known (e.g., perfectly estimated) at the Rx but unknown to the Tx. Since the Tx does not know the channel, it is logical to distribute the available power uniformly on the transmit antennas. In fact, this is the optimal way for power allotment over the M_T antennas in this case. We denote by ρ_T the total average received signal-to-noise ratio (SNR) at the receiver array $\rho_T = P_T / \sigma_n^2$, where σ_n^2 is the variance of the additive white Gaussian complex noise. The MIMO channel capacity in units of bps/Hz is

$$C = \log_2 \det \left[I_{M_R} + \frac{\rho_T}{M_T} HH^\dagger \right]. \tag{4.5}$$

Here I_{M_R} is the $(M_R \times M_R)$ identity matrix. Equation (4.5) can also be written in a different form if we consider the singular value decomposition of H:

$$H = U_H \, \Lambda_H \, V_H^\dagger \tag{4.6}$$

where † denotes complex conjugate transpose, U_H and V_H are unitary matrices of dimensions $(M_R \times M_R)$ and $(M_T \times M_T)$, respectively, and Λ_H is an $(M_R \times M_T)$ matrix containing the singular values of H. Let us define $M = \min\{M_T, M_R\}$. We denote these singular values by $\lambda_{H,i}$, $i = 1, ..., M$. The MIMO capacity can be written in the following form:

$$C = \log_2 \det\left[I_{M_R} + \frac{\rho_T}{M_T} \Lambda_H \Lambda_H^\dagger \right] = \sum_{i=1}^{M} \log_2\left(1 + \frac{\rho_T}{M_T} \lambda_{H,i}^2 \right). \qquad (4.7)$$

4.3.2.3 CSI Known to Both Tx and Rx

This is the case when the estimated CSI at the Rx is provided for the Tx, which is practically feasible when the channel varies slowly in time. Providing the CSI for the Tx can be done using a (hopefully low-bandwidth) feedback channel or via the reverse link when the communication takes place in a duplex mode. We assume that CSI is provided perfectly and without any delay to the Tx. In this case, the Tx can allot the available power on the antennas in an optimal manner in order to achieve the maximum capacity [33]. This capacity is often called *known CSI* capacity or *water-filling* (WF) capacity. For this purpose, we should weight the transmitted symbols vector x by the matrix V_H, and the received signal vector by the matrix $U_H^\dagger$ [15, 34]. This can be regarded as an *optimal beamforming* solution. By this weighting, we can in fact consider an equivalent channel H_{eq} between x and y:

$$H_{eq} = U_H^\dagger H V_H = \Lambda_H. \qquad (4.8)$$

In other words, the MIMO channel matrix H is decomposed into several parallel independent single-input single-output subchannels. The number of these subchannels is equal to rank(H), and their gain is given by the singular values of H. Let R_X be the autocorrelation matrix of x, with eigenvalues $\lambda_{X,i}$, $i = 1, \cdots, M$. The optimal WF solution consists in distributing the available power P_T, optimally over the equivalent parallel subchannels, which results in [15, 35]

$$\lambda_{X,i} = \left(\psi - \frac{\sigma_n^2}{\lambda_{H,i}^2} \right)^+, \quad i = 1, \cdots, M \qquad (4.9)$$

where $(s)^+ = s$ if $s > 0$, and 0 otherwise. Also, ψ is a constant that is determined so as to satisfy the constraint on the total transmit power,

$$\sum_{i=1}^{M} \lambda_{X,i} = P_T.$$

The WF solution imposes that we allocate more power to best subchannels and lower (or perhaps no) power to worse ones. Now, the WF capacity C_{WF} is given by [15, 35]

$$C_{\text{WF}} = \sum_{i=1}^{M} \log_2 \left(1 + \frac{\lambda_{X,i} \lambda_{H,i}^2}{\sigma_n^2} \right). \tag{4.10}$$

4.3.2.4 CSI Unknown to Both Tx and Rx

If the channel is not known to the Tx and the Rx, we can consider (4.5) as an upper bound on capacity. Let Δ be the coherence interval of the channel in units of channel uses. As Δ tends to infinity, the channel capacity approaches this upper bound, because with a greater Δ, tracking the channel variations becomes more possible for the Rx [36]. For the same reason, there is less difference between the capacity and the upper bound for higher SNR values. This difference becomes more considerable, however, for larger M_T or M_R [36]. For a fast varying channel, the capacity is far less than the Rx-perfect-knowledge upper bound, because practically there is no possibility to estimate the channel at the Rx. For M > Δ, no increase is achieved in the MIMO capacity by an increase in M [36]. It is shown in [37] that at high SNR and for a rich-scattering propagation medium, the MIMO unknown CSI capacity increases linearly with $M^*(1 - M^*/\Delta)$, where

$$M^* = \min\left\{ M_T, M_R, \left\lfloor \frac{\Delta}{2} \right\rfloor \right\}.$$

4.3.3 Some Numerical Results

We present some numerical results, excluding the case where CSI is unknown to both the Tx and the Rx. We consider the outage capacity C_{out} for an outage probability of $P_{out} = 0.01$ and will refer to it simply as capacity. Let us first consider the case where CSI is known only to the Rx. We will call this case *no-WF*. Figure 4.4 contrasts the outage capacity of MIMO, single-input multiple-output (SIMO), and multiple-input single-output (MISO) systems under the conditions of uncorrelated Rayleigh flat fading [38] and $P_{out} = 0.01$. SNR stands for ρ_T. For the MIMO system, we have $M_T = M_R = M$. Uncorrelated fading necessitates enough antenna spacings at the Tx and the Rx that depend, in turn, on the propagation conditions [39, 16]. The MIMO capacity increases linearly with M and is much more considerable than that of MISO and SIMO systems.

Now consider the case of known CSI at both the Tx and the Rx. For $M_R \geq M_T$, the improvement in capacity by performing WF, which we call the *WF gain*, is considerable for low SNR and a large number of transmit antennas [33]. But this gain is much more considerable when $M_T > M_R$. For example, Figure 4.5 shows curves of no-WF and WF capacities for two cases of $M_R > M_T = 4$ and $M_T > M_R = 4$. Notice that the WF capacity of an (M_T, M_R) system is equal to that of an (M_R, M_T) system. Also, the WF capacity of a MISO system is equal to that of the equivalent SIMO system.

The gain in capacity by WF is especially interesting for the case of correlated channels. For instance, for the case of Ricean fading [40, 41], curves of capacity versus the Ricean factor (RF) are presented in Figure 4.6, for a MIMO system with $M_T = M_R = M$ and two cases of $M = 2$ and $M = 4$. RF represents the percentage of power received from

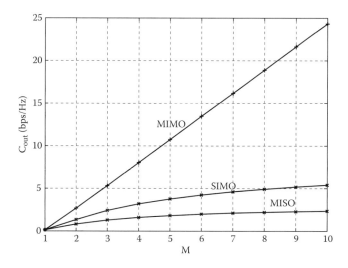

FIGURE 4.4 Capacity for MIMO, SIMO, and MISO structures; uncorrelated Rayleigh flat fading; SNR = 10 dB, P_{out} = 0.01.

the line of sight (LOS) to the total received average power [41]. For relatively high RF that can be regarded as a more correlated channel, the WF gain is quite considerable.

4.4 Iterative Signal Detection

Signal detection is a crucial part of the transmission system. Among the various detection techniques proposed for the case of MIMO systems, there are iterative (also called *turbo*) detectors. This is what we are going to focus on in this section. In effect, since the invention of turbo-codes by Berrou and Glavieux [42], who proposed iterative decoding of parallel concatenated convolutional codes, the *turbo principle* has been applied to several problems in communications, such as channel equalization [43], channel estimation [44], synchronization [45], multiuser detection [46, 47], and, of course, MIMO signal detection [48]. The turbo principle consists of the exchange of *soft* information between two different stages of the Rx, mostly including the soft channel decoder. In MIMO systems too, iterative processing has attracted special attention as it makes a good compromise between complexity and performance. Before presenting the basics of iterative detection, we have to present a brief introduction on space-time coding that is performed at the Tx. We mostly consider frequency-non-selective (flat) fading conditions and single-carrier modulation.

4.4.1 Space-Time Coding and Decoding

An important aspect in the implementation of MIMO systems is to appropriately distribute redundancy in space and in time at the Tx, what is called space-time (ST) coding [49]. To date, there has been considerable work on this subject, and a variety

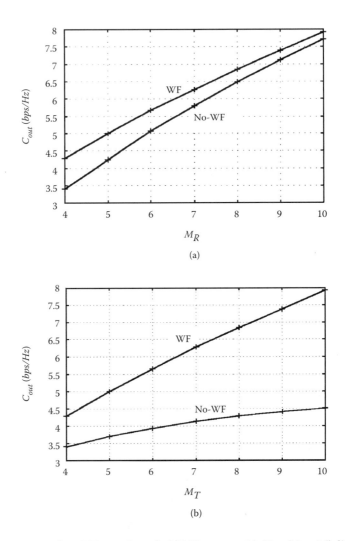

FIGURE 4.5   WF and no-WF capacities of a MIMO system with $M_R > M_T = 4$ (left) and $M_T > M_R = 4$ (right), Rayleigh flat fading, SNR = 3 dB, $P_{out} = 0.01$.

of ST schemes have been proposed for MIMO systems. The key criteria in the design of ST codes are the coding gain and the diversity gain. The first one aims at achieving high rate by capitalizing on the MIMO capacity, whereas the latter aims at profiting from the space diversity to reduce fading at the Rx. The two extreme schemes corresponding to these criteria are respectively spatial multiplexing and transmit diversity. For instance, orthogonal space-time block codes (OSTBCs) [50, 51] aim at diversity gain and some coding rate; space-time trellis codes (STTCs) [52] and linear constellation precoding (LCP) [53] aim at both coding and diversity gain; spatial multiplexing or the V-BLAST architecture maximizes the coding rate [28]; and the more general family of

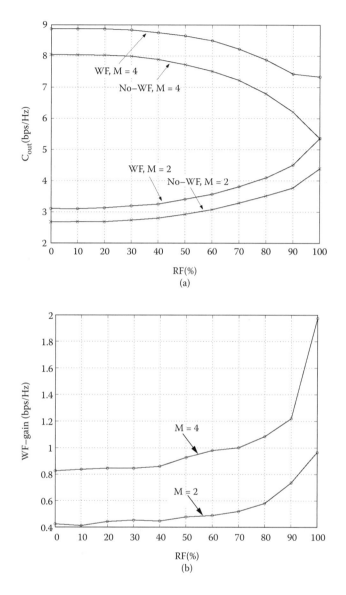

FIGURE 4.6 WF and no-WF capacities (left) and WF gain in capacity (right) for a MIMO system with $M_T = M_R = M$, Ricean flat fading, SNR = 10 dB, $P_{out} = 0.01$.

linear dispersion (LD) codes [54], which maximize the mutual information between the Tx and the Rx, allow flexible rate-diversity trade-off.

Apart from the problem of code construction, one important criterion in the choice of the appropriate ST scheme could be its decoding complexity. ST orthogonal designs like OSTBCs offer full diversity and can be decoded using an optimal decoder with linear complexity. However, these schemes suffer from low rate, especially for a large number

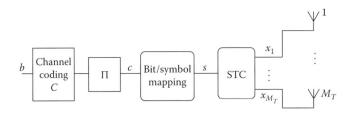

FIGURE 4.7 Block diagram of the BICM transmission scheme.

of transmit antennas. Moreover, full-rate OSTBCs exist only for a restricted number of transmit antennas and modulations [50]. Nonorthogonal schemes, on the other hand, offer higher coding rates, but their optimal decoder becomes prohibitively complex for a large number of transmit antennas and large signal constellation sets. This is especially the case for STTC schemes that, although offering high rates and good diversity gains, are complex to decode and, moreover, suffer from long decoding delays.

For nonorthogonal schemes, instead of performing complex optimal decoding, we may use suboptimal decoding based on simple linear-algebraic techniques such as sphere decoding [55] or interference-cancelling-based decoding [28, 48]. For either solution, the Rx performance can be improved considerably by performing iterative detection.

4.4.1.1 ST Coding, Tx Scheme

In addition to using a special ST scheme, we usually perform channel coding at the Tx. Let us consider bit-interleaved coded modulation (BICM) [56] for which a typical scheme is shown in Figure 4.7. The advantage of BICM is its flexibility regarding the choice of the code and the bit-symbol mapping, as well as its conformity to iterative detection. In Figure 4.7, the binary data b are encoded by a channel code C, before being interleaved (the block Π). The output bits c are then mapped to symbols according to a given constellation set. We will mostly consider QAM modulation with B bits per symbol. Power-normalized symbols s are next combined according to a given ST scheme and then transmitted on M_T antennas.

4.4.2 General Formulation of LD Codes

Before talking about ST decoding, let us present the general formulation of the LD codes from [54] that can be equally used for other ST schemes as well. Let S of dimension $(Q \times 1)$ be the vector of data symbols prior to ST coding:

$$S = \left[s_1, s_2, \cdots, s_Q \right]^t, \tag{4.11}$$

where $.^t$ denotes transposition. By ST coding, these symbols are mapped into a $(M_T \times T)$ matrix X, where T is the number of channel uses. We define the ST coding rate as $R_{STC} = Q/T$. Corresponding to an encoded matrix X, we receive the $(M_R \times T)$ matrix Y. We separate the $\Re$ and $\Im$ parts of the entries of S and X and stack them row-wise in vectors $\mathcal{S}$ of dimension $(2Q \times 1)$ and $\mathcal{X}$ of dimension $(2M_T T \times 1)$, respectively. For instance,

$$S = \left[\Re\{s_1\} \ \Im\{s_1\}, \ \cdots, \ \Re\{s_Q\} \ \Im\{s_Q\} \right]^t. \tag{4.12}$$

We have then $\mathcal{X} = \mathcal{F} S$, where the $(2M_T T \times 2Q)$ matrix $\mathcal{F}$ depends on the actual ST scheme (see [54] for more details). Let the $(M_R \times M_T)$ matrix $\boldsymbol{H}$ represent our flat channel. Similar to $\mathcal{X}$, we construct the $(2M_R T \times 1)$ vector $\mathcal{Y}$ from $\boldsymbol{Y}$. Vectors $\mathcal{X}$ and $\mathcal{Y}$ are then related through a $(2M_R T \times 2M_T T)$ matrix $\mathcal{H}$:

$$\mathcal{Y} = \mathcal{H} \mathcal{X} + \mathcal{N} \tag{4.13}$$

where $\mathcal{N}$ is the vector of real AWGN of zero mean and variance σ_n^2. Matrix $\mathcal{H}$ is composed of $(2T \times 2T)$ segments $\mathcal{H}_{ij}$, $i = 1, \cdots, M_R, j = 1, \cdots, M_T$, described below:

$$\mathcal{H}_{ij} = \begin{bmatrix} \boldsymbol{H}_{ij} & 0 & \cdots & 0 \\ 0 & \boldsymbol{H}_{ij} & \cdots & 0 \\ \vdots & & \ddots & \vdots \\ 0 & \cdots & 0 & \boldsymbol{H}_{ij} \end{bmatrix} \tag{4.14}$$

The (2×2) elements $\boldsymbol{H}_{ij}$ are obtained from each entry H_{ij} of the initial matrix $\boldsymbol{H}$:

$$\boldsymbol{H}_{ij} = \begin{bmatrix} \Re\{H_{ij}\} & -\Im\{H_{ij}\} \\ \Im\{H_{ij}\} & \Re\{H_{ij}\} \end{bmatrix}. \tag{4.15}$$

Now, we can describe the ST encoder and channel input/output relationship by considering an equivalent channel matrix $\mathcal{H}_{eq}$ of dimension $(2M_R T \times 2Q)$:

$$\mathcal{Y} = \mathcal{H} \mathcal{F} S + \mathcal{N} = \mathcal{H}_{eq} S + \mathcal{N}. \tag{4.16}$$

4.4.3 Iterative Detection for Nonorthogonal ST Schemes

We assume that $\mathcal{H}_{eq}$ and σ_n^2 are known at the Rx. Having received the vector $\mathcal{Y}$, we should extract from it the transmitted data S. As we perform channel coding together with ST coding, the idea of iterative detection comes to mind. Indeed, by profiting in this way from the channel coding gain, we can obtain a good performance after only a few iterations and approach the optimal ST decoder + channel decoder performance. This is, of course, the case for nonorthogonal ST schemes. In what follows, we explain the principle of iterative detection and explain in detail the ST decoding part.

The block diagram of such an Rx is shown in Figure 4.8. Soft-input soft-output signal detection and channel decoding are performed. For MIMO signal detection or ST

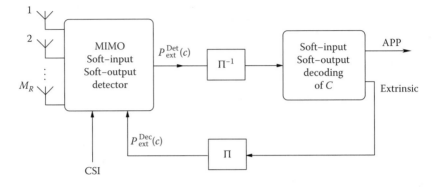

FIGURE 4.8 Block diagram of the receiver.

decoding, we can use the optimal maximum *a posteriori* (MAP) algorithm, or a suboptimal solution based on sphere decoding or interference cancelling, for example. In fact, the optimal MAP detector becomes too complex to implement in practice, especially for large Q or large signal constellations.

Soft channel decoding, on the other hand, can be done using the well-known forward-backward algorithm [57], the soft-output Viterbi algorithm (SOVA) [58], or a simplification of them [59]. For the final decision making on the transmitted data bits, we use the *a posteriori* probabilities at the decoder output. Like in any other turbo processing case, *extrinsic* information is exchanged between the two blocks of MIMO detector and channel decoder. In what follows, we explain the principle of MIMO detection based on MAP, sphere decoding, and soft interference cancelling, while focusing on the third approach.

4.4.3.1 MAP Signal Detection

We present here the formulation of the MAP detector based on probabilities. It can also be implemented using logarithmic likelihood ratios (LLRs). Remember the expression of $\mathcal{Y}$ from (4.16). The MIMO detector provides at its output extrinsic probabilities on the coded bits c. Let $\mathcal{Q}$ be the cardinality of $\mathcal{S}$ of size $q \triangleq |\mathcal{Q}| = 2^{BQ}$. Let also $c_i, i = 1, \cdots, BQ$ be the bits corresponding to a vector of symbols $\mathcal{S} \in \mathcal{Q}$. The extrinsic probability on the bit c_j at the MIMO detector output, $P_{\text{ext}}^{\text{Det}}(c_j)$, is calculated as follows [60]:

$$P_{\text{ext}}^{\text{Det}}\left(c_j = 1\right) = K \sum_{\substack{\mathcal{S} \in \mathcal{Q} \\ c_j = 1}} \exp\left(-\frac{\left\|\mathcal{Y} - \mathcal{H}_{eq}\mathcal{S}\right\|^2}{\sigma_n^2}\right) \prod_{\substack{i=1 \\ i \neq j}}^{BQ} P_{\text{ext}}^{\text{Dec}}\left(c_i\right), \tag{4.17}$$

where K is the normalization factor satisfying $P_{\text{ext}}^{\text{Det}}(c_j = 1) + P_{\text{ext}}^{\text{Det}}(c_j = 0) = 1$. The probability $P_{\text{ext}}^{\text{Dec}}(c_i)$ is in fact the *a priori* information on bit c_i, fed back from the channel decoder. At the first iteration, where no *a priori* information is available on bits c_i, $P_{\text{ext}}^{\text{Dec}}$ are set to 1/2. The summation in (4.17) is taken over the product of the conditional channel likelihood (the exponential term) given a vector S, and the *a priori* probability on this symbol, i.e., the term $\Pi\, P_{\text{ext}}^{\text{Dec}}$. In this latter term, we exclude the *a priori* probability corresponding to the bit c_j itself, so as to respect the exchange of *extrinsic* information between the channel decoder and the MIMO detector. Also, this term assumes independent coded bits c_i, which is true for random interleaving of large size.

4.4.3.2 Sphere Decoding

As it was seen, in the calculation of the extrinsic probabilities, the MAP detector considers the exhaustive list of all possibly transmitted symbol vectors. Hence, the complexity of the MAP detector grows exponentially with the number of transmit antennas M_T and the number of bits per modulation symbol B. By sphere decoding, these probabilities are calculated based on a nonexhaustive list [55]. Corresponding to a vector $\mathcal{Y}$, we only take into account those lattice points that are in a hypersphere of radius R around $\mathcal{Y}$. The main detection tasks are then setting the radius R as well as determining which lattice points are within the sphere. In this way, the average complexity of the detector under flat fading conditions becomes polynomial (often subcubic) under high SNR [61]. At low SNR, however, the detector complexity can be very high. A drawback of this method is the time variability of the detector complexity, as it depends on SNR.

4.4.3.3 Parallel Interference Cancelling-Based Detection

The block diagram of the detector based on soft-parallel interference cancellation (soft-PIC) is shown in Figure 4.9, where the detector soft outputs are considered in the form of LLR. Reformulation is trivial if we want to consider these soft outputs in the form of probability (like in Figure 4.8). We will refer to the corresponding receiver scheme as turbo-PIC. Soft-PIC is essentially composed of the three blocks of PIC detector, conversion to LLR, and soft estimation of transmit symbols.

4.4.3.3.1 PIC Detector

In order to present general expressions for the detector, let us denote by $\hat{\gamma}_p$, $p = 1, \cdots, 2Q$, the detected data at the PIC detector output, corresponding to the real or imaginary parts of s_q, $q = 1, \cdots, Q$. At the first iteration, $\hat{\gamma}_p$ are obtained via minimum mean square error (MMSE) filtering [48, 62, 63]:

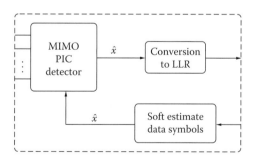

FIGURE 4.9 Block diagram of the soft-PIC detector.

$$\hat{\gamma}_p = W_p^\dagger \mathcal{Y} = h_p^\dagger \left(\mathcal{H}_{eq} \mathcal{H}_{eq}^\dagger + \sigma_n^2 I \right)^{-1} \mathcal{Y}, \tag{4.18}$$

where W_p denotes the filter, h_p is the pth column of $\mathcal{H}_{eq}$, and $(.)^\dagger$ stands for transpose complex conjugate. Note that the entries of $\mathcal{H}_{eq}$ are real values, and hence $(.)^\dagger$ is equivalent to transposition. From the second iteration, we can calculate soft estimates $\tilde{S}$ of the transmitted data using the soft decoder outputs. Using these estimates, we perform interference cancelling followed by a simple zero forcing (ZF) or MMSE filtering:

$$\tilde{\mathcal{Y}}_p = \mathcal{Y} - \mathcal{H}_p \, \tilde{S}_p, \quad \hat{\gamma}_p = W_p^\dagger \, \tilde{\mathcal{Y}}_p, \tag{4.19}$$

$$\text{ZF: } W_p = \frac{1}{h_p^\dagger h_p} h_p, \quad \text{MMSE: } W_p = \frac{1}{\left(h_p^\dagger h_p + \sigma_n^2 \right)} h_p, \tag{4.20}$$

where $\tilde{S}_p$ of dimension $((2Q - 1) \times 1)$ is $\tilde{S}$ with its pth entry removed, and $\mathcal{H}_p$ of dimension $(2M_R T \times (2Q - 1))$ is the matrix $\mathcal{H}$ with its pth column removed. Notice that, compared to the exact MMSE filtering proposed in [48], (4.20) are simplified solutions that assume almost perfect estimation of data symbols and permit a considerable reduction of the computational complexity. Thanks to iterative processing, the performance loss due to this simplification would be negligible. In the results that we present later, we will consider the simplified ZF solution.

4.4.3.3.2 Conversion to LLR

For QAM modulation with B (an even number) bits per symbol, we can attribute $m = B/2$ bits to the real and imaginary parts of each symbol. Let, for instance, the bit c_i correspond to the real (imaginary) part of the symbol s_q. Let also $a_{1,j}$ and $a_{0,j}, j = 1, \cdots, B/2$ denote the real (imaginary) part of the signal constellation points, corresponding to $c_i = 1$ and $c_i = 0$, respectively. Remember that the signal constellation points have normalized average power. The LLR corresponding to c_i is calculated as follows [62]:

$$\text{LLR}_i = \log_{10} \frac{\displaystyle\sum_{j=1}^{2^{m-1}} \exp\left(-\frac{1}{2\sigma_p^2} \left(\hat{\gamma}_p - \alpha_{1,j} \right)^2 \right)}{\displaystyle\sum_{j=1}^{2^{m-1}} \exp\left(-\frac{1}{2\sigma_p^2} \left(\hat{\gamma}_p - \alpha_{0,j} \right)^2 \right)}, \quad i = 1, \cdots, m, \tag{4.21}$$

where σ_p^2 is the variance of noise plus the residual interference (RI) that intervenes in the detection of $\hat{\gamma}_p$, and is assumed to be Gaussian. Note that as the detection is performed on blocks of Q complex symbols, or in other words, on blocks of *2Q real symbols* in our model, the RI comes in fact from $(2Q - 1)$ other real symbols in the corresponding channel use [64]. In LLR calculation, we need the variances σ_p^2. These variances can be calculated analytically as shown in [46], or estimated at each iteration and for each one of

2Q real symbols, as done in [64]. To simplify the detector further, we may neglect the RI and take into account only the noise variance. For not too large signal constellation sizes, this simplification causes a negligible performance loss [65, 66].

4.4.3.3.3 Soft Estimation of Transmit Symbols

Each element $\tilde{\gamma}_p$ of the vector $\tilde{S}$ is obtained by taking a summation over all the possible values of the real part (or imaginary part) of the signal constellation, multiplied by the corresponding probability calculated using the soft decoder output [62, 63]. It is preferable to use the *a posteriori* information from the decoder output rather than extrinsic information in the calculation of $\tilde{\gamma}_p$. This has the advantage of permitting a better and faster convergence of the Rx.

4.4.3.4 Case Study

As we focus here on the turbo-PIC detector as the suboptimal solution, we just compare the performance of this detector with that of turbo-MAP. We consider the simplified implementation of soft-PIC based on ZF filtering given in (4.20). For this comparison, we consider the simple spatial multiplexing (V-BLAST) ST scheme and the case of four transmit antennas, $M_T = 4$, while we take M_R between 1 and 4. The Tx and Rx schemes correspond to Figures 4.7 and 4.8, respectively. We consider Gray bit/symbol mapping and random interleaving, as well as the Rayleigh flat quasi-static channel model. The nonrecursive and nonsystematic convolutional (NRNSC) channel code $(5, 7)_8$ (in octal representation) is considered with rate $R_c = 1/2$. SNR is considered in the form of E_b/N_0, where E_b is the average received energy per information bit and N_0 is the unilateral noise power spectral density; E_b/N_0 includes the Rx array gain, M_R.

Curves of bit error rate (BER) versus E_b/N_0 are given in Figure 4.10. In fact, the performances of turbo-PIC and turbo-MAP are relatively close to each other for $M_R \geq M_T$. Turbo-PIC can still be used for certain values of $M_R < M_T$, mostly for $M_R > M_T/2$ [63]. So, for these M_R values where turbo-PIC converges properly, it would be preferred to turbo-MAP due to its considerably lower complexity. Better performances are obtained for turbo-PIC if the variance of the RI is taken into account in LLR calculation [63].

4.4.4 Orthogonal versus Nonorthogonal ST Schemes

In practice, to attain a desired spectral efficiency, we should adopt the most appropriate scheme by fixing the degrees of freedom of the system, that is, the signal constellation, the channel coding rate, and the ST coding scheme. The answer to the question "What is the most suitable combination?" is not obvious for moderate to high spectral efficiencies. In effect, if a low spectral efficiency is required, an OSTBC scheme together with a powerful turbo-code would be a suitable solution, as the reduction in the overall coding rate is best invested in turbo channel codes [67]. To attain high spectral efficiencies with OSTBC schemes, however, we have to use large signal constellations and increase the channel coding rate by puncturing the encoder output bits. Use of larger signal constellations complicates the tasks of synchronization and detection at the Rx and also results in a higher SNR required to attain a desired BER. On the other hand, puncturing results in a reduced channel code robustness against noise. Higher ST coding rates are offered

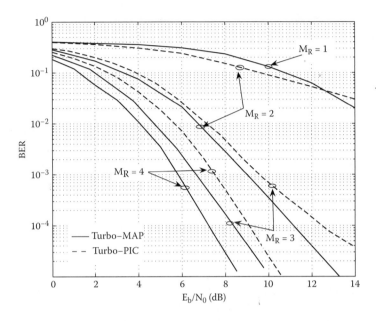

FIGURE 4.10 Comparison of turbo-PIC and turbo-MAP detectors, $M_T = 4$, $(5, 7)_8$ channel code, QPSK modulation, 64 channel uses per frame.

by nonorthogonal schemes, hence relaxing the conditions on signal constellation and channel coding. Here, a simple (suboptimal) iterative detector can be used for ST decoding, as explained in the previous subsection, and we may approach the optimal detection performance after few iterations. Nevertheless, the detector remains more complex, as compared to the OSTBC case. However, this increased Rx complexity is quite justified; using such an appropriate nonorthogonal ST scheme and iterative detection, we obtain a considerable gain in performance with respect to OSTBC choice [64–66]. Results in [65, 66] have also confirmed that the gain obtained by using nonorthogonal with respect to orthogonal schemes is still considerable, and even more important when channel estimation errors are taken into account.

4.4.4.1 Case Study

We consider the case of a (2×2) MIMO system, Gray bit/symbol mapping and random interleaving, as well as the Rayleigh flat block-fading channel model with $N_c = 32$ independent fades per frame. The number of channel uses corresponding to a frame is 768. The NRNSC channel code $(133, 171)_8$ is considered with rate $R_c = 1/2$. The ST schemes we consider are shown in Table 4.1, where η is the spectral efficiency in units of bps/Hz. As the OSTBC scheme, we consider the Alamouti code [51]. Using the formulation of LD codes that we presented in section 4.4.2, we have $Q = T = 2$, $R_{STC} = 1$, and

TABLE 4.1 Different ST Schemes for a (2×2) MIMO System with $\eta = 2$ bps/Hz

ST Scheme	R_{STC}	Modulation	R_c
Alamouti	1	16-QAM	1/2
V-BLAST	2	QPSK	1/2
GLD	2	QPSK	1/2

$$X = \begin{bmatrix} s_1 & -s_2^* \\ s_2 & s_1^* \end{bmatrix}. \tag{4.22}$$

For the OSTBC case, the decoding is performed once using (4.18). We also consider two nonorthogonal schemes. The first one is the simple V-BLAST scheme described by $X = [s_1 \ s_2]^t$, for which $Q = 2$, $T = 1$, and $R_{STC} = 2$. The second one is the optimized scheme proposed in [68] and called *Golden code*, which we denote by GLD. For this code that offers full rate and full diversity with the property of nonvanishing determinant, we have $Q = 2$, $T = 2$, and $R_{STC} = 2$:

$$X = \frac{1}{\sqrt{5}} \begin{bmatrix} \alpha(s_1 + \theta s_2) & \gamma\bar{\alpha}(s_3 + \bar{\theta} s_4) \\ \alpha(s_3 + \theta s_4) & \bar{\alpha}(s_1 + \bar{\theta} s_2) \end{bmatrix}, \tag{4.23}$$

where

$$\theta = \frac{1+\sqrt{5}}{2}, \alpha = 1 + j(1-\theta), \bar{\theta} = 1-\theta, \bar{\alpha} = 1 + j(1-\bar{\theta}), \gamma = j = \sqrt{-1}.$$

The factor $1/\sqrt{5}$ ensures normalized transmit power per channel use. For $\eta = 2$ bps/Hz, performance curves are shown in Figure 4.11, where again perfect channel knowledge is assumed at the Rx. For V-BLAST and GLD schemes, BER curves are shown for the second and fourth iterations, where almost full Rx convergence is attained. We see that, by using the V-BLAST scheme, we gain about 3.3 and 3.75 dB in SNR at BER = 10^{-4} after two and four iterations, respectively, compared to Alamouti coding. The corresponding gains by using GLD code are about 3.5 and 4.3 dB, respectively. We note that even when, for the reasons of complexity and latency, only two iterations are to be performed, the gain in SNR compared to the Alamouti scheme is still considerable.

4.5 Advanced Polarization Diversity Techniques for MIMO Systems

The initial research demonstrates that the MIMO channel capacity based on the uncorrelated channel model can be proportionally increased by increasing the number of antennas. However, in practice, the performance of the MIMO communication channel is affected by spatial correlation, which is dependent on antenna array configurations (such as radiation pattern, antenna spacing, and array geometry), and propagation channel characteristics, which are dependent on the environment (such as number of channel paths, distribution and properties of scatterers, angle spread, and cross-polarization discrimination). Thus, the antenna arrays at Tx and Rx should be properly designed to reduce the spatial correlation effects and to improve the communication performance.

However, it is possible to reduce these effects by increasing antenna array spacing, but this solution is not always suitable in some wireless applications where the array size is

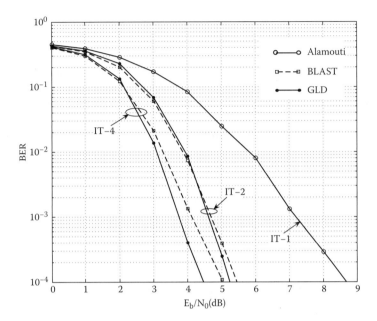

FIGURE 4.11 (2×2) MIMO system, turbo-PIC detection, $(133, 171)_8$ channel code, $\eta = 2$ bps/Hz, each frame contains 768 channel uses corresponding to 32 independent fades.

limited. Therefore, in order to eliminate the spatial correlation effects and remain with high transmission performance, there are essentially two diversity techniques, such as angular [69] and polarization diversity techniques [70]. For the pattern diversity technique, the radiation of antennas should be generated in a manner to isolate the radiation pattern. For polarization diversity techniques [71], the antennas are designed to radiate with orthogonal radiation polarizations to create uncorrelated channels across different array elements. The polarization diversity techniques could be applied in point-to-point communication systems such as intermobile base station communications, mobile satellite communications, high-resolution localization systems, military communications, etc.

Finally, there are also other diversity techniques, such as multimode diversity [72], that exploit the difference of high-order modes to obtain low correlated channels across the modes and a combination of pattern and polarization diversity techniques [73, 74] that take together the advantages of orthogonal radiation patterns and polarizations.

4.5.1   Antennas

In practice, not only the propagation environment for the multiple antenna systems has an important role in determining the transmission performance, but also the proper implementation of the antennas plays a dominant role. For example, uniform linear arrays, uniform circular arrays, and cube antenna arrays give different performances

TABLE 4.2 Patterns of Different Electric Dipoles

	E_x	E_y	E_z	E_ϕ	E_θ
$E_\theta(\theta, \phi)$	$-\cos\theta\cos\phi$	$-\cos\theta\cos\phi$	$-\sin\theta$	0	1
$E_\phi(\theta, \phi)$	$\sin\phi$	$-\cos\phi$	0	1	0

in terms of channel capacity. Moreover, different array configurations produce different correlation effects. In this part, we will analyze five types of antennas [75], x-, y-, and z-oriented dipole antennas, azimuth, and elevation isotropic antennas, applied to a uniform linear array.

The radiating patterns of the antennas are considered in the far-field case, and are also simplified by neglecting path loss and distance phase. Hence, these radiating patterns are simply dependent on the azimuth and elevation angle direction as shown in Table 4.2. A general expression of radiation patterns is given by [75]

$$E = E_\theta(\theta,\phi)\vec{\theta} + E_\phi(\theta,\phi)\vec{\phi}, \tag{4.24}$$

where $E_\theta(\theta,\phi)$ and $E_\phi(\theta,\phi)$ are the amplitudes of polarization vector at the $\vec{\theta}$- and $\vec{\phi}$-directions, and x, y, and z are the antenna orientations. As different types of antennas are employed in this chapter, it is necessary to normalize the radiation pattern when comparing all channel performances. Thus, all radiation patterns of an antenna are normalized by that of an isotropic antenna, and they can be written as

$$G = \sqrt{\frac{\iint\limits_{\Delta\phi,\Delta\theta} \sin\theta d\theta d\phi}{\iint\limits_{\Delta\phi,\Delta\theta} \left[\left|E_\theta(\theta,\phi)\right|^2 + \left|E_\phi(\theta,\phi)\right|^2 \right] \sin\theta d\theta d\phi}} \left(E_\theta(\theta,\phi)\vec{\theta} + E_\phi(\theta,\phi)\vec{\phi} \right) \tag{4.25}$$

where G is the antenna gain that is used for the computation of the channel matrix.

4.5.2 Cross-Polarization Discrimination

In wireless communications, due to the interactions of environment, such as diffractions, reflection, and refraction, the transmitted signals are generally not only attenuated but also depolarized. Depolarization is the change of the original state of the polarization of the electromagnetic wave propagated from the Tx.

Cross-polarization discrimination (XPD) is defined as the power ratio of the co-polarization and cross-polarization components of the mean incident wave. The higher the XPD, the less energy that is coupled in the cross-polarized channel. Therefore, there are two transmission cases, azimuth transmission (χ_θ) and elevation transmission (χ_ϕ), as follows:

$$\chi_\theta = \frac{E\left\{\left|E_{\theta\theta}\right|^2\right\}}{E\left\{\left|E_{\theta\phi}\right|^2\right\}}, \quad \chi_\phi = \frac{E\left\{\left|E_{\phi\phi}\right|^2\right\}}{E\left\{\left|E_{\phi\theta}\right|^2\right\}}, \tag{4.26}$$

where $E_{\theta\phi}$ denotes the θ-polarized electric field, which is propagated from a Tx and received in the φ polarization. It has the same explanation for $E_{\theta\theta}$, $E_{\phi\phi}$, and $E_{\phi\theta}$. In the literature, some measurement campaigns have been carried out, and they concluded that the XPD depended on the physical obstacles, the distance between Tx and Rx, and the delay of multipath components of each environment [76]. Therefore, for simplicity, XPD can be approximated by a Gaussian statistical distribution with average μ and variance σ^2 [77, 78]. For urban environments, the mean value of XPD can vary from 0 to 16 dB and the standard deviation can change from 3 to 9 dB.

4.5.3 Geometry-Based Stochastic Channel Models

We now focus on a useful model for simulation purposes, geometry-based stochastic modeling or geometric scattering modeling [10, 79, 80], which can be easily exploited to examine the performance of different antenna patterns and polarizations. This model is based on the assumption that scatterers around the Tx and Rx influence the direction of departure (DOD) and the direction of arrival (DOA), respectively, within transmit and receive scattering areas. Scatterers are randomly located according to a certain probability distribution. In particular, the scatterers are used to represent the depolarization and attenuation mechanism of incident waves traveling from the transmitters.

In Figure 4.12, a scattering geometry is shown. A uniform linear array of z-oriented dipole antennas at both Tx and Rx is employed. The heights of Tx and Rx are the same level. Moreover, transmit and receive scatterers are uniformly distributed within an angular region defined by $|\phi + \pi/2| \leq \Delta\phi/2$ in elevation area and $|\theta + \pi/2| \leq \Delta\theta/2$ in azimuth area. In order to determine one propagation path, from one transmit scatterer to one receive scatterer, we consider that there is a double depolarization mechanism

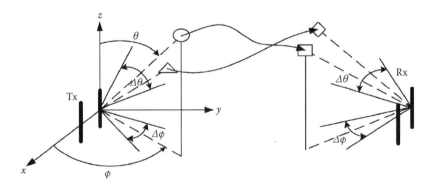

FIGURE 4.12   Geometries of MIMO channel.

replaced by one scattering matrix. One of the propagation path channels occurs when one of the transmit and one of the receive scatterers are randomly linked. To reduce the computational cost, a scatterer is used to generate only one propagation channel. Then the actual channel impulse response is established by a sum of propagation channels. We also assume that the channel coherence bandwidth is larger than the transmitted bandwidth of the signal. This channel is usually called frequency-non-selective or flat fading channel.

In case of far-field transmission without the line-of-sight channel, the flat fading transmission channel between antenna p at the Tx and antenna m at the Rx can be expressed as

$$
h_{mp}(t, f) = \frac{1}{\sqrt{N_S}} \sum_{i=1}^{N_S} a_m^{(i)} a_p^{(i)} \exp\left\{-j\vec{k}^{(i)} \cdot \vec{v}_{Rx} t - j\vec{k}'^{(i)} \cdot \vec{v}_{Tx} t + \varphi_{mp}\right\}
$$

$$
\left[G_\theta^m(\theta_i, \phi_i) \quad G_\phi^m(\theta_i, \phi_i) \right] \mathbf{S}_{mp}^{(i)} \begin{bmatrix} G_\theta^p(\theta_i, \phi_i) \\ G_\phi^p(\theta_i, \phi_i) \end{bmatrix},
$$

(4.27)

where t is time; f is frequency; N_s is the number of scatterers at the Rx and Tx; $\vec{v}_{Rx}$ and $\vec{v}_{Rx}$ are the velocity vectors of the Tx and Rx; $\vec{k}'^{(i)}$ and $\vec{k}^{(i)}$ are the vectors of wave number in the direction of the ith transmit scatterer and the ith receive scatterer, where $|\vec{k}^{(i)}| = |\vec{k}'^{(i)}| = 2\pi/\lambda$; $G_\theta^p(\theta_i, \phi_i)$ and $G_\phi^p(\theta_i, \phi_i)$ are the gain in the $\vec{\theta}$ and $\vec{\phi}$ directions of the pth transmit antenna in the direction of the ith transmit scatterer; $G_\theta^m(\theta_i, \phi_i)$ and $G_\phi^m(\theta_i, \phi_i)$ are the gain in the $\vec{\theta}$ and $\vec{\phi}$ directions of the mth receive antenna in the direction of the ith receive scatterer; $a_m^{(i)}$ is the mth element of the local vector of the receive antenna, so that the local receive vector can be expressed as $\mathbf{a}_{Rx}^{(i)} = [1 \quad \exp\{-j\vec{k}^{(i)} \cdot \vec{r}_1\} \cdots \exp\{-j\vec{k}^{(i)} \cdot \vec{r}_{M_R-1}\}]$; $a_p^{(i)}$ is the pth element of the local vector of the transmit antenna, so that the local transmit vector can be expressed as $\mathbf{a}_{Tx}^{(i)} = [1 \quad \exp\{-j\vec{k}'^{(i)} \cdot \vec{r}_1'\} \cdots \exp\{-j\vec{k}'^{(i)} \cdot \vec{r}_{M_T-1}'\}]$; and $\mathbf{S}_{mp}^{(i)}$ are a 2×2 scattering matrix for the ith transmit scatterer and the ith receive scatterer for $i = 1 \ldots N_S$ wave components. A scattering matrix contains the polarization mechanism as defined by

$$
\mathbf{S}_{mp}^{(i)} = \begin{bmatrix} \sqrt{\dfrac{\chi_\theta}{1+\chi_\theta}} \exp\left\{j\beta_{mp}^{(\theta\theta)}\right\} & \sqrt{\dfrac{1}{1+\chi_\phi}} \exp\left\{j\beta_{mp}^{(\phi\theta)}\right\} \\ \sqrt{\dfrac{1}{1+\chi_\theta}} \exp\left\{j\beta_{mp}^{(\theta\phi)}\right\} & \sqrt{\dfrac{\chi_\phi}{1+\chi_\phi}} \exp\left\{j\beta_{mp}^{(\phi\phi)}\right\} \end{bmatrix},
$$

(4.28)

where $\beta_{mp}^{(\phi\theta)}$ denotes phase offset of the ith incident wave, which changes from $\vec{\phi}$ directions to $\vec{\theta}$ directions superposing on the m-p channel, and χ_θ and χ_ϕ denote the ratio of the co-polarized average received power to the cross-polarized average received power. In [77], after sufficiently reflecting the propagation signal between transmitters and receivers, the polarization state of the signal will be independent of the transmitted signals.

4.5.4   Spatial Correlation and Angle Spread Effects

The consecutive MIMO systems based on the spatial diversity technique are directly influenced by the spatial correlation effect (or antenna correlation effect) [16]. This effect is drastically dependent on array configurations and environment characteristics. Therefore, the antenna arrays at both Tx and Rx should be properly designed or adapted to decrease spatial correlation effects.

As illustrated in Figure 4.13, the spatial correlation of a 2×2 uniform linear antenna array depends on the antenna spacing and the angle spread (AS). The general expression of spatial correlation [81] between two antenna elements can be written as

$$
\rho_{i,j} = \frac{\displaystyle\iint_{\Delta\theta,\Delta\phi} a_i \cdot a_j^* \sin(\theta)\, p\,(\theta,\phi)\, d\theta d\phi}{\sqrt{\displaystyle\iint_{\Delta\theta,\Delta\phi} \left|a_i\right|^2 \sin(\theta)\, p\,(\theta,\phi)\, d\theta d\phi \cdot \iint_{\Delta\theta,\Delta\phi} \left|a_j\right|^2 \sin(\theta)\, p\,(\theta,\phi)\, d\theta d\phi}}, \tag{4.29}
$$

where a_i is the local value of the ith transmit or receive antenna and a_j is the local value of the jth transmit or receive antenna in the local vector. The scalar $p(\theta,\phi)$ is the joint probability density function (pdf) of the angles of arrival for the receiving spatial correlations or of the angles of departure for the transmitting spatial correlations.

When the narrow angle spread of incident fields occurs in the transmitting or receiving side, the separation between antennas should be expanded in order to reduce the spatial correlation problem, as shown in Figure 4.13. This shows the spatial correlation of 2×2 MIMO z-oriented antennas in the case of a uniform distribution of the angles

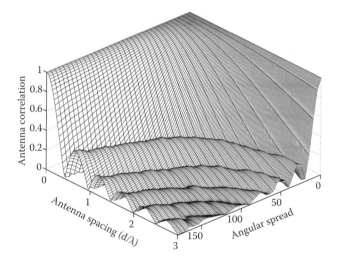

FIGURE 4.13   Spatial correlation of uniform linear antenna array for single-polarized configuration.

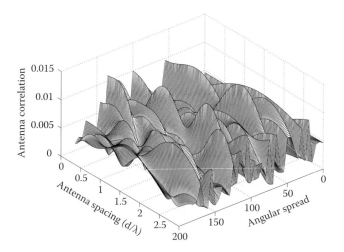

FIGURE 4.14   Spatial correlation of uniform linear antenna array for dual-polarized configuration.

of arrival within a region. It demonstrates that the received signals become uncorrelated when the antenna spacing is sufficiently increased and the angle spread is quite wide. However, wide antenna spacing may preclude implementation in some applications where size is a limitation. The use of polarized antennas is promising low spatial correlation. The spatial correlation is demonstrated in Figure 4.14, while the 2×2 polarized MIMO configuration employs a pair of y- and z-oriented dipole antennas at both Tx and Rx. The spatial correlation of dual-polarized MIMO is much lower than that of single-polarized MIMO in all simulation scenarios.

Although there are only two diversity branches, the use of polarized antennas does allow the antenna elements to be collocated without the correlation effect. However, there is considerable interest in many diversity branches by applying a combination of the pattern and polarization diversity [74, 82].

4.5.5   Capacity of Polarized Channels

Antenna polarization diversity is very useful in MIMO systems for enhancing channel capacity. Indeed, employing polarization diversity can reduce the antenna array size and also the spatial correlation; then we can obtain a better capacity. That is why the multipolarized antennas become more and more interesting in MIMO transmission systems. In this section, MIMO systems are investigated to show the potential of using multipolarized antennas for differently oriented dipole antennas.

We assume that the channel state information (CSI) is perfectly known to the Rx but unknown to the Tx. This is in theory what happens to signals propagating through an urban and an indoor environment. In the case of a random channel model, the channel matrix (4.27) is stochastic, and then the capacity given by (4.5) is also random. In this situation, the ergodic capacity can be obtained by taking the expectation of capacity over all possible channel realizations.

Capacity of 2×2 MIMO with single-polarized antennas

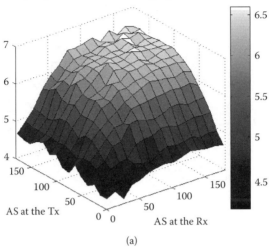

(a)

Capacity of 2×2 MIMO with dual-polarized antennas

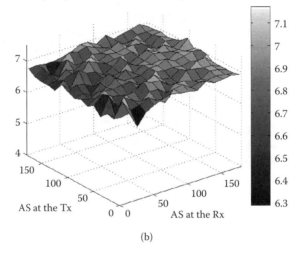

(b)

FIGURE 4.15 (2×2) MIMO channel capacity of isotropic antennas: (a) single-polarization configuration and (b) dual-polarization configuration.

Figure 4.15 demonstrates the 2×2 MIMO channel capacity of isotropic antennas with single polarization (a) and dual polarization (b). In this case, the single-polarization configuration exploits only the azimuth isotropic antenna, and the dual polarization configuration applies the azimuth and elevation isotropic antennas. The channel capacity is examined in function of angular spread (AS) with twenty scatterers distributed around the Tx and Rx. As mentioned in the previous section, the XPD is defined for

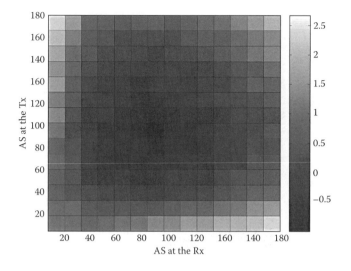

FIGURE 4.16 Difference between the triple-polarized and single-polarized channel capacities of 3 × 3 MIMO systems employing dipole antennas.

the urban case by χ_θ and $\chi_\phi \sim \mathcal{N}(0, 9)$. As shown in Figure 4.15a, the MIMO channel capacity increases as the angle spread increases at Tx and Rx for the same polarization antennas. In contrast, dual polarization improves the channel capacity due to the lower antenna correlation, as shown in Figure 4.15b. However, when the spatial correlation of the single-polarized antennas is lower, the channel capacity is proportional until 6.7 dB for AS > 80°. It should be noted that the MIMO channel capacity is significantly dependent on the antenna correlation.

Figure 4.16 shows the difference between the triple-polarized and single-polarized channel capacities ($\Delta C = C_{\text{triple-polar}} - C_{\text{single-polar}}$) of the 3 × 3 MIMO system versus the average XPD and AS. The propagation environment has the same conditions as the previous section. For the triple-polarization configuration, x-, y-, and z-oriented dipole antennas are employed. This can represent a combination of angular and polarization techniques. We found that for the triple-polarization case, the average power of the subchannel can be unfortunately lost when the angular spread is not large enough until covering all antennas. As seen in Figure 4.16, the single-polarized channel capacity can be superior to the triple-polarized channel capacity because single polarization has a low spatial correlation and triple polarization loses the subchannel power due to insufficient angular spread. However, generally speaking for most scenarios, triple polarizations can maintain a higher capacity with respect to the single-polarization case.

4.5.6 Impact of Depolarization Effect on MIMO Configurations

In this section, we investigate the impact of the depolarization effect on 4 × 4 MIMO systems with single- and dual-polarization configurations. While pattern and polarization

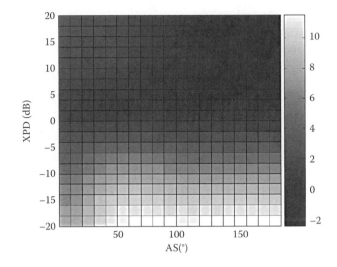

FIGURE 4.17   Difference between the dual-polarized and single-polarized channel capacities of 4 × 4 MIMO systems in the functions of XPD and AS.

diversity techniques are employed, the special correlation effect can be reduced or eliminated when there is no pattern interference. Nevertheless, the cross-polarization discrimination (XPD) becomes the most important parameter because XPD represents the ratio of the co-polarized average received power to the cross-polarized average received power. Then, for a high XPD value, less energy is coupled between the cross-polarized wireless channels. At lower XPD and higher K-factor values [83], multipolarized antenna arrays can give high capacity. However, at higher XPD and lower spatial correlation, a single-polarized antenna array can provide even better results.

Figure 4.17 explains the difference between the dual-polarized and single-polarized channel capacity ($\Delta C = C_{\text{dual-polar}} - C_{\text{single-polar}}$) of 4 × 4 MIMO systems versus XPD and AS. Same angle spreads at Tx and Rx are considered in the simulation. For high XPD and sufficiently large angle spread, we can note that the MIMO channel capacity of the single-polarized antenna is superior to that of the dual-polarized antenna because of the subchannel power loss. The Frobenius norm of the MIMO channel is used to investigate the total channel power. It confirms that with a high XPD and low spatial correlation the average transmission power of single-polarized isotropic antenna arrays is $\|H\|_F \approx MN$, but that of dual-polarized isotropic antenna arrays is always $\|H\|_F \approx M^\theta N^\theta + M^\phi N^\phi$ for high XPDs and $\|H\|_F \approx M^\theta N^\phi + M^\phi N^\theta$ for low XPDs.

Figure 4.18 shows the XPD impact on 4 × 4 MIMO systems with single- and dual-polarization configurations. We employ four azimuth isotropic antennas for the single-polarization configuration and two elevation and two azimuth isotropic antennas for the dual-polarization configurations, which are applied to a uniform linear antenna array with a λ/2 antenna separation. It shows that the channel power of single-polarization configuration augments significantly with respect to XPD until $\|H\|_F = 4 \times 4 = 16$, while that of dual-polarization configuration keeps around $\|H\|_F = 2 \times 2 + 2 \times 2 = 8$.

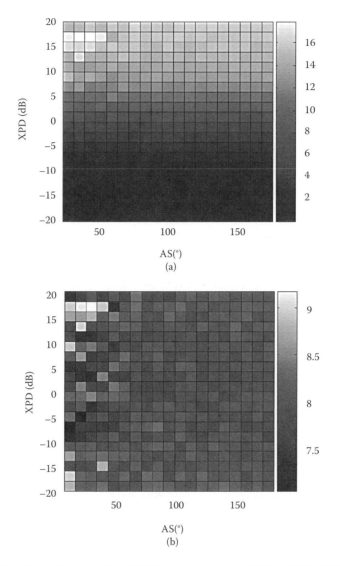

FIGURE 4.18 Frobenius norm of 4×4 MIMO systems in terms of XPD and AS with (a) a single-polarization configuration and (b) a dual-polarization configuration.

4.5.7   Adaptive MIMO Polarized-Antenna Selection Technique (AMPAS)

In the previous sections, we defined the scattering mechanisms that are used to represent not only the attenuation of traveling waves but also the polarization of the electromagnetic wave. The achieved performance in capacity is calculated under the assumption

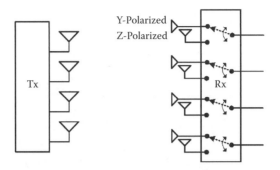

FIGURE 4.19   Adaptive polarization technique (AMPAS) at Rx.

that the average received power is normalized and the channel attenuation is neglected. Subsequently, the variation of polarization is characterized only by the XPD effects.

As shown in Figure 4.17, this phenomenon affects directly the performance of non-polarized MIMO systems. In the single-polarization communications case, low XPD causes higher losses in channel power; in other words, there is some sort of mismatch in polarization. That is why we will apply adaptive techniques to reduce this mismatch in the polarization of MIMO systems. This technique is called adaptive MIMO polarized-antenna selection technique (AMPAS). The principle of this method is to choose properly the antenna polarizations that optimize the receiving signal power while minimizing fading correlation antenna effects. In Figure 4.19, an example of an adaptive polarization system employing four z-oriented dipoles at Tx and four pairs of y- and z-oriented dipoles at Rx is illustrated. Simulation results based on three-dimensional ray-tracing techniques show that the channel capacity obtained by an adaptive polarization increases 7–13% in comparison to the single-polarization channel capacity.

Another example of an adaptive polarization system is based on the rotation of the antenna elements according to the polarization of traveling waves at Rx. The proposed MIMO system consists of P half-wavelength dipole antennas that are rotated against one another by the rotation angle $\gamma = 180°/P$ with phase centers at the same point at both Tx and Rx.

Figure 4.20 demonstrates the obtained performances of 1×1 SISO, 1×2 SIMO, and 2×2 MIMO communication systems while the receiving antenna is rotating on the y-z plane with χ_θ and $\chi_\phi \sim \mathcal{N}(0,5)$. Performance can be better enhanced if polarization at the Rx is properly matched to that of incident waves. In contrast, it can be worse if they are not well matched, as shown in Figure 4.20a (at rotational angle $(\gamma) \approx 140°$). Thus, for improving the MIMO channel capacity, the receiving antenna elements should be rotated to find maximum receiving signals while minimizing fading correlation antenna effects.

While the other systems employ only z-oriented dipoles at Tx and y- and z-oriented dipoles at Rx in the case of the same antenna position (b and c) and the $\lambda/2$-separated antenna elements (d), the rotations of two dipoles on plane y-z can provide different performances, as shown in Figure 4.20b–d. The channel performance is maximized in Figure 4.20b when the polarizations of receiving antennas are well matched, and

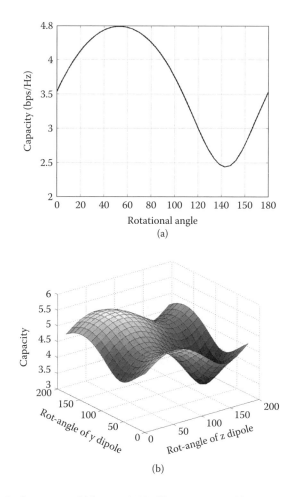

FIGURE 4.20 Performances of (a) 1 × 1 SISO, (b) 1 × 2 SIMO, (c) 2 × 2 MIMO as the same antenna position, and (d) 2 × 2 MIMO as the separated antenna position while the receiving antenna is rotating on plane *y-z*.

is minimized if they are mismatched. In Figure 4.20c, when the antennas have been rotated to nearly the same position, high correlation is produced. Subsequently, the channel capacity is reduced even if the polarizations are correctly matched. However, it has better performance in the situation of polarization diversity when the antenna rotation has the difference of 90°. Moreover, we observe the MIMO capacity with spatial diversity as illustrated in Figure 4.20d, where the antenna correlation is reduced and the channel capacity is improved.

In order to achieve a better transmission performance, the polarized antenna selection can exploit together all diversity techniques, such as pattern, spatial, and polarization diversity. Pattern diversity should be employed when a large angle spread is detected. Spatial diversity could be exploited when a high antenna correlation is observed.

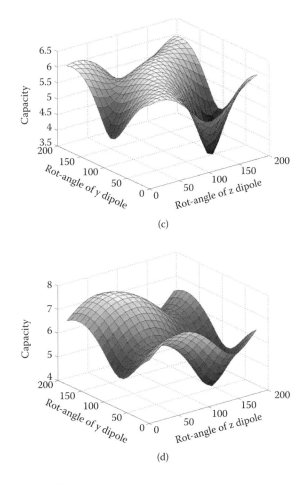

(c)

(d)

FIGURE 4.20   (continued)

Polarization diversity should be used when a low XPD occurs. Therefore, the employed diversity should be properly selected according to the propagation environment.

4.6   MIMO Applications

4.6.1   Wireless LAN-Based MIMO

In this chapter it was shown that employing MIMO systems could achieve higher performance in data transmission. MIMO signaling can increase network bandwidth, range, and reliability. Recently, many communication systems began to take advantage of this channel capacity enhancement, such as wireless local area network (WLAN) and wireless metropolitan area network (WMAN).

The IEEE 802.11 WLAN and IEEE 802.16 WMAN standards are based on orthogonal frequency division multiplexing (OFDM). OFDM is a multicarrier modulation system, reducing the required bandwidth but keeping the modulated signals orthogonal so they do not interfere with each other. An important high-data-transmission-rate extension of these standards could be based on MIMO. An advantage of these systems is that they are principally deployed in indoor environments and suburban environments that are characterized by a rich multipath.

There are also some motivations in order to improve the performance and transmission rate in MIMO-OFDM systems when the CSI is available at the transmitter. As mentioned in previous sections, the water-filling technique is used to optimize the distribution of the total transmit power over transmit antennas. Therefore, information symbols and power could be optimally allocated over space and frequency in MIMO-OFDM communication.

The number of antennas utilized in a MIMO (WLAN) router, for example, can vary; a typical MIMO router contains three or four antennas. These systems are driving the need for the next broadband revolution focused on home networking. Such systems will cover next-generation game consoles, video on demand, HDTV, and other new products. These services are creating a more sophisticated home entertainment environment, together with a high level of quality of service (QoS) to facilitate multimedia connectivity.

The intelligence behind the antenna polarization is described as adaptive polarization. The router receives feedback from the client adapter and has the ability to focus the polarization of the signals. As signal travels between an access point and wireless card, it will bounce off of walls, ceilings, and any other obstacle, resulting in multiple reflections of the original signal arriving by different paths and different polarizations. By applying adaptive antenna polarization algorithms (AMPAS), these reflections can be used to improve the signal-to-noise ratio, instead of having just one copy of the original signal.

This adaptive transmitting feature provides a more reliable signal at extreme ranges. Moreover, MIMO can eliminate dead spots, delivering reliable whole-home coverage with all the speed you need for application in the future. Today, one can say, wireless is faster than leased wire systems.

Finally, MIMO networking has the potential to increase communications data rates by 10–20 times above those of current systems. Such systems will use multipath reflections to create parallel channels in the same frequency bandwidth, thereby increasing spectral efficiency. In the next section sensor networks will be discussed as an application of a MIMO-based system.

4.6.2   MIMO for Cooperative Sensor Networks

A sensor network can be considered a self-contained circuit with its sensor and RF interface, as shown in Figure 4.21. Recent hardware advances allow more signal processing functionality to be integrated into a single chip. For example, it is possible to integrate an RF transceiver, sensor interface, and baseband processors into one device that is as small as a piece of coin and can be used as a fully functional wireless sensor node. Such wireless nodes typically operate with small batteries; that is why these sensor nodes have limited power capabilities. In many scenarios, the wireless nodes must operate without battery replacement for many years. Consequently, minimizing the energy consumption

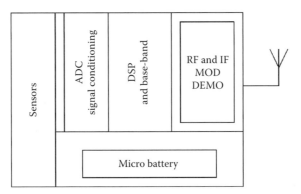

FIGURE 4.21   Typical wireless sensor node.

is a very important design consideration, and energy-efficient transmission schemes must be used for the data transfer in sensor networks.

In addition, because sensor nodes will be deployed in remote and oftentimes dangerous locations, their maintenance (in particular, battery replacement) will be unlikely [84].

Sensor networks are a new attraction for many potential applications, such as industrial, military, geolocalization, surveillance, intrusion detection, and environmental monitoring [84, 85].

Robust communications between sensor nodes are highly demanded at low power. As was shown, MIMO communication promises performance enhancements over conventional single-input single-output (SISO) technology without increasing the bandwidth consumed by the system or the total power radiated from a transmitter. MIMO technology has promising characteristics that make it a serious candidate for sensor network communication technology. Signal processing techniques that use multiple transmit and receive antennas, such as space-time coding (ST coding), have been shown to increase transmission reliability.

In a surveillance application, the ability of sensor nodes to relay data is critical to the utility and effectiveness of the sensor network.

For a given node density, nodes are more likely to be out of range, thus inhibiting communication. In a situation such as this, the extended range of MIMO is of greater importance because it enables cohesion (the ability of the sensor nodes to form a completely connected network), which guarantees the success of the final application [86].

New protocols for target reporting and a procedure for target localization that conserve energy have recently been developed [86, 87]. In [88], the authors summarize and compare several routing MIMO technologies.

Mean path length provides a measurement of the impact of MIMO communications on a wireless sensor network. Mean path length provides a rough estimate of the amount of time and energy expended in a data transmission from one node to another in the network.

Most significant mean path length reduction is provided by MIMO in the low or mid range of node densities because the internode spacing is such that MIMO can reliably form some links that SISO cannot. In the elongated region scenario this trend holds, though it is less apparent [89].

There is also an increasing need for mobile networks with distributed transmitters and receivers, typically referred to as mobile ad hoc networks (MANETs). There, transmitters and receivers do not pool their information together, either due to geographical disper-siveness, the bandwidth and resource limitation, or due to security/privacy concerns.

Recognizing that multiple antennas at the transceivers provide inherent multiplexing capability due to their spatial selectivity, it is attractive to study MIMO communication in ad hoc networks with interference transmission.

Energy-efficient communication techniques typically focus on minimizing the trans-mission energy only, which is reasonable in long-range applications where the transmis-sion energy is dominant in the total energy consumption.

In cooperative sensor networks, we allow the cooperation among sensors for infor-mation transmission or reception, so that energy consumption as well as transmission delays over some distance ranges can be reduced.

In conclusion, for the same throughput requirement, MIMO systems require less transmission energy than SISO systems. However, direct application of multiantenna techniques to sensor networks is impractical due to the limited physical size of a sensor node, which typically can only support a single antenna. If individual single-antenna nodes allowed cooperating on information transmission or reception, a cooperative MIMO system can be constructed such that energy-efficient MIMO schemes can be deployed [90]. Finally, MIMO can provide significant network performance improve-ments in power consumption, latency, and network robustness.

References

[1] E. Telatar. 1999. Capacity of multi-antenna Gaussian channels. *Eur. Trans. Tele-commun.* 10:585–95.

[2] P. Bello. 1963. Characterization of randomly time-variant linear channels. *IEEE Trans. Commun. Syst.* 11:360–93.

[3] J. D. Parsons. 2000. *The mobile radio propagation channel*. New York: Wiley.

[4] G. Matz. 2005. On non-WSSUS wireless fading channels. *IEEE Trans. Wireless Commun.* 4:2465–78.

[5] J. W. McJown and R. L. Hamilton. 1991. Ray tracing as a design tool for radio net-works. *IEEE Network Mag.* 5(6):27–30.

[6] T. Kurner, D. J. Cichon, and W. Wiesbeck. 1993. Concepts and results for 3D digi-tal terrain-based wave propagation models: An overview. *IEEE J. Selected Areas Commun.* 11:1002–12.

[7] R. Kouyoumjian and P. Pathak. 1974. A uniform geometrical theory of diffraction for an edge in a perfectly conducting surface. *Proc. IEEE* 62:1448–61.

[8] A. R. N. G. E. Athanasiadou and J. P. McGeehan. 2000. A microcellular ray-trac-ing propagation model and evaluation of its narrowband and wideband. *IEEE J. Selected Areas Commun.* 18:322–35.

[9] P. Petrus, J. Reed, and T. Rappaport. 2002. Geometrical-based statistical macrocell channel model for mobile environments. *IEEE Trans. Commun.* 50:495–502.

[10] J. C. Liberti and T. Rappaport. 1996. A geometrically based model for line-of-sight multipath radio channels. In *Proceedings of VTC*, pp. 844–48.

[11] C. Oestges, V. Erceg, and A. Paulraj. 2003. A physical scattering model for MIMO macro-cellular broadband wireless channels. *IEEE J. Selected Areas Commun.* 21:721–29.

[12] A. F. Molisch, A. Kuchar, J. Laurila, K. Hugl, and R. Schmalenberger. 2003. Geometry-based directional model for mobile radio channels—Principles and implementation. *Eur. Trans. Telecommun.* 14:351–59.

[13] K. Raoof and N. Prayongpun. 2005. Channel capacity performance for MIMO polarized diversity systems. In *Proceedings of WCNM*, pp. 1–4.

[14] G. J. Foschini and M. J. Gans. 1998. On limits of wireless communications in a fading environment when using multiple antennas. *Wireless Personal Commun.* 6:311–35.

[15] M. A. Khalighi, K. Raoof, and G. Jourdain. 2002. Capacity of wireless communication systems employing antenna arrays, a tutorial study. *Wireless Personal Commun.* 23:321–52.

[16] D. Shiu, G. J. Foschini, M. J. Gans, and J. M. Kahn. 2000. Fading correlation and its effect on the capacity of multi-element antenna systems. *IEEE Trans. Commun.* 48:502–13.

[17] J. Kermoal et al. 2002. A stochastic MIMO radio channel model with experimental validation. *IEEE J. Selected Areas Commun.* 20:1211–26.

[18] H. Özcelik et al. 2003. Deficiencies of the "Kronecker" MIMO radio channel model. In *Elec. Lett.* 39(16):1209–1210.

[19] A. Saleh and R. Valenzuela. 1987. A statistical model for indoor multipath propagation. *IEEE J. Selected Areas Commun.* 5:128–37.

[20] D. Gesbert et al. 2002. Outdoor MIMO wireless channels: Models and performance prediction. *IEEE Trans. Commun.* 50:1926–34.

[21] A. Burr. 2003. Capacity bounds and estimates for the finite scatterers MIMO wireless channel. *IEEE J. Selected Areas Commun.* 21:812–18.

[22] M. Debbah and R. Muller. 2005. MIMO channel modeling and the principle of maximum entropy. *IEEE Trans. Inf. Theory* 51(5).

[23] A. Sayeed. 2002. Deconstructing multiantenna fading channels. *IEEE Trans. Signal Processing* 50:2563–79.

[24] L. M. Correia. 2001. *Wireless flexible personalised communication.* COST 259 Final Report. New York: Wiley.

[25] L. M. Correia. 2006. *Mobile broadband multimedia networks.* COST 273 Final Report. London: Elsevier.

[26] 3GPP. 2003. *Spatial channel model for multiple input multiple output (MIMO) simulations.* 3GPP2 TR 25.996.

[27] G. J. Foschini. 1996. Layered space-time architecture for wireless communication in a fading environment when using multi-element antennas. *Bell Labs Tech. J.* 1:41–59.

[28] G. D. Golden, G. J. Foschini, R. A. Valenzuela, and P. W. Wolniansky. 1999. Detection algorithm and initial laboratory results using V-BLAST space-time communication architecture. *Electronic Lett.* 35:14–16.

[29] G. G. Raleigh and V. K. Jones. 1999. Multivariate modulation and coding for wireless communication. *IEEE J. Selected Areas Commun.* 17:851–66.

[30] E. Biglieri, J. Proakis, and S. S. Shitz. 1998. Fading channels, information-theoretic and communications aspects. *IEEE Trans. Inf. Theory* 44:2619–92.

[31] W. C. Jakes. 1998. *Microwave mobile communications.* New York: John Wiley & Sons.

[32] H. Hashemi. 1993. The indoor radio propagation channel. *Proc. IEEE* 81:943–68.

[33] M. A. Khalighi, J. M. Brossier, G. Jourdain, and K. Raoof. 2001. Water filling capacity of Rayleigh MIMO channels. In *Proceedings of PIMRC*, San Diego, vol. A, pp. 155–58.

[34] J. B. Anderson. 2000. Array gain and capacity for known random channels with multiple element arrays at both ends. *IEEE J. Selected Areas Commun.* 18:2172–78.

[35] G. G. Raleigh and J. M. Cioffi. 1998. Spatio-temporal coding for wireless communication. *IEEE Trans. Commun.* 46:357–66.

[36] T. L. Marzetta and B. M. Hochwald. 1999. Capacity of a mobile multiple-antenna communication link in Rayleigh flat fading. *IEEE Trans. Inf. Theory* 45:139–57.

[37] L. Zheng and D. N. C. Tse. 2002. Communication on the Grassmann manifold: A geometric approach to the noncoherent multiple-antenna channel. *IEEE Trans. Inf. Theory* 48:359–83.

[38] B. Sklar. 1997. Rayleigh fading channels in mobile digital communication systems. Part I. Characterization. Part II. Mitigation. *IEEE Commun. Mag.* 35:90–109.

[39] J. Salz and J. H. Winters. 1994. Effect of fading correlation on adaptive arrays in digital mobile radio. *IEEE Trans. Veh. Technol.* 43:1049–57.

[40] P. F. Driessen and G. J. Foschini. 1999. On the capacity formula for multiple input-multiple output wireless channels: A geometric interpretation. *IEEE Trans. Commun.* 47:173–76.

[41] M. A. Khalighi, J. M. Brossier, G. Jourdain, and K. Raoof. 2001. On capacity of Ricean MIMO channels. In *Proceedings of PIMRC*, San Diego, vol. A, pp. 150–54.

[42] C. Berrou and A. Glavieux. 1996. Near optimum error correcting coding and decoding: Turbo-codes. *IEEE Trans. Commun.* 44:1261–71.

[43] C. Douillard, M. Jézéquel, C. Berrou, A. Picart, P. Didier, and A. Glavieux. 1995. Iterative correction of intersymbol interference: Turbo-equalization. *Eur. Trans. Telecommun.* 6:507–11.

[44] M. A. Khalighi and J. J. Boutros. 2006. Semi-blind channel estimation using EM algorithm in iterative MIMO APP detectors. *IEEE Trans. Wireless Commun.* 5:3165–73.

[45] N. Noels, C. Herzet, A. Dejonghe, V. Lottici, H. Steendam, M. Moeneclaey, M. Luise, and L. Vandendorpe. 2003. Turbo synchronization: An EM algorithm interpretation. In *Proceedings of ICC*, Anchorage, AK, vol. 4, pp. 2933–37.

[46] X. Wang and H. V. Poor. 1999. Iterative (turbo) soft interference cancellation and decoding for coded CDMA. *IEEE Trans. Commun.* 47:1046–61.

[47] H. El Gamal and E. Geraniotis. 2000. Iterative multiuser detection for coded CDMA signals in AWGN and fading channels. *IEEE J. Selected Areas Commun.* 18:30–41.

[48] M. Sellathurai and S. Haykin. 2002. Turbo-BLAST for wireless communications: Theory and experiments. *IEEE Trans. Signal Processing* 50:2538–46.

[49] B. Vucetic and J. Yuan. 2003. *Space-time coding.* Chichester, England: John Wiley & Sons Ltd.

[50] V. Tarokh, H. Jafarkhani, and A. R. Calderbank. 1999. Space-time block codes from orthogonal designs. *IEEE Trans. Inf. Theory* 45:1456–67.

[51] S. M. Alamouti. 1998. A simple transmit diversity technique for wireless communications. *IEEE J. Selected Areas Commun.* 16:1451–58.

[52] V. Tarokh, N. Seshadri, and A. R. Calderbank. 1998. Space-time codes for high data rate wireless communication: Performance criterion and code construction. *IEEE Trans. Inf. Theory* 44:744–65.

[53] Y. Xin, Z. Wang, and G. B. Giannakis. 2003. Space-time diversity systems based on linear constellation precoding. *IEEE Trans. Wireless Commun.* 2:294–309.

[54] B. Hassibi and B. M. Hochwald. 2002. High-rate codes that are linear in space and time. *IEEE Trans. Inf. Theory* 48:1804–24.

[55] E. Viterbo and J. J. Boutros. 1999. A universal lattice code decoder for fading channels. *IEEE Trans. Inf. Theory* 45:1639–42.

[56] G. Caire, G. Taricco, and E. Biglieri. 1998. Bit-interleaved coded modulation. *IEEE Trans. Inf. Theory* 44:927–46.

[57] L. Bahl, J. Cocke, F. Jelinek, and J. Raviv. 1974. Optimal decoding of linear codes for minimizing symbol error rate. *IEEE Trans. Inf. Theory* 20:284–87.

[58] J. Hagenauer, E. Offer, and L. Papke. 1996. Iterative decoding of binary block and convolutional codes. *IEEE Trans. Inf. Theory* 42:429–45.

[59] P. Robertson, P. Hoeher, and E. Villebrun. 1997. Optimal and sub-optimal maximum a posteriori algorithms suitable for turbo decoding. *Eur. Trans. Telecommun.* 8:119–25.

[60] J. J. Boutros, N. Gresset, and L. Brunel. 2003. Turbo coding and decoding for multiple antenna channels. In *Proceedings of the International Symposium on Turbo Codes and Related Topics*, Brest, France, pp. 1–8..

[61] B. Hassibi and H. Vikalo. 2005. On the sphere-decoding algorithm I. Expected complexity. *IEEE Trans. Signal Processing* 53:2806–18.

[62] M. A. Khalighi and J. J. Boutros. 2005. Channel estimation in turbo-BLAST detectors using EM algorithm. In *Proceedings of ICASSP*, Philadelphia, vol. III, pp. 1037–40.

[63] M. A. Khalighi, J. J. Boutros, and J.-F. Hélard. 2006. Data-aided channel estimation for turbo-PIC MIMO detectors. *IEEE Commun. Lett.* 10:350–52.

[64] M. A. Khalighi and J.-F. Hélard. 2005. Should MIMO orthogonal space-time coding be preferred to non-orthogonal coding with iterative detection? In *Proceedings of ISSPIT*, Athens, pp. 340–45.

[65] M. A. Khalighi, J.-F. Hélard, and S. Bourennane. 2006. Choice of appropriate space-time coding scheme for MIMO systems employing channel coding under BICM. In *Proceedings of SPAWC*, Cannes, France, pp. 1–5.

[66] M. A. Khalighi, J.-F. Hélard, and S. Bourennane. 2006. Contrasting orthogonal and non-orthogonal space-time schemes for perfectly-known and estimated MIMO channels. In *Proceedings of ICCS*, Singapore, pp. 1–5.

[67] V. Le Nir, J.-M. Auffray, M. Hélard, J.-F. Hélard, and R. Le Gouable. 2003. Combination of space-time block coding with MC-CDMA technique for MIMO systems with two, three and four transmit antennas. In *Proceedings of the IST Mobile Communications Summit Conference*, Aveiro, Portugal.

[68] J.-C. Belfiore, G. Rekaya, and E. Viterbo. 2005. The golden code: A 2×2 full-rate space-time code with nonvanishing determinants. *IEEE Trans. Inf. Theory* 51:1432–36.

[69] R. G. Vaughan. 1998. Beam spacing for angle diversity. In *Proceedings of Globecom*, vol. 2, pp. 928–33.

[70] P. Kyritsi et al. 2002. Effect of antenna polarization on the capacity of a multiple element system in an indoor environment. *IEEE J. Selected Areas Commun.* 20:1227–39.

[71] V. Eiceg, H. Sampath, and Catreux-Erceg, S. 2006. Dual-polarization versus single-polarization MIMO channel measurement results and modeling. *IEEE Trans. Wireless Commun.* 5:28–33.

[72] T. Svantesson. 2000. On the potential of multimode antenna diversity. In *Proceedings of VTC*, pp. 2368–2372.

[73] M. A. J. T. Svantesson and J. W. Wallace. 2004. Analysis of electromagnetic field polarizations in multiantenna systems. *IEEE Trans. Wireless Commun.* 3:641–46.

[74] J. B. Andersen and B. N. Getu. 2002. The MIMO cube—A compact MIMO antenna. *Wireless Personal Multimedia Commun.* 1:112–14.

[75] C. A. Balanis. 1997. *Antenna theory.* New York: John Wiley & Sons.

[76] P. Soma, D. S. Baum, V. Ercegl, R. Krishnamoorthyl, and A. J. Paulraj. 2002. Analysis and modeling of multiple-input multiple-output (MIMO) radio channel based on outdoor measurements conducted at 2.5 GHz for fixed BWA applications. In *Proceedings of ICC*, pp. 272–76.

[77] R. G. Vaughan. 1990. Polarization diversity in mobile communications. *IEEE Trans. Veh. Technol.* 39(3).

[78] J. Lempiainen, J. K. Laiho-Steffens, and A. F. Wacker. 1997. Experimental results of cross polarization discrimination and signal correlation values for a polarization diversity scheme. In *Proceedings of VTC*, pp. 1498–502.

[79] T. Svantesson. 2002. Correlation and channel capacity of MIMO systems employing multi-mode antennas. *IEEE Trans. Veh. Technol.* 51(6).

[80] A. F. Molisch. 2004. A generic model for the MIMO wireless propagation channels in macro- and microcells. *IEEE Trans. Signal Processing* 52:61–71.

[81] S. K. Yong and J. S. Thompson. 2003. A three-dimensional spatial fading correlation model for uniform rectangular arrays. *IEEE Antennas Wireless Propagation Lett. 2.*

[82] M. R. Andrews, P. P. Mitra, and R. de Carvalho. 2001. Tripling the capacity of wireless communication using electromagnetic polarization. *Nature* 409:316–18.

[83] N. Prayongpun and K. Raoof. 2006. MIMO channel capacities in presence of polarization diversity with and without line-of-sight path. *J. WSEAS Trans. Commun.* 5:1744–50.

[84] I. E. Akyildiz et al. 2002. A survey on sensor networks. *IEEE Commun. Mag.* *40(8)*:102–14.

[85] C. Y. Chong and S. P. Kumar. 2003. Sensor networks: Evolution, opportunities, and challenges. *Proc. IEEE* 91:1247–56.

[86] J. Burdin and J. Dunyak. 2006. *Cohesion of wireless sensor networks with MIMO communications.* Technical report. MITRE Corp.

[87] L. Godard, J. Lienard, and K. Raoof. 2006. *Localisation des objets communicants.* Technical report. Laboratoire LIS.

[88] Q. Jiang and D. Manivarman. 2004. Routing protocols for sensor networks. In *Consumer Communications and Networking Conference*, pp. 93–98.

[89] J. Burdin and J. Dunyak. 2006. *Enhancing the performance of wireless sensor networks with MIMO communications.* Technical report. MITRE Corp.

[90] S. Cui, A. J. Goldsmith, and A. Bahai. 2004. Energy-efficiency of MIMO and cooperative MIMO techniques in sensor networks. *IEEE J. Selected Areas Commun* *22(6):1089–1098*.

5

Adaptive Modeling and Identification of Nonlinear MIMO Channels Using Neural Networks

Mohamed Ibnkahla
Queen's University

Al-Mukhtar
Al-Hinai
Queen's University

5.1 Introduction

Multiple-input multiple-output (MIMO) systems have been well looked at and investigated as a means of increasing channel capacity. The MIMO concept can offer significantly increased high data rate and capacity, with no additional bandwidth, when the channel exhibits rich scattering and its variations can be accurately tracked [1].

In order to achieve high data rates and fulfill the power requirement at the same time, MIMO communication systems may be equipped with high-power amplifiers (HPAs) [11, 14, 15, 18]. However, HPAs introduce nonlinearity to the system when operating near their nonlinear saturation regions [12]. This nonlinearity may restrict the communication

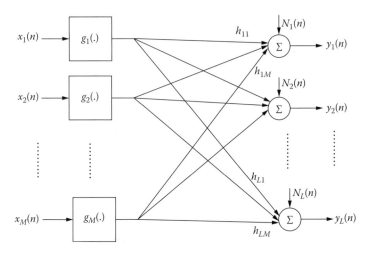

FIGURE 5.1   Nonlinear MIMO system.

system to low-order modulation schemes. In addition to the nonlinear behavior of the HPAs, a further challenge lies in the variations of the wireless fading channel [5, 17, 18].

Therefore, to reach maximum throughput, knowledge of accurate and timely channel state information (CSI) is needed by the receiver for accurate detection and demodulation.

Modeling the channel time-varying parameters and the HPA nonlinearity [5, 12] is a highly challenging task, especially when both the nonlinearity and fading parameters are unknown.

The nonlinear MIMO channel studied in this chapter is depicted in Figure 5.1. It is composed of M inputs, M memoryless nonlinearities (representing the HPAs), a linear combiner H (representing the propagation channel), and L outputs (representing the receiving antennas).

The chapter proposes a block-oriented neural network (NN) approach [4, 6] to adaptively identify the overall MIMO input-output transfer function and characterize each component of the system (i.e., the memoryless nonlinearities and the linear combiner). The proposed NN model is composed of a set of memoryless NN blocks followed by an adaptive linear combiner. Each block in the adaptive system aims at identifying the corresponding block in the unknown MIMO system.

The chapter is organized as follows. Section 5.2 presents the system to be identified and the NN algorithm. Section 5.3 presents some applications and simulation results.

5.2   System Model and Neural Network Algorithm

5.2.1   System Model

The nonlinear MIMO channel is presented in Figure 5.1. The system is assumed to be composed of M zero-mean uncorrelated inputs $x_i(n)$, $i = 1, \ldots, M$. Each input is nonlinearly transformed by a memoryless nonlinearity $g_i(.)$. The outputs of these non-linearities are then linearly combined by an $L \times M$ matrix $H = [h_{ij}]$.

The jth output of the MIMO channel can be expressed as

$$y_j(n) = \sum_{i=1}^{M} h_{ji}(n) g_i(x_i(n)) + N_j(n), \qquad (5.1)$$

where N_j is a white noise.

The system input-output relationship can be expressed in a matrix form as

$$\begin{bmatrix} y_1(n) \\ y_2(n) \\ \cdots \\ y_L(n) \end{bmatrix} = H \times \begin{bmatrix} g_1(x_1(n)) \\ g_2(x_2(n)) \\ \cdots \\ g_M(x_M(n)) \end{bmatrix} + \begin{bmatrix} N_1(n) \\ N_2(n) \\ \cdots \\ N_L(n) \end{bmatrix}. \qquad (5.2)$$

Matrix H is a propagation matrix [1] that may be time varying.

In our modeling approach, only the structure of the MIMO system is assumed known. That is, we know that the MIMO system is composed of linear memoryless blocks and a linear combining matrix, but we do not know what their values and behaviors are.

5.2.2   Neural Network Scheme

The neural network [2, 3, 5] used for modeling the nonlinear MIMO system is represented in Figure 5.2. It is composed of M neural network blocks. Each block k has a scalar input $x_k(n)$ ($k = 1, ..., M$), N neurons, and a scalar output:

$$NN_k(n) = \sum_{i=1}^{N} c_{ki} f\left(a_{ki} x_k(n) + b_{ki}\right), \quad k = 1, ..., M, \qquad (5.3)$$

where f is the NN activation function (a sigmoid transform). a_{ki}, c_{ki}, b_{ki} represent, respectively, the input weight, bias term, and output weight of the ith neuron of the kth block. The output NN_k of the kth block is connected to the jth output of the system through weight w_{jk}. The system j^{th} output is then expressed as

$$s_j(n) = \sum_{k=1}^{M} w_{jk} NN_k(n), \quad j = 1, , L. \qquad (5.4)$$

Weights w_{jk} will be put in a matrix form: $W = [w_{jk}]; j = 1, ..., L; k = 1, ..., M$.

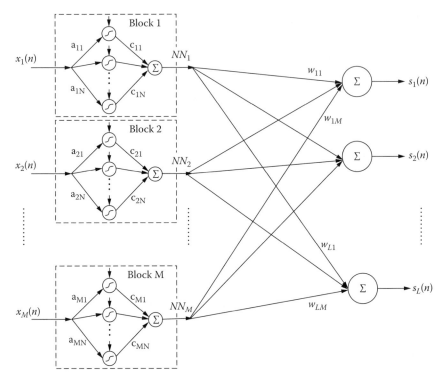

FIGURE 5.2   NN identification structure.

This can be expressed in a matrix form as

$$\begin{bmatrix} s_1(n) \\ s_2(n) \\ \text{...} \\ s_L(n) \end{bmatrix} = W \times \begin{bmatrix} NN_1(x_1(n)) \\ NN_2(x_2(n)) \\ \text{...} \\ NN_M(x_M(n)) \end{bmatrix}. \tag{5.5}$$

By choosing this structure, our goal is to characterize each part of the unknown system by the corresponding part in the modeling structure [4, 6]: each block $NN_k(.)$ would identify the corresponding nonlinearity $g_k(.)$, and matrix W would identify matrix H, to within a scaling factor for each transmitting antenna. The scaling factors can be determined by knowing the transmission antennas' average powers.

5.2.3   Learning Algorithm

In supervised learning [2, 3, 8, 9, 13], the unknown MIMO system and the proposed NN model are fed with the same input vector (Figure 5.3). At each iteration, the NN parameters are updated so that to minimize a cost function, which is defined in this

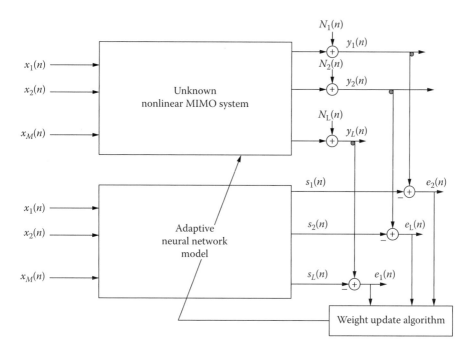

FIGURE 5.3   Adaptive system diagram.

chapter as the sum of the squared errors between the unknown system outputs and the corresponding outputs of the model:

$$J(n) = \sum_{j=1}^{L} \left(e_j(n)\right)^2,$$ (5.6)

where $e_j(n) = y_j(n) - s_j(n)$.

A gradient descent approach is used in this chapter for the updating equations:

$$w_{jk}(n+1) = w_{jk}(n) - \mu \frac{\partial J(n)}{\partial w_{jk}(n)} = w_{jk}(n) + 2\mu e_j(n) NN_k(n),$$ (5.7)

$$c_{ki}(n+1) = c_{ki}(n) - \mu \frac{\partial J(n)}{\partial c_{ki}(n)} = c_{ki}(n) + 2\mu f\left(a_{ki}x_k(n) + b_{ki}\right) \sum_{l=1}^{L} w_{lk} e_l(n),$$ (5.8)

$$a_{ki}(n+1) = a_{ki}(n) - \mu \frac{\partial J(n)}{\partial a_{ki}(n)} = a_{ki}(n) + 2\mu c_{ki}x_k(n) f'\left(a_{ki}x_k(n) + b_{ki}\right) \sum_{l=1}^{L} w_{lk} e_l(n),$$ (5.9)

$$b_{ki}(n+1)=b_{ki}(n)-\mu\frac{\partial J(n)}{\partial b_{ki}(n)}=b_{ki}(n)+2\mu c_{ki}f'\left(a_{ki}x_k(n)+b_{ki}\right)\sum_{l=1}^{L}w_{lk}e_l(n), \quad (5.10)$$

where μ is a small positive constant.

The adaptive system is initialized with a set of small random-weight values. The learning curve is defined as the evolution of the mean squared error (MSE) during the learning process. In modeling fixed (i.e., static) MIMO systems, the MSE error starts to decrease until the system reaches a steady state and only slight changes in the weights are observed. In that case, the learning process may be stopped. However, if the MIMO system is time varying, then the learning process should be maintained in order to keep tracking the time-varying parameters.

We have chosen to compare our results to the classical multilayer perceptron (MLP) [2]. The MLP is composed of M inputs, a number of nonlinear neurons in the first layer, followed by L linear combiners (Figure 5.4).

The MLP is trained using the backpropagation algorithm [2, 3]. It allows a black-box modeling [8, 9, 13] of the unknown MIMO system, i.e., the modeling of the overall nonlinear MIMO input-output transfer function, without being able to characterize the different parts of the unknown system (i.e., the memoryless nonlinearities and the combining matrix H).

For a fair comparison between the MLP structure and the proposed NN block structure, we will take the same total number of neurons in each scheme. Therefore, for the block structure, if we denote the number of neurons in each block by N, then the total number of neurons will be equal to MN. This structure will be compared to an MLP with MN neurons.

Table 5.1 compares the computational complexity of each algorithm:

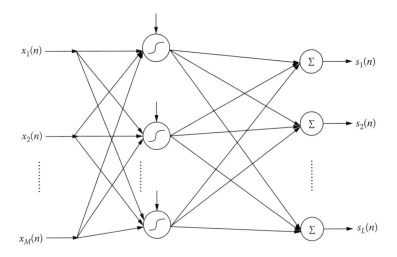

FIGURE 5.4   MLP structure.

TABLE 5.1

	Parameters	Sigmoid Transforms	Multiplications	Additions
Proposed NN	$M(3N + L)$	MN	$M(2N + L)$	$M(2N + L + 1)$
MLP	$MN(M + L + 1)$	MN	$MN(M + L)$	$MN(M + L + 1)$

The number of sigmoid transforms is the same, since we have the same number of neurons in the two structures. However, the MLP requires much more multiplications and additions. This is because in the MLP structure, all inputs are connected to all first-layer neurons (MN neurons in total), whereas in the block structure, each input is connected to one block that is composed of N neurons only.

5.3 Applications and Simulation Results

5.3.1 Modeling and Identification of MIMO Transmitters with Nonlinear Amplifiers and RF Coupling Interference

In this section, we apply the proposed NN for modeling and identification of MIMO RF transmitters that are equipped with nonlinear HPAs. Figure 5.5 shows the coupling terms generated by RF interference [10, 16]. In this application, the unknown nonlinear HPA transfer functions to be identified are taken from a family of nonlinear functions of the form

$$g_i(x) = \alpha_i x \exp\left(\frac{-\beta_i x^2}{2}\right),$$

where α_i and β_i are positive constants.

For the simulations presented in this section, we have taken the following parameters: $M = L = 2$, $\alpha_1 = \alpha_2 = 1$, $\beta_1 = 1$, $\beta_2 = 2$. For each output, the noise is taken as white Gaussian with variance σ_i^2. For the simulation, the inputs were zero-mean white Gaussian processes with unit variance.

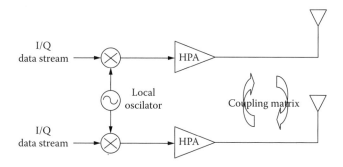

FIGURE 5.5 Example of a two-dimensional MIMO RF transmitter, including the coupling matrix between antennas.

TABLE 5.2

	Parameters	Sigmoid Transforms	Multiplications	Additions
Proposed NN	34	10	24	26
MLP	50	10	40	50

The coupling matrix simulated in this example was taken as

$$H = \begin{bmatrix} 1 & 0.3 \\ 0.3 & 1 \end{bmatrix}.$$

In this example, the proposed block-oriented NN scheme was composed of two blocks of $N = 5$ neurons each. The total number of neurons is then equal to 10. We have compared this scheme to an MLP structure composed of ten neurons. The *erf* function has been taken for the activation function.

The complexity of the algorithms is displayed in Table 5.2.

As can be seen in the table, the number of multiplications and additions in the MLP structure is almost twice that of the proposed block structure.

We have tested both algorithms for a range of μ values belonging to the interval $[10^{-5}\ 10^{-2}]$ under various initial conditions. The proposed block structure has always outperformed the MLP.

Figure 5.6 shows the learning curves for the block structure and the MLP for $\mu = 0.005$. The proposed approach shows lower MSE and faster convergence speed.

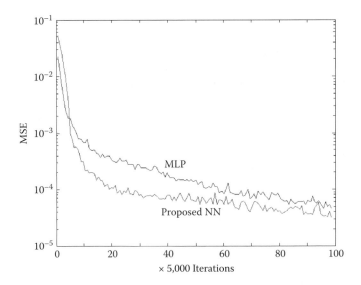

FIGURE 5.6   Smoothed MSE curves (an averaging window of 1,000 samples has been used). $\mu = 0.005$.

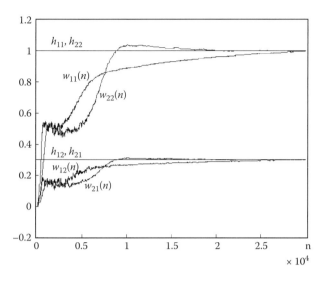

FIGURE 5.7   Evolution of the weights (normalized) during the learning process.

Moreover, the block structure allows excellent characterization of each of the nonlinearities as well as the coupling matrix H, to within scaling factors. That is, $NN_1(x)$ has converged to $\gamma_1 g_1(x)$, $NN_2(x)$ has converged to $\gamma_2 g_2(x)$, and W has converged to

$$\begin{bmatrix} \dfrac{1}{\gamma_1} & 0 \\ 0 & \dfrac{1}{\gamma_2} \end{bmatrix} \times H,$$

where γ_1 and γ_2 are real constants.

It is easy to show that if the output of block k (i.e., $NN_k(x_k)$) is multiplied by a given factor γ_k, and if the corresponding weights $\{w_{ki}\}$, $i = 1, ..., L$ are multiplied by the inverse of that factor, i.e., $1/\gamma_k$, then the overall NN system transfer function remains unchanged (equation (5.5)). This reveals that the adaptive system has several equivalent stationary points. The values of γ_k can be determined if the amplifiers' output average powers or operating points are given.

Figure 5.7 shows the transient behavior of matrix W weights (after normalization with the scaling factors). The scaling factors here were $\gamma_1 = -1.1246$ and $\gamma_2 = -0.9319$. Figures 5.8 and 5.9 show that the memoryless nonlinearities have been successfully modeled by our NN approach.

5.3.2   Application to Fault Detection

In this section, the proposed scheme is applied to fault detection in nonlinear MIMO systems.

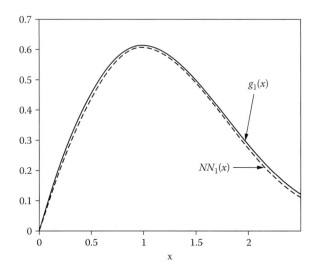

FIGURE 5.8   Unknown function $g_1(x)$ and normalized $NN_1(x)$.

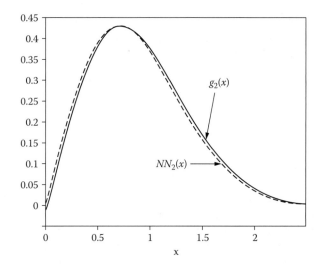

FIGURE 5.9   Unknown function $g_2(x)$ and normalized $NN_2(x)$.

Here we consider the case studied in section 5.3.1 (modeling MIMO RF transmitters). We have simulated an abrupt change in $g_1(x)$. The change occurred during the learning process at the 12,500th iteration. $g_2(x)$ and the other parameters of the MIMO system were unchanged.

The learning curves are shown in Figure 5.10 for the block NN structure and MLP. As expected, the MSE error increases at the time of change, then decreases. Here again, the block NN structure performs better than the MLP. Figure 5.11 shows that the block structure has correctly detected and characterized the change in $g_1(x)$. $g_2(x)$ and H (not

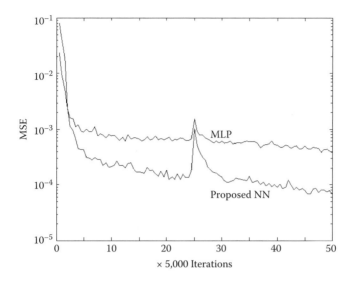

FIGURE 5.10 Fault detection: learning curves.

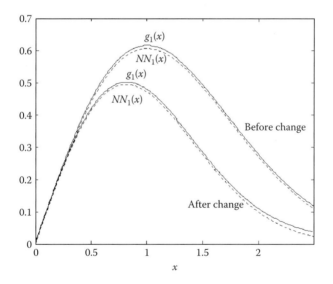

FIGURE 5.11 $g_1(x)$ and $NN_1(x)$ (normalized) before and after the change.

shown in the figure) were correctly identified as well. We have studied other MIMO systems with higher dimensions (3×3 and 4×4 systems) in which we simulated changes in more than one nonlinearity; the proposed NN model was always capable of characterizing the changes.

On the other hand, the black-box MLP structure could not characterize the change. By looking at the MLP learning curve (Figure 5.10), one could only predict that "something unusual" happened to the unknown system (or to the backpropagation algorithm) at iteration 12,500, without any further indication of what exactly happened during the learning process.

5.3.3 Tracking of Slowly Time-Varying Propagation Channels

This section deals with another application in which matrix H is slowly time varying.

In this application, we consider a multiple antenna transmitter/receiver system (Figure 5.1), with $M = L = 2$. The propagation channel H coefficients are modeled as slowly time-varying Ricean fading gains with a Ricean factor $K = 5$, and a normalized Doppler frequency of 0.0001.

Figure 5.12 shows the learning curves of the block structure and MLP. It can be noticed that the MSE errors here are higher than those of the static case studied in section 5.3.1 (Figure 5.6). This is because of the fact that gradient descent algorithms do not provide excellent tracking capabilities in time-varying environments.

As shown in Figure 5.12, our approach is faster and yields lower MSE than the MLP. Moreover, the proposed structure allows good identification of the unknown nonlinearities (Figures 5.13 and 5.14). The time-varying coefficients were correctly tracked by matrix W. An illustration is made for $w_{11}(n)$ versus $h_{11}(n)$ (Figure 5.15).

5.3.4 Nonlinear MIMO Channel Receiver Design

This section applies the NN identification scheme to MIMO receiver design [17, 18]. A V-BLAST (Vertical Bell Laboratories Layered Space-Time) receiver is proposed. It

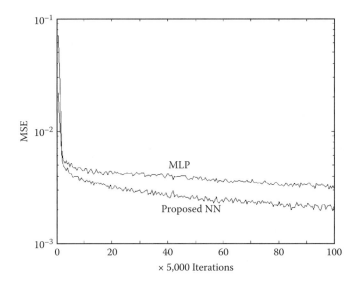

FIGURE 5.12 Learning curves: time-varying case.

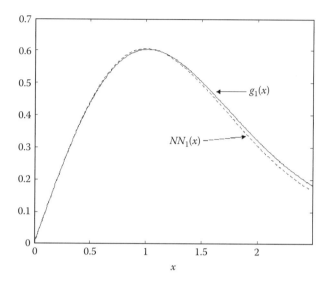

FIGURE 5.13 $g_1(x)$ and $NN_1(x)$ (normalized): time-varying case.

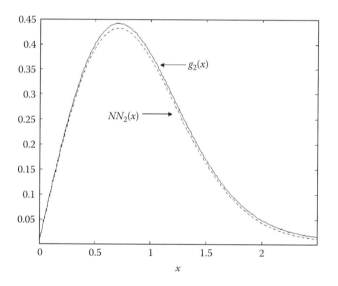

FIGURE 5.14 $g_2(x)$ and $NN_2(x)$ (normalized): time-varying case.

consists of a NN MIMO channel estimator (NNCE) and a ZF V-BLAST (zero-forcing V-BLAST) detection algorithm [1, 7, 11]. The NNCE performs an online estimation of the nonlinear MIMO channel. The estimated channel state information (CSI) is provided to the ZF V-BLAST detection algorithm, which gives an estimation of the transmitted symbols (Figure 5.16).

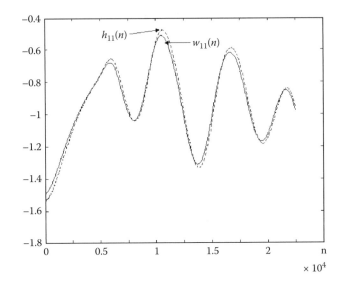

FIGURE 5.15 Identification of the time-varying channel: $h_{11}(n)$ and normalized $w_{11}(n)$.

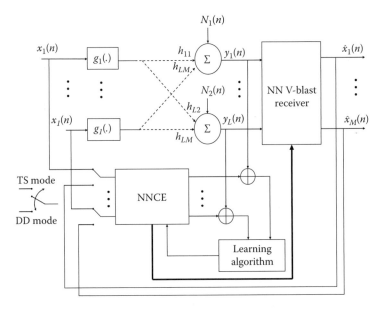

FIGURE 5.16 NN-based V-BLAST receiver.

The NNCE operation consists of two modes: training sequence (TS) mode and decision-directed (DD) mode. In the TS mode the amplitudes of the transmitted signals are used to train the NNCE. The TS mode continues until the NNCE is fully trained, then the NNCE switches to the DD mode. In the DD, the amplitudes of the detected signals $(\hat{x}_k(n), k = 1, ..., M)$ are used to train the NNCE.

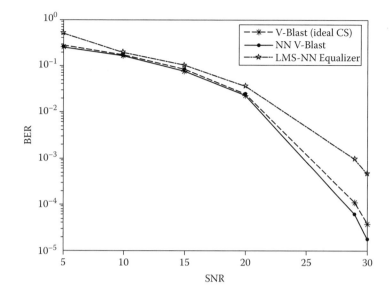

FIGURE 5.17   BER performance versus SNR.

Here the transmitted signals are 16-QAM modulated, and $M = L = 2$. The unknown propagation channel is modeled as time-varying Rician fading gains with a Rician factor $K = 5$ and normalized Doppler frequency of $f_D = 10^{-4}$. For the modeling and identification part, the noise is taken as white Gaussian with variance 0.001. However, for the detection part, the noise is varied depending on the channel signal-to-noise ratio (SNR). Each subnetwork in each block is composed of $N = 5$ neurons. It is assumed in the simulations that the HPA transmission powers are known to the receiver. The receiver is trained using a training sequence of fifty thousand transmitted symbols, after which the DD mode is activated.

Figure 5.17 shows the BER performance of our NN V-BLAST receiver. The proposed receiver achieves a bit error rate (BER) of 10^{-4} at an SNR of 29.2 dB. The NN V-BLAST receiver is compared to a least mean squares (LMS) tracker following a memoryless neural network (LMS-NN) equalizer [6]. It is composed of a linear combiner, W^{-1}, acting as an inverter to the channel matrix H, and a NN structure, acting as an inverter to the channel nonlinearities. The general structure of the LMS-NN equalizer is shown in Figure 5.18. BER simulation results show that our NN V-BLAST receiver outperforms the LMS-NN equalizer in terms of BER. We can also see that the performance of the NN V-BLAST receiver is close to that of the ideal V-BLAST receiver, which assumes perfect channel knowledge. This should not be surprising, since the different parts of the channel have been well identified.

5.4   Conclusion

The chapter proposed an NN approach for modeling nonlinear MIMO channels composed of a set of memoryless nonlinearities followed by a linear combining channel. We

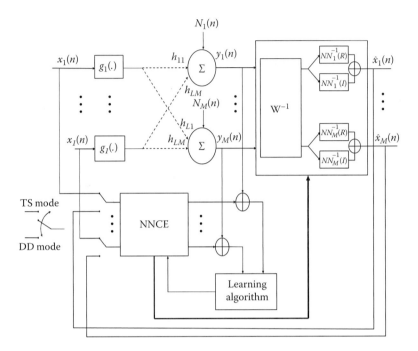

FIGURE 5.18   LMS-NN equalizer.

followed a block-oriented NN approach for the MIMO system modeling. The NN model allows excellent modeling and identification of the different parts of the unknown MIMO system; the learning process is being made by using the MIMO system input-output data. The proposed approach outperforms the black-box MLP method in terms of computational complexity, convergence speed, and MSE performance. The proposed scheme has been successfully applied to tracking of slowly time-varying MIMO channels, detection and characterization of changes in MIMO channels, and receiver design for nonlinear Rician MIMO channels.

References

[1] H. Bolcskei. 2004. MIMO systems: Principles and trends. In *Signal processing for mobile communications handbook*, ed. M. Ibnkahla, chap. 12. Boca Raton, FL: CRC Press.

[2] S. Haykin. 1999. *Neural networks: A comprehensive foundation*. Englewood Cliffs, NJ: Prentice Hall.

[3] S. Haykin. 1996. *Adaptive filter theory*. Englewood Cliffs, NJ: Prentice Hall.

[4] M. Ibnkahla, N. J. Bershad, J. Sombrin, and F. Castanié. 1998. Neural network modeling and identification of non-linear channels with memory: Algorithms, applications and analytic models. *IEEE Trans. Signal Processing* 46(5):1208–1220.

[5] M. Ibnkahla, Q. Rahman, A. Sulyman, H. Al-Asady, Y. Jun, and A. Safwat. 2004. High speed satellite mobile communications: Technologies and challenges. *Proc. IEEE* 92(2):312–39.

[6] M. Ibnkahla. 2000. Applications of neural networks to digital communications: A survey. *Signal Processing* 80:1185–215.

[7] T. Javornik, G. Kandus, S. Plevel, G. White, and A. Burr. 2003. V-BLAST algorithm performance in non-linear channel. In *IEEE Computer as a Tool Conference (EUROCON)*, vol. 1, pp. 183–87.

[8] L. Ljung. 1999. *System identification: Theory for the user.* Englewood Cliffs, NJ: Prentice Hall.

[9] K. S. Narendra and K. Parthasarathy. 1990. Identification and control of dynamical systems using neural networks. *IEEE Trans. Neural Networks* 1:4–27.

[10] J. Pedro and S. Maas. 2005. A comparative overview of microwave and wireless power amplifier behavioral modeling approaches. *IEEE Trans. Microwave Theory Tech.* 53:1150–63.

[11] G. Poitau and A. Kouki. 2004. Impact of realistic amplification models on dynamic VBLAST optimization. In *Proceedings of the IEEE Vehicular Technology Conference*, pp. 894–97.

[12] A. Saleh. 1981. Frequency-independent and frequency-dependent nonlinear models of TWT amplifiers. *IEEE Trans. Commun.* 29(11).

[13] J. Sjoberg et al. 1995. Nonlinear black box modeling in system identification: A unified overview. *Automatica* 31:1691–724.

[14] A. I. Sulyman and M. Ibnkahla. 2004. Performance analysis of nonlinearly amplified M-QAM signals in MIMO channels. In *Proceedings of IEEE ICASSP'04*, Montreal, Canada.

[15] A. I. Sulyman and M. Ibnkahla. 2005. Performance of space-time codes over nonlinear MIMO channels. In *Proceedings of the International Symposium on Signal Processing and Its Applications (ISSPA'05)*, Sydney, Australia.

[16] S. Woo, D. Lee, K. Kim, H. Hur, C. Lee, and J. Laskar. 2005. Combined effects of RF impairments in the future IEEE 802.11n WLAN systems. In *Proceedings of the IEEE Vehicular Technology Conference*, vol. 2, pp. 1346–49.

[17] S. Yang, J. Xi, and X. Mu. 2005. Decision aided joint compensation of clipping noise and nonlinearity for MIMO-OFDM systems. In *Proceedings of the IEEE International Symposium on Communications and Information Technology ISCIT 2005*, vol. 1, pp. 725–28.

[18] S. Yang, J. Xi, F. Wang, X. Mu, and H. Kobayashi. 2005. Decision aided compensation of residual frequency offset for MIMO-OFDM systems with nonlinear channel. In *Proceedings of the International Symposium on Intelligent Signal Processing and Communication Systems ISPACS 2005*, pp. 113–16.

6

Joint Adaptive Transmission and Switched Diversity Reception

Hong-Chuan Yang
University of Victoria

Young-Chai Ko
Korea University

Haewoon Nam
Motorola, Inc.

Mohamed-
Slim Alouini
*Texas A&M University
(TAMU)-Qatar*

6.1 Introduction

Future wireless communication systems have to provide multimedia services to battery-operated portable terminals. These systems must efficiently utilize the limited bandwidth and power resources to enable high-data-rate high-reliability transmission over wireless fading channels. Adaptive modulation and diversity combining are two of the most important enabling techniques for future wireless communication systems. Adaptive modulation can achieve high spectral efficiency over wireless channels [1–3]. The basic idea of adaptive modulation is to match the modulation parameters, such as constellation size and coding rate, to the prevailing fading channel conditions while maintaining the instantaneous error rate below a target value. Usually, the modulation mode is chosen based on the comparison results of received signal strength with

153

several predetermined thresholds. Diversity combining, on the other hand, can improve the reliability of wireless fading channels by appropriately combining differently faded information-bearing signals [4]. While adaptive modulation can certainly benefit from diversity combining through the improved channel quality, the design and analysis of these two techniques were usually carried out separately.

From a performance perspective, the optimal combining solution is the well-known maximum ratio combining (MRC) [5, chapter 12]. With MRC, the combiner output is obtained by coherently summing up the optimally weighted signal replicas from all diversity paths. Meanwhile, the implementation complexity of MRC is high, especially when the number of diversity paths is large. Indeed, MRC not only requires the same number of RF chains as the number of available diversity paths, but also mandates the complete and simultaneous knowledge of the channel condition for each diversity path. To reduce the implementation complexity of the MRC diversity combiner, generalized selection combining (GSC) was proposed and extensively studied (see, for example, [6–11] and references therein). The receiver with GSC combines a fixed number of best diversity paths as per the rules of the optimal MRC scheme. As one can intuitively expect, diversity combining techniques such as MRC and GSC improve the spectral efficiency of adaptive transmission systems by allowing a higher-order transmission mode. On the other hand, these combining schemes operate rather independently of the adaptive transmission system.

Recently, there has been a growing interest in adaptive combining schemes for processing power-saving purposes [12–17]. The basic idea of these power-saving schemes is to use the diversity combiner resource adaptively in a way that the output signal satisfies a certain quality requirement. For example, minimum selection GSC (MS-GSC) was proposed by Kim et al. [12] as a power-saving implementation of the GSC scheme. With MS-GSC, the receiver combines the least number of best branches such that the combined SNR exceeds a certain predetermined threshold. It has been shown [13–15] that MS-GSC can save a considerable amount of processing power by keeping fewer branches active on average, while still providing nearly the same performance as MRC. Output-threshold MRC (OT-MRC) is another adaptive combining scheme [17], which can further reduce the number of path estimations by sequentially combining additional paths if necessary. It is interesting to note that both adaptive modulation and adaptive combining concepts utilize some predetermined threshold in their operation. Very recently, based on this observation, two joint adaptive modulation and diversity combining designs, where both the transmission mode and combiner structure are adaptively determined based on the fading channel condition, have been proposed and analyzed [18, 19].

We note at this point that both MRC- and GSC-based diversity combining schemes require the implementation of multiple RF chains. While this may not be an issue for conventional cellular or wireless LAN systems, it certainly becomes very challenging or even prohibitive for future generations of wireless personal area network (PAN) systems operating over 60 GHz bands with millimeter (MM)-wave technology, which are being currently standardized by the IEEE 802.15.3c group. Only until very recently, RF-CMOS technology was able to produce chips operating over the 60 GHz frequency range, but still with considerable high cost. On the other hand, implementing multiple MM-wave antennas at the receiver is achievable at a reasonable price. In the resulting multiple antennas with a single RF chain receiver scenario, only traditional selection or switched combining schemes can be employed to explore the diversity benefit. In this chapter, we complement the previous work on joint adaptive transmission and diversity combining by extending the design and analysis in [18, 19] to the lower-complexity selection and switched combining schemes. The proposed joint design can also apply to a transmit diversity system to achieve high efficiency with low feedback load and no power spreading loss.

The selection combining (SC) scheme [4, section 6.2] has much lower complexity than MRC. The receiver with SC uses only the best diversity paths of all available ones. This nevertheless requires the estimation and comparison of the quality indicators, such as the signal-to-noise ratio (SNR), of all diversity paths. With switched combining, where switch-and-stay combining (SSC) is the popular example [20–27], the complexity is further reduced by eliminating the need for simultaneously checking the path quality of all diversity paths. In fact, SSC, where the combiner switches to the other branch only when the output SNR of the current branch is below a certain threshold, can be viewed as the origin of aforementioned adaptive combining schemes such as MS-GSC and OT-MRC. Recent applications (such as transmit diversity, for example) have also motivated studies of multibranch switch-and-examine combining (SEC) [28, 29]. In particular, with SEC, if the current path is not of acceptable quality, then the combiner switches and examines the quality of the next available path. This switching-and-examining process is repeated until either an acceptable path is found or all available diversity paths have been examined. In the latter case, the combiner either settles on the last examined path [28] or connects to the receiver the path with the best quality among all examined paths, and this SEC variant is known as SEC with post-examine selection (SECps) [29].

In this chapter, we study joint adaptive transmission and diversity combining with SC, SEC, and SECps schemes. We first examine the spectral efficiency benefit provided by the SC scheme for adaptive transmission systems. Then, capitalizing on the adaptive nature of switched combining, we develop more integrated joint designs based on the SEC and SECps schemes. In particular, the resulting system jointly selects the most appropriate transmission mode and diversity paths based on the current channel conditions and the desired bit error rate (BER) requirement. Depending on the primary objective of the joint design, we arrive at a scheme that requires a minimum number of path estimations (termed as minimum estimation scheme) and a scheme with a high bandwidth efficiency (termed as bandwidth-efficient scheme), both of which satisfy the desired BER requirement. For both schemes under consideration, we quantify through accurate analysis their processing complexity (quantified in terms of average number of paths estimated), spectral efficiency (quantified in terms of average number of transmitted bits/s/Hz), and performance (quantified in terms of average BER). Finally, some selected numerical examples are presented to illustrate the mathematical formalism.

The rest of this chapter is organized as follows. Section 6.2 contains the general description of the system and channel model under consideration. While section 6.3 addresses the design and analysis of SC-based joint adaptive modulation and diversity combining schemes, section 6.4 is dedicated to the operation and performance of switched combining–based schemes. Finally, section 6.5 provides some concluding remarks. In all sections, selected numerical examples are provided together with some related discussions and interpretations.

6.2   System and Channel Models

In this section, we first present the adaptive modulation scheme adopted in this work. After discussing the discrete-time implementation model for the proposed transmission system, we introduce the block-fading channel model under consideration.

6.2.1   Adaptive Modulation

We adopt the constant-power variable-rate uncoded M-ary quadrature amplitude modulation (M-QAM) scheme studied in [2].* With this adaptive modulation scheme, the mode selection is solely based on the fading channel condition. In particular, the SNR range is divided into N regions and the constellation size $M = 2^n$ is used during the subsequent data burst transmission if the output SNR of diversity combiner ends up being in the nth region $[\gamma_{T_n}, \gamma_{T_{n+1}})$, where $n = 2, 3, \cdots, N$. Note that in this case, n bits are carried by each symbol. The region boundaries, denoted by γ_{T_n}, are determined such that the instantaneous BER for the chosen constellation is below a certain required value, denoted by BER_0. More specifically, to meet a BER requirement of 0.1%, the thresholds can be set by solving the following equations:

$$\gamma_{T_n} = \text{BER}_n^{-1}(0.001), \tag{6.1}$$

where $\text{BER}_n^{-1}(\cdot)$ is the inverse BER expression. After applying a numerical method to solve (6.1) with the exact BER expression for square M-QAM given in [30], we obtain the exact E_s/N_0 values for the thresholds, γ_{T_n}, for the instantaneous BER requirement of 10^{-2}, 10^{-3}, and 10^{-4} cases, and these values are summarized in Table 6.1. Note that γ_{T_1} is equal to 0 or $-\infty$ dB, while $\gamma_{T_{n+1}} = +\infty$.

TABLE 6.1   SNR Thresholds (E_s/N_0) at Each Modulation Level to Satisfy 1, 0.1, and 0.01% Bit Error Rate for 2_n-ary QAM, Respectively

n	γ_{T_n} (dB) for $\text{BER}_0 = 10^{-2}$	γ_{T_n} (dB) for $\text{BER}_0 = 10^{-3}$	γ_{T_n} (dB) for $\text{BER}_0 = 10^{-4}$
2	7.33	9.64	11.35
3	11.84	13.32	16.07
4	13.90	16.63	18.23
5	17.83	19.79	22.29
6	19.73	22.86	24.30
7	23.85	25.91	28.24
8	25.43	28.94	30.23

---

* Note that the design can be easily extended to include the coding scheme, in which case we should use the proper SNR thresholds for different modulation and coding scheme combinations.

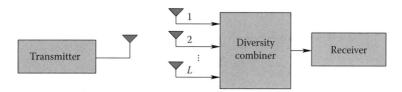

FIGURE 6.1 Block diagram of a receiver space diversity system.

6.2.2 System Model

In this work, we consider a receiver space diversity system as shown in Figure 6.1. In particular, L antennas are used at the receiver to create differently faded replicas of the transmitted signal. Because of the hardware complexity constraint that is imposed by MM-wave systems, we assume that the receiver can only process a single selected diversity branch. We adopt a discrete-time implementation for the proposed transmission system. More specifically, short guard periods are periodically inserted into the transmitted signal. During these guard periods, the receiver performs a series of operations, including diversity path estimations and their comparisons, in order to select the appropriate diversity branch and the suitable adaptive modulation mode to be used during the subsequent data burst reception. Once these decisions are made at the receiver, the adaptive modulation mode is fed back to the transmitter via an error-free reverse channel before the guard period ends. After that, the transmitter and receiver are configured accordingly throughout the subsequent data burst transmission.

6.2.3 Channel Model

We assume that the length of the guard period plus data burst is of the order of the channel coherence time, and therefore, the faded signal amplitude remains constant during each guard period–data burst pair and de-correlates after that. We also assume that the received signal on each antenna branch experiences an independent and identically distributed (i.i.d.) fading process. As such, the faded SNR, denoted by γ_i, $i = 1, \cdots, L$, on each diversity branch shares a common probability density function (PDF) and cumulative distribution function (CDF). In Table 6.2, we summarize the PDF, $p_\gamma(x)$, and the CDF, $F_\gamma(x)$, of the received SNRs under three popular fading models: Rayleigh, Rice, and Nakagami-m. In Table 6.2, $\bar{\gamma}$ is the average SNR, $\Gamma(\cdot)$ is the Gamma function [31, section 8.31], $I_0(\cdot)$ is the modified Bessel function of the first kind with zero order [31, section 8.43], $\Gamma(\cdot,\cdot)$ is the incomplete Gamma function [31, section 8.35], and $Q_1(\cdot,\cdot)$ is the first-order Marcum Q-function [32].

6.3 Joint Adaptive Modulation and Selection Combining

In this section, we consider the analysis of a joint adaptive modulation and selection combining system. We first present the mode of operation of the transmission system

TABLE 6.2 Statistics of the Faded SNR for the Three Fading Models under Consideration

Model	Rayleigh	Rice/Nakagami-n	Nakagami-m
Parameter	.	$K = n^2 \geq 0$	$m \geq \frac{1}{2}$
PDF ($p_\gamma(x)$)	$\dfrac{1}{\bar{\gamma}} e^{-\frac{x}{\bar{\gamma}}}$	$\dfrac{(1+n^2)e^{-n^2}}{\bar{\gamma}} e^{-\frac{1+n^2}{\bar{\gamma}}x} I_0\left(2n\sqrt{\dfrac{1+n^2}{\bar{\gamma}}x}\right)$	$\left(\dfrac{m}{\bar{\gamma}}\right)^m \dfrac{x^{m-1}}{\Gamma(m)} e^{-\frac{mx}{\bar{\gamma}}}$
CDF ($F_\gamma(x)$)	$1-e^{-\frac{x}{\bar{\gamma}}}$	$1-Q_1\left(n\sqrt{2},\sqrt{\dfrac{2(1+n^2)}{\bar{\gamma}}}x\right)$	$1-\dfrac{\Gamma\left(m,\dfrac{mx}{\bar{\gamma}}\right)}{\Gamma(m)}$

during guard periods. The statistics of the received SNR for the SC-based joint design are then derived before this result is applied to the performance and spectral efficiency analysis.

6.3.1 Mode of Operation and Statistics

During each guard period, the receiver performs the following operations. First, the receiver estimates the faded SNR of each antenna branch and selects the best branch after proper comparisons. The combiner output SNR is then given by $\gamma_c = \max[\gamma_1, \gamma_2, ..., \gamma_L]$. After that, the receiver determines the appropriate modulation scheme by sequentially comparing the output SNR γ_c with the threshold γ_{T_n}, $n = N, N-1, \cdots, 2$. Whenever the receiver finds that the output SNR is smaller than $\gamma_{T_{n+1}}$ but greater than γ_{T_n}, it selects the modulation mode n for the subsequent data burst and feeds back that mode index to the transmitter. If the combined SNR of all L available branches is below γ_{T_2}, the receiver may ask the transmitter to either (1) transmit using the lowest modulation mode, i.e., QPSK, in violation of the target instantaneous BER requirement (option 1), or (2) buffer the data and wait until the next guard period for more favorable channel conditions (option 2). Note that option 1 is useful in supporting delay-sensitive traffic with hard delay constraint, especially when coupled with a powerful upper-layer error correction mechanism.

Based on the mode of operation described above, we can see that the received SNR, γ_c, of the SC-based scheme is the same as the output SNR of a traditional L branch selection combiner. In other words, the CDF of the received SNR, $F_{\gamma_c}(\cdot)$, is given by

$$F_{\gamma_c}(\gamma) = \begin{cases} F_{\gamma_c}^{SC}(\gamma), & \text{for option 1;} \\ F_{\gamma_c}^{SC}(\gamma), & \gamma > \gamma_{T_2} \\ F_{\gamma_c}^{SC}(\gamma_{T_2}), & 0 < \gamma \leq \gamma_{T_2} \end{cases} \quad \text{for option 2,} \tag{6.2}$$

where $F_{\gamma_c}^{SC}(\cdot)$ denotes the CDF of the combined SNR with L branch SC, which is given for i.i.d. fading environment by

$$F_{\gamma_c}^{SC}(\gamma) = \left(F_\gamma(\gamma)\right)^L. \tag{6.3}$$

Correspondingly, the PDF of the received SNR, $p_{\gamma_c}(\cdot)$, is given by

$$p_{\gamma_c}(\gamma) = \begin{cases} p_{\gamma_c}^{SC}(\gamma), & \text{for option 1;} \\ p_{\gamma_c}^{SC}(\gamma)\mathcal{U}\left(\gamma - \gamma_{T_2}\right) + F_{\gamma_c}^{SC}(\gamma_{T_2})\delta(\gamma), & \text{for option 2,} \end{cases} \tag{6.4}$$

where $\mathcal{U}(\cdot)$ is the Heaviside unit step function, $\delta(\cdot)$ is the Dirac Delta function and $p_{\gamma_c}^{SC}(\cdot)$ denotes the PDF of the combined SNR with L branch SC, which is given for the i.i.d. fading environment by

$$p_{\gamma_c}^{SC}(\gamma) = L\left(F_\gamma(\gamma)\right)^{L-1} p_\gamma(\gamma). \tag{6.5}$$

6.3.2 Performance and Efficiency Analysis

6.3.2.1 Average Spectral Efficiency

The average spectral efficiency of an adaptive modulation system can be calculated as [2, equation (33)]

$$\eta = \sum_{n=2}^{N} n \, p_n,$$

where p_n is the probability that the nth constellation is used. For the SC-based system under consideration, it can be shown that p_n is given by

$$p_n = \begin{cases} \begin{cases} F_{\gamma_c}^{SC}\left(\gamma_{T_{n+1}}\right) - F_{\gamma_c}^{SC}\left(\gamma_{T_n}\right), & n \geq 3; \\ F_{\gamma_c}^{SC}\left(\gamma_{T_3}\right), & n = 2 \end{cases} & \text{for option 1;} \\ F_{\gamma_c}^{SC}\left(\gamma_{T_{n+1}}\right) - F_{\gamma_c}^{SC}\left(\gamma_{T_n}\right), & \text{for option 2,} \end{cases} \tag{6.6}$$

Therefore, the average spectral efficiency of the SC-based scheme is given by

$$\eta = \begin{cases} N - \sum_{n=3}^{N} F_{\gamma_c}^{SC}\left(\gamma_{T_n}\right), & \text{for option 1;} \\ N - \sum_{n=2}^{N} F_{\gamma_c}^{SC}\left(\gamma_{T_n}\right), & \text{for option 2.} \end{cases} \tag{6.7}$$

6.3.2.2   Average Error Rate

The average BER for an adaptive modulation system can be calculated as [2, equation (35)]

$$\langle BER \rangle = \frac{1}{\eta} \sum_{n=2}^{N} n \, \overline{BER}_n, \tag{6.8}$$

where $\overline{BER}_n$ is the average error rate for constellation n, and which is given by

$$\overline{BER}_n = \int_{\gamma_{T_n}}^{\gamma_{T_{n+1}}} BER_n(\gamma) \, p_{\gamma_c}^{SC}(\gamma) d\gamma, \tag{6.9}$$

where $BER_n(\gamma)$ is the conditional BER of the constellation n over an additive white Gaussian noise (AWGN) channel given that its SNR is equal to γ, which is given in [30] for uncoded M-QAM. Therefore, the average BER of the SC-based scheme under consideration can be calculated as

$$\langle BER \rangle = \begin{cases} \dfrac{\displaystyle\int_0^{\gamma_{T_3}} BER_2(\gamma) \, p_{\gamma_c}^{SC}(\gamma) \, d\gamma + \sum_{n=3}^{N} n \int_{\gamma_{T_n}}^{\gamma_{T_{n+1}}} BER_n(\gamma) \, p_{\gamma_c}^{SC}(\gamma) \, d\gamma}{N - \displaystyle\sum_{n=3}^{N} F_{\gamma_c}^{SC}\left(\gamma_{T_n}\right)}, & \text{for option 1;} \\[4ex] \dfrac{\displaystyle\sum_{n=2}^{N} n \int_{\gamma_{T_n}}^{\gamma_{T_{n+1}}} BER_n(\gamma) \, p_{\gamma_c}^{SC}(\gamma) \, d\gamma}{N - \displaystyle\sum_{n=2}^{N} F_{\gamma_c}^{SC}\left(\gamma_{T_n}\right)}, & \text{for option 2.} \end{cases} \tag{6.10}$$

6.3.3   Numerical Examples

Figure 6.2 plots the average spectral efficiency of the SC-based scheme with both options and for different numbers of receive antennas L. As we can see, as the number of receive antennas increases, the average spectral efficiency of the system improves, but with diminishing gain. If we compare the average spectral efficiency of the two options with the same number of antennas, we can see that for a high-SNR region, both options have nearly the same spectral efficiency, whereas for a low-SNR region, option 1 offers higher efficiency than option 2. This is because when the channel condition is poor, option 2 will buffer the data while option 1 continues transmission at the cost of violating the BER constraint.

The BER constraint violation can be immediately observed from Figure 6.3, where we plot the average error rate of the SC-based scheme with both options and different numbers of receive antenna L. The target BER is set to be 10^{-3} in this figure. We also observe that when the average SNR is greater than a certain value (10 dB for the $L = 4$ case), the average BER of the SC-based scheme with option 1 will always be smaller than the target BER of 10^{-3}. When the number of receive antennas increases, the SNR region where the BER constraint is satisfied with option 1 also increases. For option 2, however, more receive antennas will only lead to a better BER performance over the very high-SNR region.

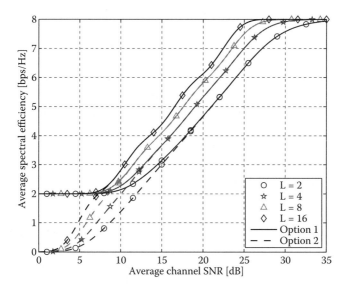

FIGURE 6.2 Average spectral efficiency of the SC-based scheme for both options and different numbers of antennas L.

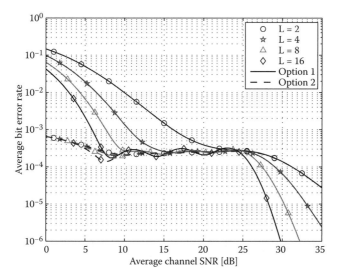

FIGURE 6.3 Average error rate of the SC-based scheme for both options and different numbers of antennas L.

6.4   Joint Adaptive Modulation
and Switched Combining

In this section, we design and analyze joint adaptive modulation and switched combining schemes. Note that switched combining schemes offer an even lower-complexity combining solution than SC since they involve less channel estimation and fewer and simpler comparisons. In what follows, we first present several switched combining schemes and then study two variants of the joint adaptive modulation and diversity combining designs based on switched combining, namely, minimum estimation schemes and bandwidth-efficient schemes.

6.4.1   Switched Combining

6.4.1.1   SSC/SEC

Dual-branch SSC is one of the most widely studied switched combining schemes. With SSC, the receiver uses the current branch until it becomes unacceptable, and then it switches and stays on the other branch regardless of its quality. The branch acceptance is determined by comparing the SNR of the current branch with a preselected fixed threshold, denoted by γ_T, which can be implemented with a threshold detector. Note that the major complexity saving of SSC over SC is that SSC needs only the SNR estimate of the current path and a simpler comparison to reach the combining decision. Because the receiver simply stays on the switch-to branch irrespective of its quality, SSC cannot benefit from additional diversity paths.

In multibranch scenario, i.e., $L > 2$, we can generalize SSC in order to take advantage of the additional diversity paths, and this leads to the SEC scheme [28]. In this case, the receiver with SEC cyclically switches between the L antenna branches. Branch switching occurs only when the received SNR of the currently used branch is below the threshold γ_T, and as such, this branch becomes unacceptable. Unlike SSC, the receiver examines the received SNR of the switch-to branch and switches again if it is found unacceptable. The receiver will repeat this process until either it finds an acceptable branch or all L antenna branches have been examined. In the latter case, it uses the last examined antenna branch for the data reception. It can be seen that the SEC scheme retains the complexity advantage of less path estimation and simple comparison.

Based on the mode of operation of SSC/SEC, it can be shown that the CDF and PDF of the output SNR with L-SEC are given by [28]

$$F_{\gamma_c}^{SEC}(\gamma) = \begin{cases} \left[F_\gamma(\gamma_T)\right]^{L-1} F_\gamma(\gamma), & \gamma < \gamma_T; \\ \sum_{j=0}^{L-1} \left[F_\gamma(\gamma) - F_\gamma(\gamma_T)\right]\left[F_\gamma(\gamma_T)\right]^j \\ \quad + \left[F_\gamma(\gamma_T)\right]^L, & x \ge \gamma_T, \end{cases} \qquad (6.11)$$

and

$$p_{\gamma_c}^{SEC}(\gamma) = \begin{cases} \left[F_\gamma(\gamma_T) \right]^{L-1} p_\gamma(\gamma), & \gamma < \gamma_T; \\ \sum_{j=0}^{L-1} \left[F_\gamma(\gamma_T) \right]^{j} p_\gamma(\gamma), & \gamma \geq \gamma_T, \end{cases}$$

(6.12)

respectively, where $F_\gamma(\cdot)$ and $p_\gamma(\cdot)$ are the common CDF and PDF of received SNR per path. Note that when $L = 2$, (6.11) and (6.12) reduce to the CDF and PDF for dual-branch SSC, respectively.

6.4.1.2 SECps

We can further improve the performance of SSC/SEC by better exploiting the available channel estimates in the worst-case scenario for which no acceptable diversity path is found. Since all available diversity paths have been examined in this case, the receiver may use the strongest one among all these unacceptable paths, instead of randomly choosing the last unacceptable path, for data reception. We term the resulting scheme SECps [29]. In particular, the receiver with SECps tries to use an acceptable path by examining as many diversity paths as necessary, as is the case with SEC. When no acceptable path is found after examining all diversity paths, unlike conventional SEC, the receiver with SECps selects the best unacceptable path, i.e., the one with the highest SNR, for data reception.

More specifically, the receiver first estimates the SNR of the currently used diversity path, denoted by γ_1, without loss of generality, and compares it with the threshold γ_T. If the current diversity path is acceptable (i.e., $\gamma_1 \geq \gamma_T$), then the receiver continues to use it for data reception (i.e., $\gamma_c = \gamma_1$). Otherwise (i.e., $\gamma_1 < \gamma_T$), the receiver tries to find an acceptable path by sequentially examining the other $L - 1$ diversity paths. This process is continued until either an acceptable path is found or all available diversity paths have been estimated. In the latter case, since all path SNRs are known, the receiver compares the estimate path SNRs and selects the one with the largest SNR for data reception. Compared to conventional SEC, SECps requires the additional complexity for the occasional selection of the best path among all unacceptable paths. However, SECps takes full advantage of the available path estimates. In particular, by using the best unacceptable path instead of a randomly chosen one (usually the last one examined) as in SEC, SECps can deliver better performance than conventional SEC.

Based on the mode of operation of SECps, it can be shown that the CDF and PDF of the output SNR with L-SECps are given by [29]

$$F_{\gamma_c}^{SEC}(\gamma) = \begin{cases} 1 - \sum_{i=0}^{L-1} \left[F_\gamma(\gamma_T) \right]^{i} \left[1 - F_\gamma(\gamma) \right], & \gamma \geq \gamma_T; \\ \left[F_\gamma(\gamma) \right]^{L}, & \gamma < \gamma_T. \end{cases}$$

(6.13)

and

$$
p_{\gamma_c}^{SEC}(\gamma) = \begin{cases} \sum_{i=0}^{L-1} \left[F_\gamma(\gamma_T) \right]^i p_\gamma(\gamma), & \gamma \geq \gamma_T; \\ L\left[F_\gamma(\gamma) \right]^{L-1} p_\gamma(\gamma), & \gamma < \gamma_T, \end{cases}
\tag{6.14}
$$

respectively.

6.4.2 Minimum Estimation Schemes

6.4.2.1 Mode of Operation

The primary objective of the minimum estimation schemes based on switched combining is to minimize the number of path estimations for path selection at the receiver. Once this primary objective is met, this scheme tries to afford the largest possible spectral efficiency while meeting the required target BER. Based on these objectives, the diversity combiner will perform just enough combining operations such that at least the lowest adaptive modulation mode, i.e., QPSK ($M = 4$), will exhibit an instantaneous BER smaller than the predetermined target value. In particular, the receiver tries to increase the output SNR Γ above the threshold for QPSK, i.e., γ_{T_2}, by performing SSC/SEC or SECps diversity. The receiver will sequentially estimate the received SNR of each diversity path and compare it with γ_{T_2}. The first acceptable path, i.e., the one with SNR greater than γ_{T_2}, will be used for data burst reception. Then, the receiver starts to determine the modulation mode to be used by checking in which interval the resulting output SNR falls. In particular, the receiver sequentially compares the output SNR with respect to the thresholds, $\gamma_{T_3}, \gamma_{T_4}, \cdots, \gamma_{T_N}$. Whenever the receiver finds that the output SNR is smaller than $\gamma_{T_{n+1}}$ but greater than γ_{T_n}, it selects the modulation mode n for the subsequent data burst and feeds back that particular modulation mode to the transmitter. If the received SNRs of all L available branches are below γ_{T_2}, the receiver may ask the transmitter to either (1) transmit using the lowest modulation mode in violation of the target instantaneous BER requirement (option 1) or (2) buffer the data and wait until the next guard period for more favorable channel conditions (option 2). For option 1, the receiver will use the last unacceptable path for data reception for the case of SSC/SEC and the strongest unacceptable path instead for the case of SECps.

6.4.2.2 Statistics of the Received SNR

Based on the mode of operation described above, we can see that the received SNR, Γ, of the minimum estimation scheme based on SSC/SEC is the same as the combined SNR of the SSC/SEC scheme with γ_{T_2} as the output threshold. In other words, the CDF of the received SNR, $F_\Gamma(\cdot)$, of the SSC/SEC-based minimum estimation scheme is given by

$$
F_\Gamma(\gamma) = \begin{cases} F_{\gamma_c}^{SEC(\gamma_{T_2})}(\gamma), & \text{for option 1;} \\ F_{\gamma_c}^{SEC(\gamma_{T_2})}(\gamma), & \gamma > \gamma_{T_2}; \\ & \hspace{2em} \text{for option 2,} \\ F_{\gamma_c}^{SEC(\gamma_{T_2})}(\gamma_{T_2}), & 0 < \gamma \leq \gamma_{T_2} \end{cases}
\tag{6.15}
$$

where $F_{\gamma_c}^{SEC(\gamma_{T2})}(\cdot)$ denotes the CDF of the combined SNR with L branch SEC and using γ_{T_2} as an output threshold, which is given for the i.i.d. fading environment in (6.11). Correspondingly, the PDF of the received SNR, $p_\Gamma(\cdot)$, is given by

$$p_\Gamma(\gamma) = \begin{cases} p_{\gamma_c}^{SEC(\gamma_{T2})}(\gamma), & \text{for option 1;} \\ p_{\gamma_c}^{SEC(\gamma_{T2})}(\gamma)\mathcal{U}(\gamma-\gamma_{T_2}) + F_{\gamma_c}^{SEC(\gamma_{T2})}(\gamma_{T_2})\delta(\gamma), & \text{for option 2,} \end{cases} \tag{6.16}$$

where $p_{\gamma_c}^{SEC(\gamma_{T2})}(\cdot)$ denotes the PDF of the combined SNR with L branch SSC/SEC and using γ_{T_2} as an output threshold, which is given for the i.i.d. fading environment by (6.12).

Similarly, the CDF and PDF of the received SNR, Γ, of the minimum estimation scheme based on SECps can be obtained as

$$F_\Gamma(\gamma) = \begin{cases} F_{\gamma_c}^{SECps(\gamma_{T2})}(\gamma), & \text{for option 1;} \\ \begin{cases} F_{\gamma_c}^{SECps(\gamma_{T2})}(\gamma), & \gamma > \gamma_{T_2}; \\ F_{\gamma_c}^{SECps(\gamma_{T2})}(\gamma_{T_2}), & 0 < \gamma \le \gamma_{T_2} \end{cases} & \text{for option 2,} \end{cases} \tag{6.17}$$

and

$$p_\Gamma(\gamma) = \begin{cases} p_{\gamma_c}^{SECps(\gamma_{T2})}(\gamma), & \text{for option 1;} \\ p_{\gamma_c}^{SECps(\gamma_{T2})}(\gamma)\mathcal{U}(\gamma-\gamma_{T_2}) + F_{\gamma_c}^{SECps(\gamma_{T2})}(\gamma_{T_2})\delta(\gamma), & \text{for option 2,} \end{cases} \tag{6.18}$$

respectively, where $F_{\gamma_c}^{SECps(\gamma_{T2})}(\cdot)$ and $p_{\gamma_c}^{SECps(\gamma_{T2})}(\cdot)$ denote the CDF and PDF of the combined SNR with L branch SECps and γ_{T_2} as an output threshold, which is given for the i.i.d. fading environment by (6.13) and (6.14), respectively.

6.4.2.3 Performance and Efficiency Analysis

With the statistics of the received SNR Γ derived for minimum estimation schemes based on both SSC/SEC and SECps, we evaluate the performance and efficiency of the system. Following an approach similar to the one used for the SC-based schemes in the previous section, we obtain the average spectral efficiency of the SSC/SEC-based minimum estimation scheme as

$$\eta = \begin{cases} N - \displaystyle\sum_{n=3}^{N} F_{\gamma_c}^{SEC(\gamma_{T2})}(\gamma_{T_n}), & \text{for option 1;} \\ N - \displaystyle\sum_{n=2}^{N} F_{\gamma_c}^{SEC(\gamma_{T2})}(\gamma_{T_n}), & \text{for option 2.} \end{cases} \tag{6.19}$$

The average BER of the SSC/SEC-based minimum estimation scheme can be similarly obtained as

$$
\langle BER \rangle =
\begin{cases}
\dfrac{\displaystyle\int_0^{\gamma_{T_3}} BER_2(\gamma)\, p_{\gamma_c}^{SEC_{(T_2)}}(\gamma)\, d\gamma + \sum_{n=3}^{N} n \int_{\gamma_{T_n}}^{\gamma_{T_{n+1}}} BER_n(\gamma)\, p_{\gamma_c}^{SEC_{(T_2)}}(\gamma)\, d\gamma}{N - \displaystyle\sum_{n=3}^{N} F_{\gamma_c}^{SEC_{(T_2)}}\left(\gamma_{T_n}\right)}, & \text{for option 1;} \\[4ex]
\dfrac{\displaystyle\sum_{n=2}^{N} n \int_{\gamma_{T_n}}^{\gamma_{T_{n+1}}} BER_n(\gamma)\, p_{\gamma_c}^{SEC_{(T_2)}}(\gamma)\, d\gamma}{N - \displaystyle\sum_{n=2}^{N} F_{\gamma_c}^{SEC_{(T_2)}}\left(\gamma_{T_n}\right)}, & \text{for option 2.}
\end{cases}
$$

$$\tag{6.20}$$

Similar results can be obtained for SECps-based schemas by using the CDF and PDF of the combined SNR with L branch SECps given in (6.13) and (6.14) instead, but these results are omitted here for conciseness.

The complexity saving of switched combining schemes over selection combining schemes manifests in the average number of path estimations. Note that SC always needs L path estimations, i.e., $N_E^{SC} = L$. On the other hand, the receiver with SSC/SEC and SECps needs to estimate additional paths only if necessary. Whenever an acceptable path is found, the receiver stops path estimation. Therefore, the average number of path estimations needed with the SSC/SEC- or SECps-based minimum selection scheme is given by

$$
N_E^{SEC/SECps} = 1 + \sum_{i=1}^{L-1} \left[P_\gamma\left(\gamma_{T_2}\right) \right]^i = \frac{1 - \left[P_\gamma\left(\gamma_{T_2}\right) \right]^L}{1 - P_\gamma\left(\gamma_{T_2}\right)}.
\tag{6.21}
$$

Therefore, the switched combining–based scheme requires less path estimations than the SC-based scheme on average, i.e.,

$$
N_E^{SEC} = N_E^{SECps} \leq N_E^{SC}.
$$

6.4.3 Bandwidth-Efficient Schemes

6.4.3.1 Mode of Operation

The primary objective of the switched combining-based bandwidth-efficient scheme is to maximize the spectral efficiency. As such, the receiver with this scheme performs as many combining operations as needed so that the highest achievable modulation mode can be used while satisfying the instantaneous BER requirement. More specifically, the receiver tries first to increase the output SNR Γ above the threshold of highest modulation mode 2^N-QAM, i.e., γ_{T_N}, by performing SSC/SEC or SECps. The receiver will sequentially estimate the received SNR of each diversity path and compare it with γ_{T_N}. Whenever a path with SNR greater than γ_{T_N} is found, the receiver stops path estimation, uses that path for data burst reception, and informs the transmitter to use 2^N-QAM as the modulation mode for the subsequent data burst. If the received SNRs of all the available paths are below γ_{T_N}, the receiver selects the modulation mode corresponding to the SNR interval in which the combined SNR falls. With SSC/SEC, the combined SNR is the SNR of the last unacceptable path, whereas with SECps, it is the SNR of the best unacceptable path. In particular, the receiver sequentially compares the output SNR with the thresholds, $\gamma_{T_{N-1}}, \gamma_{T_{N-2}}, \cdots, \gamma_{T_2}$. Whenever the receiver finds that the output SNR is smaller than $\gamma_{T_{N+1}}$ but greater than γ_{T_n}, it selects the modulation mode n for the subsequent data burst and feeds back this selected mode to the transmitter. If, in the worst case, the combined SNRs of all the available paths ends up being below γ_{T_2}, the receiver has the same two termination options as for the minimum estimation scheme (i.e., to transmit using the lowest modulation mode [option 1] or to wait until the next guard period [option 2]).

6.4.3.2 Statistics of the Received SNR

Based on the mode of operation described above, we can see that the received SNR, Γ, of the bandwidth-efficient scheme is the same as the combined SNRs of SSC/SEC or SECps diversity with γ_{T_N} as the output threshold. Therefore, the CDF of the received SNR of this bandwidth-efficient scheme based on SEC is given by

$$F_\Gamma(\gamma) = \begin{cases} F_{\gamma_c}^{SEC(\gamma_{T_N})}(\gamma), & \text{for option 1;} \\ F_{\gamma_c}^{SEC(\gamma_{T_N})}(\gamma), & \gamma > \gamma_{T_2}; \\ F_{\gamma_c}^{SEC(\gamma_{T_N})}(\gamma_{T_2}), & 0 < \gamma \leq \gamma_{T_2} \end{cases} \quad \text{for option 2,} \tag{6.22}$$

where $F_{\gamma_c}^{SEC(\gamma_{T_N})}(\cdot)$ denotes the CDF of the combined SNR with L branch SEC and γ_{T_N} as an output threshold. Correspondingly, the PDF of the received SNR is given by

$$p_\Gamma(\gamma) = \begin{cases} p_{\gamma_c}^{SEC(\gamma_{T_N})}(\gamma), & \text{for option 1;} \\ p_{\gamma_c}^{SEC(\gamma_{T_N})}(\gamma)\mathcal{U}(\gamma - \gamma_{T_2}) + F_{\gamma_c}^{SEC(\gamma_{T_N})}(\gamma_{T_2})\delta(\gamma), & \text{for option 2,} \end{cases} \tag{6.23}$$

where $p_{\gamma_c}^{SEC(\gamma_{T_N})}(\cdot)$ denotes the PDF of the combined SNR with L branch SEC and γ_{T_N} as output threshold. The CDF and PDF of received SNR with the SECps-based bandwidth-efficient scheme can be similarly obtained, but are omitted here for conciseness.

6.4.3.3 Performance and Efficiency Analysis

With the statistics of the received SNR at hand, we study the performance, efficiency, and complexity of the bandwidth-efficient scheme as we did for the minimum estimation. For conciseness, we just list the analytical results in what follows for the SSC/SEC-based scheme, while noting that those for the SECps-based scheme can be easily obtained by using the appropriate CDF and PDF instead.

Average spectral efficiency:

$$
\eta = \begin{cases}
N - \displaystyle\sum_{n=3}^{N} F_{\gamma_c}^{SEC_{(\gamma_{T_N})}}\left(\gamma_{T_n}\right), & \text{for option 1;} \\[2em]
N - \displaystyle\sum_{n=2}^{N} F_{\gamma_c}^{SEC_{(\gamma_{T_N})}}\left(\gamma_{T_n}\right), & \text{for option 2.}
\end{cases}
\tag{6.24}
$$

Average BER:

$$
\langle BER \rangle = \begin{cases}
\dfrac{\displaystyle\int_0^{\gamma_{T_3}} BER_2(\gamma)\, P_{\gamma_c}^{SEC_{(\gamma_{T_N})}}(\gamma)\, d\gamma + \sum_{n=3}^{N} n \int_{\gamma_{T_n}}^{\gamma_{T_{n+1}}} BER_n(\gamma)\, P_{\gamma_c}^{SEC_{(\gamma_{T_N})}}(\gamma)\, d\gamma}{N - \displaystyle\sum_{n=3}^{N} F_{\gamma_c}^{SEC_{(\gamma_{T_N})}}\left(\gamma_{T_n}\right)}, & \text{for option 1;} \\[3em]
\dfrac{\displaystyle\sum_{n=2}^{N} n \int_{\gamma_{T_n}}^{\gamma_{T_{n+1}}} BER_n(\gamma)\, P_{\gamma_c}^{SEC_{(\gamma_{T_2})}}(\gamma)\, d\gamma}{N - \displaystyle\sum_{n=2}^{N} F_{\gamma_c}^{SEC_{(\gamma_{T_N})}}\left(\gamma_{T_n}\right)}, & \text{for option 2.}
\end{cases}
\tag{6.25}
$$

Average number of path estimation:

$$
N_E^{SEC/SECps} = 1 + \sum_{i=1}^{L-1} \left[P_\gamma\left(\gamma_{T_N}\right) \right]^i = \frac{1 - \left[P_\gamma\left(\gamma_{T_N}\right) \right]^L}{1 - P_\gamma\left(\gamma_{T_N}\right)}.
\tag{6.26}
$$

6.4.4   Numerical Examples

In this section, we examine the different design trade-offs involved in the joint adaptive modulation and switched combining schemes through several selected numerical examples.

Figure 6.4 plots the average spectral efficiency of the SEC-based minimum estimation schemes. It is interesting to see that the average spectral efficiency of this scheme for both options and different numbers of receive antennas overlaps for the high-SNR region. This is because when the channel condition is favorable, the first path examined will always be acceptable and will be used for modulation mode selection. As a result, the system becomes equivalent to the no-diversity case. We also observe from Figure 6.4 that over the low- to medium-SNR region, the system can benefit from an increasing number of receive antennas, while this benefit is more significant for option 2 than for option 1. Note that the probability of no transmission for option 2 is reduced when the number of receive antennas increases. Figure 6.4 also shows that option 1 has a considerable spectral efficiency advantage over option 2 in the low-SNR region. This advantage

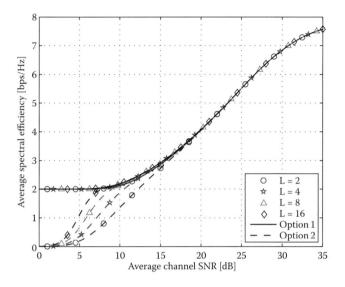

FIGURE 6.4 Average spectral efficiency of SEC-based minimum estimation schemes for both options and different numbers of antennas L.

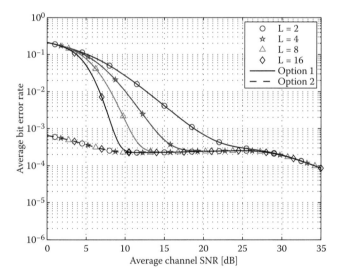

FIGURE 6.5 Average error rate of SEC-based minimum estimation schemes for both options and different numbers of antennas L.

again comes at the cost of the violation of the BER constraint, as shown in Figure 6.5. In Figure 6.5, we plot the average error rate of the SEC-based minimum estimation scheme with both options and different numbers of receive antenna L. It is clear that option 1 constantly violates the target BER requirement over the low- to medium-SNR region,

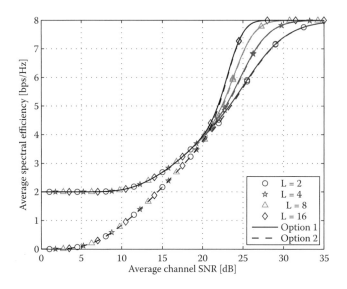

FIGURE 6.6 Average spectral efficiency of SEC-based bandwidth-efficient schemes for both options and different numbers of antennas L.

whereas option 2 always satisfies the requirement. We also notice from Figure 6.5 that while increasing the number of receive antennas can considerably alleviate the BER violation for option 1, it has nearly no effect on the BER performance of option 2.

We plot the average spectral efficiency of the SEC-based bandwidth-efficient schemes for both options and different numbers of antennas L in Figure 6.6. Unlike the minimum estimation schemes, increasing the number of receive antennas has little effect on the spectral efficiency for either option over the low- to medium-SNR region, whereas option 1 has certain spectral advantage over option 2 for the same reason given earlier. For the high-SNR region, the spectral efficiency of both options converges and benefits considerably from the additional number of receive antennas. This is because when the channel condition is favorable, having more antennas will increase the chance of using the largest possible constellation size and, as such, increase spectral efficiency. On the other hand, when the channel condition is poor, the largest constellation is not feasible, and based on the operation of the bandwidth-efficient scheme, the receiver will always choose constellation size based on the last examined path. As a result, the system becomes equivalent to the no-diversity case. Figure 6.7 shows the average BER of the SEC-based bandwidth-efficient schemes. As expected, option 1 violates the BER requirement over a large SNR range, while option 2 always satisfies the requirement. Again, increasing the number of antennas helps improve the BER performance of option 1 over the high-SNR region but affects little that of option 2.

It can be expected that the spectral efficiency of the SECps-based minimum estimation scheme is the same as that of the SEC-based scheme, and that the SECps-based scheme leads to an improved error performance for option 1 over the low-SNR region due to the occasional selection of the best unacceptable paths in the worst case. We omit the numerical examples of the SECps-based minimum estimation scheme for conciseness and focus

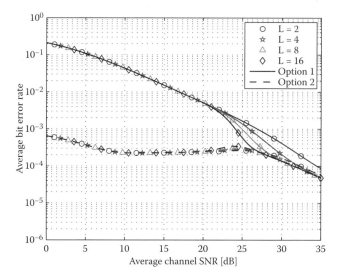

FIGURE 6.7 Average error rate of SEC-based bandwidth-efficient schemes for both options and different numbers of antennas L.

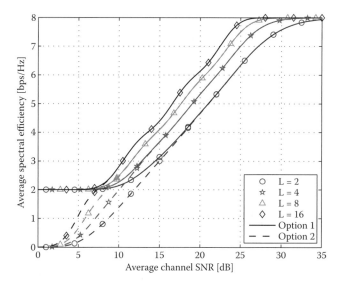

FIGURE 6.8 Average spectral efficiency of SECps-based bandwidth-efficient schemes for both options.

on those of the SECps-based bandwidth-efficient scheme here. From Figure 6.8, unlike the SEC-based bandwidth-efficient scheme, the spectral efficiency of the SECps-based scheme benefits from the increasing number of antennas over all SNR regions. Basically, when the largest constellation is not feasible, the SECps-based bandwidth-efficient

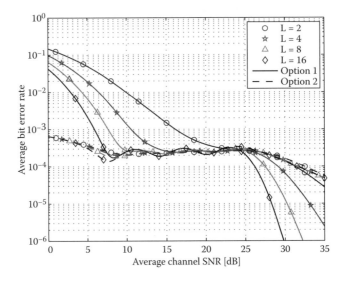

FIGURE 6.9   Average error rate of SECps-based bandwidth-efficient schemes for both options.

scheme can use a larger constellation than the SEC-based scheme since the best path is used in this case. Comparing Figures 6.2 and 6.8, we can conclude that the SECps-based bandwidth-efficient scheme achieves nearly the same spectral efficiency as the SC-based scheme. From Figure 6.9, we observe that similar to the SEC-based scheme, the BER performance of the SECps-based bandwidth-efficient scheme with option 1 shows a more significant improvement than with option 2 when the number of antennas increases. But the difference is that now the BER improvement manifests over all the SNR region. In fact, the average BER performance of the SECps-based bandwidth-efficient scheme with option 1 is very similar to that of the SC-based scheme shown in Figure 6.3.

We now compare the performances and spectral efficiencies of different schemes in Figures 6.10 and 6.11, respectively, for a fixed number of receive antennas ($L = 4$). The target BER is set to 10^{-4} in these figures. In particular, the average BER performance of different schemes with option 1 is shown in Figure 6.10. Note that we use subscript 1 to denote the minimum estimation schemes based on SEC or SECps, and subscript N to denote the bandwidth-efficient schemes. It can be seen that the SECps-based bandwidth-efficient scheme offers nearly the same error performance as the SC-based scheme. The complexity advantage of the SEC-based scheme comes at the cost of certain error performance degradation. It is worth pointing out that for the high-SNR region, the average BER of the SEC-based bandwidth-efficient scheme becomes better than that of the SECps-based minimum estimation scheme. Figure 6.11 compares the average spectral efficiencies of different schemes with option 2. Again, the SECps-based bandwidth-efficient scheme offers the same high spectral efficiency as the SC-based scheme. The SEC-based minimum estimation scheme shows a certain spectral advantage over the no-diversity case in the low-SNR region, whereas this advantage is more pronounced for

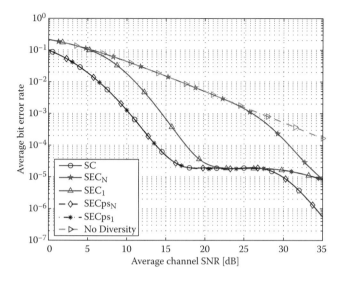

FIGURE 6.10 Comparison of the average error rate of SC-, SEC-, and SECps-based schemes (all for option 1).

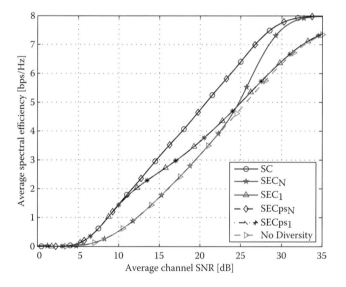

FIGURE 6.11 Comparison of the average spectral efficiency of SC-, SEC-, and SECps-based schemes (all for option 2).

the SEC-based bandwidth-efficient scheme in the high-SNR region. Finally, it is interesting to note that the spectral efficiencies of the SEC-based and SECps-based minimum estimation schemes for option 2 are exactly the same as one would intuit.

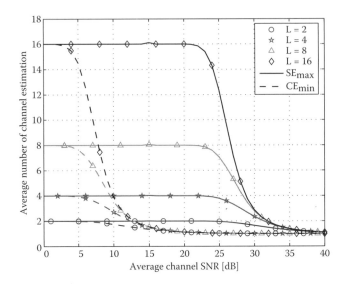

FIGURE 6.12 Comparison of the average number of channel estimation for the bandwidth-efficient and minimum estimation schemes.

Finally, we examine the average number of channel estimations required by switched combining–based joint designs in Figure 6.12. We use SE_{max} to denote the bandwidth-efficient schemes and CE_{min} the minimum estimation schemes. As we can see, as the average SNR increases, the number of channel estimations for all cases decreases from L to 1. For the low- to medium-SNR range, the minimum estimation schemes always require much less channel estimation than the bandwidth-efficient scheme, especially when L is large. Both types of schemes require less than L channel estimations, which is required by the SC-based scheme, in the high-SNR region. Therefore, the complexity advantage of switched combining-based schemes is that on average, the system needs to perform less channel estimation for diversity path and modulation mode selection.

6.5 Concluding Remarks

In this chapter, we investigated the joint design of adaptive modulation and switched diversity combining. The proposed system selects the diversity path and modulation mode jointly based on the fading channel condition and target BER requirement. Both SC and different variants of switched combining schemes, including SSC, SEC, and SECps, were considered. For the switched combining-based systems, we took into account both minimum channel estimation and maximum spectral efficiency design objectives. For the resulting schemes, we accurately quantified the performance, spectral efficiency, and complexity. It is observed from the selected numerical examples that the SECps-based bandwidth-efficient scheme can achieve nearly the same spectral efficiency and link reliability as SC–based schemes while requiring less path estimation on average. We also noticed that the SEC-based schemes offer different design trade-offs. With its low implementation complexity, we believe that the proposed joint design can be readily applied to the emerging MM-wave-based WPAN systems.

Acknowledgments

This work is supported in part by a discovery grant from NSERC Canada, in part by the Ministry of Information and Communications (MIC), Korea, under the Information Technology Research Communication (ITRC) support program supervised by the Institute of Information Technology Advancement (IITA) (IITA-2008-C1090-0801-0037), and in part by the Qatar Foundation for Education, Sciences, and Community Development, Doha, Qatar.

References

[1] A. J. Goldsmith and S.-G. Chua. 1998. Adaptive coded modulation for fading channels. *IEEE Trans. Commun.* 46:595–602.

[2] M.-S. Alouini and A. J. Goldsmith. 2000. Adaptive modulation over Nakagami fading channels. *Kluwer J. Wireless Commun.* 13:119–43.

[3] K. J. Hole, H. Holm, and G. E. Øien. 2000. Adaptive multidimensional coded modulation over flat fading channels. *IEEE J. Select. Areas Commun.* 18:1153–58.

[4] G. L. Stüber. 2000. *Principles of mobile communications.* 2nd ed. Norwell, MA: Kluwer Academic Publishers.

[5] R. Prasad. 1998. *Universal wireless personal communications.* Boston: Artech House.

[6] N. Kong and L. B. Milstein. 1999. Average SNR of a generalized diversity selection combining scheme. *IEEE Commun. Lett.* 3:57–59.

[7] M. Z. Win and J. H. Winters. 1999. Analysis of hybrid selection/maximal-ratio combining in Rayleigh fading. *IEEE Trans. Commun.* 47:1773–76.

[8] M.-S. Alouini and M. K. Simon. 2000. An MGF-based performance analysis of generalized selective combining over Rayleigh fading channels. *IEEE Trans. Commun.* 48:401–15.

[9] A. Annamalai and C. Tellambura. 2002. Analysis of hybrid selection/maximal-ratio diversity combiner with Gaussian errors. *IEEE Trans. Wireless Commun.* 1:498–512.

[10] Y. Ma and S. Pasupathy. 2004. Efficient performance evaluation for generalized selection combining on generalized fading channels. *IEEE Trans. Wireless Commun.* 3:29–34.

[11] S. W. Kim, Y. G. Kim, and M. K. Simon. 2004. Generalized selection combining based on the log-likelihood ratio. *IEEE Trans. Commun.* 52:521–24.

[12] S. W. Kim, D. S. Ha, and J. H. Reed. 2003. Minimum selection GSC and adaptive low-power RAKE combining scheme. In *Proceedings of the IEEE International Symposium on Circuits and Systems (ISCAS'03)*, Bangkok, Thailand, vol. 4, pp. 357–60.

[13] P. Gupta, N. Bansal, and R. K. Mallik. 2005. Analysis of minimum selection H-S/MRC in Rayleigh fading. *IEEE Trans. Commun.* 53:780–84.

[14] H.-C. Yang. 2006. New results on ordered statistics and analysis of minimum-selection generalized selection combining (GSC). *IEEE Trans. Wireless Commun.* 5:1876–85.

[15] R. K. Mallik, P. Gupta, and Q. T. Zhang. 2005. Minimum selection GSC in independent Rayleigh fading. *IEEE Trans. Veh. Technol.* 54:1013–21.

[16] M.-S. Alouini and H.-C. Yang. 2005. Minimum estimation and combining generalized selection combining (MEC-GSC). In *Proceedings of the IEEE International Symposium on Information Theory (ISIT'05)*, Adelaide, Australia, pp. 578–82.

[17] H.-C. Yang and M.-S. Alouini. 2005. MRC and GSC diversity combining with an output threshold. *IEEE Trans. Veh. Technol.* 54:1081–90.

[18] H.-C. Yang, N. Belhaj, and M.-S. Alouini. 2006. Bandwidth-efficient-power-greedy joint adaptive modulation and diversity combining. In *Proceedings of the International Symposium on Information Theory (ISIT '2006)*, Seattle, WA, pp. 937–941.

[19] Y.-C. Ko, H.-C. Yang, and M.-S. Alouini. 2006. Adaptive modulation and diversity combining based on output-threshold MRC. In *Proceedings of the 63rd IEEE Semiannual Vehicular Technology Conference (VTC'2006)*, Melbourne, Australia, pp. 1693–97.

[20] W. C. Jakes. 1994. *Microwave mobile communication.* 2nd ed. Piscataway, NJ: IEEE Press.

[21] M. A. Blanco and K. J. Zdunek. 1979. Performance and optimization of switched diversity systems for the detection of signals with Rayleigh fading. *IEEE Trans. Commun.* 27:1887–95.

[22] M. A. Blanco. 1983. Diversity receiver performance in Nakagami fading. In *Proceedings of the IEEE Southeastern Conference*, Orlando, FL, pp. 529–32.

[23] A. A. Abu-Dayya and N. C. Beaulieu. 1994. Analysis of switched diversity systems on generalized-fading channels. *IEEE Trans. Commun.* 42:2959–66.

[24] A. A. Abu-Dayya and N. C. Beaulieu. 1994. Switched diversity on microcellular Ricean channels. *IEEE Trans. Veh. Technol.* 43:970–76.

[25] G. Femenias and I. Furió. 1997. Analysis of switched diversity TCM-MPSK systems on Nakagami fading channels. *IEEE Trans. Veh. Technol.* 46:102–7.

[26] Y.-C. Ko, M.-S. Alouini, and M. K. Simon. 2000. Analysis and optimization of switched diversity systems. *IEEE Trans. Veh. Technol.* 49:1569–74.

[27] C. Tellambura, A. Annamalai, and V. K. Bhargava. 2001. Unified analysis of switched diversity systems in independent and correlated fading channel. *IEEE Trans. Commun.* 49:1955–65.

[28] H.-C. Yang and M.-S. Alouini. 2003. Performance analysis of multibranch switched diversity systems. *IEEE Trans. Commun.* 51:782–94.

[29] H.-C. Yang and M.-S. Alouini. 2006. Improving the performance of switched diversity with post-examining selection. *IEEE Trans. Wireless Commun.* 5:67–71.

[30] K. Cho and D. Yoon. 2002. On the general BER expression of one- and two-dimensional amplitude modulation. *IEEE Trans. Commun.* 50:1074–80.

[31] I. S. Gradshteyn and I. M. Ryzhik. 1994. *Table of integrals, series, and products.* 5th ed. San Diego: Academic Press.

[32] A. H. Nuttall. 1975. Some integrals involving the Q_M function. *IEEE Trans. Information Theory* 1:95–96.

7

Adaptive Opportunistic Beamforming in Ricean Fading Channels

Il-Min Kim
Queen's University

Zhihang Yi
Queen's University

7.1 Introduction

Fading is one of the major impairments of wireless channels, and it has traditionally been seen as an obstacle to reliable data transmission. Considerable efforts have been devoted to combat fading and to send data more reliably over wireless channels. One of the key methods to mitigate fading is to implement *diversity* techniques in wireless communication systems. There are several different ways to achieve the diversity: frequency diversity [1], time diversity [2], and spatial diversity [3–6].

Recently, a new type of diversity, so-called *multiuser diversity*, has received a lot of attention. Multiuser diversity exploits the fact that in a multiuser system with

177

independently varying channels, there is likely to be at least a single user whose channel is near its peak at any time, if the number of users is sufficiently large. By allowing only the user having the best channel to use the system resource at a given time, the total throughput of the entire system can be maximized. In [7], Knopp and Humblet focused on the uplink channel of a multiuser system. In order to maximize the total information-theoretic capacity of the system, they showed that the optimum strategy was to transmit the signals of the user with the best channel at any time. Similar results were also obtained by studying the downlink channel from the base station to the mobile users [8]. In [9], the space-time coding was combined with multiuser diversity. In [10] and [11], the interaction between multiple-antenna diversity and multiuser diversity was discussed in detail. Furthermore, the multiuser diversity technique has been implemented in practical communication systems, such as the downlink of the IS-856 system [12].

In a multiuser system exploiting multiuser diversity, channel fading can be considered a source of randomization providing multiuser diversity, as opposed to an impairment. Therefore, large multiuser diversity gain is achieved when the dynamic range of the channel fluctuation is large or the variation rate of the channel is fast. In practice, however, the channel fluctuation may not be large enough to provide satisfactory multiuser diversity. Furthermore, when the channel fading is slower than the delay constraint of a system or an application, the user cannot wait until its channel reaches the peak, and thus, the multiuser diversity gain may get smaller. Addressing these problems, in [13], Viswanath et al. proposed *opportunistic beamforming*, which artificially induced channel fluctuation when the fluctuation of the underlying physical channel was small or the fading was slow. They studied a system where the base station was equipped with multiple antennas and the same signal was transmitted from the antennas after being multiplied by pseudorandom weight coefficients. The phase and magnitude of each weight coefficient were changing in a controlled but pseudorandom fashion. By using a single pilot signal, the signal-to-noise ratio (SNR) of the overall *equivalent* channel was measured at every user and was fed back to the base station. Based on the SNR feedback, the base station picked the user with the best equivalent channel and the data only for this user were transmitted.

Over the past few years, many works have been devoted to extend the opportunistic beamforming technique. In [14], multiple weight coefficients were used at every time slot and the one producing the highest SNR was chosen. By doing this, better performance could be achieved at the expense of the increased feedback overhead. Opportunistic beamforming was combined with the water-filling method and extended to multiple-input multiple-output systems in [15]. Furthermore, several new opportunistic schemes, including opportunistic cophasing and antenna selection, were proposed and their performance was analyzed in [16].

Very recently, Kim et al. proposed an adaptive version of opportunistic beamforming in Ricean fading channels [17]. This new scheme improved the performance substantially over Ricean fading channels without introducing multiple weight coefficients or increasing the feedback overhead. Unlike the opportunistic beamforming in [13], which generated the weight coefficients in a pseudorandom fashion, the improved opportunistic beamforming generated the weight coefficients more intelligently by estimating the directions of arrival (DOAs) of the users.

The outline of the rest of this chapter is as follows. In section 7.2, the fundamental idea and theories behind multiuser diversity are reviewed. In section 7.3, opportunistic beamforming [13] is presented in detail. In section 7.4, the adaptive version of opportunistic beamforming [17] is discussed. Finally, some conclusions are drawn in section 7.5.

7.2 Multiuser Diversity

In this section we present fundamental ideas and theories of multiuser diversity. We consider the downlink of a wireless communication system where the base station with a single antenna is communicating with M users. Let $h_k(t)$ denote the channel coefficient from the base station to the kth user at time slot t, $\eta_k(t)$ the additive noise at the kth user at time slot t, and $s(t)$ the transmitted signal from the base station at time slot t. At a given time slot t, the power of $s(t)$ is P and $\eta_k(t)$ are modeled by circularly symmetric complex Gaussian random variables with zero mean and variance ρ_k^2, i.e., $\eta_k(t) \sim \mathcal{CN}(0, \rho_k^2)$. The received signal $r_k(t)$ at the kth user at time slot t is given by

$$r_k(t) = h_k(t)s(t) + \eta_k(t). \tag{7.1}$$

The instantaneous SNR $\gamma_k(t)$ of the kth user is $\gamma_k(t) = P|h_k(t)|^2/\rho_k^2$. We define the sum capacity of this downlink channel as the maximum achievable sum of the long-term average data rate transmitted to all users. It has been shown that the sum capacity is achieved by transmitting to the user with the highest $\gamma_k(t)$ at every time slot t [8].

In Figure 7.1, we compare the sum capacity of the downlink channel under Rayleigh fading and additive white Gaussian noise (AWGN) environments, which have the same average SNR. We note that, with more than two users, the sum capacity under the

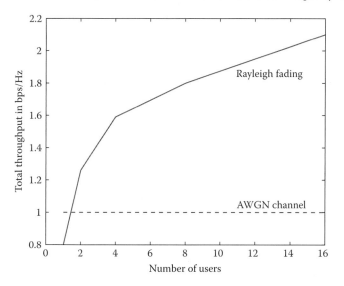

FIGURE 7.1 Sum capacity of Rayleigh fading and AWGN environments, average SNR = 0 dB. (Reproduced from Viswanath et al., 2002. © 2002, IEEE. With permission.)

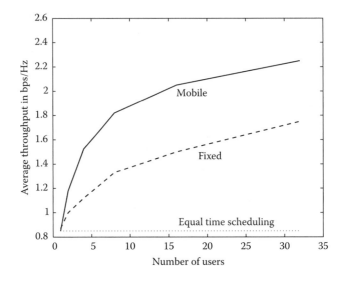

FIGURE 7.2 Multiuser diversity gain in fixed and mobile environments. (Reproduced from Viswanath et al., 2002. © 2002, IEEE. With permission.)

Rayleigh fading environment is larger than that under the AWGN environment. This is due to multiuser diversity. Figure 7.2 gives more insight into how the dynamic range and the variation rate of the channel fluctuation affect the benefits of multiuser diversity. The figure demonstrates the total throughput of the downlink channel of the IS-856 system under the following two environments: (1) fixed—users are stationary but there are objects moving around them, and (2) mobile—users move at a walking speed (3 km/h). Under both environments, the total throughput increases with the number of users, but the increase is more impressive in the mobile case. This is because the dynamic range and the variation rate of the channel fluctuation for the mobile case are larger than those for the fixed case.

In order to exploit multiuser diversity, every user needs to estimate its own instantaneous SNR $\gamma_k(t)$ by a pilot signal and feed it back to the base station. On the other hand, the base station needs to properly schedule the transmission to all the users and adapt the data rate based on the instantaneous channel condition. This technique has been adopted in some communications systems, such as IS-856 [18].

In practical communication systems, two important problems must be solved before multiuser diversity is implemented: fairness and delay. If the fading statistics of the users' channels are the same, transmitting data only to the user with the highest $\gamma_k(t)$ not only maximizes the throughput of the system, but also maximizes the throughput of every user. In reality, however, the fading statistics are not symmetrical. For example, some users may have better average SNRs, because they are closer to the base station. Furthermore, practical systems usually have a delay constraint on every user. Thus, one user cannot wait for too long until the user's instantaneous SNR becomes the highest among all the users.

In order to solve the fairness and delay problem, the proportional fair scheduling algorithm has been developed in [12]. In this algorithm, the kth user feeds back the requested data rate $R_k(t)$ to the base station, where $R_k(t)$ is the data rate the user can support at time slot t. The proportional fair scheduling algorithm also stores the average throughput $T_k(t)$ of every user in a past window of length t_c. At any time slot t, the proportional fair scheduling algorithm transmits data to the user with the largest

$$\frac{R_k(t)}{T_k(t)}. \tag{7.2}$$

In order to see how the proportional fair scheduling algorithm works, we consider a system with only two users. If the two users have identical fading statistics, the average throughput $T_k(t)$ of each user will converge to the same value. Thus, the proportional fair scheduling algorithm just picks the user with the highest $R_k(t)$, and it is fair for every user in the long term. It is possible that the first user's channel is better than that of the second user on average. Always transmitting to the user with the highest $R_k(t)$ implies that the first user will be served for most of the time and the second user will not be served in a resource fair manner. This problem is solved by using the proportional fair scheduling algorithm. Because the proportional fair scheduling algorithm selects the user based on $R_k(t)/T_k(t)$, a user is selected when its instantaneous channel condition is high relative to its own average channel condition over the time scale t_c. To an extreme case, if the second user is not served in the past window of length t_c, $T_2(t)$ becomes zero and the base station will transmit data to the second user immediately.

7.3   Opportunistic Beamforming

As we have seen in section 7.2, the dynamic range and the variation rate of the channel fluctuation determine the performance gain of multiuser diversity. Thus, if one can induce larger and faster channel fluctuation, higher gain will be achieved. This can be realized by opportunistic beamforming [13]. In this scheme, N antennas are deployed at the base station and there are M users in the system. Let $h_{n,k}(t)$ denote the channel coefficient from the nth antenna to the kth user at time slot t. The transmitted signal $s(t)$ is first multiplied by a complex weight coefficient

$$w_n(t)=\sqrt{\alpha_n(t)}e^{j\phi_n(t)}$$

and then transmitted by the nth antenna, for $n = 1, \cdots, N$. In order to satisfy the power constraint, it is assumed that

$$\sum_{n=1}^{N}\alpha_n(t)=1.$$

Thus, the received signal $r_k(t)$ at the kth user is given by

$$r_k(t) = \left(\sum_{n=1}^{N} \sqrt{\alpha_n(t)} e^{j\phi_n(t)} h_{n,k}(t) \right) s(t) + \eta_k(t). \tag{7.3}$$

The overall equivalent channel gain $\tilde{h}_k(t)$ is

$$\tilde{h}_k(t) = \sum_{n=1}^{N} \sqrt{\alpha_n(t)} e^{j\phi_n(t)} h_{n,k}(t), \tag{7.4}$$

and the overall equivalent instantaneous SNR is

$$\tilde{\gamma}_k(t) = P \frac{|\tilde{h}_k(t)|^2}{\rho_k^2}. \tag{7.5}$$

The power fraction $\alpha_n(t)$ is modeled as samples of a random variable varying from 0 to 1, and the artificial phase shift $\phi_n(t)$ is modeled as samples of a random variable uniformly distributed over $[0, 2\pi)$. At different transmit antennas, $\alpha_n(t)$ and $\phi_n(t)$ are varying independently.

The kth user feeds the equivalent instantaneous SNR $\tilde{\gamma}_k(t)$ back to the base station, and the base station selects the user by using the proportional fair scheduling algorithm. Note that the kth user does not need to estimate $h_{n,k}(t)$. Actually, the existence of multiple antennas at the base station can be completely transparent to the users. Note that the dynamic range and the variation rate of the equivalent channel $\tilde{h}_k(t)$ can be controlled by $w_n(t)$.

The performance of opportunistic beamforming is demonstrated in Figure 7.3 under the same simulation environment as in Figure 7.2. Two antennas are deployed at the base stations and $\alpha_n(t) = 1/\sqrt{2}$ at any time slot t. It can be seen that the performance of the fixed case is considerably improved by using opportunistic beamforming. This is because the channel is changing faster and the dynamic range of the channel fluctuation is larger after the use of $\alpha_n(t)$ and $\phi_n(t)$. In the following, the slow fading and fast fading environments are considered separately.

7.3.1 Slow Fading

In the slow fading environment, we assume that $h_{n,k}(t) = h_{n,k}$ in a time slot t. If only one antenna is used at the base station, the SNR $\gamma_n(t)$ of every user will remain constant at every time slot t, and hence, no multiuser diversity can be exploited. However, by using opportunistic beamforming, the equivalent channel $\tilde{h}_k(t)$ will vary in time and provide the opportunity for achieving multiuser diversity.

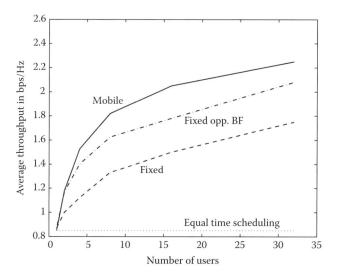

FIGURE 7.3 Amplification in multiuser diversity gain with opportunistic beamforming in a fixed environment. (Reproduced from Viswanath et al., 2002. © 2002, IEEE. With permission.)

For a particular user k, it is well known that coherent transmit beamforming can maximize $\tilde{\gamma}_n(t)$ by setting [13]

$$\alpha_n = \frac{|h_{n,k}|^2}{\sum_{n=1}^{N} |h_{n,k}|^2}, \tag{7.6}$$

$$\phi_n = -\arg(h_{n,k}), \tag{7.7}$$

which is called the beamforming configuration. In this case, however, the base station needs to know $h_{n,k}$, for $n = 1, \cdots, N$, and hence, this amount of feedback information might be prohibitive. On the other hand, opportunistic beamforming selects the user when its equivalent SNR is at the peak. Therefore, opportunistic beamforming may achieve the performance of coherent transmit beamforming, while only requiring the feedback of $\tilde{\gamma}_k(t)$. This is confirmed by the following theorem.

Theorem 7.1

[13] *"Suppose the slow fading states of the users are i.i.d and are discrete, and the joint stationary distribution of $(\alpha_1(t), \ldots, \alpha_N(t), \phi_1(t), \ldots, \phi_2(t))$ is the same as that of*

$$\left(\frac{|h_{1k}|^2}{\sum_{n=1}^{N} |h_{n,k}|^2}, \cdots, \frac{|h_{n,k}|^2}{\sum_{n=1}^{N} |h_{n,k}|^2}, -\arg(h_{1k}), \cdots, -\arg(h_{n,k}) \right) \tag{7.8}$$

for the slow fading state of any individual user k. Then, almost surely, we have

$$\lim_{M \to \infty} KT_k^{'(M)} = R_k^{bf} \tag{7.9}$$

for all k. $T_k^{(M)}$ is the average throughput of user k in a system with M users and R_k^{bf} is the instantaneous data rate that user k achieves when it is in the beamforming configuration, i.e. when its instantaneous SNR is

$$P \sum_{n=1}^{N} |h_{n,k}|^2 / \rho_k ."$$

Proof: See appendix A in [13].

When the number of users is sufficiently large, theorem 7.1 implies that, with very high probability, one user is selected when it is in its beamforming configuration and every user is allocated an equal amount of time. The stationary distributions of $\alpha_n(t)$ and $\phi_n(t)$ required by this theorem are given in closed form in [13]. In a slow Rayleigh fading environment with ten antennas at the base station, the throughput of one specific user is given in Figure 7.4. The proportional fair scheduling algorithm is employed. One can see that the throughput converges to that of coherent transmit beamforming asymptotically with the number of users. Figure 7.5 demonstrates the total throughput of all users.

7.3.2   Fast Fading

We have seen that opportunistic beamforming can considerably improve the performance when the underlying channels are slow fading. But if the underlying channels are already fast fading, can opportunistic beamforming improve the performance? The performance gain of opportunistic beamforming is from the randomization of $\alpha_n(t)$ and $\phi_n(t)$, which makes the dynamic range of the channel fluctuation larger and the variation rate of the channel faster. Therefore, if the dynamic range of the equivalent channel $\tilde{h}_k(t)$ becomes larger after using opportunistic beamforming, the system will achieve better performance.

We first consider the independent Rayleigh fading environment, where $h_{n,k}(t)$ are i.i.d circularly symmetric complex Gaussian random variables with zero mean. In this case, it is easy to see from (7.4) that the distribution of $\tilde{h}_k(t)$ is exactly the same as that of the $h_{n,k}(t)$. The use of $\alpha_n(t)$ and $\phi_n(t)$ neither makes the dynamic range of the channel fluctuation larger nor makes the variation rate of the channel faster. Therefore, in an independent fast Rayleigh fading environment, opportunistic beamforming is not able to improve the performance.

In contrast, when the underlying channels are Ricean fading channels, opportunistic beamforming can bring considerable performance gain. The Ricean channel can be modeled as [19]

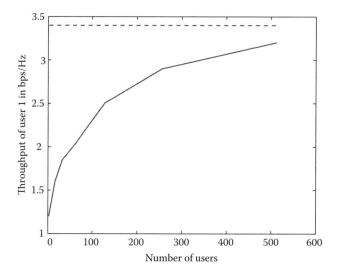

FIGURE 7.4 Throughput in bps/Hz for user 1 multiplied by number of users scheduled for slow Rayleigh fading at average SNR = 0 dB with the proportional fair scheduling algorithm. Performance of coherent transmit beamforming for user 1 and scheduled at all times is plotted as a dotted line. There are ten antennas in this experiment. (Reproduced from Viswanath et al., 2002. © 2002, IEEE. With permission.)

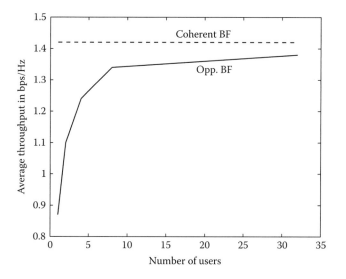

FIGURE 7.5 Total throughput in bps/Hz averaged over slow Rayleigh fading at average SNR = 0 dB with the proportional fair scheduling algorithm. Performance of coherent transmit beamforming is also plotted. (Reproduced from Viswanath et al., 2002. © 2002, IEEE. With permission.)

$$h_{n,k}(t) = \sqrt{\frac{K_k}{1+K_k}} \exp(jn\theta_k) + \sqrt{\frac{1}{1+K_k}} b_{nk}(t), \tag{7.10}$$

where K_k is the Ricean K-factor and θ_k is related to the DOA of the user. The first term in (7.10) denotes the line-of-sight (LOS) component of the channel. The second term is the diffused component and $b_{nk}(t) \sim \mathcal{CN}(0,1)$. Then the equivalent channel gain becomes

$$\tilde{h}_k(t) = \sum_{n=1}^{N} \sqrt{\frac{\alpha_k(t)K_k}{1+K_k}} \exp\left(j(n\theta_k + \phi_n(t))\right) + \sum_{n=1}^{N} \sqrt{\frac{\alpha_n(t)}{1+K_k}} \exp(j\phi_n(t))b_{nk}(t). \tag{7.11}$$

It can be seen that the randomization of $\alpha_n(t)$ and $\phi_n(t)$ will not change the dynamic range of the diffused term, which is the same as in the Rayleigh fading environment. However, the randomization of $\alpha_n(t)$ and $\phi_n(t)$ induces the fluctuation of the LOS component significantly, which leads to performance improvement. Therefore, if the LOS component is more dominant than the diffused path, larger fluctuation will be created by opportunistic beamforming and more performance gain can be achieved. To an extreme case that $K_k \to \infty$, the channel reduces to the slow fading case and opportunistic beamforming can improve the performance considerably, as we have seen in section 7.3.1. Figure 7.6 shows the total throughput for Ricean fading channels with $K_k = 10$. There is an impressive improvement in performance after using opportunistic beamforming. For comparison, the performance curve for the Rayleigh fading channels is plotted.

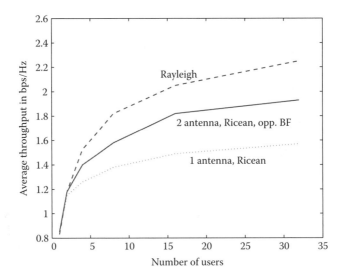

FIGURE 7.6 Total throughput as a function of the number of users under Ricean fading, with and without opportunistic beamforming. The power allocations $\alpha_n(t)$ are uniformly distributed in [0, 1] and the phases $\theta_n(t)$ are uniform in [0,2π]. (Reproduced from Viswanath et al., 2002. © 2002, IEEE. With permission.)

7.4 Adaptive Opportunistic Beamforming in Ricean Channels

In section 7.3, it was indicated that the throughput of opportunistic beamforming con-verged to that of coherent transmit beamforming, when the number of users was suf-ficiently large. However, the number of users required grows exponentially with the number of antennas. For example, in Figure 7.4, the throughput is approximately 65% of coherent transmit beamforming even with one hundred users. In order to improve the throughput up to 90% of coherent transmit beamforming, the number of users must be increased up to about four hundred, which may not be a practical number of users in a cell. With a realistic number of users, the performance of opportunistic beamforming falls much lower than that of coherent transmit beamforming. This problem can be solved by the adaptive opportunistic beamforming proposed in [17] over Ricean chan-nels. This section is devoted to presenting the adaptive opportunistic beamforming technique and its performance. In this section, we will refer to opportunistic beamform-ing in [13] as conventional opportunistic beamforming in order to distinguish it from adaptive opportunistic beamforming.

7.4.1 Adaptive Opportunistic Beamforming

The downlink of a single cell is considered, where the number of users is M and each user has a single antenna. The base station is equipped with a *linear* antenna array with N elements. The fading channel is a Ricean channel modeled by (7.10). The parameter θ_k is related to the DOA Θ_k of the kth user as follows:

$$\theta_k = \frac{2\pi d f_0 \cos\Theta_k}{c},\qquad (7.12)$$

where c is the speed of propagation of the plane wave, f_0 is the carrier frequency of the transmitted signal, and d is the spacing between two antenna elements [4].

It is assumed that a mini-slot exists at the beginning of each time slot. Through the mini-slot, a known pilot signal $s_k(t)$ is transmitted from every antenna after being multi-plied by a weight coefficient $w_n(t)$. When the transmission power is uniformly allocated to the N antenna elements, $w_n(t)$ is given by

$$w_n(t) = \frac{1}{\sqrt{N}}\exp(jn\phi(t)).\qquad (7.13)$$

The signal $r_k(t)$ received by the kth user during the mini-slot of time slot t is given by

$$r_k(t) = \tilde{h}_k(t)s_k(t) + \eta_k(t),\qquad (7.14)$$

where $\eta_k(t) \sim \mathcal{CN}(0,\rho_k^2)$. The overall equivalent channel $\tilde{h}_k(t)$ for user k is given by

$$\tilde{h}_k(t) = \sqrt{a_k} \sum_{n=1}^{N} \exp\left(jn\left(\theta_k + \phi(t)\right)\right) + B_k(t),$$ (7.15)

where

$$\sqrt{a_k} = \sqrt{\frac{K_k}{N(1+K_k)}},$$ (7.16)

$$B_k(t) = \sqrt{\frac{1}{N(1+K_k)}} \sum_{n=1}^{N} b_{nk}(t) \exp\left(jn\phi(t)\right).$$ (7.17)

It is easy to show that $B_k(t) \sim \mathcal{CN}(0, \sigma_k^2)$, where $\sigma_k^2 = 1/(1 + K_k)$.

Let $H_k(t)$ denote the magnitude of $\tilde{h}_k(t)$:

$$H_k(t) = \left|\tilde{h}_k(t)\right| = \left|\sqrt{a_k} \sum_{n=1}^{N} \exp(jn(\theta_k + \phi(t))) + B_k(t)\right|.$$ (7.18)

Each user measures only $H_k(t)$ and then determines the data rate $R_k(t)$ at which the data can be reliably transmitted from the base station to the user with a predetermined SNR threshold. The determined rates $R_k(t)$ for all the users are fed back to the base station. The requested rate $R_k(t)$ can be expressed as $R_k(t) = f(H_k(t))$, where $f(\cdot)$ is a nondecreasing function, which can be assumed to be known at the base station. Hence, the values of $H_k(t)$ are also assumed to be known at the base station in this section.

On the right-hand side of (7.18), the first term

$$\sqrt{a_k} \sum_{n=1}^{N} \exp\left(jn\left(\theta_k + \phi(t)\right)\right)$$

is related to the LOS component and the second term $B_k(t)$ to the diffused component. Let $G_k(t)$ denote the magnitude of the LOS component:

$$G_k(t) \overset{def}{=} \left|\sqrt{a_k} \sum_{n=1}^{N} \exp\left(jn\left(\theta_k + \phi(t)\right)\right)\right|.$$ (7.19)

Then one can easily see that $G_k(t)$ is maximized with $\phi(t) = -\theta_k$ and the maximum value is $\sqrt{a_k}\,N$. Motivated by this observation, the following adaptive opportunistic beamforming algorithm is proposed to improve the performance [17]:

1. As in the conventional opportunistic beamforming, the information on $H_k(t)$, or equivalently $R_k(t)$, is fed back to the base station.
2. The base station estimates the DOAs* of the users using the values of $H_k(t)$.† The estimated DOA of user k is denoted by $\hat{\theta}_k$, $k = 1, \cdots, M$.
3. The base station conducts the proportional fair scheduling to choose a user. Let us assume that user k^* is chosen.
4. When the base station transmits the data to user k^* during a time slot excluding the mini-slot, the artificial phase shift $\phi(t)$ of the weight coefficient is set to $-\hat{\theta}_{k^*}$. and this weight coefficient is multiplied to the transmitted data.

Note that this adaptive scheme forms the beams only in the directions where users really exist, as opposed to conventional opportunistic beamforming, which forms the beams blindly over the omnidirectional space. Therefore, the adaptive scheme can improve the performance without wasting resources such as time and power. In adaptive opportunistic beamforming, the LOS-related component of $H_k(t)$, which is $G_k(t)$, is maximized. Therefore, the performance heavily depends on the K-factors. Specifically, larger K-factors $\{K_k\}$ result in larger performance improvement; smaller K-factors result in smaller improvement. In an extreme case where the K-factors are zero, which is the Rayleigh case, the adaptive scheme reduces to the conventional scheme because $\phi(t)$ is chosen as a sample of a random variable uniformly distributed over $[0, 2\pi)$.

The vital step of adaptive opportunistic beamforming is to estimate the DOA as accurately as possible with the information available. In the previous publications, a number of algorithms for DOA estimation have been studied, such as the multiple signal classification (MUSIC) method [20], Root-MUSIC [21], and the estimation of signal parameters via the rotational invariance technique [22] (also see [4] and the references therein). Those previous methods, however, are not applicable to the system considered, because the base station has only very limited channel information: the magnitude values $H_k(t)$ of the equivalent channels. In the next section, a new and efficient DOA estimation algorithm requiring only $H_k(t)$ values is proposed for use in adaptive opportunistic beamforming.

7.4.2 Estimation of Users' DOAs

In this section, a maximum-likelihood (ML) estimator of $\{\theta_k\}_{k=1}^{M}$ is developed.‡ To this end, the probability density function (PDF) and cumulative density function (CDF) of $H_k(t)$ are first derived in the following theorem:

---

* To be precise, Θ_k is the DOA of user k. However, since Θ_k is uniquely determined by θ_k, we simply refer to θ_k as the DOA of user k in this section.

† The DOAs must be estimated based only on $H_k(t)$ values to ensure that the channel estimation algorithm of the receivers and the feedback overhead remain the same as in conventional opportunistic beamforming.

‡ Note that ML estimators can be asymptotically considered as minimum variance unbiased estimators [23].

Theorem 7.2

[17] "Let $A_k(t)$ be defined as follows:

$$A_k(t) \stackrel{def}{=} \begin{cases} \sqrt{a_k} + 2\sqrt{a_k} \displaystyle\sum_{n=1}^{(N-1)/2} \cos\left(n\left(\theta_k + \phi(t)\right)\right), & \text{if } N \text{ is odd} \\ 2\sqrt{a_k} \displaystyle\sum_{n=1}^{N/2} \cos\left(\left(n - \frac{1}{2}\right)\left(\theta_k + \phi(t)\right)\right), & \text{if } N \text{ is even.} \end{cases} \tag{7.20}$$

Then the PDF $p_{H_k(t)}(\cdot)$ and CDF $F_{H_k(t)}(\cdot)$ of $H_k(t)$ are given by

$$p_{H_k(t)}(x) = \frac{2x}{\sigma_k^2} \exp\left(-\frac{x^2 + A_k^2(t)}{\sigma_k^2}\right) I_0\left(\frac{2xA_k(t)}{\sigma_k^2}\right), \tag{7.21}$$

$$F_{H_k(t)}(x) = 1 - Q_1\left(\frac{\sqrt{2}A_k(t)}{\sigma_k^2}, \frac{\sqrt{2}x}{\sigma_k^2}\right), \tag{7.22}$$

where $I_0(\cdot)$ is the modified Bessel function of order zero and $Q_m(\cdot, \cdot)$ is the Marcum Q-function."

Proof: See appendix A in [17].

It is very interesting to see that (7.21) is exactly the same form as a Ricean distribution, except that $A_k(t)$ is a summation of multiple cosine functions. When $N = 1$, (7.21) reduces to the well-known Ricean PDF. As in the conventional Ricean case, we define $\Omega_k(t)$ to be the total power of (7.21) as follows:

$$\Omega_k(t) \stackrel{def}{=} E_{H_k(t)}\left[H_k^2(t)\right] = A_k^2(t) + \sigma_k^2. \tag{7.23}$$

Also, the K-factor of (7.21) can be defined as follows:

$$K_k(t) = \frac{A_k^2(t)}{\sigma_k^2}. \tag{7.24}$$

It follows that (7.21) can be rewritten as

$$p_{H_k(t)}(x) = \frac{2x\left(1 + K_k(t)\right)}{\Omega_k(t)} \exp\left(-K_k(t) - \frac{\left(1 + K_k(t)\right)x^2}{\Omega_k(t)}\right) I_0\left(2x\sqrt{\frac{K_k(t)\left(K_k(t)+1\right)}{\Omega_k(t)}}\right). \tag{7.25}$$

The DOAs are estimated using the $H_k(t)$ values during an estimation period $\mathcal{P}$, which is defined to be a set of L mini-slots. Each estimation period is divided into Q subestimation periods $\mathcal{P}_i$, $i = 1, \cdots, Q$, such that $\mathcal{P} = \mathcal{P}_1 \cup \mathcal{P}_2 \cup \cdots \cup \mathcal{P}_Q$ and $\mathcal{P}_i \cap \mathcal{P}_j = 0$, $i \neq j$. Each $\mathcal{P}_q$ is composed of L_q mini-slots with

$$ L = \sum_{q=1}^{Q} L_q. $$

In subestimation period $\mathcal{P}_q$, the artificial phase shift $\phi(t)$ of adaptive opportunistic beamforming is set to ϕ_q, i.e., $\phi(t) = \phi_q \in [0, 2\pi)$, for $t \in \mathcal{P}q$, $q = 1, 2, \cdots, Q$.* For example, we can set $\phi_q = (q/Q)2\pi$, for $q = 1, 2, \cdots, Q$. We define $A_{k,q}$, $\Omega_{k,q}$, and $\mathcal{K}_{k,q}$ as follows: $A_{k,q} = A_k(t)|_{t \in \mathcal{P}_q}$, $\Omega_{k,q} = \Omega_k(t)|_{t \in \mathcal{P}_q} = A_{k,q}^2 + \sigma_k^2$, and $\mathcal{K}_{k,q} = \mathcal{K}_k(t)|_{t \in \mathcal{P}_q}$.

Given σ_k and $A_{k,q}$ at the base station, the ML estimation of θ_k is given by

$$ \hat{\theta}_k^{ML} = \arg\max_{\theta_k \in [0,2\pi)} \prod_{q=1}^{Q} \left(\frac{2}{\sigma_k^2} \right)^{L_q} \left\{ \prod_{t \in \mathcal{P}_q} H_k(t) \exp\left(-\frac{H_k^2(t) + A_{k,q}^2}{\sigma_k^2} \right) I_0\left(\frac{2H_k(t)A_{k,q}}{\sigma_k^2} \right) \right\} \quad (7.26) $$

for $k = 1, \cdots, M$. Taking logarithm, a simpler estimator is given by

$$ \hat{\theta}_k^{ML} = \arg\max_{\theta_k \in [0,2\pi)} \sum_{q=1}^{Q} \sum_{t \in \mathcal{P}_q} \left[\ln I_0\left(\frac{2H_k(t)A_{k,q}}{\sigma_k^2} \right) - A_{k,q}^2 \right] \quad (7.27) $$

for $k = 1, \cdots, M$. In order to reduce the computational complexity further at the loss of the ML optimality, suboptimum estimators may be considered. In particular, noting that $G_k(t)$ is maximized when $\phi_q = -\theta_k$, an efficient and very simple suboptimum estimator can be given by

$$ \hat{\theta}_k^{SUB} = -\left(q^* / Q \right) 2\pi, \quad (7.28) $$

$$ q^* = \arg\max_{q=1,\cdots,Q} \frac{1}{L_q} \sum_{t \in \mathcal{P}_q} H_k^2(t). \quad (7.29) $$

7.4.3 Estimation of *K*-Factors of the Physical Channels

In the proposed DOA estimation algorithms, it was assumed that the exact $\{K_k\}_{k=1}^M$ values were known at the base station. In practical systems, however, each user needs

---

* Recall that in conventional opportunistic beamforming, $\phi(t)$ is randomly chosen from $[0, 2\pi)$. Also, note that the pilot overhead is the same for the adaptive and conventional schemes, because only one mini-slot of fixed length is used for pilot signaling in each time slot for both schemes.

to estimate the *K*-factor and feed it back to the base station. Unfortunately, this may raise some problems, such as the change of the receiver of every user and the increased feedback overhead. These problems can be completely avoided if the base station can directly estimate $\{K_k\}$ based on only $H_k(t)$ values, which are already available at the base station. To this end, we first have the following theorem, which gives the relation between the *K*-factors $\{K_k\}$ of the physical channels $h_{n,k}(t)$ and the *K*-factors $\{\mathcal{K}_{k,q}\}$ of (7.21).

Theorem 7.3

[17] "When $\phi_q = (q/Q)2\pi$, $q = 1, \cdots, Q$, the *K*-factors $\{K_k\}$ of the physical channels $h_{n,k}(t)$ can be expressed by the *K*-factors $(\mathcal{K}_{k,q})$ of (7.21) as follows:

$$K_k = \lim_{Q \to \infty} \frac{1}{Q} \sum_{q=1}^{Q} \mathcal{K}_{k,q}, \quad k = 1, \cdots, M."$$ (7.30)

Proof: See appendix D in [17].

Then the following theorem shows that $\{\mathcal{K}_{k,q}\}$ can be estimated only by using $\{\Omega_{k,q}\}$ and $\{H_k(t)\}$:

Theorem 7.4

[17] "The *K*-factors $\{\mathcal{K}_{k,q}\}$ of (7.21) can be expressed in terms of $\Omega_{k,q} = E_q[H_k^2(t)]$ and $E_q[H_k^4(t)]$:

$$\mathcal{K}_{k,q} = \frac{\sqrt{1 - \gamma_{k,q}}}{1 - \sqrt{1 - \gamma_{k,q}}},$$ (7.31)

where

$$\gamma_{k,q} = \frac{-\Omega_{k,q}^2 + E_q\left[H_k^4(t)\right]}{\Omega_{k,q}^2}$$ (7.32)

and E_q denotes the expectation with respect to $H_k(t)$ for $t \in \mathcal{P}_q$."

Proof: See proof of theorem 6 in [17].

The final step is to estimate $\Omega_{k,q}^2$ and $E_q[H_k^4(t)]$ values in (7.32). They can be estimated as in the classical single-input single-output (SISO) systems. Let $\hbar(t)$ and $\bar{L}$ denote the received signal envelope at time t and the total number of samples obtained in the SISO system, respectively. Then it is well known that

$$\left(1/\bar{L}\right)\sum_{t=1}^{\bar{L}} \hbar^2(t)$$

is the ML estimate of $E_h(t)[\hbar^2(t)]$ [24], and $E_h(t)[\hbar^4(t)]$ can be estimated simply as

$$\left(1/\bar{L}\right)\sum_{t=1}^{\bar{L}} \hbar^4(t)$$

[25]. Exactly in the same manner, one can show that $\Omega_{k,q}$ and $E_q[H_k^4(t)]$, and thus $\gamma_{k,q}$, can be estimated as follows:

$$\hat{\gamma}_{k,q} = \frac{-\left(\hat{\Omega}_{k,q}^{ML}\right)^2 + \dfrac{1}{L_q}\displaystyle\sum_{t\in P_q} H_k^4(t)}{\left(\hat{\Omega}_{k,q}^{ML}\right)^2}, \tag{7.33}$$

where $\hat{\Omega}_{k,q}^{ML}$ denotes the ML estimate of $\Omega_{k,q}$, which is given by

$$\hat{\Omega}_{k,q}^{ML} = \frac{1}{L_q}\sum_{t\in P_q} H_k^2(t). \tag{7.34}$$

It follows that the *K*-factors are estimated as follows:

$$\hat{K}_k = \frac{1}{Q}\sum_{q=1}^{Q} \hat{K}_{k,q} = \frac{1}{Q}\sum_{q=1}^{Q} \frac{\sqrt{1-\hat{\gamma}_{k,q}}}{1-\sqrt{1-\hat{\gamma}_{k,q}}}. \tag{7.35}$$

By analytically deriving the Cramer-Rao lower bound (CRLB) of the *K*-factors $\{K_k\}_{k=1}^{M}$, it has been demonstrated that this is a good estimator [17].

7.4 Performance Evaluation

In this section, the performance of the ML estimator of DOAs, the Ricean *K*-factor estimation algorithm, and the adaptive opportunistic beamforming scheme based on DOAs and the *K*-factor estimation are numerically evaluated.

7.4.1 ML Estimation of Users' DOAs

In order to concentrate on the performance evaluation of the DOA estimation, the true values of the Ricean *K*-factor are assumed to be known at the base station. Figure 7.7

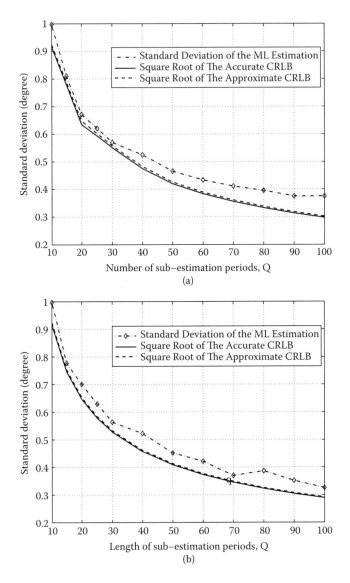

FIGURE 7.7 Comparison of CRLB and the estimation variance of the ML estimator of $\{\theta_k\}_{k=1}^{M}$ (a) $N = 8$, $K_k = 5$, $M = 1$, $L_q = 10$; (b) $N = 8$, $K_k = 5$, $M = 1$, $Q = 10$. (Reproduced from Kim et al., 2006. © 2006, IEEE. With permission.)

shows the square root of the CRLB and the standard deviation of the ML estimator of $\{\theta_k\}_{k=1}^{M}$, when $\{\theta_k\}_{k=1}^{M}$ are generated randomly from $[0,2\pi)$ and $\phi_q = (q/Q)2\pi$. One can see that the standard deviation of the DOA estimation algorithm is very close to the CRLB. This agrees with the theory that the proposed ML estimator of DOAs is asymptotically a minimum variance estimator.

7.4.2 PDF of $\max_{k=1,\cdots,M} H_k(t)$

We define $\mathcal{H}_M(t)$ as follows: $\mathcal{H}_M(t) \overset{def}{=} \max_{k=1,\dots,M} H_k(t)$. The PDFs of $\mathcal{H}_M(t)$ of conventional and adaptive opportunistic beamforming algorithms are compared in Figure 7.8. It can be seen that adaptive opportunistic beamforming considerably shifts the range of $\mathcal{H}_M(t)$

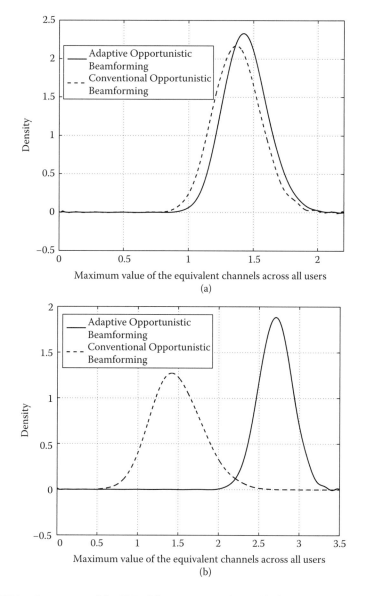

FIGURE 7.8 Comparison of the PDF of the maximum value $\mathcal{H}_k(t)$ of the equivalent channels in conventional and adaptive opportunistic beamforming: (a) $K_k = 10$, $M = 5$, $N = 2$; (b) $K_k = 10$, $M = 5$, $N = 8$. (Reproduced from Kim et al., 2006. © 2006, IEEE. With permission.)

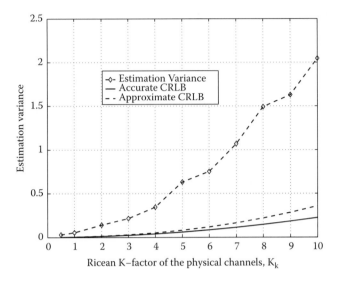

FIGURE 7.9 Comparison of the CRLB and the estimation variance of the K-factor estimator of the physical channels $h_{nk}(t)$, $N = 8$, $M = 1$, $Q = 50$, $L_q = 20$. (Reproduced from Kim et al., 2006. © 2006, IEEE. With permission.)

toward $+\infty$. This implies that the maximum channel gain exploited by multiuser diversity has increased significantly, and thus, the multiuser diversity gain can be notable. As can be seen from (7.18), adaptive opportunistic beamforming achieves more improvement when the number N of transmission antennas grows. This is because the value of $H_k(t)$ increases with N when $\phi(t) \simeq -\theta_k$, and in turn, the value of $\mathcal{H}_M(t)$ increases with N.

7.4.3   Estimation of *K*-Factors of the Physical Channels

In order to evaluate the performance of the K-factor estimator, the estimation variance and the CRLB are compared in Figure 7.9 when $\phi_q = (q/Q)2\pi$. One can see that the difference between the CRLB and the actual variance grows with the K-factor.* However, the difference itself is not very large. For example, when $K_k = 10$, the difference is approximately 2, and thus the standard deviation is approximately 1.4, which is 14% of the true K_k value. From the numerical results in the next section, it turns out that the effect of the K-factor estimation errors on the throughput performance is negligible.

7.4.4   Throughput

The final performance measure is the throughput obtained by different schemes. In slow fading channels, the length t_c of the past window is $t_c = +\infty$, and hence, the proportional fair scheduling algorithm converges to select the user with the highest $R_k(t)$. The

---

* In many previous K-factor estimators for SISO systems, the estimation variance also increases with the K-factor. For example, see [25].

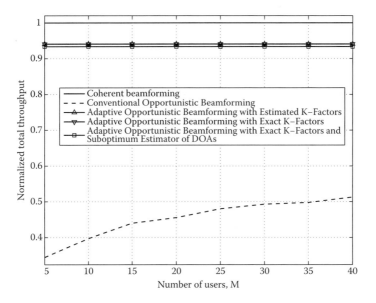

FIGURE 7.10 Normalized total throughput in the slow Ricean fading channels, $K_k = 10$, $N = 8$, $Q = 50$, $L_q = 20$. (Reproduced from Kim et al., 2006. © 2006, IEEE. With permission.)

throughput values of coherent beamforming, conventional opportunistic beamforming, and adaptive opportunistic beamforming are compared. Particularly, to see the impact of the K-factor estimation errors on the performance of adaptive opportunistic beamforming, the throughput is evaluated with exact K-factors and estimated K-factors. All throughput values obtained are normalized by the throughput of coherent transmit beamforming, which is the performance limit. The average SNR $PH_k^2(t)/\rho_k^2$ of each user is set to 0 dB, where P is the transmission power.

Figure 7.10 shows that adaptive opportunistic beamforming considerably outperforms conventional opportunistic beamforming, and actually the performance is very close to that of coherent beamforming. In particular, even with the small number of users, the adaptive scheme performs very well. Also, the throughput degradation due to the K-factor estimation errors and the use of the suboptimum DOA estimator is negligible. This implies that the estimation of users' DOAs is still very accurate with the errors of K-factor estimation, and the suboptimum DOA estimator is working very well at a very low complexity.

Figure 7.11 shows the impact of the number of antennas on the performance. As discussed before, the performance of conventional opportunistic beamforming heavily depends on the number of antennas, as it forms beams randomly and blindly. On the other hand, the adaptive scheme works much better than the conventional one with few antennas.

Finally, Figure 7.12 shows the effect of the K-factor on the performance. With smaller K-factor, the performance improvement diminishes. In the extreme case of $K = 0$, the conventional scheme and the adaptive scheme have the same performance, as expected.

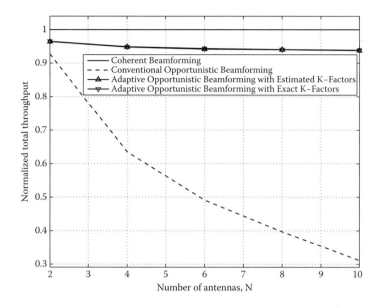

FIGURE 7.11   Normalized total throughput in the slow Ricean fading channels, $K_k = 10$, $M = 10$, $Q = 50$, $L_q = 20$. (Reproduced from Kim et al., 2006. © 2006, IEEE. With permission.)

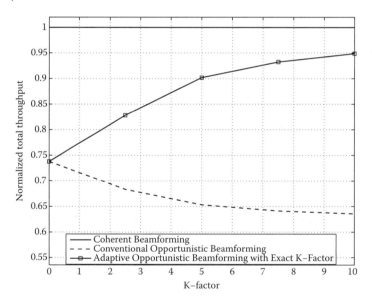

FIGURE 7.12   Normalized total throughput in the slow Ricean fading channels, $M = 10$, $N = 4$, $Q = 50$, $L_q = 20$. (Reproduced from Kim et al., 2006. © 2006, IEEE. With permission.)

On the other hand, as K grows, the performance improvement by the adaptive scheme becomes significant, because the diffused component $B_k(t)$ in (7.15) can be neglected and the LOS component becomes more dominant.

7.5 Conclusions

In the context of multiuser diversity, fading is seen as a resource that can be exploited, not an impediment that must be combated. In a multiuser system where every user has an independent fading channel, transmitting data only to the user with the best channel condition not only increases the throughput of the system but also increases the throughput of every user in the long term on average. The multiuser diversity technique has been adopted in practical systems, such as the downlink of the IS-856 system.

In order to achieve the higher multiuser diversity gain, the dynamic range of the channel fluctuation should be larger and the variation rate of the channel should be faster. When the underlying physical channels have small fluctuation and change slowly, opportunistic beamforming artificially induces the channel fluctuation by using some pseudorandom weight coefficients at the base station. By doing this, higher multiuser diversity can be achieved. However, this opportunistic beamforming requires a large number of users, especially with many transmit antennas at the base station.

Adaptive opportunistic beamforming can solve this problem very well in Ricean fading channels. The adaptive opportunistic beamforming algorithm generates the weight coefficients not randomly, but intelligently by estimating the DOAs of the users. That is, each beam is generated only in the direction where users really exist. This enables adaptive opportunistic beamforming to achieve excellent performance even with a small number of users.

References

[1] G. K. Kaleh. 1996. Frequency-diversity spread-spectrum communication system to counter bandlimited Gaussian interference. *IEEE Trans. Commun.* 44:886–93.

[2] Y. E. Dallal and S. Shamai. 1992. Time diversity in DPSK noisy phase channels. *IEEE Trans. Commun.* 40:1703–15.

[3] L. C. Godara. 1997. Applications of antenna arrays to mobile communications. Part I. Performance improvement, feasibility, and system considerations. *Proc. IEEE* 85:1031–60.

[4] L. C. Godara. 1997. Applications of antenna arrays to mobile communications. Part II. Beamforming and direction-of-arrival considerations. *Proc. IEEE* 85:1195–245.

[5] V. Tarokh, N. Seshadri, and A. R. Calderbank. 1998. Space-time codes for high data rate wireless communication: Performance criterion and code construction. *IEEE Trans. Inform. Theory* 44:744–65.

[6] J. H. Winters. 1998. Smart antennas for wireless systems. *IEEE Personal Commun.* 5:23–27.

[7] R. Knopp and P. Humblet. 1995. Information capacity and power control in single cell multiuser communications. In *Proceedings of the IEEE International Computer Conference (ICC'95)*, Seattle, WA, pp. 331–335.

[8] D. N. C. Tse. 1997. Optimal power allocation over parallel Gaussian channels. In *Proceedings of the International Symposium on Information Theory*, Ulm, Germany, p. 27.

[9] R. Gozali, R. M. Buehrer, and B. D. Woerner. 2003. The impact of multiuser diversity on space-time block coding. *IEEE Commun. Lett.* 7:213–15.

[10] J. Jiang, R. M. Buehrer, and W. H. Tranter. 2004. Antenna diversity in multiuser data networks. *IEEE Trans. Commun.* 52:490–97.

[11] E. G. Larsson. 2004. On the combination of spatial diversity and multiuser diversity. *IEEE Commun. Lett.* 8:517–19.

[12] D. N. C. Tse. 2002. Transmitter directed, multiple receiver system using path diversity to equitably maximize throughput. U.S. Patent 6449490.

[13] P. Viswanath, D. N. C. Tse, and R. Laroia. 2002. Opportunistic beamforming using dumb antennas. *IEEE Trans. Inform. Theory* 48:1277–94.

[14] I.-M. Kim, S.-C. Hong, S. S. Chassemzadeh, and V. Tarokh. 2005. Opportunistic beamforming based on multiple weighting vectors. *IEEE Trans. Wireless Commun.* 4:2683–87.

[15] J. Chung, C.-S. Hwang, K. Kim, and Y. K. Kim. 2003. A random beamforming technique in MIMO systems exploiting multiuser diversity. *IEEE J. Selected Areas Commun.* 21:848–55.

[16] N. Sharma and L. H. Qzarow. 2005. A study of opportunism for multiple-antenna systems. *IEEE Trans. Inform. Theory* 51:1808–14.

[17] I.-M. Kim, Z. Yi, D. Kim, and W. Chung. 2006. Improved opportunistic beamforming in Ricean fading channels. *IEEE Trans. Commun.* 54:2199–211.

[18] P. Bender, P. Black, M. Grob, R. Padovani, N. Sindhushayana, and A. Viterbi. 2000. CDMA/HDR: A bandwidth efficient high speed wireless data service for nomadic users. *IEEE Commun. Mag.* 38:70–78.

[19] D. Tse and P. Viswanath. 2005. *Fundamentals of wireless communication.* Cambridge: Cambridge University Press.

[20] R. O. Schmidt. 1986. Multiple emitter location and signal parameter estimation. *IEEE Trans. Antennas Propagation* 34:276–80.

[21] A. Barabell. 1983. Improving the resolution of eigenstructured based direction finding algorithms. In *Proceedings of IEEE ICASSP' 83*, pp. 336–39.

[22] R. Roy and T. Kailath. 1989. ESPRIT—Estimation of signal parameters via rotational invariance techniques. In *IEEE Trans. Acoustics Speech Signal Processing* 37:984–95.

[23] S. M. Kay. 1993. *Fundamentals of statistical signal processing.* Upper Saddle River, NJ: Prentice-Hall PTR.

[24] A. Abdi, C. Tepedelenlioglu, M. Kaveh, and G. Giannakis. 2001. On the estimation of the K parameters for the Rice fading distribution. *IEEE Commun. Lett.* 5:92–94.

[25] C. Tepedelenlioglu, A. Abdi, and G. B. Giannakis. 2003. The Ricean K factor: Estimation and performance analysis. *IEEE Trans. Wireless Commun.* 2:799–810.

8

Adaptive Beamforming for Multiantenna Communications

Alex B. Gershman
*Darmstadt University
of Technology*

8.1   Introduction

Beamforming is a versatile approach to signal spatial filtering that has found numerous applications in diverse areas, including radar, sonar, wireless communications, geophysics, speech and audio processing, ultrasonic imaging, biomedicine, radio astronomy, and other fields [1]. Early attempts of applying beamforming to wireless communications go back to the late 1970s and early 1980s [2–4]. During the last two decades, there has been a major trend to use multiantenna transceivers in wireless communication systems to facilitate the explosive growth of the number of users and meet their rapidly increasing demands for new high-data-rate services [5–20]. As a result, spatial division multiple access (SDMA) technology recently became one of the key concepts in third and higher generations of mobile communication systems. In particular, the receive (uplink) and transmit (downlink) beamforming techniques used at multiantenna base stations (BSs) have been shown to enable efficient mitigation of multiuser interference and offer substantial improvements in system capacity and performance [6, 7, 14, 19].

In this chapter, we provide an overview of fundamentals and recent advances in the field of receive and transmit beamforming for multiantenna communication systems. The chapter is organized as follows. In section 8.2, the basic receive and transmit signal models are introduced. Section 8.3 is devoted to the receive beamforming problem and methods. In the same section, applications of receive adaptive beamforming to space-time multiuser multiple-input multiple-output (MIMO) receivers are highlighted. The transmit beamforming problem and methods are overviewed in section 8.4, and conclusions are given in section 8.5.

8.2 Basic Signal Models

Let us consider a receive (transmit) M-element BS array depicted in Figure 8.1, whose sensors are weighted by the weight vector $\mathbf{w} = [w_1, w_2, \ldots, w_M]^T$, where $(\cdot)^T$ denotes the transpose. Assume that there are L single-antenna users with flat fading channels. The $M \times 1$ uplink user channel vectors (hereafter referred to as user *spatial signatures*) are denoted by $\mathbf{a}_l, l = 1, \ldots, L$, while the $M \times 1$ downlink channel vectors are denoted by $\mathbf{h}_l, l = 1, \ldots, L$. Note that for each user, the uplink and downlink channel vectors may differ from each other because the difference in the uplink and downlink frequencies in the frequency division duplex (FDD) mode and channel variability in the time division duplex (TDD) mode may violate the uplink-downlink reciprocity property [14]. In the sequel, we assume without any loss of generality that the first user is the user of interest.

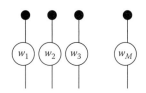

FIGURE 8.1 Receive/transmit beamformer.

8.2.1 The Uplink Case

In the uplink case, the baseband $M \times 1$ complex signal vector received at the BS array can be written as

$$\mathbf{x}(t) = \sum_{l=1}^{L} s_l(t)\mathbf{a}_l + \mathbf{n}(t) = \mathbf{As}(t) + \mathbf{n}(t), \tag{8.1}$$

where $s_l(t)$ is the baseband receive signal waveform of the lth user, $\mathbf{n}(t)$ is the $M \times 1$ vector of additive sensor noise, $\mathbf{A} \triangleq [\mathbf{a}_1, \mathbf{a}_2, \ldots, \mathbf{a}_L]$, $\mathbf{s}(t) \triangleq [s_1(t), s_2(t), \ldots, s_L(t)]^T$, and t is the time index. The additive noise is assumed to be zero mean and spatially white, that is, its correlation matrix is given by

$$E\{\mathbf{n}(t)\mathbf{n}^H(t)\}=\sigma^2\mathbf{I}, \tag{8.2}$$

where $\mathbf{I}$ is the identity matrix, $E\{\cdot\}$ denotes the statistical expectation, and $(\cdot)^H$ stands for the Hermitian transpose.

The baseband complex signal at the output of the receive beamformer can be written as

$$y(t)=\mathbf{w}^H\mathbf{x}(t)=\sum_{l=1}^{L}s_l(t)\mathbf{w}^H\mathbf{a}_l+\mathbf{w}^H\mathbf{n}(t). \tag{8.3}$$

8.2.2 The Downlink Case

Now, let us consider the transmit beamforming mode with single BS and the same signal sent to all users. The baseband signal received by the lth user can be expressed as

$$z_l(t)=s(t)\mathbf{h}_l^H\mathbf{w}+n_l(t), \tag{8.4}$$

where $s(t)$ is the transmit baseband signal, $\mathbf{w}$ is the BS weight vector, and $n_l(t)$ is additive noise at the lth user.

The latter model can be further extended to the case of K different BSs and L mobile users. Let $\mathbf{w}_l$ be the weight vector used at the BS assigned to the lth user to transmit the baseband signal $s_l(t)$ to this user. Let us also define the BS cell site index $c(l)$ as the index of the particular BS that is assigned to the lth user. Note that $c(l) = c(m)$ if both the lth and mth users are assigned to the same BS, and $c(l) \neq c(m)$ if these users are assigned to different BSs. Using these notations, the vector of signals transmitted from the kth BS can be expressed as [14]

$$\mathbf{x}_k(t)=\sum_{i\in\mathcal{G}(k)}s_i(t)\mathbf{w}_i, \tag{8.5}$$

where

$$\mathcal{G}(k)\triangleq\{i:c(i)=k\} \tag{8.6}$$

is the set of indices of all weight vectors that are used at the kth BS (or equivalently, the set of indices of all users that are assigned to this BS). Equation (8.5) implies that the kth BS transmits only to the users that are assigned to it rather than to all the users in the cellular network.

Using (8.5), the baseband signal received by the lth user can be modeled as [14]

$$z_l(t)=\sum_{k=1}^{K}\mathbf{h}_{l,k}^H\mathbf{x}_k(t)+n_l(t), \tag{8.7}$$

where $\mathbf{h}_{l,k}$ is the downlink channel vector between the kth BS and the lth user.

8.3   Receive Beamforming

The goal of receive beamforming is, given the knowledge of the spatial signature of the user of interest, to receive the signal of this user (hereafter referred to as the *desired signal*) with the maximal possible gain while suppressing the interfering users and noise as much as possible. The spatial signature of the user of interest has to be estimated in advance using, for example, traditional training-based techniques or blind approaches; see [21–25] and references therein.

8.3.1   Maximal Ratio Combining

The most traditional and simplest beamforming strategy, called *maximal ratio combining* (MRC), is to maximize the signal-to-noise ratio (SNR):

$$
\text{SNR} = \frac{\text{E}\left\{\left|s_1(t)\mathbf{w}^H\mathbf{a}_1\right|^2\right\}}{\text{E}\left\{\left|\mathbf{w}^H\mathbf{n}(t)\right|^2\right\}} = \frac{\sigma_1^2\left|\mathbf{w}^H\mathbf{a}_1\right|^2}{\sigma^2\mathbf{w}^H\mathbf{w}}
\tag{8.8}
$$

ignoring the effect of multiuser interference. Here, σ_l^2 is the power of the lth user, and as already mentioned before, the first user is assumed to be the user of interest.

From the Cauchy-Schwartz inequality it follows that the SNR in (8.8) is maximized with

$$
\mathbf{w}_{\text{MRC}} = \alpha\mathbf{a}_1,
\tag{8.9}
$$

where α is an arbitrary constant that does not affect the SNR value. It should be stressed that the MRC approach belongs to the class of nonadaptive (conventional) beamforming techniques because its weight vector does not depend on the received data.

It is well known that the MRC approach works acceptably well if

$$
\sigma_1\left\|\mathbf{a}_1\right\|^2 \gg \sigma_i\left|\mathbf{a}_1^H\mathbf{a}_i\right|, \quad i = 2, ..., L,
\tag{8.10}
$$

where $\|\cdot\|$ hereafter denotes the 2-norm of a vector or the Frobenius norm of a matrix. From (8.10), it follows that the MRC method can only be used when the users are spatially well separated (so that $\|\mathbf{a}_1\|^2 \gg |\mathbf{a}_1^H\mathbf{a}_i|$), and when the powers of interfering users, σ_i^2 ($i = 2, ..., L$), do not substantially exceed the power of the user of interest, σ_1^2. However, if the power of one or more interfering users is substantially higher than that of the user of interest, then the MRC technique is not a proper approach anymore, and we have to resort to *adaptive* beamforming methods.

8.3.2   Minimum Variance Beamforming

The main idea of adaptive beamforming is to maximize the signal-to-interference-plus-noise ratio (SINR) rather than the SNR. Therefore, in contrast to the MRC approach, the effect of multiuser interference is no longer ignored. The output SINR is given by

$$\mathrm{SINR} = \frac{\mathbf{w}^H \mathbf{R}_s \mathbf{w}}{\mathbf{w}^H \mathbf{R}_{i+n} \mathbf{w}}, \tag{8.11}$$

where $\mathbf{R}_s$ is the correlation matrix of the desired signal component in (8.1), whereas $\mathbf{R}_{i+n}$ is the correlation matrix of the interference and noise components. From (8.1) it follows that

$$\mathbf{R}_s = \sigma_1^2 \mathbf{a}_1 \mathbf{a}_1^H, \tag{8.12}$$

and, therefore, (8.11) can be rewritten as

$$\mathrm{SINR} = \frac{\sigma_1^2 \left| \mathbf{w}^H \mathbf{a}_1 \right|^2}{\mathbf{w}^H \mathbf{R}_{i+n} \mathbf{w}}. \tag{8.13}$$

To obtain the optimal weight vector that maximizes the SINR in (8.11), one can minimize the output interference-plus-noise power while maintaining the distortionless array response to the desired signal [26, 27]:

$$\min_{\mathbf{w}} \mathbf{w}^H \mathbf{R}_{i+n} \mathbf{w} \quad \text{subject to} \quad \mathbf{w}^H \mathbf{R}_s \mathbf{w} = 1. \tag{8.14}$$

Taking into account (8.12), and therefore using (8.13) instead of (8.11), the problem of maximizing SINR under the distortionless response constraint can be rewritten in a simpler form [26]:

$$\min_{\mathbf{w}} \mathbf{w}^H \mathbf{R}_{i+n} \mathbf{w} \quad \text{subject to} \quad \mathbf{w}^H \mathbf{a}_1 = 1. \tag{8.15}$$

The beamforming problems (8.14) and (8.15) are commonly referred to as the *minimum variance* (MV) problems. The solution to the general MV beamforming problem (8.14) is given by [27]

$$\mathbf{w}_{opt} = \mathcal{P} \left\{ \mathbf{R}_{i+n}^{-1} \mathbf{R}_s \right\}, \tag{8.16}$$

where $\mathcal{P}\{\cdot\}$ stands for the *principal eigenvector* of a matrix. Note that this eigenvector should be normalized to enable the resulting weight vector to satisfy the distortionless response constraint in (8.14). However, it is clear from (8.11) that any rescaling of the weight vector by a complex nonzero constant does not alter the output SINR (8.11). Hence, the aforementioned normalization is immaterial.

The solution to the simplified problem (8.15) is given by

$$\mathbf{w}_{opt} = \alpha \mathbf{R}_{i+n}^{-1} \mathbf{a}_1, \tag{8.17}$$

where α can be obtained from the distortionless response constraint in (8.15) and is equal to $\alpha = 1/\mathbf{a}_1^H \mathbf{R}_{i+n}^{-1} \mathbf{a}_1$ [26]. Again, this constant is immaterial and therefore will be dropped in the sequel.

8.3.2.1   The SMI Algorithm

In practical applications, the true matrix $\mathbf{R}_{i+n}$ is unavailable but can be estimated from the receiver data* as [6, 26, 28]

$$\hat{\mathbf{R}} = \frac{1}{N}\sum_{t=1}^{N}\mathbf{x}(t)\mathbf{x}^{H}(t), \tag{8.18}$$

where N is the number of snapshots available. The key idea of the sample matrix inversion (SMI)-based version of the MV beamformer is to replace $\mathbf{R}_{i+n}$ by (8.18) [28]. For the MV beamforming problems in (8.14) and (8.15), the SMI weight vectors are given by [26, 28]

$$\mathbf{w}_{MV} = \mathcal{P}\{\hat{\mathbf{R}}^{-1}\mathbf{R}_{s}\}, \tag{8.19}$$

$$\mathbf{w}_{MV} = \hat{\mathbf{R}}^{-1}\mathbf{a}_{1}, \tag{8.20}$$

respectively.

The use of $\hat{\mathbf{R}}$ instead of $\mathbf{R}_{i+n}$ in (8.19) or (8.20) is known to cause severe performance degradation in the case when the desired signal component is present in the beamformer snapshots used to estimate the sample correlation matrix (in the sequel, the latter case is referred to as the *signal-present* case as opposed to the *signal-free* case). Although in the signal-free case the output SINR of the SMI beamformer (8.20) rapidly converges to the optimal SINR value with increasing N [28], in the signal-present case this convergence becomes much slower [29], and the beamformer performance degrades severely even in the presence of small errors between the presumed and actual spatial signatures of the user of interest [27, 30]. Such signature errors can be caused by high user mobility and wireless channel variability, as well as by a limited amount of training symbols and the effect of multiuser interference that may prevent obtaining spatial signature estimates of acceptable quality. The effect of the beamformer performance degradation due to spatial signature errors is commonly known as *signal self-nulling*.

Another typical cause of the beamformer performance degradation in mobile communications is a highly nonstationary behavior of the propagation channel and, in particular, high mobility of interfering users [31]. The effect of such nonstationarity on the performance of receive adaptive beamformers is that the array weights may not be able to adapt fast enough to compensate for the interfering user motion [30, 31]. This phenomenon is usually referred to as *interference undernulling*.

---

* These data can be either information-bearing or training symbols. Therefore, there is no need to increase the amount of training data when using the SMI-based MV beamformers in multiantenna wireless communication systems.

8.3.3   Robust Minimum Variance Beamforming

8.3.3.1   Linearly Constrained Minimum Variance Beamformer

A classic approach to prevent signal self-nulling and improve the robustness of MV beamforming against spatial signature errors is to use additional point or derivative mainlobe constraints [26, 32, 33].

In the case of point constraints, the sample version of the MV beamforming problem (8.15) is modified as [26]

$$\min_{\mathbf{w}} \mathbf{w}^H \hat{\mathbf{R}} \mathbf{w} \quad \text{subject to} \quad \mathbf{C}^H \mathbf{w} = \mathbf{f}, \tag{8.21}$$

where $\mathbf{C}$ is the $M \times P$ matrix of P constrained steering vectors in the neighborhood of the presumed spatial signature of the user of interest, and the $P \times 1$ vector $\mathbf{f} = [1, 1, \ldots, 1]^T$. In the case of derivative constraints, the problem can be formulated in the same way as in (8.21), but the matrix $\mathbf{C}$ and the vector $\mathbf{f}$ should be defined in a different way [26]. The multiple constraints given by the equation $\mathbf{C}^H \mathbf{w} = \mathbf{f}$ in (8.21) help to stabilize the mainlobe area of the array beampattern, and therefore to prevent signal self-nulling effects. This approach is commonly referred to as linearly constrained minimum variance (LCMV) beamforming.

The solution to (8.21) is given by [26]

$$\mathbf{w}_{\mathrm{LCMV}} = \hat{\mathbf{R}}^{-1} \mathbf{C} \left(\mathbf{C}^H \hat{\mathbf{R}}^{-1} \mathbf{C} \right)^{-1} \mathbf{f}. \tag{8.22}$$

Although the LCMV beamformer is a popular technique in radar and sonar, its major shortcoming in wireless communications is that it requires exact knowledge of the array manifold, that is, it assumes the structure of the user-of-interest spatial signature to be known up to the signal direction-of-arrival parameters. In traditional array processing–based source localization techniques, the simplest free-space propagation model is typically adopted, that is, the source spatial signatures are assumed to be plane waves. However, in wireless communications, the plane wave structure of the user spatial signatures may be distorted by multipath scattering effects [34], and it is rather difficult to precisely parameterize the spatial signature vectors in terms of directions-of-arrival (DOAs). This may limit the use of the LCMV beamforming approach in wireless communications.

8.3.3.2   Diagonally Loaded Minimum Variance Beamformer

A popular approach to improve the robustness of MV beamformers is the so-called diagonal loading technique, whose idea is to *regularize* the adaptive array weights by adding a quadratic penalty term to the objective function in (8.14) or (8.15) [35–37]. The sample version of such a modified objective function can be written as

$$\mathbf{w}^H \hat{\mathbf{R}} \mathbf{w} + \gamma \mathbf{w}^H \mathbf{w} = \mathbf{w}^H (\hat{\mathbf{R}} + \gamma \mathbf{I}) \mathbf{w}, \tag{8.23}$$

where γ is a fixed loading factor that penalizes high-norm realizations of the weight vector. Minimizing this objective function under the distortionless response constraints used in (8.14) and (8.15), the following SMI-based diagonally loaded minimum variance (DLMV) beamformers can be obtained [26, 27, 35, 36]:

$$\mathbf{w}_{\text{DLMV}} = \mathcal{P}\left\{(\hat{\mathbf{R}}+\gamma\mathbf{I})^{-1}\mathbf{R}_s\right\}, \tag{8.24}$$

$$\mathbf{w}_{\text{DLMV}} = (\hat{\mathbf{R}}+\gamma\mathbf{I})^{-1}\mathbf{a}_1. \tag{8.25}$$

From (8.24) and (8.25), it can be seen that the diagonal loading operation can be viewed as injecting an artificial amount of white noise into the main diagonal of the sample correlation matrix. This warrants that, even though in the small sample case $\hat{\mathbf{R}}$ may be singular, the diagonally loaded matrix $\hat{\mathbf{R}} + \gamma\mathbf{I}$ is always positive definite, and therefore invertible.

In addition to ensuring that the correlation matrix is invertible, the diagonal loading approach is known to substantially improve the performance of the SMI-based MV beamformers (8.19) and (8.20) in the signal-present case. This improvement is especially pronounced in scenarios with mismatched spatial signature of the user of interest [30, 36, 38].

Despite the popularity of the DLMV beamforming technique, its applications may be limited by the fact that the optimal choice of the loading factor γ is scenario-dependent, and therefore, *ad hoc* choices of γ proposed in the literature [35–37] can lead to a drastic degradation of the beamformer performance.

8.3.3.3 Eigenspace-Based Beamformer

Another useful technique that helps to prevent signal self-nulling is the eigenspace-based beamformer [29, 39]. The essence of this beamformer is to reduce the spatial signature errors by projecting the presumed signature onto the estimated signal-plus-interference subspace obtained by means of the eigendecomposition of the sample correlation matrix

$$\hat{\mathbf{R}} = \hat{\mathbf{E}}\hat{\mathbf{\Lambda}}\hat{\mathbf{E}}^H + \hat{\mathbf{G}}\hat{\mathbf{\Gamma}}\hat{\mathbf{G}}^H ,$$

where the $M \times L$ matrix $\hat{\mathbf{E}}$ contains the L signal-plus-interference subspace eigenvectors of $\hat{\mathbf{R}}$, and the $L \times L$ diagonal matrix $\hat{\mathbf{\Lambda}}$ contains the corresponding eigenvalues of this matrix. Similarly, the $M \times (M - L)$ matrix $\hat{\mathbf{G}}$ contains the $(M - L)$ noise subspace eigenvectors of $\hat{\mathbf{R}}$, and the $(M - L) \times (M - L)$ diagonal matrix $\hat{\mathbf{\Gamma}}$ contains the corresponding eigenvalues. The total number of users is assumed to be known at the receiver and less than M. The weight vector of the eigenspace-based beamformer can be written as

$$\mathbf{w}_{\text{eig}} = \hat{\mathbf{R}}^{-1}\mathbf{P}_{\hat{\mathbf{E}}}\,\mathbf{a}_1, \tag{8.26}$$

where $\mathbf{P}_{\hat{\mathbf{E}}} = \hat{\mathbf{E}}(\hat{\mathbf{E}}^H\hat{\mathbf{E}})^{-1}\hat{\mathbf{E}}^H = \hat{\mathbf{E}}\hat{\mathbf{E}}^H$ is the orthogonal projection matrix onto the estimated signal-plus-interference subspace.

If the number of users is low and their SNRs are high, the eigenspace-based beam-former is known to provide a significant improvement of the robustness against spatial signature errors compared to the SMI-based MV beamformer [39]. However, the eigen-space-based approach may degrade severely if the number of users is comparable with the number of BS antennas, or if the SNRs of some of the users are low. In the latter case, subspace swap effects may cause a severe performance degradation due to the errors in estimating the user spatial signature subspace [27]. This makes it very difficult to use the eigenspace-based beamformer in mobile communications where the number of users served may be comparable to the number of BS array sensors, and some user powers may be low due to near–far effects.

8.3.3.4 MV Beamformers with Data-Dependent Beampattern Null Constraints

In mobile communications, user spatial signatures can rapidly change in time because of the user mobility and channel variability effects. As has been mentioned before, this may cause interference undernulling and, as a result, lead to a substantial performance degradation of BS receive beamformers. To preserve the beamformer performance in nonstationary scenarios, the idea of artificially broadening the adaptive beampattern nulls in interferer directions can be used [31, 40–42], where either point or derivative data-dependent constraints can be exploited to specify the array beampattern in the adaptive null areas.

One such technique developed in [41] uses derivative data-driven constraints (DDCs). The essence of this approach is to replace the sample correlation matrix in the MV or DLMV beamformers by the following modified sample correlation matrix:

$$\hat{\mathbf{R}}_{\mathrm{mod}} = \sum_{p=0}^{P} \zeta_p \mathbf{B}^p \hat{\mathbf{R}} \mathbf{B}^p, \tag{8.27}$$

where $\mathbf{B}$ is an $M \times M$ diagonal matrix whose entries are determined by the known array geometry, P is the highest order of the data-dependent constraints, and ζ_p ($p = 0, \ldots, P, \zeta_0 = 1$) determine a proper trade-off between the constraints of different orders. In practice, $P = 1$ has been demonstrated to suffice for providing a substantial robustness against interferer motion [41]. In the case $P = 1$, (8.27) can be rewritten in a simpler form:

$$\hat{\mathbf{R}}_{\mathrm{mod}} = \hat{\mathbf{R}} + \zeta \mathbf{B} \hat{\mathbf{R}} \mathbf{B}, \tag{8.28}$$

where ζ determines the required trade-off between the null depth and width. Under several mild conditions, the optimal value of ζ was shown to remain independent of the user parameters and can be straightforwardly obtained from the known BS array parameters [41].

Another approach to widen the adaptive beampattern nulls is based on data-driven point constraints [31, 40] and is commonly called the correlation matrix tapering (CMT) technique [42]. The essence of this technique is to replace the sample correlation matrix in the MV or DLMV beamformer by the so-called *tapered* sample correlation matrix:

$$\hat{\mathbf{R}}_{\text{tap}} = \hat{\mathbf{R}} \odot \mathbf{T}, \qquad\qquad (8.29)$$

where $\mathbf{T}$ is the $M \times M$ matrix taper and $\odot$ denotes the Schur-Hadamard (element-wise) matrix product. Particular designs of matrix tapers are discussed in [31], [40], and [42].

An interesting link between the DDC and CMT approaches was found in [43], where it has been proven that the matrix (8.27) can be viewed as a tapered correlation matrix (8.29) with some particular choice of $\mathbf{T}$. Hence, the DDC approach can be interpreted in terms of the CMT approach.

Although both the DDC and CMT approaches have been shown to provide excellent robustness against moving interferers with plane wavefronts, they may severely degrade when the wavefronts of the interferers deviate from the plane wavefront form [27]. This may essentially limit the application of these two approaches in wireless communications.

8.3.3.5   Worst-Case Minimum Variance Beamformers

All the previously discussed robust beamforming techniques use rather *ad hoc* ways to incorporate the robustness feature into the SMI-based MV beamforming scheme. Recently, more theoretically rigorous techniques have been proposed that improve the robustness of MV beamforming by means of *worst-case* performance optimization [38, 44–46]. Such worst-case MV beamformers are discussed in this section.

The first worst-case robust MV (RMV) beamformer was developed in [38], where the mismatch

$$\boldsymbol{\delta} \triangleq \tilde{\mathbf{a}}_1 - \mathbf{a}_1 \qquad\qquad (8.30)$$

between the actual spatial signature $\tilde{\mathbf{a}}_1$ of the user of interest and its presumed version $\mathbf{a}_1$ was assumed to be norm-bounded by some known constant ε, that is,

$$\|\boldsymbol{\delta}\| \le \varepsilon. \qquad\qquad (8.31)$$

Equation (8.31) corresponds to the case when all possible mismatched spatial signatures of the user of interest belong to a *spherical uncertainty set*. The essence of the approach proposed in [38] is to add robustness to the SMI-based MV beamforming problem by maximizing the output SINR for the worst-case spatial signature mismatch that satisfies (8.31). This is equivalent to minimizing the output beamformer power under the worst-case distortionless response constraint, which should be satisfied for all mismatched spatial signature vectors that belong to the spherical uncertainty set of (8.31). The latter problem can be written as the following modification of the sample MV problem [38]:

$$\min_{\mathbf{w}} \mathbf{w}^H \hat{\mathbf{R}} \mathbf{w} \quad \text{subject to} \quad \left| \mathbf{w}^H (\mathbf{a}_1 + \boldsymbol{\delta}) \right| \ge 1 \quad \text{for all } \|\boldsymbol{\delta}\| \le \varepsilon. \qquad (8.32)$$

At first glance, the problem in (8.32) appears to be computationally intractable because it involves minimization of a quadratic function subject to infinitely many nonconvex quadratic constraints. However, it has been found in [38] that (8.32) can be converted to a much simpler form:

$$\min_{\mathbf{w}} \mathbf{w}^H \hat{\mathbf{R}} \mathbf{w} \quad \text{subject to} \quad \mathbf{w}^H \mathbf{a}_1 \geq \varepsilon \|\mathbf{w}\| + 1, \tag{8.33}$$

where the constraint in (8.33) can be shown to hold with equality at the optimal point of (8.33). The problem in (8.33) belongs to the class of convex second-order cone programming (SOCP) problems [47–49], which can be easily solved using efficient interior point methods [50] at complexity $O(M^3)$.

An alternative way to solve the problem in (8.33) is to resort to Newton-type algorithms. Several approaches of that type (all with complexities $O(M^3)$) are described in [46–52] to solve problem (8.33) and its extensions to the ellipsoidal uncertainty case [46, 52], and to a more general class of beamformers or multiuser detectors [44, 51, 52].

In [45], the RMV beamformer of (8.33) has been generalized to account for interferer nonstationarity in addition to the spatial signature errors. The essence of the approach of [45] is, in addition to modeling the uncertainty in the spatial signature vector, to model data nonstationarity by considering an uncertainty in the data matrix

$$\mathbf{X} \triangleq [\mathbf{x}(1), \dots \mathbf{x}(N)]. \tag{8.34}$$

The sample correlation matrix can be expressed through the data matrix (8.34) as

$$\hat{\mathbf{R}} = \frac{1}{N} \mathbf{X} \mathbf{X}^H. \tag{8.35}$$

To take into account the nonstationarity of the array data, the so-called *mismatch matrix*

$$\boldsymbol{\Delta}_x \triangleq \tilde{\mathbf{X}} - \mathbf{X} \tag{8.36}$$

was introduced in [45], where $\tilde{\mathbf{X}}$ and $\mathbf{X}$ stand for the actual and presumed data matrices, respectively. Here, $\mathbf{X}$ is the data matrix acquired by the beamformer, while the actual data matrix $\tilde{\mathbf{X}}$ may differ from $\mathbf{X}$ because of a nonstationary character of sensor array snapshots. That is, the data samples in $\mathbf{X}$ can become irrelevant at the time when the beamformer is used (and when the actual, yet unknown, data matrix is $\tilde{\mathbf{X}}$ rather than $\mathbf{X}$). In such nonstationary scenarios, the actual sample correlation matrix is given by

$$\hat{\tilde{\mathbf{R}}} = \frac{1}{N} \tilde{\mathbf{X}} \tilde{\mathbf{X}}^H = \frac{1}{N} (\mathbf{X} + \boldsymbol{\Delta}_x)(\mathbf{X} + \boldsymbol{\Delta}_x)^H. \tag{8.37}$$

To make the SMI-based MV beamformer jointly robust against both the interference nonstationarity and spatial signature error effects, the authors of [45] have exploited an idea similar to that in [38]. In particular, it has been assumed that the norms of both the spatial signature mismatch $\boldsymbol{\delta}$ and the data matrix mismatch $\boldsymbol{\Delta}_x$ are norm-bounded by some known constants ε and η as

$$\|\boldsymbol{\delta}\| \leq \varepsilon, \tag{8.38}$$

$$\|\boldsymbol{\Delta}_x\| \leq \eta. \tag{8.39}$$

Then, the problem in (8.32) can be extended as [45]

$$\min_{\mathbf{w}} \max_{\|\boldsymbol{\Delta}_x\| \leq \eta} \left\| (\mathbf{X} + \boldsymbol{\Delta}_x)^H \mathbf{w} \right\| \quad \text{subject to} \quad \left| \mathbf{w}^H (\mathbf{a}_1 + \boldsymbol{\delta}) \right| \geq 1 \quad \text{for all} \quad \|\boldsymbol{\delta}\| \leq \varepsilon \tag{8.40}$$

This problem has been converted in [45] to an equivalent form:

$$\min_{\mathbf{w}} \left\| \mathbf{X}^H \mathbf{w} \right\| + \eta \|\mathbf{w}\| \quad \text{subject to} \quad \mathbf{w}^H \mathbf{a}_1 \geq \varepsilon \|\mathbf{w}\| + 1. \tag{8.41}$$

The problem in (8.41) belongs to the class of convex SOCP problems and, similar to (8.33), can be efficiently solved using standard interior point methods [50].

Another interesting approach to extend problem (8.32) to a more general class of problems and, at the same time, to turn the solution for the weight vector into a closed form has been developed in [44]. The authors of [44] have considered the so-called general-rank signal case where the array response to the user of interest is characterized by the signal correlation matrix $\mathbf{R}_s$ rather than the spatial signature $\mathbf{a}_1$, and then have relaxed the so-obtained problem to come up with a closed-form beamformer.

Following [44], let us consider the errors between the presumed and actual interference-plus-noise and signal correlation matrices. These unknown error matrices can be expressed as

$$\boldsymbol{\Delta}_s \triangleq \tilde{\mathbf{R}}_s - \mathbf{R}_s, \tag{8.42}$$

$$\boldsymbol{\Delta}_{i+n} \triangleq \tilde{\mathbf{R}}_{i+n} - \mathbf{R}_{i+n}, \tag{8.43}$$

where $\mathbf{R}_s$ and $\mathbf{R}_{i+n}$ are the presumed signal and interference-plus-noise correlation matrices, respectively, whereas $\tilde{\mathbf{R}}_s$ and $\tilde{\mathbf{R}}_{i+n}$ are the actual values of these matrices. It should be stressed that the presumed value of the interference-plus-noise correlation matrix is given by the sample array correlation matrix, that is, $\mathbf{R}_{i+n} = \hat{\mathbf{R}}$ in (8.43).

In the case of nonzero mismatches $\boldsymbol{\Delta}_s$ and $\boldsymbol{\Delta}_{i+n}$, equation (8.11) for the output SINR has to be rewritten as

$$\text{SINR} = \frac{\mathbf{w}^H \tilde{\mathbf{R}}_s \mathbf{w}}{\mathbf{w}^H \tilde{\mathbf{R}}_{i+n} \mathbf{w}}. \tag{8.44}$$

In the spirit of [38], the approach of [44] assumes that the error matrices $\boldsymbol{\Delta}_s$ and $\boldsymbol{\Delta}_{i+n}$ are norm-bounded by known constants ε and γ, that is,

$$\left\| \boldsymbol{\Delta}_s \right\| \le \varepsilon, \quad \left\| \boldsymbol{\Delta}_{i+n} \right\| \le \gamma. \tag{8.45}$$

To provide robustness against errors in the signal and interference-plus-noise correlation matrices, it has been suggested in [44] to obtain the beamformer weight vector by means of maximizing the worst-case output SINR:

$$\max_{\mathbf{w}} \min_{\boldsymbol{\Delta}_s, \boldsymbol{\Delta}_{i+n}} \frac{\mathbf{w}^H (\mathbf{R}_s + \boldsymbol{\Delta}_s) \mathbf{w}}{\mathbf{w}^H (\hat{\mathbf{R}} + \boldsymbol{\Delta}_{i+n}) \mathbf{w}} \quad \text{for all} \quad \left\| \boldsymbol{\Delta}_s \right\| \le \varepsilon, \left\| \boldsymbol{\Delta}_{i+n} \right\| \le \gamma. \tag{8.46}$$

Clearly, (8.46) may be only an approximate (relaxed) version of the worst-case SINR optimization problem in the case when the worst-case values of the matrices $\boldsymbol{\Delta}_s$ and $\boldsymbol{\Delta}_{i+n}$ lead to the matrices $\mathbf{R}_s + \boldsymbol{\Delta}_s$ and $\hat{\mathbf{R}} + \boldsymbol{\Delta}_{i+n}$ that are not positive semidefinite.* In other words, to warrant the strict equivalence of these two problems, the constraints

$$\mathbf{R}_s + \boldsymbol{\Delta}_s \succeq 0, \tag{8.47}$$

$$\hat{\mathbf{R}} + \boldsymbol{\Delta}_{i+n} \succeq 0 \tag{8.48}$$

should be added to (8.46), where $\mathbf{Q} \succeq 0$ for any matrix $\mathbf{Q}$ means that this matrix is positive semidefinite. The authors of [44], however, drop these two constraints with the motivation that they make the optimization problem much more complicated, and that ignoring them can only make the problem more conservative.

It has been shown in [44] that the worst-case error matrices $\boldsymbol{\Delta}_s$ and $\boldsymbol{\Delta}_{i+n}$ satisfying (8.45) are given by

$$\boldsymbol{\Delta}_s^{\text{WC}} = -\varepsilon \frac{\mathbf{w}\mathbf{w}^H}{\left\| \mathbf{w} \right\|^2}, \tag{8.49}$$

$$\boldsymbol{\Delta}_{i+n}^{\text{WC}} = \gamma \frac{\mathbf{w}\mathbf{w}^H}{\left\| \mathbf{w} \right\|^2}, \tag{8.50}$$

---

* Note that any correlation matrix must by definition be positive semidefinite.

and therefore, the solution to the RMV beamforming problem (8.46) can be expressed as

$$\mathbf{w}_{\text{RMV}} = \mathcal{P}\left\{(\hat{\mathbf{R}} + \gamma \mathbf{I})^{-1}(\mathbf{R}_s - \varepsilon \mathbf{I})\right\}. \tag{8.51}$$

From (8.49) and (8.51) it follows that, although $\boldsymbol{\Delta}_{\text{i+n}}^{\text{WC}}$ satisfies (8.48), the constraint (8.47) is always violated with $\boldsymbol{\Delta}_s^{\text{WC}}$. Therefore, as expected, problem (8.46) is an approximate (conservative) version of the worst-case SINR optimization problem.

Taking into account that $\mathbf{R}_s = \sigma_1^2 \mathbf{a}_1 \mathbf{a}_1^H$ and absorbing σ_1^2 in ε (that is, assuming without any loss of generality that $\sigma_1^2 = 1$), equation (8.51) can be rewritten as

$$\mathbf{w}_{\text{RMV}} = \mathcal{P}\left\{(\hat{\mathbf{R}} + \gamma \mathbf{I})^{-1}(\mathbf{a}_1 \mathbf{a}_1^H - \varepsilon \mathbf{I})\right\}. \tag{8.52}$$

From (8.51) and (8.52) it can be seen that the approach of [44] is equivalent to a combination of negative and positive diagonal loading. In the case of $\varepsilon = 0$, (8.51) and (8.52) simplify to the DLMV beamformer weight vectors (8.24) and (8.25), respectively.

An interesting interpretation of these two types of diagonal loading has been established in [44], where it has been obtained that (8.51) can be interpreted as a generalization of the DLMV beamformer whose amount of positive diagonal loading is *scenario-adaptive* rather than fixed. Similar adaptive diagonal loading interpretation of the RMV beamformer (8.33) has been found in [38], where it has been shown that the amount of diagonal loading in the latter RMV beamformer is optimally matched to the uncertainty in the spatial signature of the user of interest.

Although the worst-case RMV beamformers are known to be very robust techniques, they might be overly conservative because the actual worst operational conditions may occur in practice with a very low probability. Thus, obtaining less conservative robust alternatives to the worst-case techniques is of significant interest. This motivated the authors of [53–55] to develop an alternative, more flexible approach to RMV beamforming that can provide the robustness against spatial signature errors with a certain selected probability (i.e., using soft probabilistic constraints rather than deterministic worst-case constraints). In particular, the probabilistically constrained counterpart of problem (8.32) can be written as [53]

$$\min_{\mathbf{w}} \mathbf{w}^H \hat{\mathbf{R}} \mathbf{w} \text{ subject to } \Pr\left\{\left|\mathbf{w}^H(\mathbf{a}_1 + \boldsymbol{\delta})\right| \geq 1\right\} > p, \tag{8.53}$$

where $\boldsymbol{\delta}$ is assumed to be a random vector drawn from some known distribution, $\Pr\{\cdot\}$ is the probability operator whose explicit form can be obtained from the statistical assumptions on the steering vector error, and p is some preselected probability threshold value. It can be seen that, in contrast to the *hard* constraint used in (8.32) (that requires the distortionless response to be maintained for all norm-bounded error vectors in the uncertainty set), the *soft* constraint in (8.53) maintains the distortionless response only for the error vectors $\boldsymbol{\delta}$ whose probability is high enough, while skipping the distortionless response condition for those values of δ that are unlikely to occur. Therefore, the constraint in (8.53) can be interpreted as an *outage probability* constraint that maintains this probability acceptably low.* This makes the beamformer design in (8.53) more flexible than that in (8.32) because using soft constraints, more freedom is left for minimizing the objective function. Furthermore, the choice of the design parameter p in (8.53) is dictated by the quality-of-service (QoS) requirements (outage probability), and therefore is easier to specify than the robustness parameter ε in (8.32).

Although the problem in (8.53) describes a potentially more attractive beamformer design than (8.32), the main challenge of using (8.53) is that this problem belongs to the class of *chance programming* problems [56] that are rather difficult to solve. To enable a practical solution, several approximations of (8.53) by deterministic convex optimization problems have been considered in [53–55].

8.3.4 Extensions of Minimum Variance Beamforming to MIMO Communications

In this section, a useful extension of the theory of MV beamforming to the problem of designing multiantenna multiuser receivers for space-time block-coded multiple-input multiple-output (MIMO) communications is considered. The problem of MIMO receiver design is generally more difficult that the receive beamformer design problem because in MIMO communications, both the receive BS antenna and mobile transmitters have multiple antennas, and the information symbols may be sent in different time slots through different antennas of each

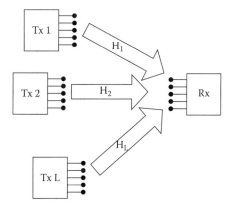

FIGURE 8.2 Uplink multiuser MIMO scenario.

transmitter using some space-time code. The problem is to design a linear receiver that enables the decoding of each symbol of the user of interest while mitigating multiaccess interference (MAI) caused by the remaining users [57–60]. To illustrate the problem, a typical multiuser MIMO uplink scenario is shown in Figure 8.2.

Let us first consider a point-to-point flat block-fading MIMO system with M_t transmit and M_r receive antennas. Assuming that the channel is used at the times 1, 2, ..., T with the block length T, the input-output relationship for such a system can be written as [9, 10, 20]

$$\mathbf{X} = \mathbf{GH} + \mathbf{N}, \tag{8.54}$$

---

* For the RMV beamforming problem (8.53), the outage probability is equal to 1 – p.

where $\mathbf{H}$ is the $M_t \times M_r$ complex channel matrix, $\mathbf{G}$ is the $T \times M_t$ complex matrix of the transmitted signals, $\mathbf{X}$ is the $T \times M_r$ complex matrix of the received signals, and $\mathbf{N}$ is the $T \times M_r$ matrix of noise.

Let us denote complex information symbols prior to space-time encoding as $s_1, \ldots, s_J$ and define the $J \times 1$ symbol vector $\mathbf{s} \triangleq [s_1, \ldots, s_J]^T$. The $T \times M_t$ matrix $\mathbf{G} = \mathbf{G(s)}$ is called an orthogonal space-time block code (OSTBC) if [16–18]:

- All elements of $\mathbf{G(s)}$ are linear functions of the J complex variables $s_1, \ldots, s_J$ and their complex conjugates.
- For any $\mathbf{s}$, it satisfies $\mathbf{G}^H(\mathbf{s})\,\mathbf{G(s)} = \|\mathbf{s}\|^2\,\mathbf{I}$.

Assuming the OSTBC signaling, the matrix $\mathbf{G(s)}$ can be written as [16, 61, 62]

$$\mathbf{G(s)} = \sum_{l=1}^{J} \left(\mathbf{C}_l \mathrm{Re}\{s_l\} + \mathbf{D}_l \mathrm{Im}\{s_l\} \right), \tag{8.55}$$

where $\mathbf{C}_l \triangleq \mathbf{G(e}_l)$, $\mathbf{D}_l \triangleq \mathbf{G}(j\mathbf{e}_l)$, $\mathbf{e}_l$ is the $J \times 1$ vector having one in the lth position and zeros elsewhere, and $j \triangleq \sqrt{-1}$. Using (8.55), the MIMO model (8.54) can be rewritten as [59, 62]

$$\underline{\mathbf{X}} = \mathcal{A}\underline{\mathbf{s}} + \underline{\mathbf{N}}, \tag{8.56}$$

where the "underline" operator for any matrix $\mathbf{P}$ is defined as

$$\underline{\mathbf{P}} \triangleq \begin{bmatrix} \mathrm{vec}\{\mathrm{Re}(\mathbf{P})\} \\ \mathrm{vec}\{\mathrm{Im}(\mathbf{P})\} \end{bmatrix}, \tag{8.57}$$

$\mathrm{vec}\{\cdot\}$ is the vectorization operator stacking all columns of a matrix on top of each other, and the $2M_rT \times 2J$ real matrix $\mathcal{A} = \mathcal{A}(\mathbf{H})$ is given by

$$\mathcal{A} = \begin{bmatrix} \underline{\mathbf{C}_1\mathbf{H}}, \ldots, \underline{\mathbf{C}_J\mathbf{H}}, \underline{\mathbf{D}_1\mathbf{H}}, \ldots, \underline{\mathbf{D}_J\mathbf{H}} \end{bmatrix}. \tag{8.58}$$

This matrix can be shown to satisfy the following orthogonality property (that, in turn, is due to the orthogonality property of the space-time code):

$$\mathcal{A}^T\mathcal{A} = \|\mathbf{H}\|^2\,\mathbf{I}. \tag{8.59}$$

From (8.58), it can be seen that the matrix $\mathcal{A}$ in (8.56) captures both the effects of the OSTBC and the channel, while the vector $\underline{\mathbf{s}}$ depends on the information symbols only. The columns of $\mathcal{A}$ can be viewed as spatio-temporal signatures that describe the receiver response to each entry of $\underline{\mathbf{s}}$ (i.e., to real or imaginary parts of each information symbol).

Obviously, there is a strong similarity between the vectorized MIMO model in (8.56) and the beamforming snapshot model in (8.1).

The established similarity between the models (8.1) and (8.56) opens an avenue for extending MV beamforming techniques to MIMO communications. A linear MIMO space-time receiver can be expressed as [57, 59]

$$\hat{\underline{s}} = \mathbf{W}^T \underline{\mathbf{X}}, \tag{8.60}$$

where $\hat{\underline{s}}$ is the receiver estimate of the vector $\underline{s}$, and $\mathbf{W} \triangleq [\mathbf{w}_1, \mathbf{w}_2, \ldots, \mathbf{w}_{2J}]$ is the $2M_rT \times 2J$ matrix of the receiver weights. Note that each entry of $\underline{s}$ requires a separate weight vector for estimation and subsequent decoding. For example, the vector $\mathbf{w}_l$ can be interpreted as the space-time receiver weight vector for the lth entry of $\underline{s}$.

The MIMO receiver counterpart of the MRC technique is commonly referred to as the matched-filter (MF) receiver, and can be written as [59, 62]

$$\hat{\underline{s}} = \frac{1}{\|\mathbf{H}\|^2} \mathcal{A}^T \underline{\mathbf{X}}. \tag{8.61}$$

The MF receiver (8.61) corresponds to the following weight matrix:

$$\mathbf{W}_{\text{MF}} = \frac{1}{\|\mathbf{H}\|^2} \mathcal{A}. \tag{8.62}$$

When followed by the simple nearest-neighbor symbol-by-symbol decoder, this receiver is known to be equivalent to the maximum likelihood (ML) space-time decoder for the single-user (point-to-point) case [62].

In the multiuser case (when multiple multiantenna transmitters simultaneously communicate with a multiantenna receiver; see Figure 8.2), the models in (8.54) and (8.56) can be straightforwardly modified to account for multiple users. For example, if all the transmitters use the same OSTBC, (8.56) can be rewritten as [59]

$$\underline{\mathbf{X}} = \sum_{l=1}^{L} \mathcal{A}_l \underline{s}_l + \underline{\mathbf{N}}, \tag{8.63}$$

where $\underline{s}_l$ is the $J \times 1$ vector of information symbols of the lth user,

$$\mathcal{A}_l = \left[\underline{\mathbf{C}_1 \mathbf{H}_l}, \ldots, \underline{\mathbf{C}_J \mathbf{H}_l}, \underline{\mathbf{D}_1 \mathbf{H}_l}, \ldots, \underline{\mathbf{D}_J \mathbf{H}_l} \right]$$

$$\triangleq [\mathbf{a}_{l,1} \ \mathbf{a}_{l,2} \cdots \mathbf{a}_{l,2J}], \tag{8.64}$$

and $\mathbf{H}_l$ is the matrix channel between the lth user and the receiver.

In the multiuser case, the MF receiver (8.62) is no longer the optimal ML decoder, and the complexity of the optimal ML decoding technique grows exponentially with the number of users. Therefore, in this case suboptimal but simple linear receivers are of particular interest [57, 59, 60].

As before, we assume without any loss of generality that the first transmitter is the user of interest. Then, using (8.60), we can express the output vector of a linear receiver as

$$\hat{\underline{\mathbf{s}}}_1 = \mathbf{W}^T \underline{\mathbf{X}}, \tag{8.65}$$

where for each user of interest, a separate matrix $\mathbf{W}$ should be used.

Given the matrix $\mathbf{W}$, the information symbols of the user of interest can be estimated as

$$\hat{\mathbf{s}}_1 = [\mathbf{I}, j\mathbf{I}]\hat{\underline{\mathbf{s}}}_1, \tag{8.66}$$

where the dimension of the identity matrices in (8.66) is $J \times J$. Using the linear estimate (8.66), the lth information symbol can be detected as the nearest to the lth entry of the $\hat{\mathbf{s}}_1$ signal constellation point.

Using the concept of MV beamforming, the receiver weight matrix $\mathbf{W}$ can be designed to maximally suppress interference while preserving a distortionless response toward the signal of the transmitter-of-interest. Specifically, for each entry of $\mathbf{s}_1$, the receiver output power has to be minimized while preserving the distortionless response for that particular entry of $\underline{\mathbf{s}}_1$. This is equivalent to solving the following optimization problem [59]:

$$\min_{\mathbf{w}_l} \mathbf{w}_l^T \hat{\mathcal{R}} \mathbf{w}_l \quad \text{subject to} \quad \mathbf{a}_{1,l}^T \mathbf{w}_l = 1 \quad \text{for all} \quad l = 1, 2, ..., 2J, \tag{8.67}$$

where

$$\hat{\mathcal{R}} = \frac{1}{N} \sum_{t=1}^{N} \mathbf{X}(t) \underline{\mathbf{X}}(t)^T \tag{8.68}$$

is the sample estimate of the $2M_r T \times 2M_r T$ correlation matrix $\mathcal{R} \triangleq E\{\underline{\mathbf{X}} \underline{\mathbf{X}}^T\}$ of the vectorized data, and $\mathbf{X}(t)$ denotes the tth data block.

Taking into account that problem (8.67) can be solved independently for each l, the solution to (8.67) can be written as [59]

$$\mathbf{w}_{\text{MV},l} = \frac{1}{\mathbf{a}_{1,l}^T \hat{\mathcal{R}}^{-1} \mathbf{a}_{1,l}} \hat{\mathcal{R}}^{-1} \mathbf{a}_{1,l}, \quad l = 1, 2, ..., 2J. \tag{8.69}$$

Although the MV receiver (8.69) is able to suppress MAI, it does not completely cancel *self-interference* that, for the lth entry of $\mathbf{s}_1$, is caused by entries of $\underline{\mathbf{s}}_1$ other than the lth one. Clearly, self-interference is treated in (8.67) in the same way as MAI. Thus, when the MAI component is strong, self-interference may not be sufficiently rejected. Note, however, that complete cancellation of self-interference is a strongly desirable property. Otherwise, the symbol-by-symbol detector loses its optimality [59].

To ensure that the self-interference component is completely cancelled, the following additional zero-forcing constraints can be added to (8.67):

$$\mathbf{a}_{1,l}^T \mathbf{w}_m = 0 \quad \text{for all} \quad m \neq l, \quad m,l = 1, 2, ..., 2J. \tag{8.70}$$

Problem (8.67) with the additional constraints (8.70) can be expressed as

$$\min_{\mathbf{W}} \text{tr}\{\mathbf{W}^T \hat{\mathcal{R}} \mathbf{W}\} \quad \text{subject to} \quad \mathcal{A}^T \mathbf{W} = \mathbf{I}, \tag{8.71}$$

where tr{·} denotes the trace of a matrix.

The solution to (8.71) is given by [59]

$$\mathbf{W}_{\text{MV}} = \hat{\mathcal{R}}^{-1} \mathcal{A} (\mathcal{A}^T \hat{\mathcal{R}}^{-1} \mathcal{A})^{-1}. \tag{8.72}$$

Clearly, this MV receiver can be interpreted as a combination of the prewhitener $\hat{\mathcal{R}}^{-1/2}$ and decorrelator receiver $\hat{\mathcal{R}}^{-1/2} \mathcal{A} (\mathcal{A}^T \hat{\mathcal{R}}^{-1} \mathcal{A})^{-1}$. Furthermore, using the property (8.59), it can be seen that in the specific case of $\hat{\mathcal{R}} \propto \mathbf{I}$, the MV receiver (8.72) simplifies to the MF receiver (8.62). The latter property is in agreement with the well-known fact that the MF receiver ignores the effect of MAI treating it as a white noise.

Several robust modifications of the MV receivers (8.69) and (8.72) have been considered in the literature to improve their robustness in the case of imperfect receive channel state information (CSI). In [59], diagonally loaded modifications of (8.69) and (8.72) have been developed. In [63], the approach of [38] has been used to design worst-case optimization-based RMV receivers that explicitly account for norm-bounded CSI errors. For example, the following worst-case modification of (8.67) has been proposed in [63]:

$$\min_{\mathbf{w}_l} \mathbf{w}_l^T \hat{\mathcal{R}} \mathbf{w}_l \quad \text{subject to} \quad \min_{\|\Delta\| \leq \varepsilon} \mathbf{w}_k^T \mathbf{a}_l (\hat{\mathbf{H}}_1 + \Delta) \geq 1 \quad \text{for all} \quad l = 1, ..., 2J, \tag{8.73}$$

where, for the sake of clarity, the user-of-interest spatio-temporal signature $\mathbf{a}_{l,1}$ is explicitly denoted as $\mathbf{a}_l(\mathbf{H}_1)$ to stress that it is a function of the user-of-interest channel $\mathbf{H}_1$. In (8.73), another version of this signature, $\mathbf{a}_l(\hat{\mathbf{H}}_1 + \Delta)$, is used, where Δ stands for the receive CSI error of the user-of-interest:

$$\Delta \triangleq \mathbf{H}_1 - \hat{\mathbf{H}}_1, \tag{8.74}$$

and $\mathbf{H}_1$ and $\hat{\mathbf{H}}_1$ denote the actual and presumed (estimated) channel matrices of the user of interest, respectively. Similar to [38], it is assumed in (8.73) that the CSI error is norm-bounded by some known constant ε, that is,

$$\left\|\Delta\right\| \leq \varepsilon. \tag{8.75}$$

The essence of the RMV receiver design problem in (8.73) is to minimize the receiver output power while maintaining the distortionless response for the worst-case CSI errors. It has been proved in [63] that for any OSTBC,

$$\left\|\Delta\right\| = \left\|\delta_l\right\| \quad \text{for all} \quad l = 1, \ldots, 2J, \tag{8.76}$$

where

$$\delta_l \triangleq \mathbf{a}_l(\mathbf{H}_1) - \mathbf{a}_l(\hat{\mathbf{H}}_1) \tag{8.77}$$

is the error between the actual spatiotemporal signature $\mathbf{a}_l(\mathbf{H}_1)$ and its mismatched (presumed) value $\mathbf{a}_l(\hat{\mathbf{H}}_1)$. Therefore, problem (8.73) can be rewritten as

$$\min_{\mathbf{w}_l} \mathbf{w}_l^T \hat{\mathcal{R}} \mathbf{w}_l \quad \text{subject to} \quad \min_{\|\delta_l\| \leq \varepsilon} \mathbf{w}_l^T (\mathbf{a}_l(\hat{\mathbf{H}}_1) + \delta_l) \geq 1 \quad \text{for all} \quad l = 1, \ldots, 2J. \tag{8.78}$$

This problem is mathematically equivalent to $2J$ decoupled RMV beamforming problems (8.32). Hence, using the results of [38], problem (8.78) can be reformulated as [63]

$$\min_{\mathbf{w}_l} \mathbf{w}_l^T \hat{\mathcal{R}} \mathbf{w}_l \quad \text{subject to} \quad \mathbf{w}_l^T \mathbf{a}_l(\hat{\mathbf{H}}_1) \geq \varepsilon \|\mathbf{w}_l\| + 1 \quad \text{for all} \quad l = 1, \ldots, 2J, \tag{8.79}$$

where the constraint can be shown to be satisfied with equality. Similar to (8.33), the problem in (8.79) belongs to the class of standard SOCP problems and can be efficiently solved using either modern convex optimization tools [50] or Newton-type algorithms [46, 51, 52].

In addition to (8.79), one more RMV receiver has been developed in [63] as a robust extension of the MV receiver (8.72). The later RMV receiver enables better cancellation of self-interference and, as a result, offers an improved performance compared to (8.79).

Using the concept of probabilistically constrained RMV beamforming, several promising extensions of multiuser RMV MIMO receivers of [63] have been developed in [55, 64–67] using soft (outage probability-based) rather that deterministic worst-case distortionless response constraints. An interesting open problem is how to extend these receiver techniques to other than orthogonal space-time codes with higher transmission rates.

8.4   Transmit Beamforming

In wireless communications, there are two basic ways to use transmit antenna arrays: for space-time data encoding [8, 9, 16] and spatial data multiplexing (beamforming) [11, 14]. Although space-time coding techniques do not require any CSI at the transmitter, beamforming methods typically require an accurate knowledge of the transmit CSI.

Certain combinations of space-time coding and beamforming at the transmitter are also possible, that may require only partial CSI [68, 69].

In the sequel, we will consider the transmit beamforming problem that includes the following (related) cases:

1. *Unicasting*. Different data streams should be delivered to different users.
2. *Broadcasting*. The same data content should be delivered to all users.
3. *Multicasting*. The same data content should be broadcasted to a selected group of users, but different data streams should be transmitted to different groups of users.

Clearly, the first two cases represent two extremes of the most general multicasting case. In particular, broadcasting and unicasting can be viewed as single-group multicasting and multicasting with single-user groups, respectively.

8.4.1 Unicast Transmit Beamforming

8.4.1.1 Traditional Techniques

Early formulations of the transmit beamforming problem were mostly developed in the context of voice services in a cellular mobile radio network, where from the operator perspective, the system should provide an acceptable QoS for each user and serve as many users as possible, while radiating as low power as possible [11, 14].

The QoS requirements can be set up in the form of the lowest admissible value of the received SINR at each mobile. Using (8.7), the receive SNR of the ith user can be expressed as [14]

$$\text{SINR}_i = \frac{\mathbf{w}_i^H \mathbf{R}_{i,c(i)} \mathbf{w}_i}{\sigma_i^2 + \sum_{m=1; m\neq i}^{L} \mathbf{w}_m^H \mathbf{R}_{i,c(m)} \mathbf{w}_m}, \tag{8.80}$$

where σ_i^2 is the noise power of the ith user, $\mathbf{R}_{i,c(m)}$ is the correlation matrix of the downlink channel between the BS serving the mth mobile and the ith mobile, and the other parameters used in (8.80) have been defined in section 8.2.

The numerator of (8.80) represents the receive signal power at the ith mobile, whereas the denominator of (8.80) contains the noise and interference powers at the same mobile. The interference terms are given by the sum of powers of transmissions that are intended for other than the ith mobile but which are received by the ith mobile. Clearly, this is a *crosstalk* type of interference that should be avoided to guarantee an acceptable quality of the voice message.

Equation (8.80) views all channels as stochastic random vectors. This representation is suitable for the case of fast fading where the downlink channel vectors themselves are unavailable at the transmitter, and only their correlation matrices are known. In the opposite case of slow channel fading, it is more natural to view the downlink channels as deterministic vectors. In the latter case, the downlink channel correlation matrix is rank one and is given by $\mathbf{R}_{i,c(m)} = \mathbf{h}_{i,c(m)} \mathbf{h}_{i,c(m)}^H$. Then, (8.80) can be rewritten in a simpler form:

$$\text{SINR}_i = \frac{\left| \mathbf{w}_i^H \mathbf{h}_{i,c(i)} \right|^2}{\sigma_i^2 + \sum_{m=1;\, m\neq i}^{L} \left| \mathbf{w}_m^H \mathbf{h}_{i,c(m)} \right|^2}. \tag{8.81}$$

The traditional approach to optimizing the beamformer weight vectors is to minimize the total transmit power under the constraints guaranteeing that an acceptable QoS is provided for each user. Mathematically, this problem can be expressed as [11, 14]

$$\min_{\{\mathbf{w}_l\}_{l=1}^{L}} \sum_{l=1}^{L} \mathbf{w}_l^H \mathbf{w}_l \quad \text{subject to} \quad \text{SINR}_i \geq \gamma_i \quad \text{for all} \quad i=1,\dots,L, \tag{8.82}$$

where γ_i is the minimum acceptable QoS for the ith user. This is a quadratic optimization problem with quadratic nonconvex constraints. Different algorithms have been proposed in the literature to solve (8.82). For example, the algorithms of [11] and [70] separate the problem into the following two subproblems: downlink power control and downlink beamforming with norm-one weight vectors. Using this representation of the problem at hand, the approaches of [11] and [70] propose to find the solution to (8.82) in an iterative way, by finding the optimal norm-one weight vectors for given power levels, and then updating the power levels based on these weight vectors.

Another powerful approach to solve (8.82) has been proposed in [71] (see also [14]) using convex optimization. The key idea of this approach is to reformulate problem (8.82) in a convex form as follows. Using the notation $\mathbf{W}_i \triangleq \mathbf{w}_i \mathbf{w}_i^H$, problem (8.82) with the user SINRs (8.80) can be transformed to [14, 71]

$$\min_{\{\mathbf{W}_l\}_{l=1}^{L}} \sum_{l=1}^{L} \text{tr}\{\mathbf{W}_l\} \quad \text{subject to} \quad \text{tr}\{\mathbf{R}_{i,c(i)}\mathbf{W}_i\} - \gamma_i \sum_{m\neq i} \text{tr}\{\mathbf{R}_{i,c(m)}\mathbf{W}_m\} \geq \gamma_i \sigma_i^2,$$

$$\mathbf{W}_i = \mathbf{W}_i^H, \ \mathbf{W}_i \succeq 0 \quad \text{for all} \quad i=1,\dots,L, \tag{8.83}$$

where the last two constraints guarantee that the matrices $\mathbf{W}_i$ are Hermitian and positive semidefinite for all $i = 1, \dots, L$. Note that the last constraint can be viewed as a convex relaxation of the nonconvex rank-one constraint rank $\{\mathbf{W}_i\} = 1$.

The problem in (8.83) belongs to the class of convex semidefinite programming (SDP) problems, and therefore, it can be efficiently solved using modern convex optimization tools [50]. Moreover, it has been proved in [14] that for this problem, $\mathbf{W}_i \succeq 0$ is exactly equivalent to rank $\{\mathbf{W}_i\} = 1$, and therefore, the replacement of the latter constraint by the former one is not a relaxation but actually an equivalent reformulation of the problem. After obtaining the optimal values for $\mathbf{W}_i$, $i = 1, \dots, L$, this property enables recovery of the optimal weight vectors $\mathbf{w}_i$, $i = 1, \dots, L$ in a simple way, from the principal eigenvectors of $\mathbf{W}_i$, $i = 1, \dots, L$.

In the slow fading case, where the downlink channel vectors are available at the transmitter and (8.81) can be used instead of (8.80), problem (8.82) can be rewritten in a SOCP form [14]:

$$\min_{\{\mathbf{w}_l\}_{l=1}^L} \sum_{l=1}^L \mathbf{w}_l^H \mathbf{w}_l \quad \text{subject to} \quad (\mathbf{w}_i^H \mathbf{h}_{i,c(i)})^2 - \gamma_i \sum_{m \neq i} \left| \mathbf{w}_m \mathbf{h}_{i,c(m)} \right|^2 \geq \gamma_i \sigma_i^2,$$

$$\mathbf{w}_i^H \mathbf{h}_{i,c(i)} \geq 0, \ \mathrm{Im}\{\mathbf{w}_i^H \mathbf{h}_{i,c(i)}\} = 0 \quad \text{for all} \quad i = 1, \ldots, L. \tag{8.84}$$

The SOCP problem (8.84) has much less optimization variables than the SDP problem (8.83) and can be solved easier than the latter problem; see [14].

An interesting and practically relevant extension of the problems (8.81) and (8.82) is, in addition to optimizing the beamformer weight vectors, to maximize the number of admitted users under the condition that the problem remains feasible. Such an extension of the transmit beamforming approach of [11] has been proposed in [72–74], where two computationally efficient joint beamforming and user admission control techniques have been developed using SDP and SOCP convex optimization approaches.

Another recent trend in transmit beamforming is to maximize the system throughput under certain power constraints rather than to minimize the total transmit power [75–79]. A growing interest in such types of transmit strategy is motivated by a higher emphasis of delay-tolerant high-rate packet data services for the third- and fourth-generation (3G and 4G) communication systems [79].

To illustrate the concept of sum-capacity maximization, let us consider the single-cell multiuser model with the downlink user channels $\mathbf{h}_l$, $l = 1, \ldots, L$ (see section 8.2). The baseband signal to be sent to the lth user at time t is denoted as $s_l(t)$, while the signal received at the lth user is denoted as $z_l(t)$. Introducing the notations

$$\mathbf{s}(t) \triangleq [s_1(t), s_2(t), \ldots, s_L(t)]^T, \tag{8.85}$$

$$\mathbf{z}(t) \triangleq [z_1(t), z_2(t), \ldots, z_L(t)]^T, \tag{8.86}$$

$$\mathbf{n}(t) \triangleq [n_1(t), n_2(t), \ldots, n_L(t)]^T, \tag{8.87}$$

$$\mathcal{H} \triangleq [\mathbf{h}_1, \mathbf{h}_2, \ldots, \mathbf{h}_L]^H, \tag{8.88}$$

$$\mathcal{W} \triangleq [\mathbf{w}_1, \mathbf{w}_2, \ldots, \mathbf{w}_L], \tag{8.89}$$

the receive signal model can be written as

$$\mathbf{z}(t) = \mathcal{H}\mathcal{W}\mathbf{s}(t) + \mathbf{n}(t). \tag{8.90}$$

The sum capacity of such a vector Gaussian channel $\mathbf{z} = \mathcal{H}\mathbf{y} + \mathbf{n}(t)$ with the total transmit power constrained to $\mathcal{P}$ is given by [79]

$$C = \max_{\mathbf{R}_y} \log\det\left\{\mathbf{I} + \mathcal{H}\mathbf{R}_y\mathcal{H}^H\right\}, \tag{8.91}$$

where $\mathbf{R}_y \triangleq \mathrm{E}\{\mathbf{y}\mathbf{y}^H\}$ and the maximization in (8.91) is performed over positive semi-definite matrices that satisfy the power constraint

$$\mathrm{tr}\{\mathbf{R}_y\} \leq \mathcal{P}. \tag{8.92}$$

The SINR of the lth user can be written as

$$\mathrm{SINR}_l = \frac{\left|\mathbf{h}_l^H\mathbf{w}_l\right|^2}{\sigma_l^2 + \sum_{m \neq l}\left|\mathbf{h}_l^H\mathbf{w}_m\right|^2}. \tag{8.93}$$

Using (8.90)–(8.93), the transmit beamforming problem can be formulated as [79]

$$\max_{\mathcal{W}} \sum_{l=1}^{L} \log(1 + \mathrm{SINR}_l) \quad \text{subject to} \quad \left\|\mathcal{W}\right\|^2 \leq \mathcal{P}. \tag{8.94}$$

The essence of (8.94) is to maximize the total system throughput under the transmit power constraint. In simple words, the maximum in (8.94) is achieved by means of exploiting *multiuser diversity* that suggests always to transmit to the strongest-channel users. Several computationally efficient algorithms to solve (8.94) have been proposed in the literature; see [77–79] and references therein.

8.4.1.2 Robust Extensions

The idea of incorporating robustness against transmitter CSI errors in the downlink beamforming problem (8.82) has been discussed in [14] and [80]. This approach uses an idea related to that used in [44] for receive RMV beamforming. More specifically, the following upper and lower bounds on downlink channel correlation matrices are considered in [80]:

$$\mathbf{R}_{i,c(m)}^{\mathrm{lower}} \preceq \mathbf{R}_{i,c(m)} \preceq \mathbf{R}_{i,c(m)}^{\mathrm{upper}}, \tag{8.95}$$

where

$$\mathbf{R}_{i,c(m)}^{\mathrm{lower}} = \mathbf{R}_{i,c(m)} - \xi\mathbf{I}, \qquad \mathbf{R}_{i,c(m)}^{\mathrm{upper}} = \mathbf{R}_{i,c(m)} + \xi\mathbf{I}, \tag{8.96}$$

and ξ determines the bound on the norm of downlink correlation matrix errors. To account for the CSI mismatch, it is proposed in [14] and [80] to enforce the QoS constraints for all possible values of the downlink channel correlation matrices that satisfy (8.95). Using this approach, the robust modification of (8.83) can be expressed as

$$\min_{\{\mathbf{W}_l\}_{l=1}^{L}} \sum_{l=1}^{L} \mathrm{tr}\{\mathbf{W}_l\} \quad \text{subject to} \ \ \mathrm{tr}\{(\mathbf{R}_{i,c(i)} - \xi\mathbf{I})\mathbf{W}_i\} - \gamma_i \sum_{m \neq i} \mathrm{tr}\{(\mathbf{R}_{i,c(m)} + \xi\mathbf{I})\mathbf{W}_m\} \geq \gamma_i \sigma_i^2,$$

$$\mathbf{W}_i = \mathbf{W}_i^H, \ \mathbf{W}_i \succeq 0 \quad \text{for all} \ \ i = 1, \dots, L. \tag{8.97}$$

Problem (8.97) is also a convex SDP problem that can be solved at the same complexity as (8.83). It can be interpreted as the worst-case modification of the nonrobust problem (8.83).

A much simpler yet suboptimal approach to robust unicast downlink beamforming has been recently proposed in [81]. The essence of this technique is, following the ideas of the algorithms of [11], [70], and [82], to split the transmit beamforming problem into the power control and weight vector optimization subproblems, and then to utilize simple decentralized algorithms for the weight vector optimization, while using a centralized worst-case optimization-based technique for adjusting the transmit powers. This approach has been shown in [81] to have a performance comparable to that of [80], while enjoying both a lower computational cost and a substantially reduced degree of the required cooperation between the network BSs compared to the algorithm of [80].

8.4.2 Broadcast Transmit Beamforming

The traditional broadcasting strategy is to radiate the transmitted power isotropically or using a fixed transmit beampattern. However, in such a case the transmit CSI about the user channels is ignored, and as a result, the users with weak channels can experience a severe QoS degradation. In future data broadcasting/multicasting applications, the CSI for all the intended users is likely to be available at the transmitter. Therefore, this CSI can be exploited to improve the performance with respect to the traditional broadcasting scheme [83].

Let a single multiantenna transmitter with the weight vector $\mathbf{w}$ broadcast the signal $s(t)$ to L single-antenna users whose channels $\mathbf{h}_l$, $l = 1, \dots, L$, are known at the transmitter. The signal received at the lth user can be modeled as (8.4), and the received SNR at the lth user is given by

$$\mathrm{SNR}_l = \frac{|\mathbf{w}^H \mathbf{h}_l|^2}{\sigma_l^2}. \tag{8.98}$$

The SNRs in (8.98) can be viewed as the user QoS values. Let γ_l be the minimum acceptable QoS value for the lth user. Then, defining the normalized channel vectors

$$\mathbf{g}_l \triangleq \frac{\mathbf{h}_l}{\sqrt{\gamma_l \sigma_l^2}}, \quad l = 1, \ldots, L, \tag{8.99}$$

the optimal broadcasting problem can be written as [83]

$$\min_{\mathbf{w}} \mathbf{w}^H \mathbf{w} \quad \text{subject to} \quad \left| \mathbf{w}^H \mathbf{g}_l \right|^2 \geq 1 \quad \text{for all} \quad l = 1, \ldots L. \tag{8.100}$$

According to (8.100), the optimal broadcast weight vector is designed by minimizing the total transmit power under the constraints that the minimum acceptable QoS is guaranteed for each user.

Unfortunately, the problem in (8.100) is NP-hard [83], and hence it cannot be solved in polynomial time. To obtain a reasonably simple approximate solution to (8.100), it has been proposed in [83] to reformulate this problem in terms of $\mathbf{Z} \triangleq \mathbf{w}\mathbf{w}^H$ and $\mathbf{Q}_l \triangleq \mathbf{g}_l \mathbf{g}_l^H$ as

$$\min_{\mathbf{Z}} \mathrm{tr}\{\mathbf{Z}\} \quad \text{subject to} \quad \mathrm{tr}\{\mathbf{Z}\mathbf{Q}_l\} \geq 1 \quad \text{for all} \quad l = 1, \ldots, L,$$
$$\mathbf{Z}^H = \mathbf{Z}, \ \mathrm{rank}\{\mathbf{Z}\} = 1, \tag{8.101}$$

and then to replace the nonconvex constraint rank$\{\mathbf{Z}\} = 1$ by its relaxed convex version $\mathbf{Z} \succeq 0$.

The so-obtained problem is a convex SDP problem. However, in contrast to the problem in (8.83), the rank and semidefinite constraints on $\mathbf{Z}$ are not equivalent to each other; that is, the matrix $\mathbf{Z}_{\mathrm{opt}}$ obtained by solving the relaxed SDP problem is not rank one in general [83]. Therefore, to recover the optimal value of weight vector from $\mathbf{Z}_{\mathrm{opt}}$, the so-called *randomization* approach was adopted in [83].

Another alternative broadcast beamforming problem setting in the case of fixed transmit power constraint is to maximize the minimum receiver SNR subject to this constraint. This problem can be written as [83]

$$\max_{\mathbf{w}} \min_{l} \left\{ \left| \mathbf{w}^H \mathbf{h}_l \right|^2 \Big/ \sigma_l^2 \right\}_{l=1}^{L} \quad \text{subject to} \quad \mathbf{w}^H \mathbf{w} \leq \mathcal{P}. \tag{8.102}$$

This problem can also be shown to be NP-hard, but it can be relaxed to a convex SDP form in nearly the same way as problem (8.100); see [83] for more details.

A useful worst-case design-based robust modification of problem (8.100) has been discussed in [84]. The authors of [84] have assumed the norm-bounded channel errors $\mathbf{v}_l \triangleq \tilde{\mathbf{g}}_l - \mathbf{g}_l; \ \|\mathbf{v}_l\| \leq \kappa$ (with $\tilde{\mathbf{g}}_l$ and $\mathbf{g}_l$ being the actual and presumed values of the normalized channel vector, respectively), and modified problem (8.100) as

$$\min_{\mathbf{w}} \mathbf{w}^H \mathbf{w} \quad \text{subject to} \quad \left| \mathbf{w}^H (\mathbf{g}_l + \mathbf{v}_l) \right|^2 \geq 1 \quad \text{for all} \quad \left\| \mathbf{v}_l \right\| \leq \kappa, \ l = 1, \ldots, L. \tag{8.103}$$

A simple and elegant algorithm has been proposed in [84] to find an approximate solution to (8.103) using a properly rescaled weight vector of the original nonrobust problem (8.100) or its convex relaxed reformulation presented in [83]. Another robust technique that solves (8.103) has been developed in [85] using the convex SDP approach.

8.4.3 Multicast Transmit Beamforming

The multicast transmit beamforming problem is a natural, yet nontrivial, extension of the broadcast problem to the case when different data streams have to be transmitted to different groups of users.

Let us modify the broadcast scenario by assuming a total of I ($1 \leq I \leq L$) multicast groups $\{\mathcal{D}_1, ..., \mathcal{D}_I\}$, where $\mathcal{D}_i$ is the index set of receivers participating in the ith group [86, 87]. It will be assumed that no user can be shared by more than one group, that is, $\mathcal{D}_i \cap \mathcal{D}_k = \varnothing$ for any $i \neq k$ and $\mathcal{D}_1 \cup \mathcal{D}_2 \cup \cdots \cup \mathcal{D}_I = \{1, 2, ..., L\}$, which also implies that

$$\sum_{i=1}^{I} |\mathcal{D}_i| = L.$$

Using these conventions, the joint design of the transmit beamformers for multiple user groups amounts to minimizing the total transmit power subject to the QoS constraints [86]:

$$\min_{\{\mathbf{w}_i\}_{i=1}^{I}} \mathbf{w}_i^H \mathbf{w}_i \quad \text{subject to} \quad \frac{\left|\mathbf{w}_k^H \mathbf{h}_m\right|^2}{\sigma_k^2 + \sum_{l \neq k} \left|\mathbf{w}_l^H \mathbf{h}_m\right|^2} \geq \gamma_m \quad \text{for all} \quad m \in \mathcal{D}_k, \quad k, l = 1, ..., I, \quad (8.104)$$

where $\mathbf{w}_i$ is the weight vector used for the ith group and, as before, γ_k and $\mathbf{h}_k$ are the minimal acceptable QoS and the downlink channel vector of the kth user.

Clearly, as (8.104) represents a generalization of (8.100), it it also NP-hard. Moreover, in contrast to (8.100), problem (8.104) can be infeasible due to crosstalk-type interference between the user groups. Using an approach similar to that used in [83] for relaxing broadcast problem (8.100) to a convex form, the authors of [86] have derived a relaxation of (8.104) to the following convex SDP problem:

$$\min_{\{\mathbf{Z}_i\}_{i=1}^{I}} \sum_{i=1}^{I} \text{tr}\{\mathbf{Z}_i\} \quad \text{subject to} \quad \text{tr}\{\mathbf{Q}_m \mathbf{Z}_k\} \geq \gamma_m \sum_{l \neq k} \text{tr}\{\mathbf{Q}_m \mathbf{Z}_l\} + \gamma_m \sigma_m^2$$

$$\text{for all} \quad m \in \mathcal{D}_k, \quad k, l = 1, ..., I, \quad (8.105)$$

$$\mathbf{Z}_i^H = \mathbf{Z}_i, \mathbf{Z}_i \succeq 0 \quad \text{for all} \quad i = 1, ..., I.$$

Similar to (8.101), problem (8.105) can be directly solved using available convex optimization tools [50].

The authors of [86] also derived a convex optimization-based algorithm to solve the multicast generalization of the max-min fair beamforming problem (8.102). The design of robust worst-case multicast beamformers is currently an open problem.

8.5 Conclusions

An overview of adaptive beamforming methods for wireless multiantenna communications has been presented. Both the receive and transmit beamforming problems and techniques have been discussed in detail. As many popular recent approaches to receive and transmit beamforming are based on modern optimization theory, our chapter has a strong emphasis on convex optimization-based techniques. More details about using convex optimization methods in wireless communications can be found in the recent tutorial papers [48] and [49].

Acknowledgment

This work was supported by the German Research Foundation (DFG) under Grant GE 1881/1-1.

References

[1] B. D. Van Veen and K. M. Buckley. 1988. Beamforming: A versatile approach to spatial filtering. *IEEE Acoust. Speech Signal Processing Mag.* 5:4–24.

[2] J. Mayhan. 1976. Nulling limitations for a multiple-beam antenna. *IEEE Trans. Antennas Propagation* 24:769–79.

[3] R. T. Compton. 1978. An adaptive array in a spread spectrum communication system. *Proc. IEEE* 66:289–98.

[4] J. Winters. 1982. Spread spectrum in a four-phase communication system employing adaptive antennas. *IEEE Trans. Commun.* 30:929–36.

[5] G. J. Foschini. 1996. Layered space-time architecture for wireless communication in a fading environment when using multielement antennas. *Bell Labs Tech. J.* 1:41–59.

[6] L. C. Godara. 1997. Application of antenna arrays to mobile communications. II. Beam-forming and direction-of-arrival considerations. *Proc. IEEE* 85:1195–245.

[7] T. S. Rapapport, ed. 1998. *Smart antennas: Adaptive arrays, algorithms, and wireless position location.* Piscataway, NJ: IEEE Press.

[8] S. M. Alamouti. 1998. A simple transmit diversity technique for wireless communications. *IEEE J. Selected Areas Commun.* 45:1451–58.

[9] V. Tarokh, N. Seshadri, and A. R. Calderbank. 1998. Space-time codes for high data rate wireless communication: Performance criterion and code construction. *IEEE Trans. Inform. Theory* 44:744–65.

[10] I. E. Telatar. 1999. Capacity of multi-antenna Gaussian channels. *Eur. Trans. Telecommun.* 10:585–96.

[11] F. Rashid-Farrokhi, K. J. R. Liu, and L. Tassiulas. 1998. Transmit beamforming and power control for cellular wireless systems. *IEEE J. Selected Areas Commun.* 16:1437–50.

[12] S. Kapoor, D. J. Marchok, and Y.-F. Huang. 1999. Adaptive interference suppression in multiuser wireless OFDM systems using antenna arrays. *IEEE Trans. Signal Processing* 47:3381–91.

[13] J. Razavilar, F. Rashid-Farrokhi, and K. J. R. Liu. 2000. Traffic improvements in wireless communication networks using antenna arrays. *IEEE J. Selected Areas Commun.* 18:458–71.

[14] M. Bengtsson and B. Ottersten. 2001. Optimal and suboptimal transmit beamforming. In *Handbook of antennas in wireless communications*, ed. L. Godara. Boca Raton, FL: CRC Press, chapter 18.

[15] C. Farsakh and J. A. Nossek. 1998. Spatial covariance based downlink beamforming in an SDMA mobile radio system. *IEEE Trans. Commun.* 46:1497–506.

[16] E. G. Larsson and P. Stoica. 2003. *Space-time block coding for wireless communications*. Cambridge, UK: Cambridge University Press.

[17] A. Paulraj, R. Nabar, and D. Gore. 2003. *Introduction to space-time wireless communications*. Cambridge, UK: Cambridge University Press.

[18] A. B. Gershman and N. D. Sidiropoulos, eds. 2005. *Space-time processing for MIMO communications*. New York: John Wiley & Sons.

[19] T. Kaiser, A. Boudroux, H. Boche, J. R. Fonollosa, J. B. Andersen, and W. Utschick, eds. 2005. *Smart antennas—State-of-the-art*. EURASIP Book Series on Signal Processing and Communications. New York: Hindawi.

[20] H. Bölcskei, D. Gesbert, C. B. Papadias, and A. J. van der Veen, eds. 2006. *Space-time wireless systems—From array processing to MIMO communications*. Cambridge, UK: Cambridge University Press.

[21] S. S. Jeng, H. P. Lin, G. Xu, and W. J. Vogel. 1995. Measurements of spatial signature of an antenna array. In *Proceedings of PIMRC'95*, Toronto, vol. 2, pp. 669–72.

[22] A. J. Weiss and B. Friedlander. 1996. Almost blind steering vector estimation using second-order moments. *IEEE Trans. Signal Processing* 44:1024–27.

[23] A. L. Swindlehurst. 1998. Time delay and spatial signature estimation using known asynchronous signals. *IEEE Trans. Signal Processing* 46:449–62.

[24] D. Astely, A. L. Swindlehurst, and B. Ottersten. 1999. Spatial signature estimation for uniform linear arrays with unknown receiver gains and phases. *IEEE Trans. Signal Processing* 47:2128–38.

[25] Y. Rong, S. A. Vorobyov, A. B. Gershman, and N. D. Sidiropoulos. 2005. Blind spatial signature estimation via time-varying user power loading and parallel factor analysis. *IEEE Trans. Signal Processing* 53:1697–710.

[26] H. L. Van Trees. 2002. *Optimum array processing*. New York: Wiley.

[27] A. B. Gershman. 2003. Robustness issues in adaptive beamforming and high-resolution direction finding. In *High-resolution and robust signal processing*, ed. Y. Hua, A. B. Gershman, and Q. Cheng. New York: Marcel Dekker.

[28] I. S. Reed, J. D. Mallett, and L. E. Brennan. 1974. Rapid convergence rate in adaptive arrays. *IEEE Trans. Aerospace Electronics Syst.* 10:853–63.

[29] D. D. Feldman and L. J. Griffiths. 1994. A projection approach to robust adaptive beamforming. *IEEE Trans. Signal Processing* 42:867–76.

[30] A. B. Gershman. 1999. Robust adaptive beamforming in sensor arrays. *AEU Int. J. Electronics Commun.* 53:305–14.

[31] J. Riba, J. Goldberg, and G. Vazquez. 1997. Robust beamforming for interference rejection in mobile communications. *IEEE Trans. Signal Processing* 45:271–75.

[32] S. P. Applebaum and D. J. Chapman. 1976. Adaptive arrays with main beam constraints. *IEEE Trans. Antennas Propagation* 24:650–62.

[33] K. Takao, M. Fujita, and T. Nishi. 1976. An adaptive antenna array under directional constraint. *IEEE Trans. Antennas Propagation* 24:662–69.

[34] K. I. Pedersen, P. E. Mogensen, and B. H. Fleury. 2000. A stochastic model of the temporal and azimuthal dispersion seen at the base station in outdoor propagation environments. *IEEE Trans. Vehicular Technol.* 49:437–47.

[35] Y. I. Abramovich. 1981. Controlled method for adaptive optimization of filters using the criterion of maximum SNR. *Radio Eng. Electronic Physics* 26:87–95.

[36] H. Cox, R. M. Zeskind, and M. H. Owen. 1987. Robust adaptive beamforming. *IEEE Trans. Acoust. Speech Signal Processing* 35:1365–76.

[37] B. D. Carlson. 1988. Covariance matrix estimation errors and diagonal loading in adaptive arrays. *IEEE Trans. Aerospace Electronic Syst.* 24:397–401.

[38] S. Vorobyov, A. B. Gershman, and Z.-Q. Luo. 2003. Robust adaptive beamforming using worst-case performance optimization: A solution to the signal mismatch problem. *IEEE Trans. Signal Processing* 51:313–24.

[39] L. Chang and C. C. Yeh. 1992. Performance of DMI and eigenspace-based beamformers. *IEEE Trans. Antennas Propagation* 40:1336–47.

[40] R. J. Mailloux. 1995. Covariance matrix augmentation to produce adaptive array pattern troughs. *IEE Electronics Lett.* 31:771–72.

[41] A. B. Gershman, U. Nickel, and J. F. Böhme. 1997. Adaptive beamforming algorithms with robustness against jammer motion. *IEEE Trans. Signal Processing* 45:1878–85.

[42] J. R. Guerci. 2000. Theory and application of covariance matrix tapers to robust adaptive beamforming. *IEEE Trans. Signal Processing* 47:977–85.

[43] M. A. Zatman. 2000. Comment on "Theory and application of covariance matrix tapers for robust adaptive beamforming." *IEEE Trans. Signal Processing* 48:1796–800.

[44] S. Shahbazpanahi, A. B. Gershman, Z.-Q. Luo, and K. M. Wong. 2003. Robust adaptive beamforming for general-rank signal models. *IEEE Trans. Signal Processing* 51:2257–69.

[45] S. Vorobyov, A. B. Gershman, Z.-Q. Luo, and N. Ma. 2004. Adaptive beamforming with joint robustness against mismatched signal steering vector and interference nonstationarity. *IEEE Signal Processing Lett.* 11:108–11.

[46] R. G. Lorenz and S. P. Boyd. 2005. Robust minimum variance beamforming. *IEEE Trans. Signal Processing* 53:1684–96.

[47] M. Lobo, L. Vandenberghe, S. P. Boyd, and H. Lebret. 1998. Applications of second-order cone programming. *Lin. Algebra Appl.* 284:193–228.

[48] Z.-Q. Luo. 2003. Applications of convex optimization in signal processing and digital communication. *Math. Programming* 97B:177–207.

[49] Z.-Q. Luo and W. Yu. 2006. An introduction to convex optimization for communications and signal processing. *IEEE J. Selected Areas Commun.* 24:1426–38.

[50] J. F. Sturm. 1999. Using SeDuMi 1.02, a MATLAB toolbox for optimization over symmetric cones. *Optim. Meth. Software* 11/12:625–53.

[51] K. Zarifi, S. Shahbazpanahi, A. B. Gershman, and Z.-Q. Luo. 2005. Robust blind multiuser detection based on the worst-case performance optimization of the MMSE receiver. *IEEE Trans. Signal Processing* 53:295–305.

[52] P. Stoica and J. Li, eds. 2006. *Robust adaptive beamforming*. Hoboken, NJ: John Wiley & Sons.

[53] Y. Rong, S. A. Vorobyov, and A. B. Gershman. 2005. Robust adaptive beamforming using probability-constrained optimization. In *Proceedings of the IEEE Workshop on Statistical Signal Processing*, Bordeaux, France, pp. 934–39.

[54] S. A. Vorobyov, Y. Rong, and A. B. Gershman. 2006. Robust minimum variance adaptive beamformers and multiuser MIMO receivers: From worst-case to probabilistically constrained designs. In *Proceedings of ICASSP'06*, Toulouse, France, vol. 5, pp. 977–80.

[55] S. A. Vorobyov, A. B. Gershman, and Y. Rong. 2007. On the relationship between the worst-case optimization-based and probability-constrained approaches to robust adaptive beamforming. In *Proceedings of ICASSP'07*, Honolulu, HI, vol. 2, pp. 977–980.

[56] A. Prékopa. 1995. *Stochastic programming*. Dordrecht, Netherlands: Kluwer Academic Publishers.

[57] H. Li, X. Lu, and G. B. Giannakis. 2002. Capon multiuser receiver for CDMA systems with space-time coding. *IEEE Trans. Signal Processing* 50:1193–204.

[58] D. Reynolds, X. Wang, and H. V. Poor. 2002. Blind adaptive space-time multiuser detection with multiple transmitter and receiver antennas. *IEEE Trans. Signal Processing* 50:1261–76.

[59] S. Shahbazpanahi, M. Beheshti, A. B. Gershman, M. Gharavi-Alkhansari, and K. M. Wong. 2004. Minimum variance linear receivers for multiaccess MIMO wireless systems with space-time block coding. *IEEE Trans. Signal Processing* 52:3306–13.

[60] A. Nordio and G. Taricco. 2006. Linear receivers for the multiple-input multiple-output multiple-access channel. *IEEE Trans. Commun.* 54:1446–56.

[61] B. Hassibi and B. M. Hochwald. 2002. High-rate codes that are linear in space and time. *IEEE Trans. Inform. Theory* 48:1804–24.

[62] M. Gharavi-Alkhansari and A. B. Gershman. 2005. Constellation space invariance of orthogonal space-time block codes. *IEEE Trans. Inform. Theory* 51:331–34.

[63] Y. Rong, S. Shahbazpanahi, and A. B. Gershman. 2005. Robust linear receivers for space-time block coded multi-access MIMO systems with imperfect channel state information. *IEEE Trans. Signal Processing* 53:3081–90.

[64] Y. Rong, S. A. Vorobyov, and A. B. Gershman. 2004. A robust linear receiver for multi-access space-time block coded MIMO systems based on probability-constrained optimization. In *Proceedings of IEEE VTC'04*, Milan, Italy, vol. 1, pp. 118–22.

[65] Y. Rong, S. A. Vorobyov, and A. B. Gershman. 2005. Robust linear receiver design for multi-access space-time block coded MIMO systems using stochastic optimization. In *Proceedings of the IEEE Workshop on Statistical Signal Processing*, Bordeaux, France, pp. 65–70.

[66] R. Wang, H. Li, and T. Li. 2006. Robust multiuser detection for multicarrier CDMA systems. *IEEE J. Selected Areas Commun.* 24:673–83.

[67] Y. Rong, S. A. Vorobyov, and A. B. Gershman. 2006. Robust linear receivers for multi-access space-time block coded MIMO systems: A probabilistically constrained approach. *IEEE J. Selected Areas Commun.* 24:1560–70.

[68] G. Jöngren, M. Skoglund, and B. Ottersten. 2002. Combining beamforming and space-time block coding. *IEEE Trans. Inform. Theory* 48:611–27.

[69] S. Zhou and G. B. Giannakis. 2002. Optimal transmitter eigen-beamforming and space-time block coding based on channel mean feedback. *IEEE Trans. Signal Processing* 50:2599–613.

[70] E. Visotsky and U. Madhow. 1999. Optimum beamforming using transmit antenna arrays. In *Proceedings of IEEE VTC'99*, Houston, TX, vol. 1, pp. 851–56.

[71] M. Bengtsson and B. Ottersten. 1999. Optimal downlink beamforming using semidefinite optimization. In *Proceedings of the 37th Annual Allerton Conference on Communications, Control, and Computing*, pp. 987–96.

[72] E. Matskani, N. D. Sidiropoulos, Z.-Q. Luo, and L. Tassiulas. Submitted. Convex approximation techniques for joint multiuser downlink beamforming and admission control. *IEEE Trans. Wireless Commun.*

[73] E. Matskani, N. D. Sidiropoulos, Z.-Q. Luo, and L. Tassiulas. 2007. Joint multiuser downlink beamforming and admission control: A semidefinite relaxation approach. In *Proceedings of ICASSP'07*, Honolulu, HI, vol. 3, pp. 585–588.

[74] E. Matskani, N. D. Sidiropoulos, Z.-Q. Luo, and L. Tassiulas. 2007. A second-order cone deflation approach to joint multiuser downlink beamforming and admission control. In *Proceedings of the IEEE Workshop on Signal Processing Advances in Wireless Communications*, Helsinki, Finland, pp. 1–5.

[75] G. Caire and S. Shamai. 2003. On the achievable throughput of a multi-antenna Gaussian broadcast channel. *IEEE Trans. Inform. Theory* 49:1691–706.

[76] S. Vishwanath, N. Jindal, and A. Goldsmith. 2003. Duality, achievable rates, and sum-rate capacity of Gaussian MIMO broadcast channels. *IEEE Trans. Inform. Theory* 49:2658–68.

[77] Z. Tu and R. S. Blum. 2003. Multiuser diversity for a dirty paper approach. *IEEE Commun. Lett.* 7:370–72.

[78] Q. H. Spencer, A. L. Swindlehurst, and M. Haardt. 2004. Zero-forcing methods for downlink spatial multiplexing in multi-user MIMO channels. *IEEE Trans. Signal Processing* 52:461–71.

[79] G. Dimić and N. D. Sidiropoulos. 2005. On downlink beamforming with greedy user selection: Performance analysis and a simple new algorithm. *IEEE Trans. Signal Processing* 53:3857–68.

[80] M. Bengtsson. 2000. Robust and constrained downlink beamforming. In *Proceedings of EUSIPCO'00*, Tampere, Finland, pp. 1433–36.

[81] M. Biguesh, S. Shahbazpanahi, and A. B. Gershman. 2004. Robust downlink power control in wireless cellular systems. *EURASIP J. Wireless Commun. Networking* 2:261–72.

[82] H. Boche and M. Schubert. 2001. A new approach to power adjustment for spatial covariance based downlink beamforming. In *Proceedings of ICASSP'01*, Salt Lake City, UT, vol. 5, pp. 2957–60.

[83] N. D. Sidiropoulos, T. N. Davidson, and Z.-Q. Luo. 2006. Transmit beamforming for physical-layer multicasting. *IEEE Trans. Signal Processing* 54:2239–51.

[84] E. Karipidis, N. D. Sidiropoulos, and Z.-Q. Luo. 2006. Convex transmit beamforming for downlink multicasting to multiple co-channel groups. In *Proceedings of ICASSP'06*, Toulouse, France, vol. 5, pp. 973–76.

[85] I. Wajid, A. B. Gershman, S. A. Vorobyov, and Y. A. Karanouh. 2007. Robust multiantenna broadcasting with imperfect channel state information. In *Proceedings of the IEEE Workshop on Computer Advances in Multi-Sensor Adaptive Processing*, U.S. Virgin Islands.

[86] E. Karipidis, N. D. Sidiropoulos, and Z.-Q. Luo. 2008. Submitted. Quality of service and max-min fair transmit beamforming to multiple co-channel multicast groups. *IEEE Trans. Signal Processing* 56:1268–1279.

[87] E. Karipidis, N. D. Sidiropoulos, and Z.-Q. Luo. 2005. Transmit beamforming to multiple co-channel multicast groups. In *Proceedings of the IEEE Workshop on Computer Advances in Multi-Sensor Adaptive Processing*, Puerto Vallarta, Mexico, pp. 109–12.

9

Adaptive Equalization for Wireless Channels*

Richard K. Martin
*Air Force Institute
of Technology*

Adaptive equalizers have been in use for about four decades. Since then, the applications and allowable complexity have changed dramatically, but the basic design approaches are largely unchanged. This chapter reviews common methodologies for designing trained and blind adaptive equalizers, including fast algorithms and slower, low-complexity algorithms. Application of these principles to modern wireless communication systems is demonstrated via discussion of popular equalizers for two newly popular modulation formats, multicarrier and ultrawideband communications. The chapter concludes with a discussion of the interaction of the adaptive equalizer with other adaptive blocks, such

---

* The views expressed in this chapter are those of the author and do not reflect the official policy or position of the U.S. Air Force, Department of Defense, or U.S. government. This document has been approved for public release, distribution unlimited.

as an adaptive carrier frequency offset estimator, an adaptive gain controller, and adaptive frequency-domain equalizers (for multicarrier systems).

9.1 Introduction and Historical Perspective

In this section we discuss the motivation for adaptive equalizers, including a discussion of notation. We then give a brief historical perspective, following the proposal of adaptive equalizers in the 1960s, to their heyday of research in the 1980s, through current research.

9.1.1 Equalizer Structure and Notation

Consider a single-input multiple-output (SIMO) wireless communication channel. This can be obtained via multiple sensors or oversampling by a factor of P at the receiver. The channel at sensor p or sampling subinterval p is often modeled as

$$y^{(p)}[n] = \sum_{k=0}^{L} h^{(p)}[n,k] x[n-k] + w^{(p)}[n], \quad p \in \{1,...,P\}, \tag{9.1}$$

where $w^{(p)}[n]$ is additive noise (typically white and Gaussian), the channel coefficient $h^{(p)}[n,k]$ gives the dependence of output n on input n-k, and L is the channel order. This is sometimes called the time-lag representation of the channel [1]. Throughout this chapter, we will use n to index the time value (or sample number or iteration) and k to index the lag, i.e., the dependence of the current output on the input that occurred k samples ago, and p will index the receive antennas (or oversampling subsequence). When abundance of indices permits, discrete time indices will be indicated in square brackets, element indices of a matrix or vector will be indicated as subscripts, and all other indices will be in superscripts with parentheses. Matrices and vectors will be in boldface. The superscripts $(\cdot)^*$, $(\cdot)^T$, and $(\cdot)^H$ denote complex conjugate, matrix transpose, and Hermitian (conjugate transpose), respectively. All other superscripts are powers.

When the channel is time invariant, it is not dependent on n and the model simplifies to a discrete convolution,

$$y^{(p)}[n] = \underbrace{\sum_{k=0}^{L} h^{(p)}[k] x[n-k]}_{\mathbf{x}^T[n]\mathbf{h}^{(p)}} + w^{(p)}[n], \quad p \in \{1,...,P\}.$$

$$\mathbf{h}^{(p)} = \left[h^{(p)}[0], ..., h^{(p)}[L] \right]^T, \tag{9.2}$$

$$\mathbf{x}[n] = \left[x[n], ..., x[n-L] \right]^T.$$

Unless otherwise noted, we will use the channel model of (9.2), although we will discuss some recent work that features the more challenging model of (9.1). Ideally, the channel is unity for a single value of the delay and zero otherwise, indicating a single propagation path from the transmitter to the receiver. However, realistically, there are multiple delayed and attenuated reflections, called multipath, leading to many other nonzero channel coefficients.

The goal of the receiver is to estimate the sequence $x[n]$ from the sequence $y[n]$. There are various ways to do this. Optimally, one would estimate the entire sequence $x[n]$ from the entire sequence $y[n]$, and the maximum likelihood solution is called maximum likelihood sequence estimation (MLSE), which is typically implemented via the Viterbi algorithm [2]. However, MLSE is typically extremely complex (of the order A^L, where A is the alphabet size of the transmitted signal). Thus, it is impractical for most wireless channels, which can span hundreds of samples, with A typically ranging from 4 to 64.

Equalization is a relatively low-complexity alternative to MLSE, although its performance is suboptimal. Equalization is the act of processing the output of the communication channel with the goal of recovering the input, often in the form of a filtering operation. This can be a linear finite impulse response (FIR) filter, a linear infinite impulse response (IIR) filter, or a nonlinear decision-feedback equalizer (DFE). Other equalization techniques exist, but due to considerations of brevity, this chapter focuses on equalization via linear filtering, as it is the most popular approach, although we briefly discuss the DFE and its variants.

The linear equalizer of order N is modeled as

$$\hat{x}[n] = \sum_{p=1}^{P} \left(\sum_{k=0}^{N} f^{(p)}[k] y^{(p)}[n-k] \right) = \mathbf{f}^T \mathbf{y}[n],$$

$$\mathbf{f} = \left[f^{(1)}[0], \ldots, f^{(1)}[N], f^{(2)}[0], \ldots, f^{(P)}[N] \right]^T, \qquad (9.3)$$

$$\mathbf{y}[n] = \left[y^{(1)}[n], \ldots, y^{(1)}[n-N], y^{(2)}[n], \ldots, y^{(P)}[n-N] \right]^T.$$

Later in this chapter, we will discuss methods of designing $\mathbf{f}$ for various communication standards that are currently enjoying widespread popularity. First, however, we discuss the reasons for making $\mathbf{f}$ adaptive, and give a brief history of equalizer design.

9.1.2 Motivation for Adaptation

Although the entirety of this book deals with adaptive processing, there is need to specifically motivate adaptive equalization. In a wireless environment, the channel model is highly dependent on the physical location of the reflectors that lead to multipath. As objects move, the channel model must change accordingly. In particular, if either the transmitter or receiver is mobile, every channel coefficient will gradually change, and the equalizer must change accordingly.

There are two ways to deal with this changing environment. One could assume the channel is static over a short time window, and then compute the optimal equalizer for that window via a batch processing algorithm [e.g., 3]. For each new block, a new equalizer could be computed from scratch, and that equalizer could be used to equalize the data in that particular block. Alternatively, an adaptive algorithm can recursively compute the equalizer by tweaking the coefficient values from the previous time step. The former approach has the advantage of optimality when the channel truly is static and the window is large enough to average out the noise. However, the complexity can be very high, since optimal solutions generally require matrix inversions, computation of generalized eigenvectors [3], or singular value decompositions [4], depending on the problem, and these must be repeated every block. The latter approach has the advantage of making use of the solution from the previous time step, and the complexity is usually limited to computation of a matrix-vector multiply, or often only vector-vector and vector-scalar multiplies, and moreover, the complexity is evenly spread out over time (rather than as a lump sum at initialization). However, convergence to a good value often requires many more data samples than optimal batch processes. Thus, adaptive equalizers are by no means universally superior, but they are often preferred when computational power is at a premium and mobility is high, e.g., in a mobile handset.

9.1.3   History of Adaptive Equalizers

Adaptive equalizers, sometimes called automatic equalizers, have been in use since the 1960s [5, 6]. Of particular note is the introduction of the least mean square (LMS) algorithm, sometimes called the Widrow-Hoff algorithm [5, 7]. LMS is still used today as a benchmark for comparison of adaptive equalizers, due to its low complexity and its convergence to the minimum mean squared error (MMSE) equalizer for a static channel. Research on adaptive equalization became more widespread in the 1970s, motivated by the need to equalize the impulse responses of telephone lines [8–10]. This research focused on the comparison of different cost functions, and on hybrid equalizer structures, such as the combination of a partial equalizer and a reduced-complexity MLSE. However, this research typically assumed the availability of a sufficiently long training signal, which reduces the channel throughput, and is not even available in surveillance environments.

A "blind" (or "self-recovering") equalizer is one that relies on known statistical properties of the transmitted signal, rather than on a training signal. A blind, adaptive equalizer was first introduced in 1975 in [11], which replaced the training signal in the LMS algorithm with the output of a decision device at the receiver. This idea was later termed decision direction (DD). However, it is dependent on the ability to make good decisions at initialization, which is not always the case. A more sophisticated blind equalizer, the constant modulus algorithm (CMA), was introduced in the early 1980s [12–14]. CMA assumes the transmitted data has a constant modulus, and the equalizer attempts to restore this property. However, CMA can be extended to non-constant modulus sources [15], in which case it may be viewed as a dispersion (or effective noise power)-minimizing algorithm. Despite the age of DD and CMA, most blind adaptive equalizers proposed more recently are rooted in these two algorithms. Exceptions often involve finding alternate signal properties to restore.

Since the mid-1980s, adaptive equalizer research has focused less on development of new algorithms and more on either characterizing popular algorithms or tweaking them for performance improvement or complexity reduction. Notable extensions for trained equalizers include application of nonlinearities to the data and error functions in the LMS algorithm [17], modifications to exploit sparsity in the equalizer [18], and consideration of alternate cost functions, such as the fourth power of the error instead of the square [19, 20], or hybrids of the two. Notable extensions for blind equalizers include the use of multiple moduli in CMA for higher-order constellations [21], and hybrid cost functions that retain the benefits of CMA but provide improved convergence speed [22].

In the past few years, there has been a surge of interest in algorithms that exploit the idea of CMA, but with a different contour in the constellation space. CMA tries to force the constellation points back onto a circular contour in the complex plane. Similarly, the square contour algorithm (SCA) [23], also called the constant square algorithm (CQA) [24], uses a square contour, which has a constant infinity norm (as opposed to the constant 2-norm of a circle). This idea was generalized to include all possible norms on the complex plane in [24] and [25], leading to a plethora of possible algorithms, including extended CMA (ECMA) [25] and the constant norm algorithm with an L6 norm (CNA-6) [24]. However, not all constellations are built on square or circle patterns; hence, [26] proposed the use of a more complicated cross-shaped contour, leading to the constant cross algorithm (CXA).

Further information on equalization of slowly time-varying channels can be obtained from a variety of survey papers. Qureshi's paper [16] surveys early work in trained adaptive equalizers (pre-1985). Reviews of blind adaptive equalization algorithms can be found in [27] and [28]. One of the more popular "encyclopedias" of adaptive filter algorithms is Haykin's book [29].

Up until around 2000, almost all adaptive equalizers were designed for the case in which the channel was frequency selective but quasi-static. That is, the channel impulse response was assumed to be relatively constant and to approximately obey the model of (9.2), but it could drift over time. This allows a gradient-descent type of equalizer to keep up. In the past few years, interest has risen in equalization of channels that vary rapidly with respect to the symbol period [1, 30–33]. In such a case, the channel is both time and frequency selective, or doubly selective, and is modeled more appropriately by (9.1) than by (9.2). Such channels cannot be equalized by gradient-descent equalizers (trained or blind) since the update rule cannot keep up with the speed of the channel variations, and more complex methods of equalization are required. One possibility is to take the two-dimensional Fourier transform of the channel convolution matrix, and then perform the equalization in this frequency domain. This is most appropriate for systems such as orthogonal frequency division multiplexing (OFDM), in which the data are already encoded in the frequency domain. For small amounts of time variation, MMSE symbol estimation can be performed by considering the interference from adjacent frequency bins due to the Doppler spread. The complexity can be kept manageable by ignoring the small interference coefficients at large Doppler spreads [31] or by windowing in the time domain to restrict the Doppler spread [1]. However, [1] and [31] were derived specifically in the context of OFDM, for which the frequency-domain input signal is discrete, so this approach is not applicable to other modulation schemes. A more general approach is to model both the channel and the equalizer in the time domain as matrices rather than as vectors [30]. This explicitly accounts for the time variations in the model, and allows for far more degrees of freedom in the equalizer. The challenge then becomes accurately determining all of the channel coefficients from limited observable data, and of computing the equalizer with a moderate amount of complexity. These are difficult goals, and are the primary drawbacks of this approach [1], although this can be mitigated somewhat by explicitly exploiting the sparse structure of the channel matrix [33]. Another approach that requires less degrees of freedom is to use a basis expansion model (BEM) [32]. A BEM models the doubly selective channel using basis signals that allow for both time and frequency selectivity, by substituting

$$h^{(p)}\left[n,k\right]= \sum_{q=-Q/2}^{Q/2} e^{j2\pi qn/N} h^{(p,q,k)}, \quad k\in\left\{0,\dots,L\right\} \tag{9.4}$$

into (9.1), where q indexes the Doppler spread and $h^{(p,q,k)}$ is the strength of Doppler component q at lag k for receive antenna p. For each time instant, the channel is modeled as a sum of complex exponentials with different Doppler spreads; hence, the amount of Doppler that is modeled is controlled (as opposed to the matrix model of [30], which can in principle allow arbitrarily large Doppler). The equalizer can also be modeled using the BEM, and hence it is also effectively a time-varying filter.

This chapter covers adaptive equalizers, and the equalizers discussed above for doubly selective channels are not adaptive in the classical sense of a recursive update rule. However, they do perform equalization by explicitly accounting for the time variations in the channel model, which is the essence of adaptive equalization. As mobility becomes more pervasive in communications and as data rates are pushed higher, many adaptive equalizers may need to be recast in more complex forms, such as BEMs.

9.2   Adaptive Equalizer Algorithm Formulation

This section covers adaptive algorithm design approaches, in the context of traditional digital communication systems (later sections discuss emerging digital modulation formats). We begin with a discussion of methods of trained and blind equalizer design, with a focus on the design philosophy rather than a specific algorithm. We then cover methods of improving algorithm performance, by either accelerating convergence or exploiting the sparse structure in wireless channels.

9.2.1   Trained Adaptive Algorithm Design Methodologies

Most trained adaptive equalizers take the form of a stochastic gradient descent of a cost function. By far, the most popular choice of cost function is the mean squared error (MSE), where the error is the difference of a desired signal $d[n]$ and the filter output. However, the point at which the error is measured can vary from application to application, leading to different algorithms.

In a traditional single-carrier communication system, the input-output relation is adequately modeled by (9.2) with no additional preprocessing at the transmitter. The data sequence $x[n]$ is in general complex valued and from a discrete signal constellation. When training (knowledge of $x[n]$ for short, intermittent time intervals) is available, the desired signal in wireless communications is almost always a delayed version of the input, $d[n] = x[n - \Delta]$. Then the error signal is

$$e\left[n\right]= \underbrace{x\left[n-\Delta\right]}_{d\left[n\right]}-\underbrace{\mathbf{f}^T\mathbf{y}\left[n\right]}_{\hat{x}\left[n-\Delta\right]}. \tag{9.5}$$

The cost function to be minimized is

$$J = E\left[\left|e\left[n\right]\right|^q\right] \tag{9.6}$$

with $q = 2$ leading to the MSE cost function, although $q = 1$ and $q = 4$ are sometimes encountered in the literature, leading to the mean absolute error (MAE) and mean fourth error (MFE) cost functions, respectively.

Computing a stochastic gradient descent of (9.6) requires computing the gradient with respect to $\mathbf{f}$ and removing the expectation. Then the new equalizer is additively adjusted in the direction of the negative gradient by a small step size μ,

$$\mathbf{f}\left[n+1\right]=\mathbf{f}\left[n\right]-\mu\,\nabla_{\mathbf{f}} J. \tag{9.7}$$

For $q = 2$, this leads to the LMS algorithm,

$$\mathbf{f}\left[n+1\right]=\mathbf{f}\left[n\right]+\mu\,e\left[n\right]\mathbf{y}^*\left[n\right] \tag{9.8}$$

Due to the vector structure of $\mathbf{f}$ and $\mathbf{y}[n]$, (9.8) accounts for the SIMO channel model. Similarly, $q = 1$ leads to the error sign LMS algorithm [34], and $q = 4$ leads to the least mean fourth (LMF) algorithm [19].

Almost all adaptive equalizers in the literature follow this same structure: a cost function is proposed (usually in the form of (9.6) or a hybrid of several such cost functions), and a stochastic gradient descent update is computed. In sections 9.3 and 9.4, we will talk about modifications that can be applied to most gradient descent algorithms to improve performance.

9.2.2   Blind Adaptive Algorithm Design Methodologies

The biggest challenge in adaptive equalization comes when training is unavailable. Even in standards that include training, there is typically a long interval of data between training symbols. If this is the case, the error signal of (9.5) cannot be computed for use in the cost function.

Blind algorithms often turn to known statistical properties of the source signal in lieu of training. For example, when the source data comes from a finite constellation, then a pseudo-training signal can be formed from the nearest constellation point. Denoting $Q\{\cdot\}$ as the decision or quantization function that selects the nearest constellation point, an error signal and associated cost function can be computed as

$$e\big[n\big]=Q\big\{\hat{x}\big[n\big]\big\}-\hat{x}\big[n\big],$$

$$J=E\Big[\big\|e\big[n\big]\big\|^2\Big].$$

(9.9)

The associated stochastic gradient descent algorithm is called decision-directed LMS (DD-LMS). It is identical to (9.8) except for the new error signal. DD-LMS is commonly used due to its simplicity, but it is actually a member of a far broader class of algorithms, called Bussgang algorithms [35], which use the error signal

$$e\big[n\big]=g\big\{\hat{x}\big[n\big]\big\}-\hat{x}\big[n\big],$$

(9.10)

and MSE cost function (as in (9.9)), where $g\{\cdot\}$ is a nonlinear function that estimates the source signal from the equalizer output using Bayesian techniques.

If reliable decisions cannot be made due to excessive noise, an alternative is CMA. CMA uses the error signal and cost function

$$e\big[n\big]=\big|\hat{x}\big[n\big]\big|^2-\gamma,$$

(9.11)

and MSE cost function (as in (9.9)). If the transmitted signal is not constant modulus, CMA can still be used with a performance penalty, or the multiple modulus algorithm (MMA) can be used [21]. Table 9.1 compares the error signals, cost functions, and parameters of the most common trained and blind algorithms, from which most algorithm variants are derived. Note that it is possible to manipulate the form of the CMA error term to show that it is a special case of the Bussgang algorithm, and hence some sources list it as such [29], but it is not really in the same spirit as other Bussgang algorithms (i.e., it does not make a nonlinear estimation of the source symbol), and hence it is listed separately in Table 9.1.

In emerging communications standards, the properties exploited in this section are often not readily available. In this case, the algorithm designer must search for other properties. For example, known zero padding or redundancy (such as repetition) in the transmitted signal may be exploited. Sections 9.3 and 9.4 discuss specific examples in more detail.

9.2.3 Algorithm Acceleration Techniques

The primary drawback of LMS and particularly CMA is that they take a very large number of symbols to converge. Thus, much recent research has focused on accelerating

TABLE 9.1 A Comparison of Common Trained (T) and Blind (B) Adaptive Equalizer Errors, Costs, and Parameters, When the Algorithm Can Be Written as a Stochastic Gradient Descent

	Error Signal	Cost Function	Parameters	Cost Name	Algorithm Name	Reference
T	$e[n] = x[n] - \hat{x}[n]$	$E\left[\left|e[n]\right|^{q}\right]$	$q = 1$	MAE	Error sign LMS	[34]
T			$q = 2$	MSE	LMS	[5]
T			$q = 4$	MFE	LMF	[19]
B	$e[n] = g\{\hat{x}[n]\} - \hat{x}[n]$					
B	$g\{x\} = \text{sign}(x)$	$E\left[\left|e[n]\right|^{q}\right]$	$q = 2$		Sato	[11]
B	$g\{x\} = Q\{x\}$	$E\left[\left|e[n]\right|^{q}\right]$	$q = 2$	DD	DD-LMS	[29]
B	$g\{x\} = \text{Bayesian estimator of source}$	$E\left[\left|e[n]\right|^{q}\right]$	$q = 2$		Bussgang	[35]
B	$e[n] = \|\hat{x}[n]\|_{l}^{p} - \gamma$	$E\left[\left|e[n]\right|^{q}\right]$	$(p,q,l) = (1,1,2)$			[13]
B			$(p,q,l) = (2,2,2)$	CM	CMA	[13]
B			$(p,q,l) = (\text{var},2,2)$	Godard	Godard	[12]
B			$(p,q,l) = (1,2,\infty)$		SCA or CQA	[25]
B			$(p,q,l) = (2,2,4)$		ECMA	[25]
B			$(p,q,l) = (2,2,6)$		CNA-6	[24]

Note: The delay Δ is omitted for simplicity. In each section of the table, the general form of the algorithm is given in the first line, and the next few lines give more specific cases. The variables $x[n], \hat{x}[n], e[n]$ are the source signal, its estimate, and the error signal used by the equalizer, respectively.

their convergence speed. There are techniques such as recursive least squares (RLS) that improve the speed at the cost of an increase in the number of computations (e.g., $O\{N^2\}$ per iteration for RLS, as opposed to $O\{N\}$ per iteration for LMS). However, herein we focus on acceleration techniques that have only a minimal increase in computational overhead, as is appropriate for a mobile handset or any other wireless device with power and computational restrictions.

One type of acceleration technique is to exploit the fact that wireless channels are often very sparse. This is reflected in the fact that most wireless standards define "test channels" via ray models with only about four rays over the span of many symbol intervals.

One difficulty in exploiting sparsity comes from the fact that even when the channel impulse response is sparse, the corresponding equalizer is not necessarily sparse. This is shown in Figure 9.1, which plots the magnitudes of the impulse response coefficients for (a) an example sparse channel, (b) the corresponding equalizer, which is only

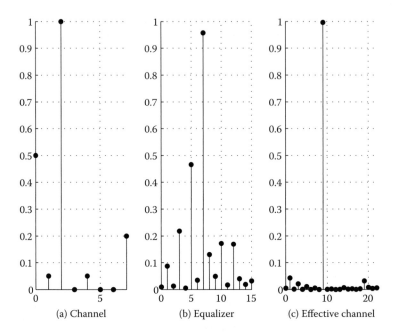

FIGURE 9.1 Even when a multipath channel is sparse, the corresponding linear equalizer need not be sparse. (a) Magnitudes of channel coefficients. (b) Magnitudes of zero-forcing equalizer coefficients. (c) Magnitudes of coefficients of resulting effective channel.

somewhat sparse, and (c) the resulting effective channel. This means there are several ways to exploit sparsity in adaptive equalization:

- Form an adaptive equalizer that is only somewhat sparse, as in Figure 9.1(b).
- Form an adaptive channel identifier that exploits sparsity, then periodically use the current estimate to compute the equalizer.
- Use an alternate equalizer structure that exploits sparsity, such as a partial feedback equalizer (PFE) [36], as shown in Figure 9.2, and adapt it via an algorithm that exploits sparsity.

We now discuss these adaptive algorithms for sparse linear filters (identifiers or equalizers).

An algorithmic paradigm that has received much attention recently due to its fast convergence for sparse adaptive filters is the use of proportionate adaptation [37], as proposed by Duttweiler [18]. The idea is to update large taps more quickly than small taps, since they are more important. The many small taps do little to reduce MSE, and if the filter is known *a priori* to be sparse, many of the small taps may actually be zero, and their updates are noise driven. Although recent advances in proportionate adaptation have been motivated by echo cancellation, the algorithms are equally applicable to wireless systems in which the channels or equalizers are known to be sparse, for example, digital television channels. As shown in [38], a histogram of measured channel coefficients for digital television (U.S. standard) follows an inverse power law distribution,

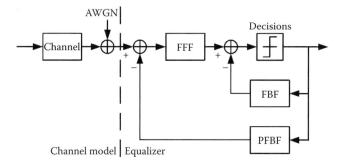

FIGURE 9.2 Block diagram of a partial feedback equalizer (PFE). The partial feedback filter (PFBF) cancels the sparse channel coefficients before the feedforward filter (FFF) smears them out, which cannot be done in a conventional DFE. Then the FFF and the feedback filter (FBF) operate as a normal DFE would. The PFBF can be long and sparse, allowing the FFF and FBF to be short.

whereas histograms of the coefficients of the feedforward and feedback filters in the optimal DFE follow decaying exponential distributions. Thus, the probability of a tap being large is very small.

The proportionate normalized LMS (PNLMS) algorithm [18] is the most widely referenced proportionate adaptation algorithm. The update rule, generalized from [18] to complex-valued signals, is

$$\gamma_{\min}\left[n\right]=\rho\max\left\{\delta_0,\left|\mathbf{f}_0\left[n\right]\right|,\ldots,\left|\mathbf{f}_N\left[n\right]\right|\right\}$$

$$\boldsymbol{\gamma}_k\left[n\right]=\max\left\{\gamma_{\min}\left[n\right],\left|\mathbf{f}_k\left[n\right]\right|\right\},\quad k\in\left\{0,\ldots,N\right\}$$

$$\mathbf{g}_k\left[n\right]=\frac{\boldsymbol{\gamma}_k\left[n\right]}{\left(\frac{1}{N+1}\right)\sum_{l=0}^{N}\boldsymbol{\gamma}_l\left[n\right]},\quad k\in\left\{0,\ldots,N\right\} \tag{9.12}$$

$$\mathbf{G}\left[n\right]=\mathrm{diag}\left[\mathbf{g}_0\left[n\right],\ldots,\mathbf{g}_N\left[n\right]\right]$$

$$\mathbf{f}\left[n+1\right]=\mathbf{f}\left[n\right]+\frac{\mu\,\mathbf{G}\left[n\right]\mathbf{y}^*\left[n\right]e\left[n\right]}{\mathbf{y}^T\left[n\right]\mathbf{G}\left[n\right]\mathbf{y}^*\left[n\right]+\delta},$$

where the error signal $e(n)$ is as in (9.5), and the set of constants $\{\rho,\delta,\delta_0\}$ are called regularizers, mollifiers, or biases, and serve to keep the numerator and the denominator of the update term from going to zero. Ignoring the mollifiers, the kth diagonal element of the matrix $\mathbf{G}[n]$ is the fraction of the 1-norm of the equalizer that occurs in filter tap k. Thus, the product $\mu\,\mathbf{G}[n]$ can be thought of as a vector step size with an average value of μ, but whose elements are proportionate to the corresponding equalizer tap magnitudes.

The denominator $\mathbf{y}^T[n]\mathbf{G}[n]\mathbf{y}^*[n] + \delta$ normalizes by the (tap-weighted) power of the filter input, which essentially makes the scale of the update insensitive to the scale of the filter input signal, and this aspect of PNLMS is taken from the normalized LMS (NLMS) algorithm [29]. Note that PNLMS and its successors were originally developed for echo cancellation, for which this normalization term was needed. However, in this chapter, we discuss equalization; hence, the power of the input signal is less variable and the normalization term is not necessary. Thus, henceforth we omit it.

By weighting the update terms by the current tap magnitudes, PNLMS forces the large taps to converge quickly at the expense of the small taps. Since many of the small taps should be zero anyway, this mitigates the amount of misadjustment (jitter-induced excess MSE) [7] while still allowing the large (and thus more important) taps to converge quickly. Thus, NLMS is an adaptation rule that distributes the update strength equally across all taps whether they need it or not, whereas PNLMS is a "capitalist" adaptation rule that allows the rich (moderate, partially converged values) to get richer (large, converged values). The sparser the optimal filter is, the more the large taps can benefit from the minimal adaptation of the small taps; hence, PNLMS is ideally suited to sparse channels.

Much recent literature has focused on analyzing and improving PNLMS. Deng and Doroslovački [39] have extended the idea of proportionate adaptation to find the optimal weighting function. That is, instead of populating the matrix $\mathbf{G}[n]$ with the normalized filter tap magnitudes, which is essentially a linear weight, they considered applying an arbitrary function to the filter taps before normalizing. They then computed the function that led to the fastest convergence, in terms of having *all* filter taps reach values within some small ε of their optimal values. They concluded that the optimal function was logarithmic rather than linear, so that the first three lines of (9.12) can be replaced by

$$\mathbf{g}_k[n] = \frac{\ln\left(\left|\mathbf{f}_k[n]\right|/\varepsilon\right)}{\left(\frac{1}{N+1}\right)\sum_{l=0}^{N}\ln\left(\left|\mathbf{f}_l[n]\right|/\varepsilon\right)}, \tag{9.13}$$

where ε serves as a mollifier when filter taps are small. The resulting algorithm is called μ-law PNLMS (MPNLMS). The natural log can be replaced by a piecewise-linear function with a minimal effect on performance [39].

Another approach is to combine the advantages of NLMS and PNLMS. The simplest approach is to alternate between the two [40, 41]. A more sophisticated approach was presented in [42], in which an improved PNLMS (IPNLMS) algorithm was created that combines NLMS and PNLMS in every update, and does not require the true weight vectors to be sparse. This is especially important for wireless equalization, since as stated above, sparse channels do not always lead to sparse equalizers. The motivation for IPN-LMS is that the max function in (9.12) is very harsh, and that if the parameter estimates are inaccurate, the proportionality can amplify the error and have a detrimental effect on performance. Thus, they modify the weights in (9.12) to

$$\boldsymbol{\gamma}_k\left[n\right]=\frac{(1-\alpha)}{2}\left(\frac{1}{N+1}\right)\left\|\mathbf{f}\left[n\right]\right\|_1+\frac{(1+\alpha)}{2}\left|\mathbf{f}_k\left[n\right]\right|,$$

$$\mathbf{g}_k\left[n\right]=\frac{\boldsymbol{\gamma}_k\left[n\right]}{\left(\frac{1}{N+1}\right)\sum_{l=0}^{N}\boldsymbol{\gamma}_l\left[n\right]}, \quad k\in\left\{0,...,N\right\}, \qquad (9.14)$$

where $\|\cdot\|_1$ is the 1-norm of a vector and $\alpha \in [-1,1]$ is a user-defined weighting parameter. Observe that $\|\boldsymbol{\gamma}[n]\|_1 = \|\mathbf{f}[n]\|_1$; for $\alpha = -1$ we get the NLMS algorithm, and for $\alpha = 1$ we get the PNLMS algorithm.

Proportionate adaptation rules are compared in Figures 9.3 to 9.5. In Figure 9.3, NLMS, PNLMS, MPNLMS, and IPNLMS are compared for channel identification, in which just over half of the channel coefficients are zero, and most of the rest are small. In Figures 9.4 and 9.5, a similar comparison is made for equalization. The magnitudes of the optimal equalizer taps are presented in Figure 9.4, sorted for easy analysis of the level of sparsity, and the learning curves are presented in Figure 9.5. Observe that for both the identification problem and the equalization problem, all proportionate adaptation algorithms see a performance gain. However, PNLMS and MPNLMS were originally developed for echo cancellation, in which as many as 90% of the channel taps may be essentially zero. In wireless identification or equalization, the optimal filters are not

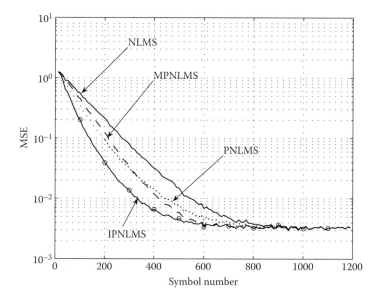

FIGURE 9.3 A demonstration of the convergence improvement of adaptive identification of wireless channels due to the use of proportionate adaptation. The algorithms considered are all in the normalized LMS family: NLMS [29], PNLMS [18], MPNLMS [39], and IPNLMS [42]. The channel had five of eleven nonzero taps, and thus was only somewhat sparse.

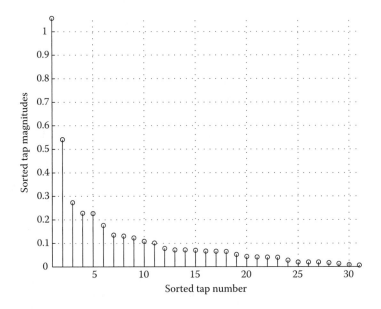

FIGURE 9.4 The tap magnitudes of an optimal equalizer, sorted in descending order. Note that although there are no zero values, the equalizer is still somewhat sparse, since two-thirds of the values are below 10% of the peak value.

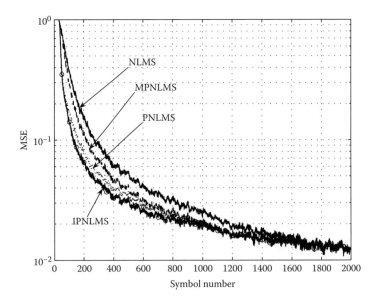

FIGURE 9.5 A demonstration of the convergence improvement of adaptive equalization of wireless channels due to the use of proportionate adaptation. The algorithms considered are all in the normalized LMS family: NLMS [29], PNLMS [18], MPNLMS [39], and IPNLMS [42]. The optimal equalizer tap magnitudes are shown in Figure 9.4.

nearly so sparse; hence, the gains are only on the order of a factor of 2 in convergence time. IPNLMS was designed to work for filters that are both sparse and nonsparse, and thus it outperforms the other algorithms in this case.

9.2.4 Complexity Reduction Techniques for Sparse Adaptive Equalizers

There are two benefits to *a priori* knowledge that the ideal filter is sparse. First, as discussed in the previous section, knowledge of sparsity can be incorporated into the adaptation rule to accelerate convergence of the important taps while suppressing (often advantageously) the adaptation of the less important taps. Second, the adaptation rule itself can be implemented sparsely, saving computations. Adaptation rules that are themselves sparse are generally referred to as partial-update algorithms. Sometimes they may incur a performance penalty, but if the system truly is sparse, they may actually improve performance. Whether or not the true system is actually sparse, they will always reduce the computational load of the equalizer by a fixed, known amount. Hence, they are very attractive for implementation in adaptive equalizers in low-power battery-operated devices.

As the name implies, partial-update algorithms only update a subset of M of the equalizer coefficients (out of the total of $N + 1$ coefficients) at each iteration. The distinctions between different partial-update algorithms are primarily the number of taps updated per iteration and the method by which the taps to be updated are chosen. Computational complexity is reduced by roughly a factor of M/N, but the exact amount depends upon the algorithm that is being partially updated.

One approach is to simply cycle through the taps to be updated at each iteration, as in sequential or periodic partial-update LMS [43]. For this baseline approach, one would expect the algorithm to take N/M times longer to converge. More sophisticated partial-update algorithms attempt to choose the M taps to be updated such that the best update possible can be formed under this constraint. However, care must be taken to ensure that the computational burden of the tap selection mechanism does not begin to balance out the computational savings of the partial update. Max-NLMS [44] chooses a single tap to update at each iteration, chosen as the tap with largest corresponding input. M-max-NLMS [45] generalizes this to update the M taps with the M largest inputs. The rationale is that the taps with the largest inputs will have the largest innovations, since the updates in NLMS algorithms are proportionate to the tap inputs. Heuristically, the largest innovations should lead to the largest improvement in the cost function. By selecting the M taps that yield the best improvement to the cost function, the update term removes the least useful steps and focuses on the most useful steps. This mitigates the misadjustment due to adapting the unnecessary taps, which can improve the performance of the algorithm, depending on how necessary or unnecessary each tap is. Since choosing the largest inputs only requires a small number of comparisons (essentially a subtraction operation each) and no multiplies, this selection criterion has very little complexity overhead.

All of these partial-update algorithms can be cast into the generic gradient descent form of

$$\mathbf{g}_j[n] = \begin{cases} 1, & \left(\text{select } M \text{ values of } j \text{ by some method}\right) \\ 0, & \text{else} \end{cases}$$

$$\mathbf{G}[n] = \text{diag}\big[\mathbf{g}[n]\big] \qquad\qquad (9.15)$$

$$\mathbf{f}[n+1] = \mathbf{f}[n] + \mu \, \mathbf{G}[n] \mathbf{y}^*[n] e[n],$$

where $\mathbf{G}[n]$ is a diagonal matrix of ones and zeros that turn on or off the updates to each coefficient. (For applications such as echo cancellation, an NLMS normalization term as in the denominator of (9.12) can be explicitly included, or it can be included implicitly by dividing the step size μ by the normalization term.) With this structure, the partial update of (9.15) has the same form as the proportionate update of (9.12), and in either case, $\mu \, \mathbf{G}[n]$ effectively forms a vector step size.

Specific tap selection rules for (9.15) have the mathematical form

$$\mathbf{g}_j^{\text{sequential}}[n] = \mathbf{g}_{(j-1)_{\text{mod}(N+1)}}^{\text{sequential}}[n-1]$$

$$\mathbf{g}_j^{\text{maxLMS}}[n] = \begin{cases} 1, & j = \arg\max_{0 \le k \le N} \big|\mathbf{y}_k[n]\big| \\ 0, & \text{else} \end{cases} \qquad (9.16)$$

$$\mathbf{g}_j^{\text{MmaxLMS}}[n] = \begin{cases} 1, & j \in \left\{\text{indices of } M \text{ largest } \big|\mathbf{y}_k[n]\big|\right\} \\ 0, & \text{else} \end{cases}.$$

Observe that max-NLMS and M-max-NLMS are better suited for equalization than channel identification when the application is digital communications. This is because digital communication channels often have inputs drawn from a constant modulus source (e.g., $\pm 1 \pm j$), in which case all of the inputs to the channel model would have the same magnitude, and no selection could be made, whereas the equalizer inputs contain intersymbol interference and noise, with a wide range of magnitudes.

Other recent work has investigated extensions to M-max-NLMS that further reduce the complexity. Selective-block NLMS [46] is like M-max NLMS, but in terms of blocks of taps rather than individual taps. Similarly, the short-sort algorithm [47] recognizes that in many cases, the significant taps are grouped together. The algorithm identifies a contiguous block of significant taps and always updates those, and performs M-max NLMS on the remaining taps, which reduces the sorting overhead. A slightly more sophisticated approach is to monitor an activity measure for each tap [48] (instead of simply comparing input magnitudes), and then favor those taps when updating. Table 9.2 lists popular proportionate adaptation and partial update algorithms.

Proportionate adaptation rules and partial-update rules were explicitly designed to exploit sparsity in the filter impulse response. PNLMS and its variants and M-max-LMS

TABLE 9.2 Proportionate Adaptation (PA) and Partial-Update (PU) Algorithms that Can Be Written in the Form $\mathbf{f}[n] = \mathbf{f}[n] + \mu\, e[n]\mathbf{G}[n]\mathbf{y}^*[n]$, with $\mathbf{G}[n] = \mathrm{diag}[\mathbf{g}[n]]$

Type	Name	Value of $\mathbf{g}[n]$	Reference
	NLMS	$\mathbf{g}[n]=\underbrace{\left[1,1,...,1\right]^{T}}_{N+1 \text{ of them}}$	[29]
PA	PNLMS	$\boldsymbol{\gamma}_k[n]=\max\left\{\left|\mathbf{f}_k[n]\right|,\rho\max\left\{\delta_0,\left\|\mathbf{f}[n]\right\|_\infty\right\}\right\}$ $\mathbf{g}_k[n]=\dfrac{\boldsymbol{\gamma}_k[n]}{\left(\frac{1}{N+1}\right)\sum_{l=0}^{N}\boldsymbol{\gamma}_l[n]},\quad k\in\left\{0,...,N\right\}$	[18]
PA	IPNLMS	$\boldsymbol{\gamma}_k[n]=\frac{(1-\alpha)}{2}\left(\frac{1}{N+1}\right)\left\|\mathbf{f}[n]\right\|_1+\frac{(1+\alpha)}{2}\left|\mathbf{f}_k[n]\right|,\quad -1\le\alpha\le1$ $\mathbf{g}_k[n]=\dfrac{\boldsymbol{\gamma}_k[n]}{\left(\frac{1}{N+1}\right)\sum_{l=0}^{N}\boldsymbol{\gamma}_l[n]},\quad k\in\left\{0,...,N\right\},$	[42]
PA	MPNLMS	$\mathbf{g}_k[n]=\dfrac{\ln\left(\left\|\mathbf{f}_k[n]\right\|/\varepsilon\right)}{\left(\frac{1}{N+1}\right)\sum_{l=0}^{N}\ln\left(\left\|\mathbf{f}_l[n]\right\|/\varepsilon\right)},\quad k\in\left\{0,...,N\right\}$	[39]
PU	Sequential PU NLMS	$\mathbf{g}_j^{\text{sequential}}[n]=\mathbf{g}_{(j-1)_{\text{mod}(N+1)}}^{\text{sequential}}[n-1]$ $\mathbf{g}[0]=\left\{\text{user-defined vector of } M \text{ ones, } N\text{-}M+1 \text{ zeros}\right\}$	[43]
PU	Max-LMS	$\mathbf{g}_j^{\text{maxLMS}}[n]=\begin{cases}1, & j=\arg\max_{0\le k\le N}\left|\mathbf{y}_k[n]\right|\\0, & \text{else}\end{cases}$	[44]
PU	M-max-LMS	$\mathbf{g}_j^{\text{MmaxLMS}}[n]=\begin{cases}1, & j\in\left\{\text{indices of }M\text{ largest }\left|\mathbf{y}_k[n]\right|\right\}\\0, & \text{else}\end{cases}$	[45]

Note: The table lists the composition of the vector $\mathbf{g}[n]$ for each algorithm.

and its variants were primarily motivated by channel identification for acoustic echo cancellation, although they have been applied to wireless channels as well. If we wish to apply them to equalization of wireless channels, it makes sense to use an equalizer structure that preserves the sparsity of the channel in its own impulse response. Motivated primarily by complexity reduction, the PFE and several related equalizer structures have been recently proposed that satisfy this goal [36]. In the PFE, as shown in Figure 9.2, a partial feedback filter (PFBF) cancels the sparse channel coefficients before the feedforward filter (FFF) smears them out, which cannot be done in a conventional DFE. Then the FFF and the feedback filter (FBF) operate as a normal DFE would. The PFBF can be long and sparse, allowing the other filters to be short.

9.3   Cyclic-Prefixed Communication Systems

Cyclic-prefixed systems come in two varieties: multicarrier and single-carrier cyclic prefixed. Multicarrier modulation is arguably the most popular choice of modulation format for communication standards that have emerged in the past decade. Examples include wireless local area networks (IEEE802.11a, HIPERLAN/2 in Europe, and MMAC in Japan), European Digital Video/Audio Broadcast (DVB/DAB), digital subscriber loops (DSL), the terrestrial repeaters in Sirius and XM satellite radio, power line communications, and the proposed "multiband" standard for ultrawideband communications, among others.

Given that the signal structure of cyclic-prefixed modulation differs markedly from that of single-carrier modulation, it provides a good framework for demonstrating the principles of adaptive equalizer design. Moreover, adaptive equalizers for cyclic-prefixed systems are fairly recent research products, and are of interest in and of themselves. In this section, we first review the signal structure, and then discuss trained and blind adaptive equalizers. Emphasis is placed on design methodology rather than on the algorithms themselves, since communications standards fall in and out of favor, but the design principles are equally applicable to other modulation formats.

The idea of cyclic-prefixed communications is to perform the equalization in the frequency domain. By considering a frequency-selective channel as a bank of parallel narrowband flat fading channels, equalization can be performed via a one-tap complex scalar on each tone. The collection of these scalars is called a frequency-domain equalizer (FEQ). However, in order to use the fast Fourier transform (FFT) to convert between time and frequency domains, the signal is necessarily discrete in time and block processed. Thus, some additional mathematical tricks are required in order to equalize in the frequency domain. Specifically, element-wise multiplication of the channel and the input data in the frequency domain is equivalent to their *circular* convolution in the time domain, but multipath propagation as modeled by (9.2) is a *linear* convolution. In order to make the convolution appear periodic (equivalently, circular), the input signal is made periodic over a short duration. A cyclic prefix (CP) is inserted as an extension to the start of each transmitted block by making a copy of the last ν samples. Thus, for a block initially of length N_{FFT}, the cyclically extended block of length $\bar{N} = N_{FFT} + \nu$ is periodic with period N_{FFT}. Provided that the channel order satisfies $L \leq \nu$, then the last N_{FFT} outputs of each block will appear to be a periodic (circular) convolution of the channel and the original N_{FFT} inputs of that block. Under this condition, equalization can be accomplished by taking an FFT, multiplying element-wise by the FEQ, and then taking an inverse FFT (IFFT) of the result. Since the FFT inherently induces block-based processing, this section uses the time index b to denote the block number, in addition to the previously introduced time index n that denotes the sample number.

Block diagrams of the two existing cyclic-prefixed systems are shown in Figure 9.6, namely, single-carrier cyclic-prefixed (SCCP) modulation (sometimes called single-carrier frequency-domain equalization [SC-FDE]) and OFDM. The distinction is that SCCP systems store the data in the time domain, so they follow the format discussed above, whereas multicarrier systems store the data in the frequency domain, so an additional

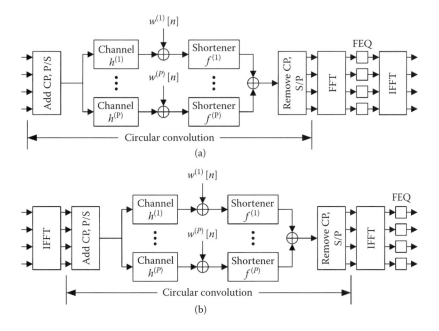

FIGURE 9.6 Block diagrams of (a) single-carrier cyclic-prefixed (SCCP) modulation and (b) multicarrier modulation. (I)FFT denotes (inverse) fast Fourier transform, P/S denotes parallel-to-serial conversion, and S/P the reverse.

IFFT is needed before transmission and the final IFFT discussed above is redundant and thus omitted. Multicarrier systems are older, and thus far more common in both the technical literature and industry standards. However, their transmitted IFFT outputs follow a Gaussian distribution (via the central limit theorem). This means that multicarrier systems tend to have a high peak-to-average power ratio, whereas SCCP systems do not.

In general, the channel order L is unknown, which often causes problems for adaptive equalizers. In the case of cyclic-prefixed systems, the CP length ν must be set without knowledge of L, in which case the condition $L \leq \nu$ may not be satisfied. Moreover, the CP introduces redundancy, reducing the data rate by a factor of $N_{FFT}/(N_{FFT} + \nu)$, so ν may be set small to keep the redundancy small even if the channel order may be larger than the maximum tolerable ν. Either way, the excess channel length induces interference and should be mitigated. This is often done via a channel-shortening equalizer (CSE). (Most literature refers to this as a time-domain equalizer [TEQ], but we avoid that term here since all the equalizers discussed in the previous sections also operate in the time domain, and the term *CSE* avoids confusion.) The CSE shortens the effective channel to the CP length, and then the FEQ can perform equalization. Like a standard equalizer, the CSE is modeled as in (9.3), and it is applied to the channel output as shown in Figure 9.6.

The next two sections discuss adaptive algorithms for the FEQ and CSE, in turn.

9.3.1   Adaptive FEQ Algorithms

Adaptation rules for the FEQ are easily formed as analogous to or generalizations of the commonly accepted adaptation rules for traditional equalizers, although there are some nuances due to modulation format. For the time being, we assume that the CSE is operating perfectly. The model of the remaining receiver processing of block b is

$$\mathbf{y}[b] = \left[y\left[(b-1)\bar{N} + v + 1 + \Delta \right], \dots, y\left[b\bar{N} + \Delta \right] \right]^T$$

$$\tilde{\mathbf{y}}[b] = \mathbf{F}\mathbf{y}[b]$$

$$\tilde{\mathbf{x}}[b] = \mathbf{d} \odot \tilde{\mathbf{y}}[b] \tag{9.17}$$

$$\hat{\mathbf{x}}_{mc}[b] = Q\{\tilde{\mathbf{x}}[b]\}$$

$$\hat{\mathbf{x}}_{scp}[b] = Q\{\mathbf{F}^{-1}\tilde{\mathbf{x}}[b]\},$$

where Δ accounts for the propagation delay as in (9.5), $\mathbf{F}$ denotes the Fourier transform of size N_{FFT} with inverse transform $\mathbf{F}^{-1} = \mathbf{F}^H$, $\mathbf{d}$ is the FEQ as a length N_{FFT} vector, $\odot$ denotes Hadamard (element-wise) multiplication, and $Q\{\cdot\}$ quantizes the output to the nearest constellation point.

First, consider trained adaptation rules. In many multicarrier standards, training is not available on every tone. For example, in DVB operating in 2K mode, the FFT size and number of subcarriers is 2,048. Of these, 1,705 are used and the rest are left null as a guard band, to mitigate adjacent channel interference. Of the used subcarriers, 45 are continual pilot tones (transmitting a pseudo-random binary sequence as training). Also, of these 1,705, every third tone is used as a pilot, but only once every four symbols, in alternating fashion [49]. This arrangement of pilots across time and frequency is shown in Figure 9.7. Thus, at best, every third coefficient of the channel in the frequency domain has training (and not all of them at once), which complicates trained adaptation. This can be dealt with by assuming the channel coherence is high enough that the remaining coefficients can be interpolated.

On a subcarrier that has training, a simple one-tap LMS or RLS adaptation rule can be created. For time or frequency indices when training is unavailable, a simple one-tap DD-LMS or CMA adaptation rule can be implemented, or the FEQ value can be interpolated across time and frequency from indices where training is available.

Adaptation of the FEQ for SCCP systems is slightly more complicated, since the outputs of the FEQ are not expected to be finite alphabet until after the final IFFT. Also, since SCCP has not yet been deployed in an industry standard, there are no set formats for training availability. Instead of indexing the transmitted signal by block number and subcarrier number, an SCCP transmission can be indexed by block number and sample number within that block, and training could be staggered in some fashion. What is critical for a simple FEQ adaptation rule is that the output of the FEQ be comparable to some known data. Since the FEQ is in the frequency domain, this requires that an

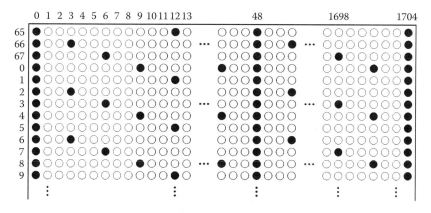

FIGURE 9.7 Diagram of the continual and staggered pilot tones (training) in the European standard for Digital Video Broadcast. The column index k is the frequency or subcarrier index, and the row index b is the block or symbol index. Shaded circles represent training data. The continual pilots are not regularly spaced in frequency; the full list is available in [49].

entire time-domain block contain training for any given frequency coefficient to be considered known. In this case, a simple one-tap LMS or RLS adaptation rule can be implemented for each FEQ coefficient. Similarly, one-tap DD-LMS and CMA rules can be formed when training is not available. In vector form, the LMS, DD-LMS, and CMA FEQ updates for an SCCP system are [50]

$$\mathbf{d}_{LMS}\big[b+1\big]=\mathbf{d}_{LMS}\big[b\big]+\mu\,\tilde{\mathbf{y}}^*\big[b\big]\odot\big(\mathbf{F}\mathbf{x}\big[b\big]-\tilde{\mathbf{x}}\big[b\big]\big)$$

$$\mathbf{d}_{DD}\big[b+1\big]=\mathbf{d}_{DD}\big[b\big]+\mu\,\tilde{\mathbf{y}}^*\big[b\big]\odot\big(\mathbf{F}Q\{\mathbf{F}^{-1}\tilde{\mathbf{x}}\big[b\big]\}-\tilde{\mathbf{x}}\big[b\big]\big)$$

$$\mathbf{d}_{CMA}\big[b+1\big]=\mathbf{d}_{CMA}\big[b\big]-\mu\,\tilde{\mathbf{y}}^*\big[b\big]\odot\big(\mathbf{F}\mathbf{e}_{CMA}\big[b\big]\big) \tag{9.18}$$

$$\mathbf{e}_{CMA}\big[b\big]=\Big(\big(\mathbf{F}^{-1}\tilde{\mathbf{x}}\big[b\big]\big)^*\odot\big(\mathbf{F}^{-1}\tilde{\mathbf{x}}\big[b\big]\big)-\gamma\Big)\odot\big(\mathbf{F}^{-1}\tilde{\mathbf{x}}\big[b\big]\big),$$

and the RLS FEQ update with forgetting factor ρ is

$$\mathbf{R}\big[b\big]=\rho\Big(\mathbf{R}\big[b-1\big]+\mathrm{diag}\big[\tilde{\mathbf{y}}^*\big[b\big]\odot\tilde{\mathbf{y}}\big[b\big]\big]\Big)$$

$$\mathbf{P}\big[b\big]=\rho\Big(\mathbf{P}\big[b-1\big]+\tilde{\mathbf{y}}^*\big[b\big]\odot\big(\mathbf{F}\mathbf{x}\big[b\big]\big)\Big) \tag{9.19}$$

$$\mathbf{d}_{RLS}\big[b\big]=\mathbf{R}^{-1}\big[b\big]\mathbf{P}\big[b\big].$$

Due to the diagonal structure of **R**, the RLS update requires only marginally more computations and converges much faster than LMS [50]; hence it is preferable of the two.

9.3.2   Adaptive CSE Algorithms

Now we return to the problem of CSE adaptation in cyclic-prefixed systems. As mentioned in the discussion of FEQ adaptation, training in multicarrier systems may be spread across time and frequency, and it may be allocated in the frequency domain. In order to train based on the equalizer output, we first assume that time-domain training is available, which is equivalent to assuming that training is available on all subcarriers in a given symbol. In this case, the obvious choice is to use an LMS-like equalization rule. However, this is complicated by the fact that the goal of the CSE is not to make the effective channel simply a delay, so we cannot form an error signal by comparing to the delayed training data as in (9.5). Instead, the goal of the CSE is to make the effective channel short, but (to a first-order approximation) we do not care about what shape the effective channel takes so long as it is short. Falconer and Magee [10] solved this problem in the context of CSE design in conjunction with MLSE by forming an additional filter, **b**, of length $v + 1$, called the target impulse response (TIR). Then the CSE and TIR can each be adapted in turn by an LMS algorithm.

Mathematically, the error function and cost are

$$e\left[n\right] = \mathbf{b}^T \mathbf{x}\left[n-\Delta\right] - \mathbf{f}^T \mathbf{y}\left[n\right],$$

$$J = E\left[\left|e\left[n\right]\right|^q\right], \tag{9.20}$$

where Falconer and Magee chose $q = 2$, corresponding to an LMS algorithm, yet others have considered $q = 4$, corresponding to an LMF algorithm [51]. Since the trivial setting $\mathbf{b} = \mathbf{0}, \mathbf{f} = \mathbf{0}$ permits a zero-cost solution, a constraint must be enforced, such as a unit-tap or unit-norm filter. Iterating through the adaptation of each filter and maintenance of the constraint, the MMSE adaptive CSE algorithm is

$$\mathbf{f}\left[n+1\right] = \mathbf{f}\left[n\right] + \mu e\left[n\right]\mathbf{y}^*\left[n\right]$$

$$\hat{\mathbf{b}}\left[n+1\right] = \mathbf{b}\left[n\right] - \mu e\left[n\right]\mathbf{x}^*\left[n-\Delta\right] \tag{9.21}$$

$$\mathbf{b}\left[n+1\right] = \hat{\mathbf{b}}\left[n+1\right] \Big/ \left\|\hat{\mathbf{b}}\left[n+1\right]\right\|_2.$$

The choice of a larger q will add factors in the first two lines, and the choice of an alternate constraint will require a different projection in the third line of (9.21). For $v = 0$, **b** becomes a scalar, and (9.21) reduces to the LMS algorithm for **f** alone.

The coupling of **b** and **f**, along with the constraint on **b**, makes formulation of an RLS adaptation rule more complicated than for a traditional equalization problem. Use of the unit-norm constraint leads to an optimal solution for **b** as an eigenvector rather than a least squares solution, and hence RLS does not apply. However, if a unit-tap constraint is used, the optimal **b** can be written as the solution to either a generalized eigen problem [52] or a least squares problem [50]. Although this constraint does lead to a suboptimal

MSE compared to the unit-norm constraint [52], it does permit the use of an RLS algorithm operating on the concatenated parameter vector

$$\mathbf{v} = \left[\mathbf{w}^T, \mathbf{b}_0, ..., \mathbf{b}_{i_0-1}, \mathbf{b}_{i_0+1}, ..., \mathbf{b}_v \right]^T, \qquad (9.22)$$

where $\mathbf{b}_{i_0}$ is the tap that is constrained to unity. The full RLS algorithm is given in [50].

We will return to the issue of trained CSE adaptation in the case of training on intermittent subcarriers momentarily. However, this requires a discussion of blind adaptation first. Since the CSE output is not expected to be finite alphabet, DD and constant modulus methods cannot immediately be used. However, the FEQ output is expected to be finite alphabet. Thus, we have two options: form a blind cost at the FEQ output, and propagate its dependence on the CSE back through the FEQ and FFT; or apply the "property restoral" concept for development of alternate blind cost functions and algorithms. We withhold the former discussion until section 9.5, since the FEQ is adaptive as well, which leads to a coupling of adaptive blocks. For now, we focus on properties of cyclic-prefixed systems that can be restored by the CSE.

A property common to both multicarrier and SCCP systems is the redundancy induced by the CP. This redundancy has been exploited for adaptive carrier frequency offset estimation [53, 54] as well as CSE adaptation [50]. To make use of this property, we form an error signal and cost function as

$$e[b] = \mathbf{f}^T \mathbf{y} \left[(b-1)\bar{N} + v + \Delta \right] - \mathbf{f}^T \mathbf{y} \left[b\bar{N} + \Delta \right],$$

$$J = E \left[\left| e[b] \right|^q \right], \qquad (9.23)$$

along with a constraint (again, unit norm or unit tap) to avoid the trivial solution $\mathbf{f} = \mathbf{0}$. Due to the linearity of the error with respect to $\mathbf{f}$, when $q = 2$ we get a simple LMS-like update rule, although a periodic projection is required to enforce the constraint, as in (9.21). The resulting algorithm is called multicarrier equalization by restoration of redundancy (MERRY), and is given by

$$\tilde{\mathbf{y}}[b] = \mathbf{y} \left[(b-1)\bar{N} + v + \Delta \right] - \mathbf{y} \left[b\bar{N} + \Delta \right]$$

$$e[b] = \mathbf{f}^T[b] \tilde{\mathbf{y}}[b],$$

$$\hat{\mathbf{f}}[b+1] = \mathbf{f}[b] - \mu \, e[b] \tilde{\mathbf{y}}^*[b] \qquad (9.24)$$

$$\mathbf{f}[b+1] = \hat{\mathbf{f}}[b+1] / \left\| \hat{\mathbf{f}}[b+1] \right\|_2.$$

More details and an RLS-like implementation are available in [50].

Another property that can be exploited in multicarrier (but not SCCP) systems is the guard band in the frequency domain, which was mentioned in section 9.3.1. Often, the subcarriers on the band edges are left as zeros at the transmitter in order to limit adjacent channel interference. These are often referred to as null tones. If the channel is benign, then at the receiver, the null tones should also be zero. Thus, we can form a cost function as

$$J = \sum_{j \in \text{null tones}} \beta_j E\left[\left\|\bar{\mathbf{y}}[b]\right\|^2\right],$$ (9.25)

where the weights β_j would typically be all ones. This is very similar to a decision-directed algorithm, with only one valid constellation point (i.e., zero). The difference is that (9.25) is invariant to magnitude and phase distortion, since $0 \cdot Ae^{j\theta} = 0$. Thus, the effects of the FEQ can be ignored, which would not be the case for a DD-LMS algorithm operating on the non-null tones. A gradient descent of (9.25) was proposed in [55] and further analyzed in [56], and it is sometimes referred to as the carrier nulling algorithm (CNA; not to be confused with the constant norm algorithm of the same acronym, discussed in the previous section). It also has an LMS-like structure, although with part of an FFT involved to relate the values on the null tones to the CSE output:

$$\mathbf{Y}_{\text{null}} = \mathbf{F}_{null}\mathbf{R}\mathbf{f}[b],$$

$$\hat{\mathbf{f}}[b+1] = \mathbf{f}[b] - \mu\, \mathbf{R}^H \mathbf{F}_{\text{null}}^H \mathbf{Y}_{\text{null}}$$ (9.26)

$$\mathbf{f}[b+1] = \hat{\mathbf{f}}[b+1]\big/\left\|\hat{\mathbf{f}}[b+1]\right\|_2,$$

where $\mathbf{R}$ is the $N \times L$ Toeplitz matrix:

$$\mathbf{R} = \begin{bmatrix} r\big[(b-1)\bar{N}+v+1+\Delta\big], & \cdots, & r\big[(b-1)\bar{N}+v+1+\Delta-L\big] \\ \vdots & & \vdots \\ r\big[b\bar{N}+\Delta\big], & \cdots, & r\big[b\bar{N}+\Delta-L\big] \end{bmatrix},$$ (9.27)

and where $\mathbf{F}_{null}$ is obtained by taking the rows of the FFT matrix indexed by the null tones, and $\mathbf{Y}_{null}$ is similarly obtained from the FFT output vector.

Now we can reconsider the case in which training is intermittently provided on some subcarriers. Let S_{null}, $S_{pilot}[b]$, and $S_{data}[b]$ be the sets of subcarriers containing null tones, pilot tones, or unknown data (respectively) during block b, with $S_{null} \cap S_{pilot}[b] \cap S_{data}[b] = \{0, \ldots, N-1\}$. Then we can form the semiblind cost function

$$J = \tag{9.28}$$

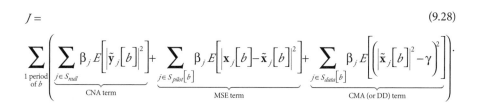

$$\underbrace{\sum_{\substack{1 \text{ period} \\ \text{of } b}}\left(\underbrace{\sum_{j \in S_{null}} \beta_j E\left[\left|\tilde{\mathbf{y}}_j[b]\right|^2\right]}_{\text{CNA term}} + \underbrace{\sum_{j \in S_{pilot}[b]} \beta_j E\left[\left|\mathbf{x}_j[b]-\tilde{\mathbf{x}}_j[b]\right|^2\right]}_{\text{MSE term}} + \underbrace{\sum_{j \in S_{data}[b]} \beta_j E\left[\left(\left|\tilde{\mathbf{x}}_j[b]\right|^2-\gamma\right)^2\right]}_{\text{CMA (or DD) term}}\right).}$$

The weights β_j control the relative use of the trained MSE term and the blind CNA and CMA terms. Note that a DD cost could be used instead of or in addition to the CMA cost.

A comparison of the learning curves of trained and blind adaptive CSE algorithms is shown in Figure 9.8. The simulated system had parameters consistent with IEEE 802.11a: an FFT size of 64 with 12 null tones, a CP of length 16, and complex baseband channels with Rayleigh-distributed taps and an approximately exponential delay profile. The SIMO channel model had one transmit antenna and two receive antennas. The performance metric is the BER, and a combination of 4-ary quadrature amplitude modulation (QAM) per tone and differential encoding was used to remove the need for and effects of an adaptive FEQ.

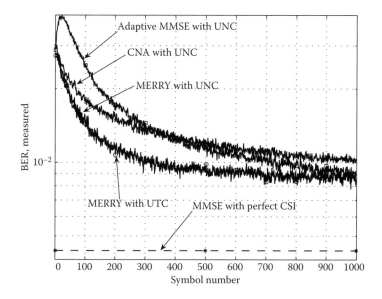

FIGURE 9.8 A performance comparison of trained and blind adaptive channel-shortening equalizers. The algorithms are MMSE [10], MERRY [50], and CNA [55]. MERRY is shown with both a unit-norm and a unit-tap constraint (UNC, UTC), and since the difference is negligible, MMSE and CNA only use a UNC.

The algorithms compared in Figure 9.8 are the adaptive MMSE algorithm [10], MERRY [50], and CNA [55, 56]. The optimal nonadaptive MMSE design with perfect channel state information (CSI) is also shown. In principle, all of these could use either a unit-norm or unit-tap constraint. Both are shown for the case of MERRY; however, since there is no appreciable difference, only a single constraint is shown for the remainder. Also, the blind algorithms can only update once per block (OFDM symbol), but the MMSE design can update as often as once per sample. For a fair comparison, the MMSE design was only updated once per block, so that the update complexity per symbol of the various algorithms would be comparable.

Observe that the various algorithms considered all had comparable learning rates for a given allowable complexity. However, the trained MMSE algorithm could in principle update $\bar{N}$ times more often (with a corresponding increase in complexity).

To recap, trained and blind adaptive equalizers for multicarrier systems can be designed using the same principles as more conventional communication systems. Trained algorithms are generally based on an MMSE framework, and blind algorithms are generally based on a property restoral framework. The unique aspects of adaptive algorithms in multicarrier systems include the need for an adaptive target impulse response for trained MMSE adaptation, the need for a constraint for most trained and blind adaptive algorithms, and the trade-offs associated with operating at the block rate versus the sample rate.

9.4 Ultrawideband Communication Systems

Like multicarrier systems, ultrawideband (UWB) systems have received much attention in recent literature due to the fact that their modulation format is very distinct from single-carrier amplitude modulation formats, although it is not in as widespread implementation as multicarrier modulation. This section reviews UWB modulation and discusses the application of adaptive equalizer design principles to this modulation format.

A UWB system is often defined as one in which the bandwidth divided by the carrier frequency exceeds ¼. The idea is that a very large bandwidth is used, but the total power is moderate—hence the power per Hertz is on the order of the noise floor. This is akin to spread-spectrum modulation in its motivation, but the process of modulation differs. Presently, there are two popular formats of UWB in the literature: pulse position modulation (PPM) and a multiband format, much like OFDM. In this chapter, we focus on PPM, due to its implications for adaptive equalizer design.

In the PPM version of UWB, information is transmitted by very short, infrequent pulses. By transmitting an impulse-like pulse, the bandwidth is very large. In some cases, the need for RF hardware can be avoided completely, since an impulsive transmission can spread the data up to very high frequencies without upconversion. Thus, PPM UWB is sometimes called impulse radio.

Consider a PPM system that transmits one of M possible symbols each time slot. The time slot is divided into $K = M/2$ chips. A symbol consists of transmitting a +1 in one chip and zeros in the other $K - 1$ chips (so there are K of these to choose from), or transmitting a −1 in one chip and zeros in the other $K - 1$ chips (so there are K of these as well). One bit of information is contained in the sign of the transmission, and the other $\log_2(K)$ bits of information are contained in the position of the transmission. This means that like multicarrier modulation, the receiver must process the data in block format. After equalization and symbol (block) synchronization, blocks of K samples are parallelized and fed into a decision device, then a single decision is made to produce the estimate of the transmitted symbol.

9.4.1 Trained Adaptation

Since PPM operates on a block basis, the MSE cost function should be block based as well. Defining the actual and desired equalizer output for block b as

$$\mathbf{z}[b] = \left[z[Kb], ..., z[Kb - K + 1] \right]^T$$
$$\mathbf{x}[b] = \left[x[Kb - \Delta], ..., x[Kb - K + 1 - \Delta] \right]^T, \tag{9.29}$$

the MSE cost function defined in [57] is

$$J = E\left[\left\| \mathbf{z}[b] - \mathbf{x}[b] \right\|_2^2 \right]. \tag{9.30}$$

Because $\mathbf{z}[b]$ is a vector of samples, it is formed by the multiplication of a *matrix* of equalizer inputs,

$$\mathbf{Y}[b] = \begin{bmatrix} y[Kb] & \cdots & y[Kb - K + 1] \\ \vdots & \ddots & \vdots \\ y[Kb - L + 1] & \cdots & y[Kb - L - K + 2] \end{bmatrix}, \tag{9.31}$$

and the equalizer, i.e., $\mathbf{z}[b] = \mathbf{Y}^T[b]\mathbf{f}$. This means that an LMS-style gradient descent of (9.30) is not simply a vector update rule; rather, there is a matrix-vector product involved. Specifically, a trained, adaptive LMS equalizer for PPM systems takes the form [57, 58]

$$\mathbf{f}[b + 1] = \mathbf{f}[b] - \mu \mathbf{Y}[b]\left(\mathbf{z}[b] - \mathbf{x}[b] \right). \tag{9.32}$$

Thus, the approach is similar to trained, adaptive equalization in traditional communication systems, but the inherent block structure introduces a slightly more complicated update rule, in terms of structure and computations.

An alternate approach would be to use a DFE. Trained zero-forcing DFE algorithms are discussed in [59, 60].

9.4.2 Blind Adaptation

The simplest blind equalizer is DD-LMS. However, even DD-LMS has a slight twist in the case of PPM. The DD cost function and update rule are

$$J = E\left[\left\|\mathbf{z}[b] - \mathbf{Q}\{\mathbf{z}[b]\}\right\|_2^2\right]$$

$$\mathbf{f}[b+1] = \mathbf{f}[b] - \mu \mathbf{Y}[b]\big(\mathbf{z}[b] - \mathbf{Q}\{\mathbf{z}[b]\}\big).$$

(9.33)

The twist is that the decision function $\mathbf{Q}\{\cdot\} : \mathbb{R}^K \mapsto \mathbb{R}^K$ operates in vector form, rather than element-wise. The vector argument is considered in its entirety and is mapped to the nearest vector in the signal constellation [57, 58].

A variant of CMA can also be derived for PPM, which will be called linear transversal equalizer adaptation for biorthogonal modulation, blindly (LTBOMB). Like DD-LMS for PPM, the distinction from traditional adaptive algorithms lies in the block structure of PPM. The cost function and algorithm are [57, 58]

$$J = E\left[\left(\left\|\mathbf{z}[b]\right\|_2^2 - 1\right)^2\right]$$

$$\mathbf{f}[b+1] = \mathbf{f}[b] - \mu\big(\mathbf{z}^T[b]\mathbf{z}[b] - 1\big)\mathbf{Y}[b]\mathbf{z}[b].$$

(9.34)

However, as implied by the time indices, the algorithm only updates once per block, i.e., once every K chips. Again, note the matrix-vector structure of the update rule, as opposed to standard CMA, which uses a scalar times vector update rule.

An alternative to CMA for blind, adaptive equalization of traditional communication systems is the Shalvi-Weinstein algorithm (SWA) [61]. SWA is similar to CMA insofar as it looks at higher-order statistics of the equalizer output, but SWA attempts to maximize the magnitude of the kurtosis. In [58] and [62], the SWA philosophy is used to create a blind, adaptive equalizer for PPM, called the recovery of M-ary biorthogonal signals via p-norm equivalence (TROMBONE). The cost function and algorithm are

$$J = E\left[\left\|\mathbf{z}[b]\right\|_2^4 - \left\|\mathbf{z}[b]\right\|_4^4\right], \quad \text{such that } \left\|\mathbf{z}[b]\right\|_2^2 = 1$$

$$\hat{\mathbf{f}}[b+1] = \mathbf{f}[b] - \mu \mathbf{Y}[b]\left(\mathbf{z}^T[b]\mathbf{z}[b]\mathbf{I}_K - \big(\mathrm{diag}\big[\mathbf{z}[b]\big]\big)^2\right)\mathbf{z}[b]$$

(9.35)

$$\mathbf{f}[b+1] = \frac{\hat{\mathbf{f}}[b+1]}{\left\|\hat{\mathbf{f}}[b+1]\right\|_2},$$

where $\mathbf{I}_K$ is the $K \times K$ identity matrix. The last line implements the constraint (similar to the MMSE, MERRY, and CNA algorithms for multicarrier systems) to constrain the equalizer away from the trivial all-zero solution. As with the DD-LMS and CMA variants of (9.33) and (9.34), the SWA variant of (9.35) only updates once per block.

To recap, traditional methods of trained and blind adaptive equalizer design can be modified for use in PPM-based UWB systems. However, the cost functions and algorithms must be modified to account for the block structure. The resulting algorithms only update once per block, they generally require matrix-vector products in the update term, as opposed to the more traditional scalar-vector products, and the decision function is a vector-to-vector mapping rather than a scalar-to-scalar mapping.

9.5 Interaction of Equalizer with Other Adaptive Blocks

Many equalizer designs are produced in isolation: the equalizer explicitly or implicitly assumes that all other adaptive blocks in the system are working perfectly. However, even in a simple single-carrier receiver, the signal amplitude, timing frequency/phase, equalizer, and carrier frequency/phase are all interdependent [15, Table 9.2]. In newer modulation formats with more or different blocks, there can be additional interdependence. We conclude this chapter with a discussion of the merits of joint analysis and design of the equalizer and other blocks in an adaptive receiver, though a full treatment of this subject is beyond the current state of the art.

One of the classical examples of this dependence is the dependence of an adaptive equalizer on accurate carrier frequency offset (CFO) estimation. For a trained, LMS algorithm, the equalizer inputs and outputs will not match their model if there is a residual CFO. At the same time, a trained CFO estimator needs ISI-free received data in order to form a good estimate. This can be a chicken-and-egg problem. However, the CMA equalizer uses a cost function that depends only on the magnitude of the equalizer output, and not the phase. In the presence of a residual CFO, the received data will have a linearly increasing additive phase, causing the signal constellation to spin about the origin. Since CMA does not care about the phase, it can remove the ISI even in the presence of the CFO-induced spinning. Then the CFO estimator can operate on the ISI-free equalizer output, and remove the CFO [15].

This method of delaying the need for CFO correction cannot be as easily incorporated into multicarrier systems, since as discussed in section 9.3, blind multicarrier equalizers do make use of a constant modulus cost function in the time domain. An alternate method of dealing with the coupling of an adaptive equalizer and another adaptive block is to jointly adapt the two. In the case of multicarrier equalization and CFO correction, the CNA, MERRY, and DD cost functions used for equalization can also be used to adjust the CFO [53, 54, 63, 64]. By forming a single cost function at the output of the two adaptive blocks in series, both algorithms may converge to their optimal setting [63].

As mentioned in section 9.3, in multicarrier systems, the CSE and FEQ must both be adapted simultaneously. Algorithms such as DD-LMS and CMA cannot directly be used for the CSE since the CSE output is not expected to be finite alphabet. However, the final FEQ output is. By forming a decision-directed or constant modulus cost at the FEQ output, both the CSE and FEQ can adapt based on this cost. If the FEQ uses a relatively higher step size than the CSE, then the CSE will adapt slowly, and the FEQ will track it and quickly reach its best value for the current CSE setting. This concatenation of adaptive equalizers allows the use of traditional cost functions. The penalty is that the FEQ step size has an upper bound for stability, and the CSE step size must be much lower, so adaptation cannot be very swift. This need for step-size-based timescale separation is present for most coupled adaptive systems adapting over a single cost function, for example, the adaptive CSE and target response in the MMSE filter [10, 65].

9.6   Summary

This chapter has discussed the design of adaptive equalizers. We began with a historical perspective and a discussion of the need for an adaptive equalizer. We then discussed popular methods of creating adaptation rules, making them converge quickly, and reducing their computational load, all of which are necessary if the equalizer is part of a small wireless device. As specific examples of creating adaptation rules, we discussed recent literature on adaptive equalization in two currently popular communication standards, multicarrier and ultrawideband. We concluded with a discussion of the interaction of an adaptive equalizer and other adaptive blocks within the receiver.

References

[1] Schniter, P. 2004. Low-complexity equalization of OFDM in doubly-selective channels. *IEEE Trans. Signal Processing* 52:1002–11.

[2] Viterbi, A. J. 1967. Error bounds for convolutional codes and an asymptotically optimal decoding algorithm. *IEEE Trans. Inf. Theory* 13:260–69.

[3] Tong, L., Xu, G., and Kailath, T. 1994. Blind identification and equalization based on second-order statistics: A time domain approach. *IEEE Trans. Inf. Theory* 40:340–49.

[4] Moulines, E., Duhamel, P., Cardoso, J., and Mayrargue, S. 1995. Subspace methods for the blind identification of multichannel FIR filters. *IEEE Trans. Signal Processing* 43:516–25.

[5] Widrow, B., and Hoff, Jr., M. E. 1960. Adaptive switching circuits. In *Proceedings of the WESCON Conference Recordings*, Part 4, pp. 96–104.

[6] Lucky, R. W. 1965. Automatic equalization for digital communication. *Bell Syst. Tech. J.* 44:547–88.

[7] Widrow, B., et al. 1976. Stationary and nonstationary learning characteristics of the LMS adaptive filter. *Proc. IEEE* 64:1151–62.

[8] Gitlin, R. D., and Mazo, J. E. 1973. Comparison of some cost functions for automatic equalization. *IEEE Trans. Commun.* 21:233–37.

[9] Qureshi, S., and Newhall, E. 1973. An adaptive receiver for data transmission over time-dispersive channels. *IEEE Trans. Inf. Theory* 19:448–57.

[10] Falconer, D. D., and Magee, F. R. 1973. Adaptive channel memory truncation for maximum likelihood sequence estimation. *Bell Syst. Tech. J.* 1541–62.

[11] Sato, Y. 1975. A method of self-recovering equalization for multilevel amplitude-modulation systems. *IEEE Trans. Commun.* 23:679–82.

[12] Godard, D. N. 1980. Self-recovering equalization and carrier tracking in two-dimensional data communication systems. *IEEE Trans. Commun.* 28:1867–75.

[13] Treichler, J. R., and Agee, B. G. 1983. A new approach to multipath correction of constant modulus signals. *IEEE Trans. Acoustics Speech Signal Processing* 31:459–72.

[14] Foschini, G. J. 1985. Equalizing without altering or detecting data. *AT&T Tech. J.* 64:1885–911.

[15] Treichler, J. R., Larimore, M. G., and Harp, J. C. 1998. Practical blind demodulators for high-order QAM signals. *Proc. IEEE* 86:1907–26.

[16] Qureshi, S. 1985. Adaptive equalization. *Proc. IEEE* 73:1349–87.

[17] Sethares, W. A. 1992. Adaptive algorithms with nonlinear data and error functions. *IEEE Trans. Signal Processing* 40:2199–206.

[18] Duttweiler, D. L. 2000. Proportionate normalized least-mean-squares adaptation in echo cancellers. *IEEE Trans. Speech Audio Processing* 8:508–18.

[19] Walach, E., and Widrow, B. 1984. The least mean fourth (LMF) adaptive algorithm and its family. *IEEE Trans. Inf. Theory* 30:275–83.

[20] Hübscher, P. I., Nascimento, V. H., and Bermudez, J. C. M. 2003. New results on the stability analysis of the LMF (least mean fourth) adaptive algorithm. In *Proceedings of the International Conference on Acoustics, Speech, and Signal Processing*, vol. VI, pp. 369–72.

[21] Sethares, W. A., Rey, G. A., and Johnson, Jr., C. R. 1989. Approaches to blind equalization of signals with multiple modulus. In *Proceedings of the International Conference on Acoustics, Speech, and Signal Processing*, vol. 2, pp. 972–75.

[22] Banovic, K., Abdel-Raheem, E., and Khalid, M. A. S. 2005. Hybrid methods for blind adaptive equalization: New results and comparisons. In *Proceedings of the IEEE Symposium on Computers and Communications*, pp. 275–80.

[23] Thaiupathump, T., and Kassam, S. A. 2003. Square contour algorithm: A new algorithm for blind equalization and carrier phase recovery. In *Proceedings of the 37th Asilomar Conference on Signals, Systems, and Computers*, vol. 1, pp. 647–51.

[24] Goupil, A., and Palicot, J. 2007. New algorithms for blind equalization: The constant norm algorithm family. *IEEE Trans. Signal Processing* 55:1436–44.

[25] Li, X.-L., and Zhang, X.-D. 2006. A family of generalized constant modulus algorithms for blind equalization. *IEEE Trans. Commun.* 54:1913–17.

[26] Abrar, S., and Qureshi, I. M. 2006. Blind equalization of cross-QAM signals. *IEEE Signal Processing Lett.* 13:745–48.

[27] Johnson, Jr., C. R., et al. 1998. Blind equalization using the constant modulus criterion: A review. *Proc. IEEE* 86:1927–50.

[28] Johnson, Jr., C. R., et al. 2000. The core of FSE-CMA behavior theory. In *Unsupervised adaptive filtering: Blind deconvolution*, ed. S. Haykin, pp. 13–112. Vol. II. New York: Wiley.

[29] Haykin, S. 2002. *Adaptive filter theory.* 4th ed. Upper Saddle River, NJ: Prentice-Hall.

[30] Stamoulis, A., Diggavi, S. N., and Al-Dhahir, N. 2002. Intercarrier interference in MIMO OFDM. *IEEE Trans. Signal Processing* 50:2451–64.

[31] Cai, X., and Giannakis, G. B. 2002. Low-complexity ICI suppression for OFDM over time- and frequency-selective Rayleigh fading channels. In *Proceedings of the Asilomar Conference on Signals, Systems, and Computers*, pp. 1822–1826.

[32] Barhumi, I., Leus, G., and Moonen, M. 2005. Time-varying FIR equalization for doubly selective channels. *IEEE Trans. Wireless Commun.* 4:202–14.

[33] Ahmed, S., Sellathurai, M., Lambothoran, S., and Chambers, J. 2006. Low complexity iterative method of equalization for single carrier with cyclic prefix in doubly selective channels. *IEEE Signal Processing Lett.* 13:5–8.

[34] Verhoeckx, N. A. M., Van dem Elzer, H., Snijders, F. A. M., and Van Gerwer, P. J. 1979. Digital echo cancellation for baseband data transmission. *IEEE Trans. Acoust. Speech Signal Processing* 27:768–81.

[35] Bellini, S. 1986. Bussgang techniques for blind equalization. In *Proceedings of the IEEE Global Telecommunications Conference*, pp. 1634–40.

[36] De, P., Bao, J., and Poon, T. 1999. A calculation-efficient algorithm for decision feedback equalizers. *IEEE Trans. Consumer Electronics* 45:526–632.

[37] Chen, Z., Haykin, S., and Gay, S. L. 2003. Proportionate adaptation: New paradigms in adaptive filters. In *Least-mean-square adaptive filters*, ed. S. Haykin and B. Widrow. New York: John Wiley & Sons, pp. 293–334.

[38] Martin, R. K., Sethares, W. A., Williamson, R. C., and Johnson, Jr., C. R. 2002. Exploiting sparsity in adaptive filters. *IEEE Trans. Signal Processing* 50:1883–94.

[39] Deng, H., and Doroslovački, M. 2006. Proportionate adaptive algorithms for network echo cancellation. *IEEE Trans. Signal Processing* 54:1794–803.

[40] Gay, S. L. 1998. An efficient, fast converging adaptive filter for network echo cancellation. In *Proceedings of the Asilomar Conference on Signals, Systems, and Computers*, pp. 394–98.

[41] Nekuii, M., and Atarodi, M. 2004. A fast converging algorithm for network echo cancellation. *IEEE Signal Process. Lett.* 11:427–30.

[42] Benesty, J., and Gay, S. L. 2002. An improved PNLMS algorithm. In *Proceedings of the IEEE International Conference on Acoustics, Speech, and Signal Processing*, vol. 2, pp. 1881–84.

[43] Douglas, S. C. 1997. Adaptive filters employing partial updates. *IEEE Trans. Circuits Syst. II* 44:209–16.

[44] Douglas, S. C. 1995. Analysis and implementation of the max-NLMS adaptive filter. In *Proceedings of the 29th Asilomar Conference on Signals, Systems, and Computers*, Pacific Grove, CA, vol. 1, pp. 659–63.

[45] Aboulnasr, T., and Mayyas, K. 1999. Complexity reduction of the NLMS algorithm via selective coefficient update. *IEEE Trans. Signal Processing* 47:1421–24.

[46] Schertler, T. 1998. Selective block update of NLMS type algorithms. In *Proceedings of the IEEE International Conference on Acoustics, Speech, and Signal Processing*, Seattle, vol. 3, pp. 1717–20.

[47] Naylor, P., and Sherliker, W. 2003. A short-sort M-max NLMS partial update adaptive filter with applications to echo cancellation. In *Proceedings of the IEEE International Conference on Acoustics, Speech, and Signal Processing*, Hong Kong, vol. 5, pp. 373–76.

[48] Homer, J. 1998. Detection guided LMS estimation of sparse channels. In *Proceedings of the IEEE Global Communications Conference*, Sydney, vol. 6, pp. 3704–9.

[49] European Telecommunications Standards Institute. 2001. *Digital video broadcasting (DVB); Framing structure, channel coding and modulation for digital terrestrial television*. ETSI EN 300 744 V1.4.1.

[50] Martin, R. K. 2007. Fast-converging blind adaptive channel shortening and frequency-domain equalization. *IEEE Trans. Signal Processing* 54:102–10.

[51] Yap, K. S., and McCanny, J. V. 2002. Improved time-domain equalizer initialization algorithm for ADSL modems. In *Proceedings of the 6th International Symposium on DSP for Communication Systems*, Sydney-Manly, Australia, pp. 253–58.

[52] Al-Dhahir, N., and Cioffi, J. M. 1996. Efficiently computed reduced-parameter input-aided MMSE equalizers for ML detection: A unified approach. *IEEE Trans. Inf. Theory* 42:903–15.

[53] van de Beek, J.-J., Sandell, M., and Borjesson, P. O. 1997. ML estimation of time and frequency offset in OFDM systems. *IEEE Trans. Signal Processing* 45:1800–5.

[54] Choi, Y., Voltz, P. J., and Cassara, F. A. 2001. ML estimation of carrier frequency offset for multicarrier signals in Rayleigh fading channels. *IEEE Trans. Veh. Technol.* 50:644–54.

[55] de Courville, M., Duhamel, P., Madec, P., and Palicot, J. 1996. Blind equalization of OFDM systems based on the minimization of a quadratic criterion. In *Proceedings of the IEEE International Conference on Communications*, Dallas, TX, pp. 1318–21.

[56] Romano, F., and Barbarossa, S. 2003. Non-data aided adaptive channel shortening for efficient multi-carrier systems. In *Proceedings of the IEEE International Conference on Acoustics, Speech, and Signal Processing*, Hong Kong, vol. 4, pp. 233–36.

[57] Klein, A. G., Johnson, Jr., C. R., and Duhamel, P. 2005. On blind equalization of M-ary bi-orthogonal signaling. In *Proceedings of the IEEE International Conference on Acoustics, Speech, and Signal Processing*.

[58] Klein, A. G., Johnson, Jr., C. R., and Duhamel, P. 2007. On blind equalization of biorthogonal signaling. *IEEE Trans. Signal Processing* 55:1421–35.

[59] Barry, J. R. 1994. Sequence detection and equalization for pulse-position modulation. In *Proceedings of the IEEE International Conference on Communications*, vol. 3, pp. 1561–656.

[60] Varanasi, M. K. 1997. Equalization for multipulse modulation. In *Proceedings of the IEEE International Conference on Personal Wireless Communications*, pp. 48–51.

[61] Shalvi, O., and Weinstein, E. 1990. New criteria for blind deconvolution of non-minimum phase systems (channels). *IEEE Trans. Inf. Theory* 36:312–21.

[62] Klein, A. G., Johnson, Jr., C. R., and Duhamel, P. 2006. A blind algorithm based on difference of norms for equalization of biorthogonal signals. In *Proceedings of the IEEE International Conference on Acoustics, Speech, and Signal Processing*.

[63] Martin, R. K. 2006. Joint blind adaptive synchronization and channel shortening. *IEEE Trans. Signal Processing* 54:4194–203.

[64] Liu, Z., and Weng, B. 2003. Finite-alphabet based blind carrier frequency offset estimation for differentially coded OFDM. In *Proceedings of the IEEE Global Tele-communications Conference*, vol. 1, pp. 327–31.

[65] Chow, J. S., Cioffi, J. M., and Bingham, J. A. C. 1993. Equalizer training algorithms for multicarrier modulation systems. In *Proceedings of the IEEE International Conference on Communications*, pp. 761–65.

10

Adaptive Multicarrier CDMA Space-Time Receivers

Besma Smida
Harvard University

Sofiène Affes
University of Quebec

10.1   Motivation of the Chapter

One important challenge for future wireless networks is the design of appropriate transceivers that can reliably transmit high data rates at a high bandwidth efficiency. Multicarrier code division multiple access (MC-CDMA) systems in particular have received considerable attention, because they have the attractive feature of high spectral efficiency and because they can be easily implemented using fast Fourier transform (FFT) without significantly increasing the transmitter and receiver complexities [1, 2].

The multicarrier systems include different combinations of multicarrier modulation (accomplished by orthogonal frequency division multiplex [OFDM]) and direct-sequence code division multiple access (DS-CDMA). This combination provides both high-data-rate transmission and multiple access capabilities. An excellent overview of the different multicarrier CDMA systems is found in [2] and [3]. They can be divided into two categories of multicarrier CDMA: one combines multicarrier modulation with frequency-domain spreading, and the other transmits several DS-CDMA waveforms in parallel with the spreading operation performed in time. The transmitter proposed here belongs to the second group, and it can be divided into MC-DS-CDMA and multi-tone (MT) CDMA, the difference between the two being the subcarrier frequency separation.

However promising, challenges remain before multicarrier CDMA can achieve its full potential. One open area is the design of transceivers that will enable the future upgrade of current wireless networks beyond the third generation (3G). Transceivers selected for early implementation need to achieve high spectrum efficiency in realistic propagation channels while being robust to imperfections such as time and frequency mismatch. Multicarrier CDMA, similar to other multicarrier schemes, is sensitive to the signal distortion generated by the imperfect frequency downconversion at the receiver due to local oscillator frequency offset. It has been found that carrier frequency offset (CFO) gives rise to a reduction of the useful signal power and to the intercarrier interference (ICI) [5]. Furthermore, one of the major obstacles in detecting multicarrier CDMA signals is interference. The multiple access interference (MAI) and the intersymbol interference (ISI), which are inherited from conventional DS-CDMA, affect likewise the performance of multicarrier CDMA systems.

The subject of this work is the design of a new adaptive multicarrier CDMA space-time receiver that provides solutions to these problems, with particular emphasis on the comparison of the MC-DS-CDMA– and MT-CDMA–air interface configurations. The proposed receiver, named MC-ISR* (multicarrier interference subspace rejection), will hence (1) perform blind channel identification and equalization as well as fast and accurate joint synchronization in time and frequency, and (2) mitigate the full interference effect. In addition, the assessment of this new receiver is oriented toward an implementation in a future, real-world wireless system.

---

* This work was presented in part at the IEEE SPAWC 2005 and IEEE ISSPA 2005 conferences and accepted for publication in *IEEE Transactions on Vehicular Technology* [6].

10.2 Overview of Multicarrier CDMA for Wireless Communications

In this section, we review concisely the class of multicarrier CDMA schemes, which have been discussed in the literature. Specifically, we discuss their parameters, spectral characteristics, advantages, and disadvantages in terms of design and structure. For more detailed information on the different multicarrier CDMA schemes, the reader is referred to the excellent monographs by Hanzo et al. [3].

The multicarrier CDMA schemes are categorized mainly in two groups. One spreads the original data stream using a given spreading code, and then modulates a different subcarrier with each chip (i.e., the spreading is in the frequency domain), and the other spreads the serial-to-parallel (S/P) converted data streams using a given spreading code, and then modulates a different subcarrier with each of the data streams (spreading is in the time domain).

10.2.1 Frequency-Domain Spreading Multicarrier CDMA: FD-MC-CDMA

Figure 10.1a shows the spectrum of the multicarrier CDMA scheme associated with frequency-domain spreading [7–10]. We refer to this scheme as FD-MC-CDMA. The FD-MC-CDMA transmitter spreads the original data stream over N_c subcarriers using a given spreading code in the frequency domain. This scheme does not include serial-to-parallel data conversion, and there exists no spreading modulation on each subcarrier. Therefore, the data rate on each of the N_c subcarriers is the same as the input data rate. However, by spreading each data bit across all of the N_c subcarriers, the fading effects of multipath channels are mitigated. In this FD-MC-CDMA system, the subcarrier frequencies are chosen to be orthogonal to each other, i.e., the subcarrier frequencies satisfy the following condition:

$$f_i - f_j = \frac{n|i-j|}{T},$$

where $n \in \mathbb{N}$ and T is the symbol duration. Therefore, the minimum spacing Δ between two adjacent subcarriers satisfies $1/T$, which is a widely used assumption [7–10] and is also the case employed in Figure 10.1a. If no overlap is assumed, then the minimum spacing Δ between two adjacent subcarriers is $2/T$.

Yee et al. [7] have considered an FD-MC-CDMA system, in which the subcarriers' frequency separation is higher than the coherence bandwidth of the channel, and therefore the individual subcarriers experience independent fading. As a result, the frequency diversity is maximized. This is the main advantage of the FD-MC-CDMA scheme over other multicarrier CDMA schemes [2]. However, this system may require a considerable transmission bandwidth. Besides that, a large delay spread per subcarrier would lessen this bandwidth requirement. But in a frequency-selective fading channel, different

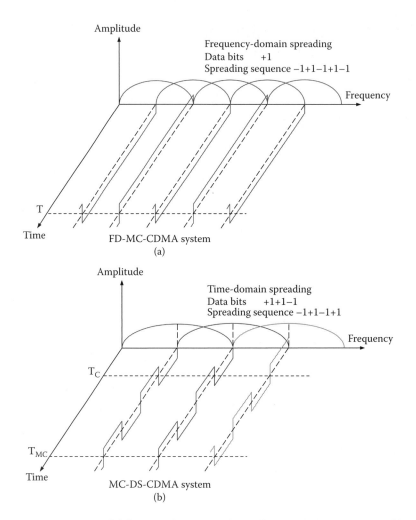

FIGURE 10.1　Spectra of different multicarrier CDMA schemes: (a) FD-MC-CDMA, (b) MC-DS-CDMA, (c) MT-CDMA.

subcarriers may encounter different amplitude attenuations and phase shifts, which can consequently destroy the orthogonality of the subcarriers.

10.2.2　Time-Domain Spreading Multicarrier CDMA: MC-DS-CDMA

In [11], a multicarrier DS-CDMA system named MC-DS-CDMA has been proposed. The MC-DS-CDMA transmitter spreads the serial-to-parallel converted data streams using a given spreading code in the time domain so that the resulting spectrum of each subcarrier can satisfy the orthogonality condition with the minimum frequency separation [11]. This scheme was originally proposed for an uplink communication system, because

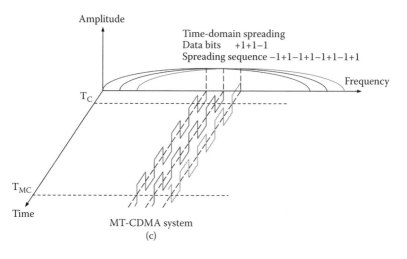

Time-domain spreading
Data bits   +1+1−1
Spreading sequence −1+1−1+1−1+1−1+1

MT-CDMA system
(c)

FIGURE 10.1 (continued).

this characteristic is effective for establishing a quasi-synchronous channel. In [12], Kondo and Milstein proposed a similar transmitter, except that band-limited subcarrier signals and larger subcarrier separation are employed. This scheme yields both frequency diversity improvement and narrowband interference suppression. The spectrum of the MC-DS-CDMA signal with 50% overlap having three subcarriers is shown in Figure 10.1b. In the MC-DS-CDMA system, the subcarrier frequencies are usually chosen to be orthogonal to each other after spreading, which can be formulated as

$$f_i - f_j = \frac{n|i - j|}{T_c},$$

where $n \in \mathbb{N}$ and T_c is the chip duration. Therefore, the minimum spacing Δ between two adjacent subcarriers satisfies $1/T_c$.

Frequency diversity in MC-DS-CDMA systems can be achieved by repeating the transmitted signal in the frequency (F) domain with the aid of several subcarriers [12–14]. Alternatively, in MC-DS-CDMA systems the F-domain repetition can be replaced by F-domain spreading [15] using a spreading code. One of the advantages of using F-domain spreading instead of F-domain repetition in MC-DS-CDMA systems is that frequency diversity can be achieved without reducing the maximum number of users supported by the system [15, 16]. The MC-DS-CDMA scheme can provide the following advantages [13]. First, the spreading processing gain is increased compared to the corresponding single-carrier DS-CDMA scheme. Second, the effect of multipath interference is mitigated because of DS spreading. Third, frequency/time diversity can be achieved. Finally, a longer chip duration may lead to more relaxed synchronization schemes.

10.2.3   Time-Domain Spreading Multicarrier CDMA: MT-CDMA

The multitone CDMA (MT-CDMA) scheme was proposed by Vandendorpe in [17, 18]. The MT-CDMA transmitter spreads the serial-to-parallel converted data streams using a given spreading code in the time domain, so that the spectrum of each subcarrier prior to the spreading operation can satisfy the orthogonality condition with the minimum frequency separation [18]. Therefore, there exists a strong spectral overlap among the different subcarrier signals after DS spreading. The spectra associated with three subcarriers for an MT-CDMA signal are shown in Figure 10.1c. In an MT-CDMA system, the subcarrier frequencies are chosen to be orthogonal to each other with the minimum frequency separation before spreading, which can be formulated as

$$f_i - f_j = \frac{n|i-j|}{T_{MC}},$$

where $n \in \mathbb{N}$ and T_{MC} is the symbol duration after S/P. It can be shown that the minimum spacing of the subcarrier frequencies is $1/T_{MC}$.

Unfortunately, the MT-CDMA scheme suffers from intercarrier interference because of the strong spectral overlap among the different subcarriers. However, the capability to use longer spreading codes results in the reduction of self-interference and multiple access interference, compared to the spreading codes assigned to a corresponding single-carrier DS-CDMA scheme. The MT-CDMA scheme uses longer spreading codes than the corresponding single-carrier DS-CDMA scheme [2], where the relative code-length extension is in proportion to the number of subcarriers. Therefore, the MT-CDMA system can accommodate more users. Simulation results will later show the advantages of MT-CDMA in increasing throughput and bandwidth efficiency.

10.2.4   Multicarrier CDMA Transmitter Selection

We had briefly outlined the features of a number of multicarrier CDMA systems, which have been studied in the literature. So far, many reports are dedicated to the BER performance comparisons of DS-CDMA with multicarrier CDMA systems. These works show that all multicarrier CDMA schemes—MC-CDMA [19–21], MC-DS-CDMA [13, 14], and MT-CDMA [22]—outperform DS-CDMA.

In addition, it can be shown that there are trade-offs associated with each multicarrier CDMA scheme considered. Each technique has different benefits and drawbacks, depending on the intended applications [2, 14, 23]. Yang and Hanzo showed in [24] that MC-DS-CDMA has the highest degree of freedom in the family of CDMA schemes that can be beneficially exploited during the system design and reconfiguration procedures. The MC-DS-CDMA constitutes a trade-off between DS-CDMA and MC-CDMA in the context of the system's architecture and performance. By employing multiple subcarriers, MC-DS-CDMA typically requires lower-chip-rate spreading codes than DS-CDMA. It necessitates a lower number of subcarriers than MC-CDMA due to imposing DS spreading on each subcarrier's signal. Consequently, MC-DS-CDMA typically requires lower-rate signal processing than DS-CDMA and has lower worst-case peak-to-average power fluctuation than MC-CDMA.

Therefore, we will study the MC-DS-CDMA–air interface in this chapter. In addition, we broaden our analysis by considering the family of generalized MC-DS-CDMA transceivers, defined in [4]. This generalized scheme includes the subclasses of MT-CDMA and MC-DS-CDMA as special cases. Simulation results will later show the advantages of MT-CDMA in increasing throughput and bandwidth efficiency.

In the following sections, we will adopt this general view and simply refer to it as MC-CDMA in the remainder of the chapter unless otherwise required. We also assume the uplink of an asynchronous multicellular multicarrier CDMA system with C in-cell active users. For the sake of simplicity, we assume that all users use the same subcarriers and transmit with the same modulation at the same rate.

10.2.5   MC-CDMA Transmission Model

This section explains in more detail the MC-CDMA scheme adopted in this chapter. The block diagram of the MC-CDMA transmitter is shown in Figure 10.2. The input information sequence of the uth user is first converted into $N_c = 2K + 1$ parallel* data sequences $b^u_{-K,n}, \ldots, b^u_{0,n}, \ldots, b^u_{K,n}$, where n is the time index. The datum $b^u_{K,n} \in C_\mathcal{M}$ is $\mathcal{M}$-PSK modulated and differentially† encoded at rate $1/T_{MC}$, where $T_{MC} = N_c \times T$ is the symbol duration after S/P conversion, T is the symbol duration before S/P, and $C_\mathcal{M} = \{\ldots, e^{j2\pi m/\mathcal{M}}, \ldots\}$, $m \in \{0, \ldots, \mathcal{M} - 1\}$. The resulting S/P converter output is then spread with a random spreading code $c_u(t)$ at a rate $1/T_c$. The spreading factor, defined as the ratio between the chip rate and the symbol rate, is $L = T_{MC}/T_c$. We write the spreading-code segment over the nth period T_{MC} as

$$c^u_n(t) = \sum_{l=0}^{L-1} c^u_{l,n} \phi\left(t - 1T_c - nT_{MC}\right), \tag{10.1}$$

where $c^u_{l,n} = \pm 1$ for $l = 0, \ldots, L - 1$ is a random sequence of length L and $\phi(t)$ is the chip pulse. We consider square-root raised cosine chip pulses with roll-off factor β (see appendix). Closed-loop power control is taken into account at the transmitter by the amplification factor $a_u(t)$. All the data are then modulated in baseband by the inverse discrete Fourier transform (IDFT) and summed to obtain the multicarrier signal. No guard interval is inserted. Indeed, the channel identification and equalization are achieved by MC-STAR (multicarrier spatiotemporal array receiver) [32], and simulation results have shown that the guard interval length does not affect the link-level performance. MC-STAR exploits the intrinsic channel diversity by combining and equalizing the multipath signals. We

---

* We selected an odd number of subcarriers to have a central frequency, but the model can easily be rearranged to operate with an even number of subcarriers.

† We can also use pilot symbols for coherent modulation and detection [25], but that is beyond the scope of this chapter.

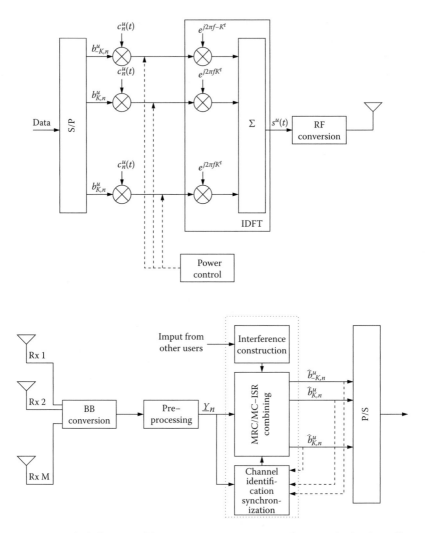

FIGURE 10.2 Block diagram of the MC-CDMA transmitter and receiver (pulse shape filtering is implemented at both transceiver ends).

hence eliminate the guard interval. Finally, the signal is transmitted after radio frequency upconversion.

The modulated subcarriers are orthogonal over the symbol duration T_{MC}. The frequency corresponding to the kth subcarrier is $f_k = \lambda \times k/T_{MC}$. The transmitter belongs to the family of MT-CDMA if λ is set to 1, and to the class of MC-DS-CDMA if λ is set to L (see resulting signal spectra in Figure 10.3). Indeed, in an MT-CDMA system, the subcarrier frequencies are chosen to be orthogonal harmonics with minimum frequency separation before spreading. By contrast, in MC-DS-CDMA, the subcarrier frequencies are chosen to satisfy the orthogonality condition with minimum possible frequency after spreading. The transmitted signal of the uth user is given by

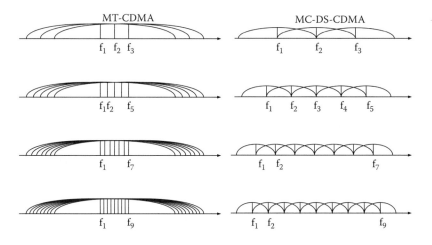

FIGURE 10.3 Different configurations of MT-CDMA and MC-DS-CDMA spectra within the same bandwidth.

$$s^u(t) = \sum_{k=-K}^{K} \sum_{n=-\infty}^{\infty} a^u(t) b_{k,n}^u c_n^u(t) e^{j2\pi f_k t}. \tag{10.2}$$

The transmitted signal's bandwidth is

$$BW = \frac{(N_c - 1)\lambda}{T_{MC}} + \frac{(1+\beta)}{T_c}. \tag{10.3}$$

10.2.6 Channel Model for Multicarrier Transmission

We consider an uplink transmission to M receiving antennas at the base station. The channel is assumed to be a slowly varying frequency-selective Rayleigh channel with delay spread $\Delta\tau$. For each kth subcarrier of user u, the key channel parameter is the number of resolvable paths, P_k^u, which is given by

$$P_k^u = \left\lfloor \frac{\Delta\tau}{T_c} \right\rfloor + 1, \tag{10.4}$$

where T_c is the chip duration. In practice, the number of multipaths depends also on the choice of the noise threshold used to differentiate between the received multipath components and the thermal noise. Typical delay-spread values are in the range of 0.4–4 µs in outdoor mobile radio channels, and the number of multipaths P_k^u varies between 2 and 5 with 3.84 megachips per second (Mcps) resolution [26]. The M-dimensional complex low-pass equivalent vector representation of the impulse response experienced by subcarrier k of the uth user, for a receiver equipped with M antennas, is

$$H_k^u(t)=\frac{\rho_k^u(t)}{\left(r^u\right)^e(t)}\sum_{p=1}^{P_k^u}\mathcal{G}_{k,p}^u(t)\delta\!\left(t-\tau_{k,p}^u(t)\right).\tag{10.5}$$

where $\rho_k^u(t)$ and $(r^u)^e(t)$ model the effects of shadowing and path loss, respectively, $r^u(t)$ is the distance from the uth user to the base station, and e is the path loss exponent. We assume their variations in time to be very slow, and hence nearly constant over several symbol durations. The M-dimensional complex vector $\mathcal{G}_{k,p}^u(t)^*$ denotes the fading and the array response from the user to the antenna elements of the receiver, and $\tau_{k,p}^u(t)$ represents the propagation time delay along the pth path. We note here that the large-scale path loss that includes free-space path loss and shadowing is the same for all subcarriers of the same user. Moreover, the number of resolvable paths and their propagation time delays depend on the reflecting objects and scatterers and can be assumed equal for all subcarriers [28]. Therefore, we omit the index k from ρ, P, and τ ($\rho_k^u=\rho^u$, $P_k^u=P^u$, and $\tau_{k,p}^u=\tau_p^u$) and reformulate equation (10.5) as

$$H_k^u(t)=\frac{\rho^u(t)}{\left(r^u\right)^e(t)}\sum_{p=1}^{P^u}\mathcal{G}_{k,p}^u(t)\delta\!\left(t-\tau_p^u(t)\right).\tag{10.6}$$

A frequency-domain channel model for a multicarrier system can be characterized by the coherence bandwidth† [28]:

$$B_c\simeq\frac{1}{2\pi\Delta\tau}.\tag{10.7}$$

When the frequency separation λ/T_{MC} is less than B_C, the MC-CDMA system is subject to correlated fading over different subcarriers. Fades across taps (multipaths) are mutually independent for the same carrier. However, fading for the same tap across different carriers is correlated. The envelope correlation coefficient between subcarrier k and subcarrier k' for user u is [28]

$$\rho_{k,k'}^u=E\!\left[\left|\mathcal{G}_{k,p,m}^u(t)\right|\left|\mathcal{G}_{k',p,m}^u(t+\tau)\right|\right]=\frac{\left(1+\lambda_{k,k'}\right)\tilde{E}\!\left(\dfrac{2\sqrt{\lambda_{k,k'}}}{1+\lambda_{k,k'}}\right)-\dfrac{\pi}{2}}{2-\dfrac{\pi}{2}},\tag{10.8}$$

---

* We may characterize $\mathcal{G}_{k,p}^u(t)$ in a space manifold parameterized by angles of arrivals [27]. However, a space characterization requires perfect antenna calibration and adequate sensor positioning.

† The coherence bandwidth is defined as the bandwidth over which the envelope correlation is above 0.5 [28].

with

$$\lambda_{k,k'} = \frac{J_0\left(2\pi f_D \tau\right)}{\sqrt{1+\left[2\pi\left(f_k - f_{k'}\right)\Delta\tau\right]^2}},$$ (10.9)

where E is the expectation function, $\mathcal{G}^u_{k,p,m}(t)$ is the fading and the antenna response from user u to antenna m of the receiver along the pth path, $\tilde{E}$ is the complete elliptic integral of the second kind, J_0 is the zeroth-order Bessel function, and f_D is the maximum Doppler frequency. We adopt the approach proposed in [29, 30] to generate correlated Rayleigh channels across subcarriers. We also assume that the received channel multipath components across the M antennas are independent.

10.2.7 Received Signal

For a multicellular MC-CDMA system with C in-cell users and $N_c = 2K + 1$ carriers, the received signal is the superposition of signals from all users and all subcarriers. Hence, the M-dimensional observation vector received, after downconversion, by the antenna array can be expressed as follows:

$$X(t) = \sum_{u=1}^{C} \sum_{k=-K}^{K} \sum_{n=-\infty}^{\infty} H_k^u(t) \otimes a^u(t) b_{k,n}^u c_n^u(t) e^{j2\pi(f_k + \Delta f^u)t} + N(t),$$

$$= \sum_{u=1}^{C} \sum_{k=-K}^{K} \sum_{n=-\infty}^{\infty} X_{k,n}^u(t) + N(t),$$ (10.10)

where $\otimes$ denotes time convolution and Δf^u models the carrier frequency offset (CFO), which is assumed equal for all subcarriers. This is a realistic assumption since there is only one oscillator per transmitter (see Figure 10.2). On the downlink, the CFO is even equal for all in-cell users (i.e., $\Delta f^u = \Delta f \; \forall \; u \in \{1, \ldots, C\}$). The noise term $N(t)$ includes the thermal noise received at the antennas as well as the out-cell interference. The contribution $X_{k,n}^u(t)$ of the nth data symbol over the kth carrier of user u to the received vector $X(t)$ is given by

$$X_{k,n}^u(t) = H_k^u(t) \otimes a^u(t) b_{k,n}^u c_n^u(t) e^{j2\pi(f_k + \Delta f^u)t}$$

$$= \psi_k^u(t) b_{k,n} \sum_{p=1}^{p^u} \mathcal{G}_{k,p}^u(t) \varepsilon_{k,p}^u(t) c_n^u\left(t - \tau_p^u\right) e^{j2\pi(f_k + \Delta f^u)(t - \tau_p^u)}.$$ (10.11)

Along the pth path, $\mathcal{G}_{k,p}^u(t) = (\sqrt{M}/\|\mathcal{G}_{k,p}^u(t)\|)\mathcal{G}_{k,p}^u(t)$ is the propagation vector over the kth subcarrier of the uth user with norm $\sqrt{M}$, and $(\varepsilon_{k,p}^u)^2(t) = \|\mathcal{G}_{k,p}^u(t)\|^2/\sum_{p=1}^{P_u}\|\mathcal{G}_{k,p}^u(t)\|$ is the fraction of the total received power on the kth subcarrier of user u:

$$\left(\psi_k^u\right)^2(t) = \left(\frac{\rho^u(t)}{\left(r^u\right)^e(t)}\right)^2 \left(a^u\right)^2(t) \sum_{p=1}^{p^u} \frac{\left\|\mathcal{G}_{k,p}^u(t)\right\|^2}{M}. \tag{10.12}$$

10.2.8 Interference Analysis

We define the matched-filtered observation vector of frame number n over a time interval $[0, T_{MC})$ as

$$Y_n(t) = \frac{1}{T_c} \int_{D_\phi} X\left(nT_{MC} + t + t'\right)\phi\left(t'\right)dt', \tag{10.13}$$

where D_ϕ denotes the temporal support* of $\phi(t)$. After sampling at a multiple of the chip rate, we frame the observation into overlapping blocks of constant length N_p. The oversampling ratio k_s is defined as the number of samples per chip. In DS-CDMA and MT-CDMA systems we need no more than one sample per chip ($k_s = 1$). In contrast, in an MC-DS-CDMA system, a higher sampling frequency is necessary for the receiver. Indeed, the sampling frequency has to satisfy the Nyquist sampling theorem, which states that the sampling interval must be smaller than the inverse of the double-sided bandwidth of the sampled signals. Hence, the smallest number greater than the number of subcarriers N_c is an adequate oversampling ratio for MC-DS-CDMA. The resulting processing block duration $T_p = N_p(T_c/k_s)$ is equal to $T_{max} + \overline{\Delta\tau}$. The processing period $T_{max} = LT_c$ contains N_c carrier symbols targeted for detection. The frame overlap $\overline{\Delta\tau} < T_{max}$, which is larger than the delay spread, allows multipath tracking [31]. Hence, we obtain the $M \times N_p$ matched-filtered observation matrix:

$$\mathbf{Y}_n = \left[Y_n(0), Y_n\left(T_c/k_s\right), \dots, Y_n\left(\left(N_p - 1\right)T_c/k_s\right)\right]. \tag{10.14}$$

It can be expressed as

$$\mathbf{Y}_n = \sum_{u=1}^{C} \sum_{k=-K}^{K} \sum_{n'=-\infty}^{\infty} \mathbf{Y}_{n',k,n}^u + N_n, \tag{10.15}$$

where the baseband preprocessed thermal noise and the out-cell interference contribute N_n, and where symbol n' of carrier k of user u contributes its observation matrix $\mathbf{Y}_{n',k,n}^u$, obtained by

---

* For a rectangular pulse, D_ϕ is $[0, T_c]$. In practice, and as assumed in this chapter, it is the temporal support of a truncated square-root raised cosine, $D_\phi = [-N_{src}T_c, N_{src}T_c]$, where N_{src} stands for the truncation span of the shaping pulse in chip samples around 0.

$$\mathbf{Y}_{n',k,n}^{u}=\left[Y_{n',k,n}^{u}(0),\,Y_{n',k,n}^{u}\!\left(T_c/k_s\right),\,\ldots,\,Y_{n',k,n}^{u}\!\left((N_p-1)T_c/k_s\right)\right],\qquad(10.16)$$

where

$$Y_{n',k,n}^{u}\left(t\right)=\frac{1}{T_c}\int_{D_\phi}X_{k,n'}^{u}\!\left(nT_{MC}+t+t'\right)\phi\!\left(t'\right)dt'.\qquad(10.17)$$

As a result of the stationarity assumptions stated in section 10.2.6, $Y_{n',k,n}^{u}(t)$ can be developed into

$$Y_{n',k,n}^{u}\left(t\right)\simeq e^{j2\pi\Delta f^{u}nT_{MC}}\,\psi_k^{u}\!\left(nT_{MC}\right)b_{k,n'}^{u}\frac{1}{T_c}\sum_{p=1}^{P}\int_{D_\phi}G_{k,p}^{u}\!\left(nT_{MC}+t+t'\right)$$

$$\varepsilon_{k,p}^{u}\!\left(nT_{MC}+t+t'\right)e^{j2\pi f_k\left(nT_{MC}+t+t'+\tau_p^{u}\right)}c_{n'}^{u}\!\left(nT_{MC}+t+t'-\tau_p^{u}\right)\phi\!\left(t'\right)dt'$$

$$\simeq e^{j2\pi\Delta f^{u}nT_{MC}}\,\psi_k^{u}\!\left(nT_{MC}\right)b_{k,n'}^{u}V_{n',k,n}^{u}\left(t\right)\qquad(10.18)$$

$$\simeq \psi_k^{u}\!\left(nT_{MC}\right)b_{k,n'}^{u}U_{n',k,n}^{u}\left(t\right),$$

where the spread channel vector without CFO $V_{n',k,n}^{u}(t)$ is obtained by

$$V_{n',k,n}^{u}\left(t\right)=\frac{1}{T_c}\int_{D_\phi}H_k^{u}\!\left(nT_{MC}+t+t'\right)\otimes c_{n'}^{u}\!\left(nT_{MC}+t+t'\right)e^{j2\pi f_k\left(nT_{MC}+t+t'\right)}\phi\!\left(t'\right)dt',\quad(10.19)$$

and the spread channel vector is

$$U_{n',k,n}^{u}\left(t\right)=e^{j2\pi\Delta f^{u}nT_{MC}}V_{n',k,n}^{u}\left(t\right).\qquad(10.20)$$

We assumed in the development of equation (10.18) that $\psi_k(nT_G + t + t')$ is constant during the interval $t' \in D_\phi$. We also considered that the frequency offset is small compared to the symbol rate ($\Delta f^{u}T_{MC} \ll 1$); thus, $e^{j2\pi\Delta f^{u}(nT_{MC}+t+t')} \simeq e^{j2\pi\Delta f^{u}nT_{MC}}$ for $t' \in D_\phi$ and $t \in [0, T_{MC})$. Substituting equation (10.18) in equation (10.15) gives

$$\mathbf{Y}_n=\sum_{u=1}^{C}\sum_{k=-K}^{K}\sum_{n'=-\infty}^{\infty}b_{k,n'}^{u}\psi_{k,n}^{u}e^{j2\pi\Delta f^{u}nT_{MC}}\mathbf{V}_{n',k,n}^{u}+\mathbf{N}_n$$

$$=\sum_{u=1}^{C}\sum_{k=-K}^{K}\sum_{n'=-\infty}^{\infty}b_{k,n'}^{u}\psi_{k,n}^{u}\mathbf{U}_{n',k,n}^{u}+\mathbf{N}_n,$$

(10.21)

where $\psi_{k,n}^{u} = \psi_k^{u}(nT_{MC})$. Due to asynchronism and multipath propagation, each user's carrier observation matrix carries information from the current as well as the previous and future symbols of the corresponding user's carrier. We therefore have

$$\mathbf{Y}_n = \sum_{u=1}^{C}\sum_{k=-K}^{K}\sum_{n'=n-1}^{n+1} b_{k,n'}^{u}\,\psi_{k,n}^{u}\,\mathbf{U}_{n',k,n}^{u} + \mathbf{N}_n. \qquad (10.22)$$

Without loss of generality, let us focus on the detection of the nth symbol carried by the kth carrier of the desired user assigned index $d \in \{1, \ldots C\}$, i.e., b_{k_d,n_d}^{d}. Using equation (10.22) and defining a vector $\underline{V}$ as a matrix $\mathbf{V}$ reshaped column-wise, we can rewrite the observation matrix for a desired user d with respect to its nth symbol of carrier k targeted for detection in the following simpler vector form:

$$\underline{Y}_n = \underbrace{s_{k,n}^{d}\underline{U}_{n,k,n}^{d}}_{\text{desired signal}} + \underbrace{\sum_{\substack{u=1 \\ u \ne d}}^{C}\sum_{k'=-K}^{K}\sum_{n'=n-1}^{n+1} s_{k',n'}^{u}\underline{U}_{n',k',n}^{u}}_{\underline{I}_{MAI,k,n}^{d}}$$

$$+ \underbrace{\sum_{\substack{k'=-K \\ k'\ne k}}^{K}\sum_{n'=n-1}^{n+1} s_{k',n'}^{d}\underline{U}_{n',k',n}^{d}}_{\underline{I}_{ICI,k,n}^{d}} + \underbrace{\sum_{\substack{n'=n-1 \\ n'\ne n}}^{n+1} s_{k,n'}^{d}\underline{U}_{n',k,n}^{d}}_{\underline{I}_{ISI,k,n}^{d}} + \underline{N}_n, \qquad (10.23)$$

$$= s_{k,n}^{d}\underline{U}_{n,k,n}^{d} + \underline{I}_{k,n}^{d} + \underline{N}_n,$$

where $s_{k',n'}^{u} = \psi_{k',n}^{u}b_{k',n'}^{u}$ and $(\psi_{k,n}^{u})^2$ are the n'^{th} signal component of the k'^{th} carrier of user u, and the received power of user u over carrier k, respectively. The total interference $\underline{I}_{k,n}^{d}$ includes three types of interference: (1) the multiple access interference $\underline{I}_{MAI,k,n}^{d}$ is the interference due to the N_c carriers from the other in-cell users $u \ne d$; (2) the inter-carrier interference $\underline{I}_{ICI,k,n}^{d}$ is the interference due to the other carriers, $k' \ne k$, from the same user d; and (3) the intersymbol interference $\underline{I}_{ISI,k,n}^{d}$ is the interference due to the same carrier k from the same user d. The noise vector $\underline{N}_n$, which comprises the pre-processed thermal noise and the interference due to out-of-cell users, is assumed to be uncorrelated both in space and time with variance σ_N^2.

In previous work [32], we proposed a receiver named MC-STAR. MC-STAR assumes the interference $\underline{I}_{k,n}^{d}$ as another contribution to the noise $\underline{N}_n$. Hence, the signal component of the desired user's carrier is extracted by spatiotemporal maximum ratio combining (MRC) as follows:

$$\hat{s}_{k,n}^{d} = \underline{W}_{MRC,k,n}^{d^{H}}\underline{Y}_n = \frac{\hat{\underline{U}}_{n,k,n}^{d^{H}}\underline{Y}_n}{\left\|\hat{\underline{U}}_{n,k,n}^{d}\right\|^2}, \qquad (10.24)$$

where anywhere in the chapter the notation $\hat{\alpha}$ stands for an estimate of a given variable α, $(.)^H$ is the Hermitian operator, and $\underline{W}_{MRC,k,n}^{d}$ is the MRC beamformer. Equation (10.23) shows that the net interference increases with the number of interferers and subcarriers, which severely limits the capacity of the MC-CDMA system with simple MRC receivers. Therefore, in the next section, we shall use the data decomposition of equation (10.23) to formulate the interference suppression problem and propose a new MC-CDMA receiver with full interference suppression capabilities.

10.2.9   Multiuser Detection Techniques for MC-CDMA Systems

Conventional multicarrier CDMA detectors—such as the matched filter, the RAKE combiner, and the MC-STAR receiver—are optimized for detecting the signal of a single desired user. RAKE combiners exploit the inherent multipath diversity in CDMA, since they essentially consist of matched filters for each resolvable path of the multipath channel. The outputs of these matched filters are then coherently combined according to a diversity combining technique, such as maximum ratio combining, equal gain combining, or selection diversity combining [13, 54]. Unlike RAKE-type receivers, which assume perfect knowledge of the channel [3], we proposed in previous work a full space-time receiver solution, named MC-STAR [43], that jointly implements adaptive channel identification and synchronization in both time and frequency.* These conventional single-user detectors are inefficient, because the interference is treated as noise and there is no utilization of the available knowledge about the mobile channel or the spreading sequences of the interferers.

In order to mitigate the problem of MAI, [33] proposed and analyzed the optimum multiuser detector for asynchronous Gaussian multiple access channels. This optimum detector significantly outperforms the conventional detector, and it is near–far resistant, but unfortunately its complexity grows exponentially with the number of interfering users. Following this work, numerous suboptimum multiuser detectors have been proposed for a variety of channels, data modulation schemes, and transmission formats. Since MC-CDMA systems also contain a DS-CDMA component, traditional suboptimum multiuser detection techniques can be performed on each carrier with some form of adaptation. A variety of linear multiuser receivers have been investigated for MC-CDMA systems such as the minimum mean square error (MMSE) detector [34] and the combination of MMSE and the decorrelator detector [35]. Interference cancellation (IC) schemes constitute another variant of multiuser detection that has been applied to MC-CDMA systems. They can be broadly divided into two categories: parallel cancellation (PIC) [37, 38] and successive cancellation (SIC) [36]. At each stage in the detector, the estimates of all the other users from the previous stage were used for reconstructing an estimate of the MAI, and this estimate was then subtracted from the interfered signal representing the wanted bit. A novel class of multicarrier multiuser detectors, referred to as subspace blind detectors, was proposed by [39] and [40], where only the spreading sequence and the delay of the desired user were known at the receiver. Based on this knowledge, a blind subspace tracking algorithm was developed for estimating the data of the desired user.

---

* MC-STAR is our starting receiver; hence, we will provide a short overview of this receiver in section 10.3.

Most of these multiuser receivers have focused on multiple access interference while ignoring the ICI. In addition, important system design issues such as carrier frequency offset recovery (CFOR) have often been neglected. In multiuser detection, the CFO of one user not only degrades the detection of that user itself, but also makes the receiver based on the ideal carrier frequency acquisition no longer optimal, thus degrading the detection of the other users [41]. An alternative multiuser detection technique, denoted interference subspace rejection (ISR), has been proposed for DS-CDMA [42]. This technique offers different modes. Each mode characterizes the interference vector in a different way and accordingly suppresses it. The flexibility and robustness inherent to ISR make its exploitation in multicarrier systems of great interest.

10.3 Proposed Adaptive Multicarrier CDMA Receiver: MC-ISR

This section is dedicated to the description, performance analysis, and implementation of the proposed MC-ISR receiver [6]. After a short overview of MC-STAR [43], which is our starting single-user receiver, we will describe and evaluate the adaptive interference rejection procedure that characterizes the proposed MC-ISR receiver.

10.3.1   The General Concept of MC-STAR

The adaptive receiver MC-STAR implements joint space-time-frequency processing over the despread data to improve the spectrum efficiency of the MC-CDMA system. The adaptive blind channel identification and equalization as well as the acquisition and tracking of multipaths and CFO are carried out on each subcarrier. However, their modules are interconnected to ensure proper information exchange and joint processing over carriers.* Mathematical details of the different adaptive procedures and their connections are provided in [43]. In this section, we explain the advantages of MC-STAR by describing the intermediate stages in our development that led to this receiver.

At the beginning, we extend original STAR, proposed for DS-CDMA [31], to a multicarrier system by placing STAR on each subcarrier. This extension requires a modification of the time-delay tracking procedure. Indeed, we introduce an intermediate transformation of the time response to reallow estimation of the multipath delays by simple linear regression.

Multi-carrier CDMA systems are very sensitive to the CFO. Therefore, we further introduce joint time-delay and frequency synchronization. The effect of the CFO on the performance of the spatiotemporal array receiver was not addressed in [31]. The space-time separation of the channel enables us to decouple time and carrier frequency synchronization. We can hence estimate the CFO by linear regression (LR) of the phase variation of each fading coefficient. Once an estimate of the carrier frequency offset

---

* The complexity of MC-STAR, which is approximately the complexity of STAR multiplied by the number of subcarriers, can be assessed using the results established in [44]. The latter suggests that MC-STAR can be implemented today on a single FPGA.

estimate $\widehat{\Delta f}$ is available by exploiting diversity in space, time, and frequency, we implement carrier frequency offset recovery (CFOR) in an adaptive closed-loop structure, where we feed back the estimate of the frequency offset to the input of the receiver. The CFOR reduces the time variations in the spatiotemporal propagation channel due to Δf to much weaker fluctuations due to the residual $\delta f = \Delta f - \widehat{\Delta f}$. It results in much weaker identification errors and enables further reduction of the carrier frequency estimation error δf.

At this stage, the receiver still consists of independent modules on each subcarrier. The purpose of the last step is to improve the performance of the overall receiver by interconnecting these modules and performing joint multicarrier processing. We exploit the intercarrier correlation, intrinsic to a multicarrier system, as a type of *frequency gain* to improve the performance by joint multicarrier channel identification and synchronization operations. Indeed, in the context of a multicarrier system, the adjacent subcarriers are exposed to correlated fading, especially if the delay spread of the channel is relatively low, resulting in relatively large coherence bandwidth. Hence, averaging the adjacent subcarrier channel parameters should improve the BER performance when transmitting over such low-dispersive fading channels. Along this perspective, the parameters common to all subcarriers can be estimated more accurately by averaging their estimates over all subcarriers. These parameters include the number of multipaths, their corresponding time delays, and the frequency offset. Other channel parameters, such as the channel fading coefficients, are correlated but not identical over all subcarriers. Therefore, combining them may not achieve the expected performance enhancement. We thus introduce a moving average technique over subcarriers with high correlation. The fact that subcarriers are highly correlated implies similar or identical channel parameters over subcarriers. Yet, the noise is uncorrelated across subcarriers, and hence the similar/common parameters can be estimated more accurately by averaging their estimates over all subcarriers, yielding the so-called frequency gain. The variance of the resulting estimation error is lower than the variance of the estimation error without frequency gain. Please bear in mind that we used the term *frequency gain* and not *frequency diversity*, which relies on the fact that the fading is different over different subcarriers.

10.3.2 Multicarrier Interference Subspace Rejection (MC-ISR)

Provided that an instantaneous estimate of the total interference $\hat{I}^d_{k,n} = \hat{I}^d_{\text{MAI},k,n} + \hat{I}^d_{\text{ICI},k,n} + \hat{I}^d_{\text{ISI},k,n}$ is made available at the receiver (see section 10.3.4), we can eliminate it and yet achieve distortionless response to the desired signal by imposing the following simple constraints to the combiner $\underline{W}^d_{k,n}$:

$$\begin{cases} \underline{W}^{d^H}_{k,n} \hat{\underline{U}}^d_{n,k,n} = 1, \\ \underline{W}^{d^H}_{k,n} \hat{\underline{I}}^d_{k,n} = 0, \end{cases} \Rightarrow \begin{cases} \underline{W}^{d^H}_{k,n} \hat{\underline{U}}^d_{n,k,n} = 1, \\ \underline{W}^{d^H}_{k,n} \left(\hat{\underline{I}}^d_{\text{MAI},k,n} + \hat{\underline{I}}^d_{\text{ICI},k,n} + \hat{\underline{I}}^d_{\text{ISI},k,n} \right) = 0. \end{cases} \quad (10.25)$$

The first constraint guarantees a distortionless response to the desired signal, while the second directs a null to the total interference realization and thereby cancels it.

Exploiting the general framework developed in [42], the solution to the specific optimization problem in equation (10.25) is the MC-ISR combiner $\underline{W}^d_{k,n}$ given as follows:

$$
\underline{W}^d_{k,n} = \frac{\mathbf{\Pi}^d_{k,n} \hat{\underline{U}}^d_{n,k,n}}{\hat{\underline{U}}^{d^H}_{n,k,n} \mathbf{\Pi}^d_{k,n} \hat{\underline{U}}^d_{n,k,n}},
\tag{10.26}
$$

$$
Q_n = 1 \left/ \left(\hat{\underline{I}}^{d^H}_{k,n} \hat{\underline{I}}^d_{k,n} \right) = 1 \left/ \left\| \hat{\underline{I}}^d_{k,n} \right\|^2 \right.\right. ,
\tag{10.27}
$$

$$
\mathbf{\Pi}^d_{k,n} = \mathbf{I}_{N_T} - \hat{\underline{I}}^d_{k,n} \hat{\underline{I}}^{d^H}_{k,n} \times Q_n,
\tag{10.28}
$$

where $N_T = M \times N_p$ is the total space dimension and $\mathbf{I}_{N_T}$ denotes an $N_T \times N_T$ identity matrix. First, we form the projector $\mathbf{\Pi}^d_{k,n}$ orthogonal to the total interference realization. Second, we project the estimated response vector $\hat{\underline{U}}^d_{n,k,n}$ and normalize it to derive the combiner. We use this combiner instead of MRC (used by MC-STAR) to extract the nth signal component of the kth carrier of the desired user as

$$
\hat{s}^d_{k,n} = \underline{W}^{d^H}_{k,n} \underline{Y}_n.
\tag{10.29}
$$

Unlike most of the multiuser receivers proposed for MC-CDMA, which focus on multiple access interference while ignoring the intercarrier interference, MC-ISR fully suppresses the total interference resulting from MAI, ISI, and ICI by simple yet efficient nulling.* In addition, the interference suppression is made adaptive to track the current situation of the wireless channel and the interference. Simulation results will later show that ICI is not negligible and that full adaptive interference suppression is required to improve the MC-CDMA system performance.

10.3.3   Link/System-Level Performance Analysis

This section is dedicated to the performance analysis of the MC-ISR receiver based on the Gaussian assumption (GA). We exploit the analysis results of DS-CDMA ISR recently developed in [45] at the link level and extend them to MC-ISR. Additionally, we broaden the scope of the analysis to the system level.

10.3.3.1   Link-Level Performance

For the sake of simplicity, we assume temporarily perfect channel identification and perfect CFO estimation and recovery. Later in the simulations, we will use the channel

---

* The formulation of MC-ISR can be extended to MMSE-type criteria [42].

and CFO estimates provided by MC-STAR* [32]. The postcombined signal can be formulated as

$$\hat{s}_{k,n}^d = \underline{W}_{k,n}^{d^H} \underline{Y}_n = s_{k,n}^d + \delta_{MAI,k,n}^d + \delta_{ICI,k,n}^d + \delta_{ISI,k,n}^d + \underline{W}_{k,n}^{d^H} \underline{N}_n, \tag{10.30}$$

where $\delta_{MAI,k,n}^d$, $\delta_{ICI,k,n}^d$, and $\delta_{ISI,k,n}^d$ are the combining residuals of $\underline{I}_{MAI,k,n}^d$, $\underline{I}_{ICI,k,n}^d$, and $\underline{I}_{ISI,k,n}^d$, respectively. We assume here that the interference rejection residuals $\delta_{MAI,k,n}^d$, $\delta_{ICI,k,n}^d$, and $\delta_{ISI,k,n}^d$ are Gaussian random variables with zero mean. Hence, we only need to evaluate their variances. Note that the residuals would be null (i.e., $\delta_{MAI,k,n}^d = \delta_{ICI,k,n}^d = \delta_{ISI,k,n}^d = 0$) if the reconstruction of the interference were perfect (i.e., $\hat{\underline{I}}_{k,n}^d = \underline{I}_{k,n}^d$), and hence $\hat{s}_{k,n}^d = s_{k,n}^d + \underline{W}_{k,n}^{d^H} \underline{N}_n$ would be corrupted only by the residual noise, which is Gaussian with zero mean and variance:

$$\mathrm{Var}\left[\underline{W}_{k,n}^{d^H} \underline{N}_n\right] = \bar{\kappa}\sigma_N^2, \tag{10.31}$$

where

$$\bar{\kappa} = \mathrm{E}\left[\left\|\underline{W}_{k,n}^d\right\|^2\right] = \frac{ML-1}{ML-2},$$

is a measure of the enhancement of the white noise compared to MRC ($\bar{\kappa} = 1$ for MRC) [45]. However, in practice the interference vector is reconstructed erroneously due to wrong tentative data decisions and power control errors, and hence $\hat{s}_{k,n}^d$ is further corrupted by non-null residual interference rejection components. Therefore, we introduce the error indicating variables

$$\xi_{k,n}^u = \hat{b}_{k,n}^{u*} b_{k,n}^u$$

and

$$\lambda_{k,n}^u = \hat{\psi}_{k,n}^{u*} \psi_{k,n}^u \Big/ \left\|\hat{\psi}_{k,n}^u\right\|^2$$

where $(.)^*$ means complex conjugate. $\xi_{k,n}^u$ models the symbol estimation error provided by MRC at the initial stage. $\lambda_{k,n}^u$ characterizes the power control error. $\xi_{k,n}^u$ and $\lambda_{k,n}^u$ equal 1 when the estimated data symbol and the power control are perfect; otherwise, they are complex numbers. Since

$$\underline{Y}_{n',k',n}^u = s_{k',n'}^u \underline{U}_{n',k',n}^u = b_{k',n'}^u \psi_{k',n}^u \underline{U}_{n',k',n}^u = \xi_{k',n'}^u \lambda_{k',n}^u \hat{b}_{k',n'}^u \hat{\psi}_{k',n}^u \underline{U}_{n',k'n}^u = \xi_{k',n'}^u \lambda_{k',n}^u \hat{\underline{Y}}_{n',k',n}^u, \dagger$$

---

* Simulations will show little deviation from analysis in the operating BER region.
† Here we assume perfect time and frequency synchronization.

we can rewrite equation (10.23) as

$$
\underline{Y}_n = \underline{Y}^d_{n,k,n} + \sum_{\substack{u=1 \\ u-d}}^{C} \sum_{k'=-K}^{K} \sum_{n'=n-1}^{n+1} \xi^u_{k',n'} \lambda^u_{k',n} \underline{\hat{Y}}^u_{n',k',n}
$$

$$
+ \sum_{\substack{k'=-K \\ k'\neq k}}^{K} \sum_{n'=n-1}^{n+1} \xi^d_{k',n'} \lambda^d_{k',n} \underline{\hat{Y}}^d_{n',k',n} + \sum_{\substack{n'=n-1 \\ n'\neq n}}^{n+1} \xi^d_{k,n'} \lambda^d_{k,n} \underline{\hat{Y}}^d_{n',k,n} + \underline{N}_n.
$$

(10.32)

The signal after MC-ISR combining is then

$$
\underline{W}^{d^H}_{k,n} \underline{Y}_n = s^d_{k,n} + \sum_{\substack{u=1 \\ u\neq d}}^{C} \sum_{k'=-K}^{K} \sum_{n'=n-1}^{n+1} \xi^u_{k',n'} \lambda^u_{k',n} \underline{W}^{d^H}_{k,n} \underline{\hat{Y}}^u_{n',k',n}
$$

$$
+ \sum_{\substack{k'=-K \\ k'\neq k}}^{K} \sum_{n'=n-1}^{n+1} \xi^d_{k',n'} \lambda^d_{k',n} \underline{W}^{d^H}_{k,n} \underline{\hat{Y}}^d_{n',k',n} + \sum_{\substack{n'=n-1 \\ n'\neq n}}^{n+1} \xi^d_{k,n'} \lambda^d_{k,n} \underline{W}^{d^H}_{k,n} \underline{\hat{Y}}^d_{n',k,n} \quad (10.33)
$$

$$
+ \underline{W}^{d^H}_{k,n} \underline{N}_n.
$$

The MC-ISR combiner $\underline{W}^d_{k,n}$ satisfies the optimization property in equation (10.25), and thus

$$
\underline{W}^{d^H}_{k,n} \underline{\hat{I}}^d_{k,n} = 0 \;\Rightarrow\; \mathrm{Var}\left[\underline{W}^{d^H}_{k,n} \left(\underline{\hat{I}}^d_{\mathrm{MAI},k,n} + \underline{\hat{I}}^d_{\mathrm{ICI},k,n} + \underline{\hat{I}}^d_{\mathrm{ISI},k,n} \right) \right] = 0. \quad (10.34)
$$

This result allows the derivation of the variance of the interference rejection residuals, as shown in the appendix. Let $\bar{\psi}^2_D = E[(\bar{\psi}^d_k)^2]$ be the average power of the kth carrier of the desired user and $\bar{\psi}^2_I$ be the average interference power on each interfering carrier. The variances of the residual $\underline{I}^d_{\mathrm{MAI},k,n}$ can be written as

$$
\mathrm{Var}\left[\delta^d_{MAI,k,n} \right] = (C-1) \frac{\bar{\psi}^2_I}{L} \left[\varsigma(\beta) + \chi_k(\beta) \right] (1 + \rho_\lambda - \rho_\xi) \bar{\kappa}, \quad (10.35)
$$

where $\varsigma(\beta) = 1 - \beta/4$ and

$$
\chi_k(\beta) = \begin{cases} \dfrac{\beta}{8}, & \text{if } k = -K \text{ or } K, \\[2mm] \dfrac{\beta}{4}, & \text{if } k = -K+1, \ldots, K-1, \end{cases} \quad (10.36)
$$

for MC-DS-CDMA ($\lambda = L$ and $f_k = k/T_c$) and

$$\chi_k(\beta) = \sum_{\substack{k'=-K \\ k'\neq k}}^{K} \vartheta\left(\left|k-k'\right|\right),\tag{10.37}$$

where

$$\vartheta(x) = \begin{cases} 1 - \dfrac{\beta}{2} - \dfrac{x}{2L} + \dfrac{3\beta}{4\pi}\sin\left(\dfrac{\pi x}{\beta L}\right) + \left(\dfrac{\beta}{4} - \dfrac{x}{4L}\right)\cos\left(\dfrac{\pi x}{\beta L}\right), & \text{if } 0 \le x/L \le \min\left(\beta, 1-\beta\right) \\[4mm] 1 - \dfrac{x}{L} & \text{if } \beta \le x/L \le 1-\beta \text{ and } \beta < 0.5 \\[4mm] \dfrac{3}{4} - \dfrac{\beta}{4} - \dfrac{x}{4L} + \dfrac{3\beta}{4\pi}\sin\left(\dfrac{\pi x}{\beta L}\right) + \left(\dfrac{\beta}{4} - \dfrac{x}{4L}\right)\cos\left(\dfrac{\pi x}{\beta L}\right) + & \\[2mm] \quad \dfrac{3\beta}{8\pi}\sin\left(\dfrac{\pi x}{\beta L} - \dfrac{\pi}{\beta}\right) - \left(\dfrac{x}{8L} - \dfrac{1-\beta}{8}\right)\cos\left(\dfrac{\pi x}{\beta L} - \dfrac{\pi}{\beta}\right) & \text{if } 1-\beta \le x/L \le \beta \text{ and } \beta > 0.5 \\[4mm] \dfrac{3}{4} + \dfrac{\beta}{4} - \dfrac{3x}{4L} + \dfrac{3\beta}{8L}\sin\left(\dfrac{\pi x}{\beta L} - \dfrac{\pi}{\beta}\right) - \left(\dfrac{x}{8L} - \dfrac{1-\beta}{8}\right)\cos\left(\dfrac{\pi x}{\beta L} - \dfrac{\pi}{\beta}\right) & \text{if } \max\left(\beta, 1-\beta\right) \le x/L \le 1 \end{cases}$$

$$\tag{10.38}$$

for MT-CDMA ($\lambda = L$ and $f_k = k/T_{Mc}$). The expressions of

$$\rho_\xi = \mathrm{E}\left[\xi_{k',n'}^u \lambda_{k',n}^u \xi_{k',n'}^{u*} \lambda_{k',n}^{u*}\right]$$

and

$$\rho_\lambda = \mathrm{E}\left[\left(\lambda_{k,n}^u\right)^2\right] - 1$$

are derived for a Rayleigh fading channel with P paths to yield

$$\rho_\xi = \left(1 - \left(1 - \cos\left(2\pi/\mathcal{M}\right)\right) S_{rec}\right)^2,$$

$$\rho_\lambda = \frac{4\pi^2 \left(f_D \times \tau_{PC}\right)^2}{P-1},$$

$$\tag{10.39}$$

where S_{rec} is the symbol error rate in the previous MC-ISR stage, f_D is the maximum Doppler frequency, and τ_{PC} is the power control feedback delay. The variances of the residuals $\underline{I}^d_{\text{ICI},k,n}$ and $\underline{I}^d_{\text{ISI},k,n}$ can be written as

$$\text{Var}\left[\delta^d_{ICI,k,n}\right] = \frac{\overline{\Psi}^2_D}{L} \delta_{is} \chi_k(\beta)\left(1+\rho_\lambda - \rho_\xi\right)\overline{\kappa},$$

$$\text{Var}\left[\delta^d_{ISI,k,n}\right] = \frac{\overline{\Psi}^2_D}{L} \delta_{is} \varsigma(\beta)\left(1+\rho_\lambda - \rho_\xi\right)\overline{\kappa},$$

(10.40)

where $\delta_{is} = (P-1)/P$ is a measure of the relative impact of the interference generated by the other paths on a given path of the desired user (for a Rayleigh fading channel with P equal paths). The SINR on the kth carrier can be estimated as

$$\text{SINR}_{ISR,k} = \frac{\overline{\Psi}^2_D}{\text{Var}\left[\delta^d_{MAI,k,n}\right] + \text{Var}\left[\delta^d_{ICI,k,n}\right] + \text{Var}\left[\delta^d_{ISI,k,n}\right] + \overline{\kappa}\sigma^2_N}.$$

(10.41)

Note that the SINR expression above applies to MRC as well by setting $\overline{\kappa} = 1$ and $\rho_\lambda = \rho_\xi = 0$ in equations (10.35) and (10.40). Note also that in [46], we provide the variance of the interference for an MC-CDMA system with a rectangular pulse. In this chapter, we improve the analytical performance evaluation by deriving the variance of the interference with a more practical band-limited square-root raised cosine waveform. The BER performance on the kth carrier is then given as follows:

$$P^k_e = \Omega\left(\text{SINR}_{ISR,k}\right),$$

(10.42)

where Ω represents the single-user bound (SUB), which is classically defined as a conditional Gaussian Q-function over ψ_D and ψ_I. When using this classical representation, the average BER is derived by first finding the probability density functions (pdfs) of ψ_D and ψ_I and then averaging over those pdfs. Since it is difficult to find a simple expression for the pdfs of ψ_D and ψ_I that takes into consideration antenna diversity, imperfect power control, and imperfect channel identification, we may consider an approximative pdf. In this analysis, we choose to simulate Ω without imposing any pdf approximation. For each multicarrier configuration, we run single-user and single-carrier link-level simulations. We reproduced as much as possible most of the real-world operating conditions: time and frequency synchronization, imperfect power control, channel identification errors, antenna diversity, etc. These link-level simulations gave a realistic Ω : $BER = \Omega(SNR)$. The simulations will later consider a multiuser and multicarrier environment. The average BER performance of the MC-ISR receiver is given by

$$P_e = \frac{1}{2K+1}\sum_{k=-K}^{K} P^k_e.$$

(10.43)

10.3.3.2   System-Level Performance

In order to compare the different MC-ISR configurations, the link-level curves provide a good picture of the performance of each system. But limiting comparisons to the

BER performance is not sufficient because the data rate is not equal for all configurations. Hence, we translate the link-level results into system-level results in terms of total throughput under the following three assumptions: (1) all users are received with an equal average power (i.e., $\bar{\psi}_D^2$ and $\bar{\psi}_i^2$) [4]; (2) all the cells have the same average load of C users per cell; and (3) the out-cell to in-cell interference ratio f is set to 0.6 [47]. Given these assumptions in an interference-limited system (thermal noise is low compared to interference), the link-level SIR at the base station antennas (ignoring ISI for simplicity) is

$$SIR_{ISR} = \frac{1}{(C-1)\alpha + \frac{1}{L}\delta_{is}\chi(\beta)(1+\rho_\lambda - \rho_\xi)\bar{\kappa} + Cf\lambda}, \tag{10.44}$$

where

$$\chi(\beta) = \max_k \left[\chi_k(\beta)\right],$$

$$\alpha = \frac{1}{L}\left(\varsigma(\beta) + \chi(\beta)\right)\left(1 + \rho_\lambda - \rho_\xi\right)\bar{\kappa}, \tag{10.45}$$

$$\gamma = \frac{1}{L}\left(\varsigma(\beta) + \chi(\beta)\right)\bar{\kappa}.$$

$(C-1)\alpha$ is the normalized variance of the residual MAI ($u \neq d$), $\frac{1}{L}\delta_{is}\chi(\beta)(1 + \rho_\lambda - \rho_\xi)\bar{\kappa}$ is the normalized variance of the residual ICI ($k' \neq k$ and $u = d$), and $Cf\gamma$ is the normalized variance of the out-cell interference. Note that equation (10.44) is derived from equation (10.41) and the assumption of negligible thermal noise and ISI.

The maximum number of users that can access the system C_{max} can be hence calculated by the simple procedure illustrated in Table 10.1. After initialization, this procedure increments the capacity C, until the SIR_{ISR} given by equation (10.44) no longer exceeds the required SNR_{req}. The SRN_{req} is the required SNR, derived from link-level simulations, to meet a BER of 5% in order to achieve a QoS of 10^{-6} after channel decoding. In step 2.2 of Table 10.1, we use the fact that the SIR expression applies to MRC by setting $\bar{\kappa} = 1$ and $\rho_\lambda = \rho_\xi = 0$ in equation (10.44). In step 2.3, we evaluate the symbol error rate S_{MRC} after the MRC stage as follows:

$$S_{MRC} = \Omega\left(SIR_{MRC}\right), \tag{10.46}$$

where Ω represents the single-user bound (SUB). Note that multistage MC-ISR is considered in step 2.5. The total throughput is hence $\mathcal{T}_{max} = C_{max} \times R_b = C_{max} \times R_s \times \log_2(\mathcal{M})$ where R_b and R_s are the bit rate and symbol rate over all subcarriers, respectively.

10.3.4 MC-ISR Receiver Implementation

As mentioned in section 10.3.2, the proposed MC-ISR receiver requires accurate channel parameter estimates and data decisions to reconstruct the total interference $\hat{I}_{k,n}^d$ and

TABLE 10.1   Capacity Computation Procedure

1. Initialize capacity $C = 0$.
2. Start computation loop:
2.1 Increment capacity $C = C + 1$.
2.2 Compute the SIR with MRC:

$$SIR_{MRC} = \frac{L}{(C-1)\left(\varsigma(\beta)+\chi(\beta)\right)+\chi(\beta)\delta_{is}+fC\left(\varsigma(\beta)+\chi(\beta)\right)}.$$

2.3 Compute the symbol error rate (SER) after MRC stage

$$S_{MRC} = \Omega\left(SIR_{MRC}\right).$$

2.4 Compute ρ_λ and ρ_ξ.
2.5 Compute the SIR:

$$SIR_{ISR_1} = \frac{1}{(C-1)\alpha+\frac{1}{L}\delta_{is}\chi(\beta)\left(1+\rho_\lambda-\rho_\xi\right)\overline{\kappa}+Cf\gamma}.$$

   If number of stages $S > 1$, start the loop; else go to 2.6.
     For $s = 2 : S$.
     Compute the symbol error rate (SER) after the $s - 1$ stage:

$$S_{ISR_{s-1}} = \Omega\left(SIRISR_{s-1}\right)$$

     Compute ρ_ξ.
     Compute the SIR:

$$SIR_{ISR_s} = \frac{1}{(C-1)\alpha+\frac{1}{L}\delta_{is}\chi(\beta)\left(1+\rho_\lambda-\rho_\xi\right)\overline{\kappa}+Cf\gamma}.$$

     End.
2.6 If $SIR_{ISR_s} > SNR_{req}$ go to 2.1; else exit.
3. Decrement capacity $C = C - 1$.

null it reliably. Unlike previous works on interference suppression or multiuser detection [34, 38] that assume perfect knowledge of the channel, we propose here an adaptive receiver that uses runtime information about the channel and interference. Indeed, MC-ISR jointly implements channel identification and synchronization in both time and frequency, using MC-STAR [32], as well as signal combining with full interference suppression capabilities. Figure 10.2 shows the block diagram of the proposed receiver implementation, divided in four main modules. The first module is a preprocessor that downconverts the received signal to baseband, then passes it through the chip-matched filter before sampling and data block framing. The second module is a signal combiner that provides symbol estimates from the data observation, first by MRC in an initial iteration, then by MC-ISR in one or more iterative stages. The third module is an adaptive channel identifier and synchronizer from MC-STAR that implements closed-loop CFOR and estimates all the channel parameters (multipath time delays and their phases and amplitudes, received power, CFO). The fourth module is a null-constraint generator common to all in-cell users. It gathers the data decisions and channel parameter estimates from the second and third modules dedicated to each in-cell user-carrier pair in order to reconstruct the total in-cell signal vector $\underline{I}_n$. Then to each combiner, say of the desired user-carrier pair as illustrated in Figure 10.1b, it passes on the associated null constraint (i.e., $\underline{I}_{k,n}^d = \underline{I}_n - s_{k,n}^d \underline{U}_{n,k,n}^d$) calculated with the least computations by simple subtraction from $\underline{I}_n$ of the desired signal contribution from the corresponding user-carrier pair.

The implementation of an adaptive closed-loop CFOR* jointly with multicarrier and multiuser detection (here by MC-ISR) requires careful attention regarding the order in which these two tasks should be processed. Indeed, conventional operation of CFOR at an early processing stage† prior to interference suppression would require (on the uplink only) as many independently CFO-compensated observations and interference null constraints as received in-cell users, thereby resulting in a tremendous complexity increase. Here we develop an efficient post-interference-suppression CFOR scheme by splitting the MC-ISR combining operation of equation (10.29) into two steps, an observation-cleaning projection and an MRC combining, and by inserting CFO compensation in between as follows:

$$\underline{Y}_{\Pi,k,n}^d = \Pi_{k,n}^d \underline{Y}_n, \tag{10.47}$$

$$\widehat{\Delta f}_n^d = \widehat{\Delta f}_{n-1}^d + \widehat{\delta f}_n^d, \tag{10.48}$$

$$\dot{\underline{Y}}_{\Pi,k,n}^d = \underline{Y}_{\Pi,k,n}^d e^{-j2\pi \widehat{\Delta f}_n^d nT}, \tag{10.49}$$

$$\dot{\hat{\underline{V}}}_{\Pi,k,n}^d = \Pi_{k,n}^d \hat{\underline{V}}_{\Pi,k,n}^d, \tag{10.50}$$

$$\hat{s}_{k,n}^d = \frac{\hat{\underline{V}}_{\Pi,k,n}^{d^H} \dot{\underline{Y}}_{\Pi,k,n}^d}{\left\| \hat{\underline{V}}_{\Pi,k,n}^d \right\|^2}. \tag{10.51}$$

The cleaning projection of equation (10.47) results in an almost interference-free observation $\underline{Y}_{\Pi,k,n}^d$ and allows for CFO estimation and compensation in equations (10.48) and (10.49), respectively, using the CFOR module of the single-user MC-STAR (refer to [32] and [48] for details on how to estimate the CFO adjustment term in equation (10.49)), and for MRC combining in equation (10.51) using the projected estimate of the

---

* In contrast to open-loop structures, closed-loop CFOR reduces the channel time variations and greatly improves their tracking [32].

† Usually CFOR is embedded in the RF chain or plugged to the preprocessor output.

spread channel vector without CFO $\hat{\underline{Y}}^d_{\Pi,k,n}$. To the best of our knowledge, we are the first to report on and address this issue and to propose an efficient scheme for closed-loop CFOR in a multiuser detection context. It is important to mention here that if $\Delta f^u = \Delta f \; \forall \; u \in \{1, \ldots, C\}$ (i.e., downlink), then there is no need to estimate the CFO for the MC-ISR to null the in-cell interference. Indeed, the MC-ISR combiner $\underline{W}^d_{k,n}$ satisfies the optimization property in equation (10.25). Thus, it is not affected by the CFO of other users, i.e.,

$$\underline{W}^{d^H}_{k,n}\hat{\underline{I}}^d_{k,n} = 0 \;\; \Rightarrow \;\; e^{j2\pi\Delta f n T_{MC}}\left[\underline{W}^{d^H}_{k,n}\hat{\underline{I}}^d_{k,n}\right] = 0. \tag{10.52}$$

Once the MC-ISR projection is performed in equation (10.47) after reconstruction of $\hat{\underline{I}}^d_{k,n}$ without CFO, we implement the same CFOR scheme implemented in part by equations (10.48) and (10.49). Hence, like the near–far resistant detector proposed in [41], the multiuser CFOR problem can be transformed on the downlink into a single-user CFOR problem, and conventional single-user methods can therefore be used to estimate the frequency offset.*

To validate the efficiency of the proposed CFOR strategy in a multicarrier and multiuser detection scheme on the uplink, we consider a multiuser DBPSK MT-CDMA system with seven subcarriers, a spreading factor of 96, and five in-cell users ($N_c = 7$, $L = 96$, $C = 5$). We select the setup that will be introduced in section 10.4.1. The frequency offset normalized by the subcarrier separation ($\Delta f \times T_{MC}$) is set to 0.005 (i.e., $\Delta f = 200$ Hz).[15] Figure 10.4 shows the link-level results of MC-ISR with and without CFOR. Results suggest that a CFO of 200 Hz has a serious impact on the performance of MC-CDMA, and that the link-level gain with the proposed CFOR is in the range of 1 dB at a BER of 5% before channel decoding. By comparing the link-level curves of MC-ISR with CFOR and MC-ISR without a frequency offset (i.e., CFO=0 Hz), we notice that CFOR compensates almost completely the performance loss due to the frequency offset. These results confirm the need for and the efficiency of the proposed CFOR in a multicarrier and multiuser detection context.

10.4   Simulation Results

10.4.1   Simulation Setup

We consider an MC-CDMA system operating at a carrier of 1.9 GHz with maximum bandwidth of 5 MHz. We select a frequency offset Δf of 200 Hz, the maximum error tolerated by 3G standards† ($\equiv 0.1$ ppm) for the frequency mismatch between the mobile and the base station [49]. We assume a frequency-selective Rayleigh fading channel with $P^u = P$ propagation paths with exponentially decreasing powers. The channel is

---

* The study of the CFOR performance is provided in [43].

† We select $\Delta f = 200$ Hz to show that even CFO residuals below the maximum value tolerated by 3G standards result in significant losses in performance.

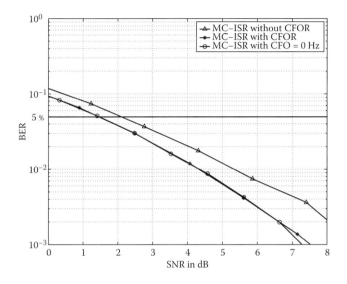

FIGURE 10.4 BER versus SNR for MT-CDMA MC-ISR, $L = 96$, $N_c = 7$, $C = 5$, with and without CFOR.

correlated across subcarriers and varying in time with Doppler shift f_D. We suppose a low Doppler situation $f_D = 8.8$ Hz unless otherwise mentioned. We consider that time delays vary linearly in time with a delay drift of 0.049 ppm. The receiver has $M = 2$ antennas. We implement closed-loop power control operating at 1,600 Hz and adjusting the power in steps of ±0.25 dB. An error rate on the power control bit of 5% and a feedback delay of 0.625 ms are simulated. The simulation parameters common to all multicarrier system configurations are listed in Table 10.2.

Table 10.3 shows the parameters specific to each multicarrier CDMA configuration. We choose as a reference the 3G DS-CDMA ($N_c = 1$) system with spreading factor $L = 32$ and a chip rate of 3.84 Mcps. We assume frequency-selective fading with $P = 3$ propagation paths. One of the features of MT-CDMA is that for a constant bandwidth the ratio between the spreading factor L and $2K = N_c - 1$ is constant. We hence maintain the same chip rate (3.840 Mcps) by changing the spreading factor and the number of subcarriers, as shown in Figure 10.3. We consider four MT-CDMA configurations. Since they use the same chip rate, there are three paths in each MT-CDMA subcarrier. For a fair comparison among different configurations of MC-DS-CDMA, the bandwidth should be the same. By reducing the chip rate, we varied the number of subcarriers while maintaining the orthogonality between them, as illustrated in Figure 10.3. Due to the reduction in bandwidth, each subcarrier in MC-DS-CDMA has either two paths (i.e., $P = 2$) or one path (i.e., $P = 1$, frequency-non-selective fading) for $N_c = 3$ and $N_c \geq 5$, respectively. The main performance criterion is the link-level SNR required per carrier to meet a BER of 5% in order to achieve a QoS of BER = 10^{-6} after channel decoding and the resulting system-level throughput. The user's data rate is calculated by summing the data rates over all subcarriers.

TABLE 10.2   Simulation Parameters

Parameter	Value	Comment
BW_{max}	5 MHz	Maximum bandwidth
M	2	Number of antennas
f_c	1.9 GHz	Central carrier frequency
f_D	8.8 Hz	Doppler frequency (5 kmph)
Δf	200 Hz	Frequency offset
f_{PC}	1,600 Hz	Frequency of PC updating
Δ_{PC}	±0.25 dB	Power control adjustment
PC_{min}^{max}	±30 dB	Power control stage
BER_{PC}	5%	Simulated PC bit error rate
$\delta\tau/\delta t$	0.049 ppm	Time-delay drift
$\Delta\tau$	4 chips	Delay spread
L_g	0	Guard interval length
N_s	2	Number of multistages
β	0.22	Roll-off factor
$2N_{src} + 1$	9	Number of pulse samples

10.4.2   Validation of the Performance Analysis

In this section, we investigate the accuracy of the analytical performance analysis in section 10.3.3 under realistic channel conditions. Indeed, we do not assume perfect channel identification; instead, we use the channel estimate provided by MC-STAR [32]. We validate the Gaussian approximation (GA) of the residual interference by comparison with simulation results. Since the SUB (Ω) is not known explicitly in this case, it has been obtained from extensive simulations. We consider two configurations: DBPSK MT-CDMA (L=64, N_c = 3) and DBPSK MC-DS-CDMA (L=32, N_c = 3). Figure 10.5 shows the link-level performances. It is seen, not surprisingly, that the GA is accurate in the presence of moderate background noise. The accuracy of GA increases at larger loads or at lower Doppler situations (speed of V = 5 kmph). Despite the realistic channel model employed and the channel estimate errors, there is a very good match between the analytical and simulation results for both MT-CDMA and MC-DS-CDMA in the target BER region (5%). This suggests that the analytical evaluation is accurate in a low Doppler situation.

10.4.3   Advantage of Full Interference Suppression

The imperfect frequency downconversion due to the instability of local oscillators combined with the multipath effect destructs the subcarriers' orthogonality and hence causes ICI. In this section we evaluate the advantage of full interference suppression on the link-level performance of MC-CDMA. We plot the link-level performances of MT-CDMA (L = 64, N_c = 3, C = 8) and MC-DS-CDMA (L = 32, N_c = 3, C = 8) with MC-MRC (i.e., multicarrier receiver with MRC combining) and MC-ISR in Figure 10.6. It is clear that MC-ISR performs better than MC-MRC. Indeed, in low Doppler (speed of

TABLE 10.3 Parameters of Each Multicarrier System Configuration

Parameter	DS-CDMA	MT-CDMA					MC-DS-CDMA					Comment
		1					L					
λ	—	1										Subcarrier spacing parameter
N_c	1	3	5	7	9	11	3	5	7	9	11	Number of subcarriers
L	32	64	128	192	256	320	32	32	32	32	32	Spreading factor
R_c in Mcps	3.840	3.840	1.4549	0.8975	0.6488	0.5081	0.4175					Chip rate
P	3	3	2	1	1	1	1					Number of paths per subcarrier
R_s in kbaud	120	180	150	140	135	132	136.4	140.2	141.9	142.9	143.5	Symbol rate over all subcarriers
R_b for DBPSK in kbps	120	180	150	140	135	132	136.4	140.2	141.9	142.9	143.5	Peak rate of DBPSK
R_b for D8PSK in kbps	360	540	450	420	405	396	409.2	420.6	425.7	428.7	430.5	Peak rate for D8PSK
BW_{nor}	1	1.026	1									Bandwidth normalized vs. DS-CDMA

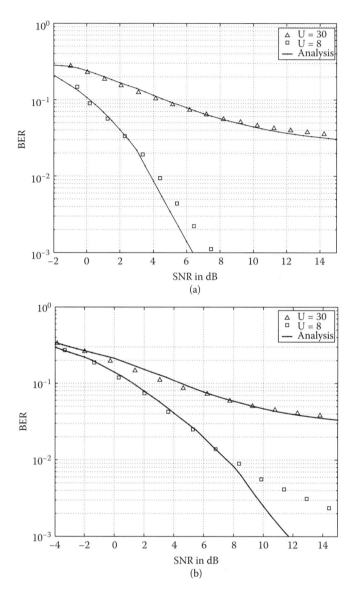

FIGURE 10.5 Analytical and simulated BER of MC-ISR versus SNR in dB for (a) MT-CDMA, $L = 64$, $N_c = 3$, DBPSK, and (b) MC-DS-CDMA $L = 32$, $N_c = 3$, DBPSK.

$V = 5$ kmph) we report 1.85 and 2 dB gains in SNR for MT-CDMA and MC-DS-CDMA, respectively. Note that the SNR gains are more important in high Doppler situations (speed of $V = 50$ kmph). At a bit error rate of 5%, MC-ISR performs 5.5 and 3 dB better than MC-MRC for MT-CDMA and MC-DS-CDMA, respectively.

In order to evaluate the specific impact of ICI on the link-level performance, we compare the BER curves of MT-CDMA ($L = 64$, $N_c = 3$, $C = 8$) MC-ISR with and without full

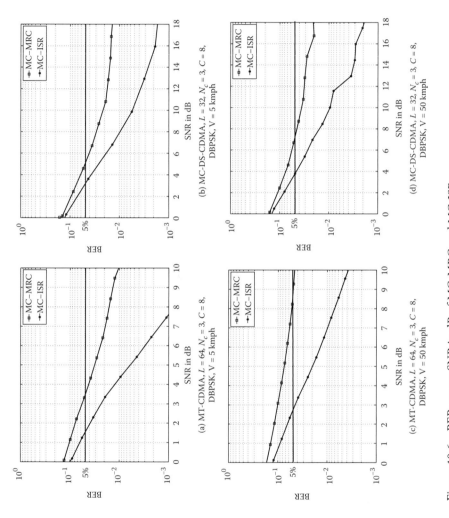

Figure 10.6   BER versus SNR in dB of MC-MRC and MC-ISR.

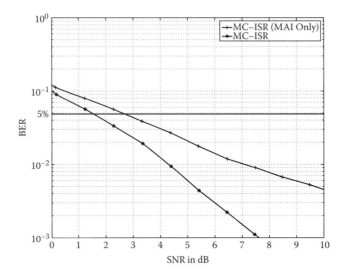

FIGURE 10.7   BER versus SNR in dB of MT-CDMA MC-ISR, $L = 64$, $N_c = 3$, $C = 8$, with and without full interference suppression (i.e., with and without ICI suppression).

interference suppression (i.e., with and without ICI suppression). Figure 10.7 shows that MC-ISR with full interference rejection is required to improve the system performance. Indeed, at a bit error rate of 5%, MC-ISR with full interference suppression performs 1.2 dB better than MC-ISR with MAI suppression only. In order to capture in more detail the gains achieved by ICI suppression in MC-CDMA, we proceed in Figure 10.8 to additional comparisons between the link-level BER performances of MC-ISR and MC-MRC in a single-user context (i.e., $C = 1$, no MAI, only ICI, and negligible ISI). Starting from the reference situation of Figure 10.8a with $L = 64$, $N_c = 7$, and DBSPK where the reported SNR gain due to ICI suppression is about 0.5 dB, the results suggest that ICI suppression is even more advantageous at higher-rate transmissions, and increasingly so when we move to the scenarios of Figures 10.8b–d; i.e., when we increase the number of carriers to $N_c = 11$ (SNR gain is about 1 dB), reduce the processing gain to $L = 32$ (SNR gain is about 3 dB), or increase the modulation order to D8PSK (SNR gain far exceeds 5 dB if not infinite), respectively. These results further confirm the benefits of ICI rejection in a full interference suppression scheme using MC-ISR.*

10.4.4   MT-CDMA, MC-DS-CDMA, and DS-CDMA Performance Comparison

This section is dedicated to the performance comparison of the proposed MC-ISR receiver with two potential next-generation multicarrier CDMA–air interface configurations: MT-CDMA and MC-DS-CDMA. Single-carrier ISR [42] for 3G DS-CDMA–air

---

* Simulation results performed in the framework of this contribution and reported in [50] show that the ICI rejection is even more beneficial in the case of one receiving antenna ($M = 1$).

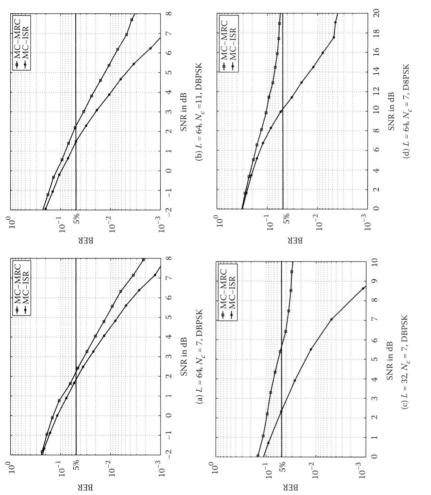

FIGURE 10.8   BER versus SNR in dB of MC-ISR and MC-MRC with single-user MT-CDMA.

interface is also considered as a reference. In addition, in order to provide a more detailed picture of the aggregate gain of the proposed MC-ISR receiver, we also compared its performance with that of MC-MRC over the same two multicarrier CDMA–air interface configurations, as well as with that of single-carrier MRC over current 3G DS-CDMA. First, we derive the SNR_{req} from link-level simulations. Then, we translate the link-level results into system-level results using the procedure in Table 10.1. In Table 10.4, we provide the required SNR and the total throughput of DBPSK and D8PSK modulated data for DS-CDMA, MT-CDMA, and MC-DS-CDMA. For DBPSK modulation, we observe that we can improve the system performance by increasing the number of subcarriers. Indeed, the total throughput continues to increase despite the increase in the number of carriers. But a gain saturation is encountered as the number of subcarriers increases. Note, however, that the throughput increase is more important with MC-ISR due to ICI suppression. Table 10.4 also shows that MT-CDMA outperforms MC-DS-CDMA with DBPSK modulation because it uses longer spreading sequences and exploits the subcarrier correlation. Moreover, due to the reduced subcarrier bandwidth, MC-DS-CDMA has less frequency diversity, while MT-CDMA is better able to exploit path diversity, and hence achieves better performance. Note also that MC-DS-CDMA is more robust against ICI, but in applying MC-ISR, this advantage over MT-CDMA becomes obsolete and the performance gap between MT-CDMA and MC-DS-CDMA increases.

Next, we compare different configurations with D8PSK modulation. We notice a link-level deterioration for MT-CDMA as the number of subcarriers increases. Indeed, higher-order modulation is more sensitive to the residual ICI. MC-DS-CDMA is much less affected by this phenomenon because it is much more robust to ICI thanks to the higher subcarrier spacing. Therefore, with high-order modulation, MC-DS-CDMA outperforms MT-CDMA when the number of subcarriers is high enough. We notice also that with the MC-MRC combiner, D8PSK MC-DS-CDMA outperforms D8PSK MT-CDMA even with a small number of subcarriers. It is clear, however, that D8PSK is less efficient than DBPSK modulation for all air interface configurations. In Table 10.4 we highlight the most spectrum-efficient MC-ISR–air interface configuration for each modulation. For both modulations MT-CDMA has the best link-level performance and the highest throughput (for a tested number of carriers less than or equal to 11). MT-CDMA with nine subcarriers and DBPSK modulation outperforms all other configurations and provides a throughput about 115% higher than that achievable with single-carrier MRC over a 3G DS-CDMA–air interface. The net benefits due to the proposed MC-ISR combiner and to the potential migration to a next-generation MT-CDMA–air interface are about 80 and 15%, respectively.

10.5   Conclusions

In this chapter we propose an adaptive multicarrier CDMA space-time receiver with full interference suppression capabilities named MC-ISR. First, we derived a complete model of the interference that takes into account MAI, ISI, and ICI in a multipath fading

TABLE 10.4 Required SNR and Maximum Throughput of DS-CDMA, MT-CDMA, and MC-CDMA for DBPSK and D8PSK (best performance values for each modulation are in bold)

MC-STAR Configuration	DS-CDMA	MT-CDMA					MC-DS-CDMA				
N_c	1	3	5	7	9	11	3	5	7	9	11
Modulation					DBPSK						
SNR_{req} in dB with MC-MRC	0.76	0.74	0.76	0.59	0.57	0.69	3.7	2.8	2.7	2.55	3
τ_{max} in kbps with MC-MRC	2,040	2,160	2,100	2,240	2,295	2,244	1091.2	1402	1419	1,571.9	1,435
SNR_{req} in dB with MC-ISR	0.76	0.74	0.75	0.55	0.5	0.62	3.62	2.8	2.6	2.48	3
τ_{max} in kbps with MC-ISR	3,600	3,960	4050	4,200	4,320	4,224	2,182.4	2,523.6	2,554.2	2,572.2	2,439.5
Modulation					DSPSK						
SNR_{req} in dB with MC-MRC	8.57	8	9.28	9	10	11	11	10.25	10.34	10.05	10.8
τ_{max} in kbps with MC-MRC	1,080	1,080	900	840	810	792	818.4	841.2	851.4	857.4	861
SNR_{req} in dB with MC-ISR	8.57	7.86	8.8	8.9	9.4	10.5	10.94	10.25	10.2	10.02	10.8
τ_{max} in kbps with MC-ISR	1,800	2,160	1,350	1,260	1,215	792	1,227.6	1,261.8	1,277.1	1,286.1	861

channel with timing and frequency mismatch. Based on this model, we proposed a new adaptive multicarrier interference subspace rejection receiver. We incorporated the least complex and more practical ISR interference rejection mode to simultaneously suppress MAI, ISI, and ICI at the signal combining step. We also proposed a realistic implementation of the new MC-ISR receiver that includes an efficient strategy for carrier offset recovery in a multicarrier and multiuser detection scheme. MC-ISR supports both the MT-CDMA– and MC-DS-CDMA–air interfaces. Furthermore, the assessment of the new MC-ISR receiver was oriented toward an implementation in a future, real-world wireless system. Indeed, we analyzed the performance of MC-ISR in an unknown time-varying Rayleigh channel with multipath, carrier offset, and cross-correlation between subcarrier channels and took into account all channel estimation errors. As another contribution in this work, we derived a link/system-level performance analysis of MC-ISR based on the Gaussian assumption (GA) and validated it by simulations. Under realistic propagation conditions and in the presence of channel estimation errors, simulation results validated the performance analysis and confirmed the net advantage of the full interference suppression capabilities of MC-ISR. With two receiving antennas and nine MT-CDMA subcarriers in 5 MHz bandwidth, MC-ISR provides about 4,320 kbps at low mobility for DBPSK, i.e., an increase of 115% in throughput over current 3G DS-CDMA with MRC.

Appendix

Derivation of the Interference Variance after MC-ISR Combining

Our goal is to estimate the variances:

$$\text{Var}\left[\delta^d_{MAI,k,n}+\delta^d_{ICI,k,n}+\delta^d_{ISI,k,n}\right]=\text{Var}\left[\sum_{\substack{u=1\\u\neq d}}^{C}\sum_{k'=-K}^{K}\sum_{n'=n-1}^{n+1}\xi^u_{k',n'}\lambda^u_{k',n}\underline{W}^{d^H}_{k,n}\underline{\hat{Y}}^u_{n',k',n}\right.$$
$$+\sum_{\substack{k'=-K\\k'\neq k}}^{K}\sum_{n'=n-1}^{n+1}\xi^d_{k',n'}\lambda^d_{k',n}\underline{W}^{d^H}_{k,n}\underline{\hat{Y}}^d_{n',k',n} \qquad (10.53)$$
$$+\left.\sum_{\substack{n'=n-1\\n'\neq n}}^{n+1}\xi^d_{k,n'}\lambda^d_{k,n}\underline{W}^{d^H}_{k,n}\underline{\hat{Y}}^d_{n',k,n}\right].$$

Let us consider the general problem of deriving the variance of the sum of random complex variables. We first introduce the variables $x_\alpha, \alpha\in\{1,...,N_t\}$ and $\xi_\alpha, \alpha\in\{1,...,N_t\}$, with the following properties: $E[\xi_\alpha\xi_{\alpha'}^*]=M_\xi, \forall\alpha\neq\alpha', E[\xi_\alpha\xi_\alpha^*]=V_\xi, E[x_\alpha]=0$, and $\text{Var}[\Sigma_{\alpha=1}^{N_t}x_\alpha]=0$. Then we assume that ξ_α and x_α are independent. Thus, we derive the variance as follows:

$$\text{Var}\left[\sum_{\alpha=1}^{N_t}\xi_\alpha x_\alpha\right]=\sum_{\alpha=1}^{N_t}\text{Var}\left[\xi_\alpha x_\alpha\right]+\sum_{\alpha=1}^{N_t}\sum_{\alpha'=1\alpha'\neq\alpha}^{N_t}\text{E}\left[\xi_\alpha\xi_{\alpha'}x_\alpha x_{\alpha'}\right]$$

$$=\sum_{\alpha=1}^{N_t}V_\xi\text{Var}\left[x_\alpha\right]+\sum_{\alpha=1}^{N_t}\sum_{\alpha'=1\alpha'\neq\alpha}^{N_t}\text{E}\left[\xi_\alpha\xi_{\alpha'}\right]\text{E}\left[x_\alpha x_{\alpha'}\right] \quad (10.54)$$

$$=\sum_{\alpha=1}^{N_t}V_\xi\text{Var}\left[x_\alpha\right]+\sum_{\alpha=1}^{N_t}\sum_{\alpha'=1\alpha'\neq\alpha}^{N_t}\rho_\xi\text{E}\left[x_\alpha x_{\alpha'}\right].$$

From $\text{Var}[\sum_{\alpha=1}^{N_t}x_\alpha]=0$ we have

$$\text{Var}\left[\sum_{\alpha=1}^{N_t}x_\alpha\right]=\sum_{\alpha=1}^{N_t}\text{Var}\left[x_\alpha\right]+\sum_{\alpha=1}^{N_t}\sum_{\alpha'=1\alpha'\neq\alpha}^{N_t}\text{E}\left[x_\alpha x_{\alpha'}\right]=0$$

$$\Rightarrow\sum_{\alpha=1}^{N_t}\sum_{\alpha'=1\alpha'\neq\alpha}^{N_t}\text{E}\left[x_\alpha x_{\alpha'}\right]=-\sum_{\alpha=1}^{N_t}\text{Var}\left[x_\alpha\right].$$

(10.55)

Then, by substituting equation (10.55) in equation (10.54) we obtain

$$\text{Var}\left[\sum_{\alpha=1}^{N_t}\xi_\alpha x_\alpha\right]=\left(V_\xi-\rho_\xi\right)\sum_{\alpha=1}^{N_t}\text{Var}\left[x_\alpha\right]. \quad (10.56)$$

Now we apply the same procedure to derive the variance of $\delta^d_{MAI,k,n}+\delta^d_{ICI,k,n}+\delta^d_{ISI,k,n}$. We substitute ξ_α with $\xi^u_{k',n'}\lambda^u_{k',n}$ and x_α with $\underline{W}^{dH}_{k,n}\hat{\underline{Y}}^u_{n',k'n}$. The MC-ISR combiner $\underline{W}^d_{k,n}$ satisfies the optimization property in equation (10.25); thus,

$$\underline{W}^{d^H}_{k,n}\hat{\underline{I}}^d_{k,n}=0 \quad\Rightarrow\quad \text{Var}\left[\underline{W}^{d^H}_{k,n}\left(\hat{\underline{I}}^d_{MAI,k,n}+\hat{\underline{I}}^d_{ICI,k,n}+\hat{\underline{I}}^d_{ISI,,k,n}\right)\right]=0. \quad (10.57)$$

Then,

$$\text{Var}\left[\delta^d_{MAI,k,n}+\delta^d_{ICI,k,n}+\delta^d_{ISI,k,n}\right]=\left(V_\xi-\rho_\xi\right)\sum_{\substack{u=1\\u\neq d}}^{C}\sum_{k'=-K}^{K}\sum_{n'=n-1}^{n+1}\text{Var}\left[\underline{W}^{d^H}_{k,n}\hat{\underline{Y}}^u_{n',k',n}\right]$$

$$+\left(V_\xi-\rho_\xi\right)\sum_{\substack{k'=-K\\k'\neq k}}^{K}\sum_{n'=n-1}^{n+1}\text{Var}\left[\underline{W}^{d^H}_{k,n}\hat{\underline{Y}}^d_{n',k',n}\right] \quad (10.58)$$

$$+\left(V_\xi-\rho_\xi\right)\sum_{\substack{n'=n-1\\n'\neq n}}^{n+1}\text{Var}\left[\underline{W}^{d^H}_{k,n}\hat{\underline{Y}}^d_{n',k,n}\right].$$

We consider $E[\|\underline{W}_{k,n}^d\|^2] = \bar{\kappa}$, which is a measure of the enhancement of the white noise compared to the MRC combiner [45]. We also assume that the combiners $\underline{W}_{k,n}^d$ and $\hat{\underline{Y}}_{n',k'n}^u$ Thus, we derive the variance of the residual interference as follows:

$$\mathrm{Var}\left[\delta_{MAI,k,n}^d + \delta_{ICI,k,n}^d + \delta_{ISI,k,n}^d\right] = \left(V_\xi - \rho_\xi\right)\bar{\kappa}\sum_{\substack{u=1 \\ u \neq d}}^{C}\sum_{\substack{k'=-K}}^{K}\sum_{n'=n-1}^{n+1}\mathrm{Var}\left[\hat{Y}_{n',k',n}^u\right]$$

$$+ \left(V_\xi - \rho_\xi\right)\bar{\kappa}\sum_{\substack{k'=-K \\ k' \neq k}}^{K}\sum_{n'=n-1}^{n+1}\mathrm{Var}\left[\hat{Y}_{n',k',n}^d\right]$$

$$\hspace{7cm}(10.59)$$

$$+ \left(V_\xi - \rho_\xi\right)\bar{\kappa}\sum_{\substack{n'=n-1 \\ n' \neq n}}^{n+1}\mathrm{Var}\left[\hat{Y}_{n',k,n}^d\right],$$

$$= \left(V_\xi - \rho_\xi\right)\bar{\kappa}\,\mathrm{Var}\left[\hat{I}_{k,n}^d\right].$$

In the developments of equation (10.59), we exploited the fact that we transmit different data sequences over distinct subcarriers for a given user, and hence assumed that the cross-correlation terms from different subcarriers are zero.

In the following, we will derive the values of V_ξ and ρ_ξ under the following three assumptions: (1) the error indicating variables $\xi_{k',n'}^u$ and $\lambda_{k',n}^u$ are independent; (2) all the random sequence variables $(\xi_{k',n'}^u)$ and $(\lambda_{k',n}^u)$ are independent and identically distributed; and (3) $E[\lambda_{k',n}^u]$. Given these assumptions we derive V_ξ as follows:

$$V_\xi = E\left[\xi_{k',n'}^u \lambda_{k',n}^u \xi_{k',n'}^{u*} \lambda_{k',n}^{u*}\right] = E\left[\xi_{k',n'}^{u*} \xi_{k',n'}^u\right] E\left[\lambda_{k',n}^u \lambda_{k',n}^{u*}\right]$$

$$= E\left[\lambda_{k',n}^u \lambda_{k',n}^{u*}\right] = \left[1 + \rho_\lambda\right].$$

$$\hspace{7cm}(10.60)$$

In order to evaluate V_ξ, we exploit the expression of the variance of the power control error in [51]. Hence, ρ_λ varies with the maximum Doppler frequency f_D (equation (10.51) in [51]), yielding:

$$\rho_\lambda = \frac{4\pi^2\left(F_D \times \tau_{PC}\right)^2}{P-1},\hspace{3cm}(10.61)$$

where τ_{PC} is the power control feedback delay. Below we derive the expectation

$$\rho_\xi = E\left[\xi_{k',n'}^u \lambda_{k',n}^u \xi_{k',n'}^{u*} \lambda_{k',n}^{u*}\right] = E\left[\xi_{k',n'}^u\right] E\left[\xi_{k',n'}^{u*}\right] E\left[\lambda_{k',n}^u\right] E\left[\lambda_{k',n}^{u*}\right] = E\left[\xi_{k',n'}^u\right]^2.$$

If $S_{rec} \ll 1$, the value of ρ_ξ can be derived as follows [45]:

$$\rho_\xi \approx \left(1 - \left(1 - \cos\left(2\pi/M_i\right)\right)S_{rec}\right)^2.\hspace{3cm}(10.62)$$

The interference $\underline{I}^d_{k,n}$ is approximated as a Gaussian distributed random variable with zero mean. Only its variance needs to be evaluated to derive the variance of the residual interference in equation (10.59).

Derivation of the Interference Variance for Band-Limited MC-CDMA

The chip waveform has been noted to be an important system parameter for DS-CDMA and MC-DS-CDMA. Hence, the performances of DS-CDMA and MC-DS-CDMA with various time-limited and band-limited chip waveforms have been investigated. However, for all the MT-CDMA systems found in the literature, a time-limited waveform is generally employed [4, 18, 52, 53]. Since we consider a practical square-root raised cosine pulse, the focus of this appendix is to derive the variance of the interference of MC-CDMA (including MC-DS-CDMA and MT-CDMA) with a band-limited square-root raised cosine waveform. Let $G(f)$ be the Fourier transform of the raised cosine filter:

$$G(f) = \begin{cases} T_c, & 0 \le |f| \le \dfrac{1-\beta}{2T_c} \\[3mm] \dfrac{T_c}{2}\left\{1+\cos\left[\dfrac{\pi T_c}{\beta}\left(|f|-\dfrac{1-\beta}{2T_c}\right)\right]\right\}, & \dfrac{1-\beta}{2T_c} \le |f| \le \dfrac{1+\beta}{2T_c} \\[3mm] 0, & |f| > \dfrac{1+\beta}{2T_c} \end{cases} \tag{10.63}$$

Let $\bar{\psi}_D^2 = E[(\psi_k^d)^2]$ be the average power of the kth carrier of the desired user and $\bar{\psi}_I^2$ be the average power on each interfering carrier (assumed equal for all u and all k). Using the general results in [54], one has

$$\mathrm{Var}\left[\underline{I}^d_{MAI,k,n}\right] = (C-1)\frac{\bar{\psi}_I^2}{L}\left[\varsigma(\beta)+\chi_k(\beta)\right], \tag{10.64}$$

where

$$\varsigma(\beta) = \frac{1}{T_c}\int_{-\infty}^{\infty} G^2(f)\,df \tag{10.65}$$

and

$$\chi_k(\beta) = \sum_{\substack{k'=-K \\ k'\neq k}}^{K} \frac{1}{T_c}\int_{-\infty}^{\infty} G(f)G\big(f-(f_k-f_{k'})\big)\,df. \tag{10.66}$$

It is easy to obtain $\varsigma(\beta) = 1 - \beta/4$. To obtain $\chi_k(\beta)$, we need to consider the MC-DS-CDMA and MT-CDMA systems separately. After mathematical evaluations of the integral, we obtain:

$$\chi_k(\beta) = \begin{cases} \dfrac{\beta}{8}, & \text{if } k=-K \text{ or } K, \\[2mm] \dfrac{\beta}{4}, & \text{if } k=-K+1,\ldots,K-1, \end{cases} \tag{10.67}$$

for MC-DS-CDMA ($\lambda = L$ and $f_k = k/T_c$) and

$$\chi_k(\beta) = \sum_{\substack{k'=-K \\ k' \neq k}}^{K} \vartheta\left(\left|k-k'\right|\right), \tag{10.68}$$

where

$$\vartheta(x) = \begin{cases} 1 - \dfrac{\beta}{2} - \dfrac{x}{2L} + \dfrac{3\beta}{4\pi}\sin\left(\dfrac{\pi x}{\beta L}\right) + \left(\dfrac{\beta}{4} - \dfrac{x}{4L}\right)\cos\left(\dfrac{\pi x}{\beta L}\right), & \text{if } 0 \leq x/L \leq \min(\beta, 1-\beta) \\[3mm] 1 - \dfrac{x}{L} & \text{if } \beta \leq x/L \leq 1-\beta \text{ and } \beta < 0.5 \\[3mm] \begin{aligned} & \dfrac{3}{4} - \dfrac{\beta}{4} - \dfrac{x}{4L} + \dfrac{3\beta}{4\pi}\sin\left(\dfrac{\pi x}{\beta L}\right) + \left(\dfrac{\beta}{4} - \dfrac{x}{4L}\right)\cos\left(\dfrac{\pi x}{\beta L}\right) + \\[2mm] & \dfrac{3\beta}{8\pi}\sin\left(\dfrac{\pi x}{\beta L} - \dfrac{\pi}{\beta}\right) - \left(\dfrac{x}{8L} - \dfrac{1-\beta}{8}\right)\cos\left(\dfrac{\pi x}{\beta L} - \dfrac{\pi}{\beta}\right) \end{aligned} & \text{if } 1-\beta \leq x/L \leq \beta \text{ and } \beta > 0.5 \\[3mm] \dfrac{3}{4} + \dfrac{\beta}{4} - \dfrac{3x}{4L} + \dfrac{3\beta}{8L}\sin\left(\dfrac{\pi x}{\beta L} - \dfrac{\pi}{\beta}\right) - \left(\dfrac{x}{8L} - \dfrac{1-\beta}{8}\right)\cos\left(\dfrac{\pi x}{\beta L} - \dfrac{\pi}{\beta}\right) & \text{if } \max(\beta, 1-\beta) \leq x/L \leq 1 \end{cases} \tag{10.69}$$

for MT-CDMA ($\lambda = L$ and $f_k = k/T_{Mc}$). The variances of the residual ICI and ISI interferences from the same user can be written as

$$\text{Var}\left[I_{ICI,k,n}^d\right] = \frac{\overline{\Psi}_D^2}{L}\delta_{is}\chi_k(\beta),$$

$$\text{Var}\left[I_{ISI,k,n}^d\right] = \frac{\overline{\Psi}_D^2}{L}\delta_{is}\varsigma(\beta), \tag{10.70}$$

where $\delta_{is} = (P-1)/P$ is a measure of the relative impact of the interference generated by the other paths on a given path of the desired user.

References

[1] L. Hanzo, T. Killer, M. S. Munster, and B. J. Choi. 2003. *OFDM and MC-CDMA for broadband multiuser communications, WLANs and broadcasting.* New York: John Wiley & Sons.

[2] S. Hata and R. Prasad. 1997. Overview of multicarrier CDMA. *IEEE Commun. Mag.* 35:126–33.

[3] L. Hanzo, L.-L. Yang, E.-L. Kuan, and K. Yen. 2003. *Single- and multi-carrier CDMA multi-user detection, space-time spreading, synchronisation and standards.* New York: John Wiley & Sons.

[4] Y. Lie-Liang and L. Hanzo. 2002. Performance of generalized multicarrier DS-CDMA over Nakagami-m fading channels. *IEEE Trans. Commun.* 50:956–66.

[5] H. Steendam and M. Moeneclaey. 2001. The effect of carrier frequency offsets on downlink and uplink MC-DS-CDMA. *IEEE J. Selected Areas Commun.* 19:2528–36.

[6] B. Smida, S. Affes, K. Jamaoui, and P. Mermelstein. 2008. A multicarrier-CDMA space-time receiver with full interference suppression capabilities. *IEEE Trans. Veh. Technol.* 57(1):363–379.

[7] N. Yee, J. P. Linnarz, and G. Fettweis. 1993. Multi-carrier CDMA in indoor wireless radio networks. In *Proceedings of the IEEE International Symposium on Personal, Indoor and Mobile Radio Communications (PIMRC 1993)*, pp. 109–13.

[8] A. Chouly, A. Brajal, and S. Jourdan. 1993. Orthogonal multicarrier techniques applied to direct sequence spread spectrum CDMA systems. In *Proceedings of the IEEE Global Telecommunications Conference (GLOBECOM 1993)*, vol. 3, pp. 1723–28.

[9] K. Fazel and L. Papke. 1993. On the performance of convolutionally-coded CDMA/OFDM for mobile communication system. In *Proceedings of the IEEE Personal, Indoor and Mobile Radio Communications Symposium (PIMRC 1993)*, pp. 468–72.

[10] A. Chouly, A. Brajal, and S. Jourdan. 1993. Orthogonal multicarrier technique applied to direct sequence spread spectrum CDMA systems. In *Proceedings of the IEEE GLOBECOM '93*, pp. 1723–28.

[11] V. DaSilva and E. S. Sousa. 1993. Performance of orthogonal CDMA codes for quasi-synchronous communication systems. In *Proceedings of the IEEE International Conference on Universal Personal Communications (ICUPC 1993)*, vol. 2, pp. 995–99.

[12] S. Kondo and L. B. Milstein. 1993. On the use of multicarrier direct sequence spread spectrum systems. In *Proceedings of the IEEE Military Communinations Conference (MILCOM 1993)*, vol. 1, pp. 52–56.

[13] E. A. Sourour and M. Nakagawa. 1996. Performance of orthogonal multicarrier CDMA in a multipath fading channel. *IEEE Trans. Commun.* 44:356–67.

[14] X. Gui and T. S. Ng. 1999. Performance of asynchronous orthogonal multicarrier CDMA system in frequency selective fading channel. *IEEE Trans. Commun.* 47:1084–91.

[15] L. L. Yang and L. Hanzo. 2002. Broadband MC DS-CDMA using space-time and frequency-domain spreading. In *Proceedings of the IEEE VTC'2002*, pp. 1632–36.

[16] Z. Luo, J. Liu, M. Zhao, Y. Liu, and J. Gao. 2006. Double-orthogonal coded space-time-frequency spreading CDMA scheme. *IEEE J. Selected Areas Commun.* 24:1244–55.

[17] L. Vandendorpe. 1993. Multitone direct sequence CDMA system in an indoor wireless environment. In *Proceedings of the IEEE First Symposium on Communications and Vehicular Technology in the Beneluz*, pp. 4.1.1–4.1.8.

[18] L. Vandendorpe. 1995. Multitone spread spectrum multiple access communications system in a multipath Rician fading channel. *IEEE Trans. Veh. Technol.* 44:327–37.

[19] S. Kaiser. 1995. OFDM-CDMA versus DS-CDMA: Performance evaluation for fading channels. In *Proceedings of the IEEE International Conference on Communications (ICC 1995)*, vol. 3, pp. 1722–26.

[20] J.-Y. Oh and M.-S. Lim. 1999. The bandwidth efficiency increasing method of multi-carrier CDMA and its performance evaluation in comparison with DS-CDMA with RAKE receivers. In *Proceedings of the IEEE Vehicular Technology Conference (VTC 1999)*, vol. 1, pp. 561–65.

[21] S. B. Slimane. 1999. Bandwidth efficiency of MC-CDMA signals. *IEEE Electronics Lett.* 35:1797–98.

[22] X. Zhang, T.-S. Ng, and J. Wang. 1999. Capacity comparison of single-tone and multitone CDMA systems. In *Proceedings of the IEEE Vehicular Technology Conference (VTC 1999)*, vol. 1, pp. 243–47.

[23] S. Hata and R. Prasad. 1996. An overview of multi-carrier CDMA. In *Proceedings of the IEEE International Symposium on Spread Spectrum Techniques and Applications*, vol. 1, pp. 107–14.

[24] L. L. Yang and L. Hanzo. 2003. Multicarrier DS-CDMA: A multiple access scheme for ubiquitous broadband wireless communications. *IEEE Commun. Mag.* 41:116–24.

[25] S. Affes and P. Mermelstein. 2003. Adaptive space-time processing for wireless CDMA. In *Adaptive signal processing: Application to real-world problems*, ed. J. Benesty and A. H. Huang. Berlin: Springer, chapter 10.

[26] 3GPP. 1997. *Universal Mobile Telecommunications System (UMTS); Selection procedures for the choice of radio transmission technologies of the UMTS.* 3GPP TR 101 112 V3.1.0.

[27] S. Affes, D. Feng, L. Ge, and P. Mermelstein. 2002. Does direction-of-arrival estimation help channel identification in multi-antenna CDMA receivers? In *IEEE ISWC'2002*, pp. 167–68.

[28] W. C. Jakes, ed. 1974. *Microwave mobile communications.* New York: Wiley.

[29] B. Natarajan, C. R. Nassar, and V. Chandrasekhar. 2000. Generation of correlated Rayleigh fading envelopes for spread spectrum applications. *IEEE Commun. Lett.* 4:9–11.

[30] R. B. Ertel and J. H. Reed. 1998. Generation of two equal power correlated Rayleigh fading envelopes. *IEEE Commun. Lett.* 2:276–78.

[31] S. Affes and P. Mermelstein. 1998. A new receiver structure for asynchronous CDMA: STAR the spatio-temporal array-receiver. *IEEE J. Selected Areas Commun.* 16:1411–22.

[32] B. Smida, S. Affes, J. Li, and P. Mermelstein. 2005. Multicarrier-CDMA STAR with time and frequency synchronization. In *IEEE ICC 2005*, vol. 4, pp. 2493–99.

[33] S. Verdu. 1986. Minimum probability of error for asynchronous Gaussian multiple access channels. *IEEE Trans. Information Theory* 32:85–96.

[34] X. Weiping and L. B. Milstein. 1998. MMSE interference suppression for multicarrier DS-CDMA in frequency selective channels. In *IEEE GLOBECOM 1998*, vol. 1, pp. 259–64.

[35] H. Lie-Liang, H. Wei, and L. Hanzo. 2003. Multiuser detection in multicarrier CDMA systems employing both time-domain and frequency-domain spreading. In *IEEE PIMRC 2003*, vol. 2, pp. 1840–44.

[36] F. Lin and L. B. Milstein. 2000. Successive interference cancellation in multicarrier DS/CDMA. *IEEE Trans. Commun.* 48:1530–40.

[37] W. Huahui, K. W. Ang, K. Yen, and Y. H. Chew. 2003. An adaptive PIC receiver with diversity combining for MC-DS-CDMA system. In *IEEE VTC 2003*, vol. 2, pp. 1055–59.

[38] W. Huahui, K. W. Ang, K. Yen, and Y. H. Chew. 2000. Performance analysis of an adaptive PIC receiver for asynchronous multicarrier DS-CDMA system. In *IEEE PIMRC 2003*, vol. 2, pp. 1835–39.

[39] W. Nabhane and H. V. Poor. Blind joint equalization and multiuser detection in dispersive MCCDMA/MC-DS-CDMA/MT-CDMA channels. In *Proceedings of the 2002 IEEE Military Communications Conference*, Anaheim, CA, vol. 2, pp. 814–819.

[40] J. Namgoong, T. F. Wong, and J. S. Lehnert. 2002. Subspace multiuser detection for multicarrier DS-CDMA. *IEEE Trans. Commun.* 48:1892–908.

[41] M. Peng, Y. J. Guo, and S. K. Barton. 2001. Multiuser detection of asynchronous CDMA with frequency offset. *IEEE Trans. Commun.* 49:952–60.

[42] S. Affes, H. Hansen, and P. Mermelstein. 2002. Interference subspace rejection: A framework for multiuser detection in wideband CDMA. *IEEE J. Selected Areas Commun.* 20:287–302.

[43] B. Smida, S. Affes, J. Li, and P. Mermelstein. 2007. A spectrum-efficient multicarrier CDMA array-receiver with diversity-based enhanced time and frequency synchronization. *IEEE Trans. Wireless Commun.* 6:2315–27.

[44] S. Jomphe, K. Cheikhrouhou, J. Belzile, S. Affes, and J.-C. Thibault. 2005. Area-efficient advanced multiuser WCDMA receiver. In *IEEE Northeast Workshop on Circuits and Systems NEWCAS'05*, pp. 296–99.

[45] H. Hansen, S. Affes, and P. Mermelstein. 2005. *Mathematical derivation of interference subspace rejection.* Technical Report INRS-EMT, EMT-014-0105.

[46] B. Smida and S. Affes. 2005. On the performance of interference subspace rejection for next generation multicarrier CDMA. In *IEEE ISSPA'2005*.

[47] T. S. Rappaport. 1999. *Wireless commmunications: Principles and practice.* Upper Saddle River, NJ: Prentice Hall PTR, pp. 417–435.

[48] B. Smida, S. Affes, and P. Mermelstein. 2003. Joint time-delay and frequency offset synchronization for CDMA array-receivers. In *IEEE SPAWC 2003*, pp. 499–504.

[49] 3rd Generation Partnership Project (3GPP). 2002. *Technical specification group (TSG), radio access network (RAN), UE radio transmission and reception (FDD).* 3GPP TS 25.101 V5.4.0.

[50] B. Smida, S. Affes, K. Jamaoui, and P. Mermelstein. 2005. A multicarrier-CDMA receiver with full interference suppression and carrier frequency offset recovery. In *IEEE SPAWC'05*, pp. 435–439.

[51] A. Abrardo and D. Sennati. 2000. On the analytical evaluation of closed-loop power-control error statistics in DS-CDMA cellular systems. *IEEE Trans. Veh. Technol.* 49:2071–80.

[52] Y. Lie-Liang and L. Hanzo. 2003. Performance of generalized multicarrier DS-CDMA using various chip waveforms. *IEEE Trans. Commun.* 51:748–52.

[53] Q. M. Rahman and A. B. Sesay. 2001. Performance analysis of MT-CDMA system with diversity combining. *In IEEE MILCOM 2001*, vol. 2, pp. 1360–1364.

[54] S. Kondo and B. Milstein. 1996. Performance of multicarrier DS CDMA systems. *IEEE Trans. Commun.* 44:238–246.

11

Cooperative Communications in Random Access Networks

Y.-W. Peter Hong
National Tsing Hua University

Shu-Hsien Wang
National Tsing Hua University

Chun-Kuang Lin
National Tsing Hua University

Bo-Yu Chang
National Tsing Hua University

11.1 Introduction

Multiple-input multiple-output (MIMO) systems [1–4] have been studied extensively in the literature to combat fading in wireless communication systems. By mounting multiple antennas on each communication terminal and by having them placed sufficiently apart on the device, each user is able to gain access to multiple independent fading paths, which can be exploited to improve communication efficiency. With access to the independent fading paths, diversity and multiplexing gains [5] can be achieved by transmitting the same message on different paths [2] or by exploiting the additional degrees of freedom to construct multiple parallel transmission channels [3]. The gains increase with the number of transmit and receive antennas. However, in practice, it is difficult to place multiple antennas on a single terminal while trying to maintain independence among the different fading paths, not to mention the cost of the devices. This is especially true as the dimension of the devices decreases. For this reason, cooperative communications have been proposed to achieve MIMO gains without requiring multiple antennas on each terminal.

Cooperative communications [e.g., 6–12] allow users in the system to cooperate by relaying each other's messages to the destination. With only one antenna on each device, the users may cooperatively form a distributed antenna array to emulate the spatial diversity and multiplexing gains in conventional MIMO systems. Most works in the literature on cooperative communications focus on the physical layer designs, such as the coding, modulation, and diversity-combining techniques. However, it is equally important to study the subject from a medium access control (MAC) layer perspective, in which case the system throughput and the efficiency of resource allocation are considered instead of the outage or error probabilities. In fact, the major drawback of cooperative systems is the loss of bandwidth efficiency due to the need for interuser coordination since the antennas are deployed at distributed locations. That is, the messages exchanged for coordination occupy the bandwidth resources that otherwise could have been used for noncooperative transmissions. Therefore, MAC protocols designed for cooperative systems, i.e., protocols that allocate the resources for cooperative transmissions or interuser coordination, may have a notable impact on the system performance. In this chapter, we provide a comprehensive survey of MAC protocols for cooperative random access networks. In contrast to most works in the literature, where a perfect centralized scheduling is usually assumed, we focus on the random access scenario where users decide locally when to transmit their own packets and when to cooperate.

Random access protocols allow users in the system to access the network without complex interactions with a centralized control. Essentially, users access the channel based only on some simple local rule whenever they have a packet to transmit and devise strategies to deal with interference or collisions after they occur. This is in contrast to centralized scheduling where collision and interference are prevented with the additional cost of communication overhead. Random access protocols are particularly useful for cooperative systems that span over a large network, such as that of sensor and ad hoc networks, where centralized coordination may be costly. However, the advantages of cooperation are not evident in random access networks since no centralized control is available to coordinate their cooperative operations.

In this chapter, we shall discuss the use of cooperative transmissions in four different random access systems: (1) the slotted ALOHA system [13, 14], (2) CSMA/CA in IEEE 802 legacy systems [15, 16], (3) the random access system with collision resolution [17, 18], and (4) the distributed ad hoc network [9, 19, 20]. Specifically, for the slotted ALOHA system, cooperative transmission and queuing schemes have been proposed in [13, 14] where each user is allowed to record the packet transmitted by its partners if the partner failed in its previous transmission attempt. The user recording the packet may then relay the message to the destination in subsequent time slots. The queue stability is used as the performance measure in these works. In [15, 16], similar cooperative relaying mechanisms have also been considered for the IEEE 802 legacy system. These schemes exploit the benefits of cooperative relaying, but do not consider the receiver's ability to perform signal combining. In [18], the authors proposed advanced signal processing methods to resolve collision resolution at the destination by modeling the problem as a multiuser detection or source separation problem. The performance of this scheme is further enhanced with cooperative relaying as proposed in [17]. Furthermore, in [9, 19, 20], cooperation is used to facilitate the broadcasting operations in large-scale distributed ad hoc networks. Specifically, with users cooperatively relaying the same signal (i.e., broadcasting), the aggregate signal arriving at each receiver can be simply treated as a multipath signal and exploited through diversity-combining techniques. Through these discussions, we demonstrate that, in random access networks, cooperation can be performed opportunistically rather than being fully coordinated.

11.2 Fundamentals of Cooperative Communications

The basic operations of a cooperative communication system can be described with the canonical two-user example illustrated in Figure 11.1. Suppose we have two users, user 1 and user 2, that are transmitting to the common destination D. At any instant in time, one of the users will serve as the source, while the other user cooperates by relaying the message from the source to the destination. As a result, the destination receives two packets encoded from the same message, one directly from the source and the other from the relay. The destination fails to decode the message only if the channels on both paths are unreliable at the time of transmission.

Based on relaying, many cooperation schemes have been proposed in the literature, such as amplify-and-forward (AF), decode-and-forward (DF) [7], coded cooperation [10, 21], quantize-and-forward, compress-and-forward [22], etc. The most intuitive and widely applied methods are the AF and DF schemes. In the AF scheme, the users, which play the role of the relays, simply amplify the signal that they receive without explicitly decoding the messages. On the other hand, in the DF scheme, the relays first decode and retransmit either a repetition of the original message or a re-encoded message to improve the coding efficiency. If the message is re-encoded (with forward error correction capabilities), this scheme is then referred to as the coded cooperation scheme. All of these methods can be employed with the addition of distributed space-time coding or

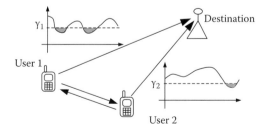

FIGURE 11.1 Cooperative communications with two users and one destination.

antenna selection techniques. Moreover, when the channel state information (CSI) can be further exploited at the transmitters, methods such as power allocation, precoding, or transmit beamforming can also be applied [11]. Details of these methods can be found in [6–11] and the references therein.

These cooperation methods can be easily extended to a network with $N > 2$ number of users. One approach is to divide the network into cooperating pairs that cooperate according to the mechanics shown in Figure 11.1. Another approach is to have, at each instant in time, one user serve as the source while the remaining users relay cooperatively the messages to the destination. In the first case, the cooperating pairs can be viewed as a super-user contributing a sum traffic to the network, while in the latter case, the relaying users effectively form a distributed antenna array possibly providing diversity and multiplexing gains for the source [12], similar to that in conventional multiple-antenna systems. Any combination of these two approaches can be taken as well.

Before each cooperative transmission, a coordination phase is required to synchronize the participating users. Inevitably, additional bandwidth and power resources are consumed in this process, as compared to the conventional MIMO system. In fact, this is the main cause of inefficiency in cooperative systems, but is often overcome by the significant gains in reliability and throughput. Two coordination methods are considered in the literature: coordination through direct interuser communication and coordination using feedback from the destination. In the first case, the users simply transmit their messages to each other and indicate the coding or modulation methods that should be used [7–9, 19, 23]. In the latter case, the feedback from destination is used to control and coordinate the messages transmitted by the users [15, 18, 24, 25].

In the past, most work on cooperative communications focused largely on the physical layer signal processing, such as the coding, modulation, diversity-combining schemes, power control, etc. A tutorial on these cooperative communication schemes can be found in [10, 11, 26]. However, most of the systems proposed in the literature assume strict synchronization between users or require a high degree of coordination, which is hard to achieve in practice. In the following, we provide examples of how cooperative communications can be applied to random access networks with reduced synchronization requirements.

11.3   Cooperation in Slotted ALOHA Random Access Networks

In random access networks (RAN), the users' transmissions are based only on simple local rules, with minimal interference from the central control. The lack of centralized control may result in collision between users and, in turn, reduce the system throughput. However, the simplicity, scalability, and robustness of random access protocols make RAN a favorable choice for low-traffic local area networks or multihop ad hoc networks. While cooperation is shown to be advantageous with perfect synchronization, it is not obvious whether cooperation can be helpful in random access networks and, if so, by how much. In fact, since the users behave in a decentralized manner, cooperation can only be achieved opportunistically in these scenarios. In this section, we study the effect of cooperation on one of the most fundamental random access protocols—the slotted ALOHA system [27].

11.3.1   Slotted ALOHA in a Wireless Network

In this section, let us start by reviewing the conventional slotted ALOHA system before incorporating the concept of cooperation in later sections.

Consider a network with N users, denoted by the set $\mathcal{U} = \{1, 2, \cdots, N\}$, transmitting to an access point (AP) through the wireless fading channel, as shown in Figure 11.2. In the slotted ALOHA system, the time is divided into equal-length time slots with the duration equal to the transmission time of one packet. The beginning and ending of each time slot are synchronized for all users, and a transmission can only occur at the beginning of a time slot. Let us assume, without loss of generality, that the length of each time slot is equal to 1 such that the transmission in the mth time slot occurs during the time $t \in [m, m+1)$. If a user, say user i, has a packet to transmit, it will transmit in the current time slot with probability p_i, which is independent from all users and over all time slots. If more than one user is transmitting in the same time slot, a collision will occur and no packet will be successfully received at the AP.

In contrast to conventional slotted ALOHA systems, where a transmission is assumed to be successful as long as there is no collision, the transmission of a packet in a wireless system may fail due to channel fading. Suppose that the channel experienced by user i in the mth time slot is parameterized by the variable $\gamma_i[m]$. For example, $\gamma_i[m]$ may be the signal-to-noise ratio (SNR) of the signal received at the AP corresponding to the transmission by user i in the mth time slot. Assume that $\gamma_i[m]$ is independent and identically distributed (i.i.d.) over time with the distribution function F_{γ_i}. Given that user i is the only user transmitting in the current time slot, the probability that the destination receives the packet correctly can be modeled by the probability $\Psi(\gamma_i[m])$, which is a function of the channel state. Let $\{H_i[m]\}_{m=0}^{\infty}$ be a sequence of i.i.d. Bernoulli random variables with probability $\psi_i = E_{\gamma_i}[\Psi(\gamma_i[m])]$, where the event $H_i[m] = 1$ indicates the event of a successful transmission. After each time slot, the access point will feedback $\{0, 1, e\}$ information to the users, where **0** indicates an idle slot, **1** indicates a success, and *e* indicates an error, due to either collision or channel fading.

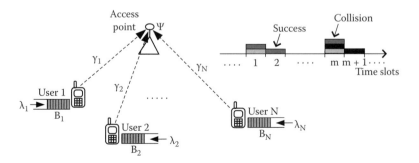

FIGURE 11.2   Illustration of the slotted ALOHA system.

Suppose that the arrival of messages at each user, say user i, is modeled as a sequence of i.i.d. Bernoulli random variables, $\{A_i[m]\}_{m=0}^{\infty}$, with rate λ_i packets per time slot. Let B_i, $i = 1, \cdots, N$, be the buffer at user i that stores the packets awaiting transmission and assume that there is no limit on the occupancy of the buffer. The number of packets in the buffer B_i at the beginning of the mth time slot is denoted by $Q_i[m]$, which is referred to as the *queue state*. Let $\{\tilde{V}_i[m]\}_{m=0}^{\infty}$ be a sequence of Bernoulli random variables with $\tilde{V}_i[m] \in \{0,1\}$ and probability $p_i = \Pr(\tilde{V}_i[m] = 1)$, where $\tilde{V}_i[m] = 1$ indicates the event that user i attempts a transmission in the mth time slot regardless of whether it has a packet to transmit. The event that user i actually transmits is then denoted by $V_i[m] = \tilde{V}_i[m] \cdot \chi_{\mathbb{N}}(Q_i[m])$, where $\mathbb{N}$ is the set of natural numbers 1, 2, 3, …, and $\chi_{\mathbb{E}}(x)$ is the indicator function, defined as

$$\chi_{\mathbb{E}}(x)=\begin{cases}1, & x \in \mathbb{E} \\ 0, & x \notin \mathbb{E}.\end{cases}$$

The evolution of the queue states can then be written as

$$Q_i[m+1]=(Q_i[m]-D_i[m])^+ + A_i[m], \quad \forall i=1,\cdots,N, \qquad (11.1)$$

where

$$D_i[m]= H_i[m]V_i[m]\prod_{j\neq i}\left(1-V_j[m]\right)$$

is the departure process and $(x)^+ = \max\{x,0\}$. It follows that $Q_i[m + 1]$ is independent of $Q_i[m - 1]$ if given the values of $\{Q_i[m], \forall i\}$, and thus $\mathbf{Q}[m] = [Q_1[m], \cdots, Q_N[m]]$ for $m = 0$, 1, 2, …, forms a discrete-time Markov chain.

In the discussions on MAC protocols, we are often concerned with the maximal stable throughput of the system, which is defined as the maximum arrival rate that each user can accommodate without causing the queues to become unstable.

Definition 11.1

The system is stable under the arrival rates $(\lambda_1, \lambda_2, \cdots, \lambda_N)$ if there exists a set of transmission probabilities $p_1, p_2, \cdots, p_N$ such that

$$\lim_{m\to\infty}\Pr\{Q_i[m]<x\}=G_i(x) \text{ and } \lim_{x\to\infty}G_i(x)=1, \forall i.$$

Our goal is then to characterize the possible values of arrival rates for which the system remains stable, which is defined as the *stability region* of the system. However, finding the stability region of the slotted ALOHA system has been a difficult task. Over the past 30 years, only the stability region of a two-user [27] or three-user [28] system has been completely characterized, although inner and outer bounds have been derived for the case where $N > 3$. In the following, we shall compare the two-user stability region of the noncooperative and cooperative cases to illustrate the cooperative advantages in the slotted ALOHA system.

11.3.2 Stability Region of the Noncooperative Slotted ALOHA System

To derive the stability region of the slotted ALOHA system, Rao and Ephremides [27] proposed the use of an auxiliary hypothetical system that has the same arrival process and follows the same transmission policy as the original system, but is less likely to be stable. This system is called a *dominant system*. The stability region of the dominant system may be easier to derive if it is set up properly, but is only guaranteed to be an inner bound of the true stability region. However, in certain cases, the stability region of both systems may actually coincide, as is the case in the following example. Even if they do not coincide, the dominant system still provides useful insights on the behavior of the original system.

A dominant system can be constructed by assuming that a subset of users in the network are *fully loaded*, i.e., they always have a packet to transmit regardless of the actual queue state in the original system. Under this assumption, the departure rate of each user will be no larger than that in the original system since contention between users increases.

Interestingly, following the same procedures as in [27], we can show that the systems considered in the following satisfy the conditions of Loynes' formulation, and thus, by applying Loynes' theorem [29], we can say that the system is stable when the arrival rates $\lambda_1, \cdots, \lambda_N$ are smaller than the service rates $\mu_1, \ldots, \mu_N$, for a given set of transmission probabilities $p_1, \cdots, p_N$, i.e., $\lambda_i < \mu_i$ for all i. The service rate of user i refers to the average number of user i's packets that can be served in each time slot. On the other hand, the system is *unstable* if there exists i such that $\lambda_i > \mu_i$. In this case, our task reduces to finding the departure rate of each system.

To derive the stability region of the two-user system, let us first consider the case where user 1 is fully loaded. In this case, a packet from user 2 can be successfully received at the base station if and only if the reception is successful, which occurs with probability ψ_2, and user 1 does not transmit. Therefore, the service rate of user 2, which is a function of p_1 and p_2, is equal to $\mu_2^{1*}(p_1, p_2) = \psi_2 p_2(1 - p_1)$. If user 2 transmits at a rate λ_2 strictly less than the service rate $\mu_2^{1*}(p_1, p_2)$, it follows from Little's theorem [30] that the probability the buffer of user 2 is non-empty is equal to

$$\frac{\lambda_2}{\psi_2 p_2(1-p_1)},$$

and the service rate of user 1 is equal to

$$\mu_1^{1*}(p_1, p_2) = \psi_1 p_1 \left[1 - \frac{\lambda_2}{\psi_2(1-p_1)} \right].$$

Hence, given p_1 and p_2, the stability region obtained by assuming that user 1 is fully loaded is given by

$$\mathcal{R}^{1^*}(p_1,p_2)=\left\{(\lambda_1,\lambda_2):\lambda_1<\mu_1^{1^*}(p_1,p_2),\lambda_2<\mu_2^{1^*}(p_1,p_2)\right\}. \tag{11.2}$$

For $\lambda_2<\psi_2 p_2(1-p_1)$, the bounds for λ_1 are not only inner bounds, but in fact coincide with that of the true system. More specifically, when λ_1 is greater than $\mu_1^{1^*}(p_1,p_2)$, user 1 will be unstable in the dominant system and the queue length will go to infinity without emptying with finite probability, which means that there are sample paths that do not return to zero infinitely often. For such sample paths, the queue state will not return to zero after a certain time, in which case the dominant system and the original system will be identical. This shows that the original system will be unstable for these values of λ_1 as well.

Similarly, by assuming that user 2 is fully loaded, the stability region can be expressed as

$$\mathcal{R}^{2^*}(p_1,p_2)=\left\{(\lambda_1,\lambda_2):\lambda_1<\mu_1^{2^*}(p_1,p_2),\lambda_2<\mu_2^{2^*}(p_1,p_2)\right\}, \tag{11.3}$$

where $\mu_1^{2^*}(p_1,p_2)=\psi_1 p_1(1-p_2)$ and

$$\mu_2^{2^*}(p_1,p_2)=\psi_2 p_2\left[1-\frac{\lambda_1}{\psi_1(1-p_2)}\right].$$

By taking the union of $\mathcal{R}^{1^*}(p_1,p_2)$ and $\mathcal{R}^{2^*}(p_1,p_2)$ for all possible transmission probabilities, the stability region of the two-user slotted ALOHA system is given by

$$\mathcal{R}=\bigcup_{(p_1,p_2)\in[0,1]^2}\left[\mathcal{R}^{1^*}(p_1,p_2)\bigcup\mathcal{R}^{2^*}(p_1,p_2)\right]=\left\{(\lambda_1,\lambda_2):\sqrt{\frac{\lambda_1}{\psi_1}}+\sqrt{\frac{\lambda_2}{\psi_2}}<1\right\}. \tag{11.4}$$

The boundary of the stability region is plotted in Figure 11.3 for the cases where $\psi_1 = 0.75$ and $\psi_2 = 0.5$ (solid curve) and $\psi_1 = \psi_2 = 0.25$ (dash-dotted curve). The case with no channel errors (dotted curve), $\psi_1 = \psi_2 = 1$, is also plotted as a reference.

11.3.3 Cooperation in a Two-User Slotted ALOHA System

In the conventional slotted ALOHA system, each user transmits independently of others, and thus, the stable throughput of each user is limited by its local channel quality. It is possible that the users experiencing bad channels on average will be seriously backlogged, while the others, i.e., the ones that experience good channels, have nothing to transmit. Intuitively, if the users are willing to cooperate, those that are more capable can help by relaying the messages from the less capable users and thereby increase their stable throughput.

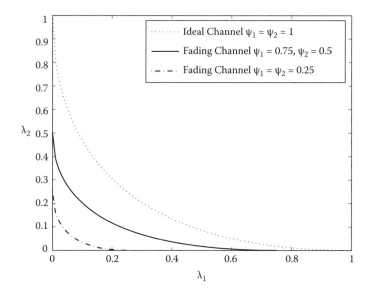

FIGURE 11.3 The two-user slotted ALOHA system without cooperation.

Let us consider a two-user slotted ALOHA system as shown in Figure 11.4. Following the same mechanics as the slotted ALOHA system, we consider the simplest class of cooperation based on decode-and-forward relaying. In the cooperative system, each idle user (i.e., users that have no packet to transmit) will continuously listen for a transmission from the other user and help by relaying the packet to the AP if the original transmission failed in its attempt to reach the AP. In order to do this, we assume that each user is equipped with an additional buffer, which will be used to store the messages transmitted by the other user. To distinguish from buffer B_i, we denote this buffer by CB_i for user i and refer to it as the *cooperative buffer*. In this case, B_i will be called the *source buffer*. Under the half-duplex assumption, a user is able to receive the message from the other user only if it is not transmitting in the same time slot. Moreover, we assume that

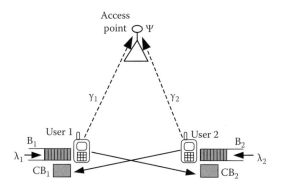

FIGURE 11.4 The two-user slotted ALOHA system with cooperation.

the message will always be correctly received by the cooperating users as long as the transmission is collision-free. This basically assumes that the users are sufficiently close to each other.

Suppose that user 1 is transmitting a certain packet for the first time from its source buffer while user 2 remains silent. During this time slot, user 2 will be able to over-hear the message transmitted by user 1 and record the message in its cooperative buffer (i.e., CB_2). If an e feedback is received from the access point, the packet will remain at the head of line (HOL) and both users will be allowed to retransmit the packet in later time slots until one of them succeeds. When a transmission is eventually successful, i.e., a **1** feedback is received from the access point, the packet will be dropped from both users. Since no new packet is transmitted by the source user until the HOL packet is successfully transmitted, the length of the cooperative queue cannot exceed 1 and, thus, the stability of this queue will not be a concern. Let $CQ_1[m]$, $CQ_2[m] \in \{0, 1\}$ and $Q_1[m]$, $Q_2[m] \in \{0, 1, 2, \ldots\}$ represent the number of packets stored in the buffers CB_1, CB_2, B_1, and B_2, respectively, at the beginning of the mth time slot.

Let p_i be the transmission probability of user i and let $A_i[m]$, $V_i[m]$, $\tilde{V}_i[m]$, and $H_i[m]$ be as defined in the previous subsection. If user i decides to transmit in the current slot, and both B_i and CB_i are nonempty, user i will choose the packet from B_i with probability q_i, and CB_i with probability $\bar{q}_i = 1 - q_i$. To incorporate this into the queue evolution equation, we introduce the new Bernoulli random process $\{U_i[m]\}_{m=0}^{\infty}$, where $\Pr\{U_i[m] = 1\} = q_i$ and $U_i[m] = 1$ indicate the event that user i transmits a packet from B_i instead of CB_i when both the buffers are nonempty. A packet from the source buffer B_i will be transmitted either if (1) CB_i is empty and B_i is not or if (2) B_i and CB_i are both nonempty but user i chooses to transmit from B_i, i.e., $U_i[m] = 1$. Therefore, the event that a transmission attempt is made by a packet in B_i is represented with

$$V_i^{coop}[m] = V_i[m] \cdot \chi_{\{0\}}(CQ_i[m]) + V_i[m] \cdot U_i[m] \cdot \chi_{\mathbb{N}}(CQ_i[m]), \qquad (11.5)$$

where $V_i[m] = \tilde{V}_i[m] \cdot \chi_{\mathbb{N}}(Q_i[m])$. Notice that $V_i^{coop}[m] = 1$ if a packet is transmitted from B_i, and $V_i^{coop}[m] = 0$ otherwise. On the other hand, a packet from the cooperative buffer CB_i is transmitted if user i attempts a transmission and the packet transmitted is not from B_i. Therefore, we have

$$CV_i^{coop}[m] = \tilde{V}_i[m][1 - \chi_{\{0\}}(Q_i[m]) \cdot \chi_{\{0\}}(CQ_i[m])] - V_i^{coop}[m]. \qquad (11.6)$$

More specifically, a packet from the cooperative buffer is transmitted if $CV_i^{coop}[m] = 1$; otherwise, $CV_i^{coop}[m] = 0$. Since the successful transmission of a packet can be achieved either by the source itself or by the cooperating user, the departure process of user i is

$$D_i[m] = V_i^{coop}[m] H_i[m](1 - V_j^{coop}[m] - CV_j^{coop}[m])$$

$$+ CV_j^{coop}[m] H_j[m](1 - V_i^{coop}[m] - CV_i^{coop}[m]), \text{ for } i = 1, 2 \text{ and } i \neq j. \qquad (11.7)$$

The evolution of the source queues is given in (11.1), while that of the cooperative queue is

$$CQ_i[m+1]=(CQ_i[m]-D_j[m])^+ +V_j^{coop}[m](1-H_j[m])(1-V_i^{coop}[m])\chi_{\{0\}}(CQ_i[m]). \quad (11.8)$$

The queue states $\{CQ_1[m],CQ_2[m],Q_1[m],Q_2[m]\}_{m=0}^{\infty}$ form a discrete-time Markov chain.

11.3.4 Stability Region of the Two-User Cooperative Slotted ALOHA System

Due to the complex interactions among the four queues $Q_1[m]$, $Q_2[m]$, $CQ_1[m]$, and $CQ_2[m]$, the stability region of the cooperative system is difficult to obtain. However, we are able to derive an inner bound by analyzing a dominant system where we let both users be fully loaded, i.e., they always have a packet available for transmission. The stability region of this dominant system will be referred to as the *fully loaded region* in the following.

When the users are fully loaded, the packet in CB_i will always be transmitted with probability $\bar{q}_i$ as long as it is nonempty, i.e., $CQ_i[m] = 1$. This is to say that when the users are fully loaded, the states of the cooperative buffers are independent of the states of the source buffers. In this case, $\{CQ_1[m],CQ_2[m]\}_{m=0}^{\infty}$ forms a finite-state Markov chain with four states: $S_0 = (0,0)$, $S_1 = (0,1)$, $S_2 = (1,0)$, and $S_3 = (1,1)$. For a given set of transmission probabilities p_1, p_2, q_1, and q_2, the Markov chain yields a steady-state distribution denoted by the probabilities π_0, π_1, π_2, and π_3, where π_i is the steady-state probability of state S_i. The service rates for users 1 and 2, respectively, are given by

$$\mu_1(p_1,p_2,q_1,q_2)=(\pi_1+\pi_3)\bar{q}_2 p_2(1-p_1)\psi_2 +[1-(\pi_2+\pi_3)q_1]p_1(1-p_2)\psi_1,$$

$$\mu_2(p_1,p_2,q_1,q_2)=(\pi_2+\pi_3)q_1 p_1(1-p_2)\psi_1 +[1-(\pi_1+\pi_3)\bar{q}_2]p_2(1-p_1)\psi_2.$$

The stability region of the fully loaded system is

$$\mathcal{R}_{coop} = \bigcup_{\substack{(p_1,p_2)\in[0,1]^2 \\ (q_1,q_2)\in[0,1]^2}} \mathcal{R}_{coop}(p_1,p_2,q_1,q_2), \quad (11.9)$$

where $\mathcal{R}_{coop}(p_1,p_2,q_1,q_2) = \{(\lambda_1,\lambda_2): \lambda_1 < \mu_1(p_1,p_2,q_1,q_2), \lambda_2 < \mu_2(p_1,p_2,q_1,q_2)\}$ is the fully loaded region given fixed values of p_1,p_2,q_1, and q_2. The region in (11.9) serves as an inner bound to the stability region of the cooperative slotted ALOHA system.

11.3.5 Cooperative Slotted ALOHA with Channel Awareness

The cooperation considered up to this point exploits the diversity gains to improve the throughput of the system. Since the same packet may be transmitted by both users in the network, the probability of success will not be limited by the local channel quality of any single source. However, the performance can be further improved by exploiting knowledge of the local channel state information (CSI) at the users [13]. For example, if a user experiences bad channels, it should reduce its transmission probability to avoid collision with the other users. This is the subject of this section.

In a multiuser system, it is well known that, with knowledge of the CSI, the throughput of the system can be increased by scheduling users with the best channel to transmit in each time slot. This advantage is referred to as multiuser diversity in the literature [31]. To exploit this in a random access system, one can adopt a channel-aware transmission control function, as proposed in [32, 33], that determines the transmission probability of each user based on its local channel state in each time slot. Specifically, instead of transmitting with a fixed transmission probability, we assume that each user, say user i, adjusts its transmission probability based on its local channel state $\gamma_i[m]$ according to the function $s_i(\gamma_i[m])$. The average transmission probability is defined as

$$p_i = \int s_i(\gamma_i)dF_{\gamma_i}(\gamma_i),$$

where $F_{\gamma_i}(\gamma_i)$ is the distribution function of the channel state γ_i.

Consider a cooperative network that consists of two users and assume that both users are fully loaded. Similar to the case without channel awareness, the cooperative queue states $\{CQ_1[m], CQ_2[m]\}_{m=0}^{\infty}$ form a four-state Markov chain where the stationary distribution can also be derived. The service rates under the fully loaded assumption are given by

$$\mu_{1,CSI}(s_1,s_2,q_1,q_2)=(\pi_1+\pi_3)\overline{q}_2\beta_2(1-p_1)+[1-(\pi_2+\pi_3)\overline{q}_1]\beta_1(1-p_2),$$

$$\mu_{2,CSI}(s_1,s_2,q_1,q_2)=(\pi_2+\pi_3)\overline{q}_1\beta_1(1-p_2)+[1-(\pi_1+\pi_3)\overline{q}_2]\beta_2(1-p_1),$$

(11.10)

for users 1 and 2, respectively, where $\beta_i = \mathbf{E}_{\gamma_i}[s_i(\gamma_i)\Psi(\gamma_i)]$. By Loynes' theorem, the fully loaded region for fixed s_1, s_2, q_1, and q_2 is

$$\mathcal{R}_{coop,CSI}(s_1,s_2,q_1,q_2)=\{(\lambda_1,\lambda_2): \lambda_1 <\mu_{1,CSI}(s_1,s_2,q_1,q_2), \lambda_2 <\mu_{2,CSI}(s_1,s_2,q_1,q_2)\}. \quad (11.11)$$

For fixed $s_1(\gamma_1)$, $s_2(\gamma_2)$, we take the union of the regions in (11.11) over q_1, q_2 to obtain

$$\mathcal{R}_{coop,CSI}(s_1,s_2)=\{(\lambda_1,\lambda_2): \lambda_1 < \frac{\beta_2(1-p_1)+\beta_1(1-p_2)}{\beta_2(1-p_1)+ p_1(1-p_2)} p_1(1-p_2)\equiv V_{1,x}(s_1,s_2),$$

$$\lambda_2 < \frac{\beta_1(1-p_2)+\beta_2(1-p_1)}{\beta_1(1-p_2)+ p_2(1-p_1)} p_2(1-p_1)\equiv V_{2,y}(s_1,s_2), \quad (11.12)$$

$$\lambda_1+\lambda_2 <\beta_1(1-p_2)+\beta_2(1-p_1)\equiv k(s_1,s_2)\}.$$

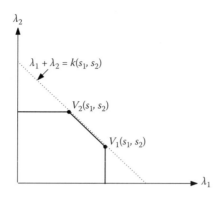

FIGURE 11.5 Stability region for given $s_1(\gamma_1)$ and $s_2(\gamma_2)$.

The region given above takes on the shape illustrated in Figure 11.5, where $V_1(s_1,s_2) = (V_{1,x}(s_1,s_2), k(s_1,s_2) - V_{1,x}(s_1,s_2))$ and $V_2(s_1,s_2) = (k(s_1,s_2) - V_{2,y}(s_1,s_2), V_{2,y}(s_1,s_2))$. The maximum service rate of user 1 is achieved by setting $q_1 = 1$ and $q_2 = 0$ (i.e., user 1 always transmits from the source buffer, and user 2 always transmits from the cooperative buffer), and similarly for user 2. By taking the union of the pentagons in Figure 11.5 over all s_1, s_2, we obtain the fully loaded region for the cooperative channel-aware system.

11.3.6 Performance Comparisons

In Figure 11.6, we plot the boundaries of three regions: (1) the stability region of the conventional slotted ALOHA system, (2) the inner bound of the cooperative system, and (3) the inner bound of the cooperative system with channel-aware transmission control. We consider, as an example, the reception model where

$$\Psi(\gamma) = \begin{cases} 1, & \text{if } \gamma \geq \gamma_{th} \\ 0, & \text{otherwise,} \end{cases}$$

which says that the packet is perfectly received if the channel state exceeds a certain threshold γ_{th} and fails otherwise. For the case with channel awareness, we assume that the channel-aware transmission control functions $s_i(\cdot)$, for all i, are step functions such that the transmission probability is 1 if the local channel state exceeds a certain value, and 0 otherwise. This was shown to be optimal in [13, 32].

Figure 11.6(a) is plotted for the case where $\psi_1 = E_{\gamma_1}[\Psi(\gamma_1)] = 0.1$ and $\psi_2 = E_{\gamma_2}[\Psi(\gamma_2)] = 0.1$, and Figure 11.6(b) is plotted for the case where $\psi_1 = 0.2$ and $\psi_2 = 0.8$. As shown in the figures, cooperation enlarges the stability region, especially when one of the users has a much better channel than the other. With cooperation, the user that on average experiences a bad channel can utilize help from its partner to increase the throughput of its own packets. When both users experience bad channels, little advantage can be gained through pure cooperation, as shown in Figure 11.6(a). However, a significant gain is

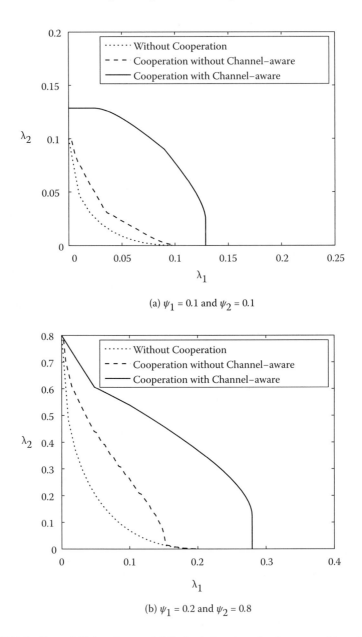

(a) $\psi_1 = 0.1$ and $\psi_2 = 0.1$

(b) $\psi_1 = 0.2$ and $\psi_2 = 0.8$

FIGURE 11.6 The stability regions and fully loaded regions for the cases with and without cooperation and channel awareness.

achieved when the users are able to adjust their transmission probabilities according to their local channel state. The channel-aware transmission control allows the users to exploit the multiuser diversity and increase the throughput of both cooperative users.

Although significant gains are already demonstrated in the above examples, it is worthwhile to notice that the form of cooperation discussed above does not utilize diversity-combining techniques at the AP. However, these techniques can be easily incorporated into our model to further improve the system performance. The analysis of these systems is subject to future investigation. The discussions for the case of N users can be found in [34]. Furthermore, we can also consider the use of multiple-packet reception in the cooperative slotted ALOHA system. Readers are referred to [35] for detailed discussions.

11.4 Cooperation with CSMA/CA in IEEE 802 Legacy Systems

Cooperation in its simplest form can be easily incorporated into the IEEE legacy systems. In fact, Liu, Tao, and Panwar [15] proposed a cooperative MAC protocol for wireless local area networks (WLANs) by making small changes to the IEEE 802.11b MAC. The idea is to use, in addition to the *request-to-send* and *clear-to-send* (RTS/CTS) messages, the *helper request-to-send* (HTS) message to help reserve the channel for the duration of the cooperative transmission and to coordinate the transmission between cooperating users.

There are two MAC operations in IEEE 802.11b: the distributed coordination function (DCF) and the point coordination function (PCF). The DCF is a contention-based method that provides asynchronous data transfers, and the PCF is a schedule-based method used to provide connection-oriented file transfers. Cooperation is less problematic in the latter case since a centralized control is used to schedule the users' transmissions. The cooperative MAC protocol proposed in [15] is based on the contention-based method, where no centralized control from the access point is required, and therefore is suitable for the ad hoc scenario. The DCF is designed based on the *Carrier Sensing Multiple Access with Collision Avoidance* (CSMA/CA) protocol. In DCF, a user must first sense whether a transmission by a neighboring user is in progress before transmitting its own data, i.e., *physical carrier sensing*. However, this does not avoid the hidden terminal problem, that is, the event that a user in the vicinity of the destination is transmitting simultaneously with the source and interfering with the reception at the destination. To avoid this problem, RTS and CTS messages are broadcast by the source and the destination before the data packet is actually transmitted to inform the neighboring users that a transmission with a certain duration is to follow. This is referred to as *virtual carrier sensing*.

For example, suppose that user 1 intends to transmit a data packet to user 2, as shown in Figure 11.7. Before transmission of the packet, user 1 first senses (physically) whether the channel is idle. If it is, user 1 will send an RTS message to the destination (i.e., user 2), which will reserve the channel from its neighboring users for a duration indicated in the network allocation vector (NAV) of the RTS message. If the destination is ready to receive the packet, that is, no other user is transmitting in its vicinity, it will reply with a CTS message to confirm the transmission and to reserve the channel from users in its vicinity. With multi-rate modulation and coding schemes available in the IEEE 802.11 standard, the users are able to adapt their transmission rates based on their local

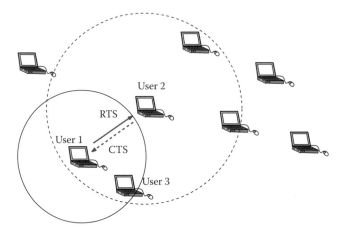

FIGURE 11.7   Carrier Sensing Multiple Access with Collision Avoidance (CSMA/CA).

channel quality. The time duration reserved by the RTS/CTS messages is then adjusted accordingly. When only a low data rate is achievable between two distant users, the transmission channel may be occupied for a long period of time but with low channel utilization, which significantly reduces the system throughput. To improve upon this, the source can ask a close-by user to relay the message to the destination, assuming that the source-relay and the relay-destination channels have sufficiently high transmission rates to reduce the total transmission time.

Cooperation can also be considered with slight modifications to the CSMA/CA protocol, as illustrated in Figure 11.8. In this case, each user is required to maintain a helper table that records the information about the potential helpers in its vicinity. The table contains four fields (e.g., see Figure 11.8(a)): the helper ID (i.e., MAC address), the time that a packet from the helper is last received, the rate between the helper and the destination (i.e., R_{hd}), and the rate between the source and the helper (i.e., R_{sh}). Before each transmission, the user searches this table for the best helper and determines whether cooperation is actually advantageous in its case.

For example, suppose that the source node (i.e., user 1) wants to transmit an N-bit packet to the destination node (i.e., user 2), as illustrated in Figure 11.8(b). With R_{sd} being the rate of the source-destination link and R_{sr}, R_{rd} the rate of the source-relay and relay-destination links, each user can compute the time that it takes for both direct and cooperative transmissions. Specifically, for direct transmission, the time that it takes is equal to N/R_{sd}, and for cooperative transmissions with relay r, the time is equal to

$$\left(\frac{N}{R_{sr}} + \frac{N}{R_{rd}} \right).$$

Before each transmission, the source first computes the cooperative transmission time corresponding to each helper and selects the user with the shortest time as the potential helper. The source then compares this with the time required for direct transmission

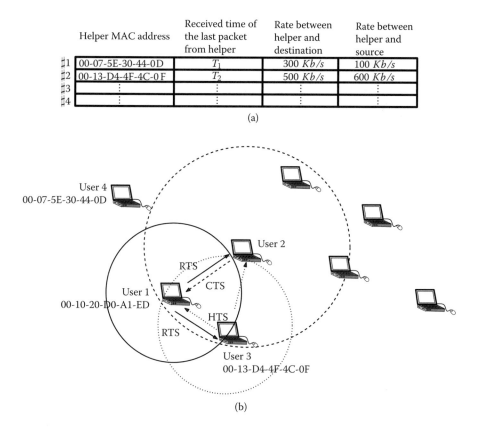

Helper MAC address	Received time of the last packet from helper	Rate between helper and destination	Rate between helper and source
♯1 00-07-5E-30-44-0D	T_1	300 Kb/s	100 Kb/s
♯2 00-13-D4-4F-4C-0 F	T_2	500 Kb/s	600 Kb/s
♯3	⋮	⋮	⋮
♯4	⋮	⋮	⋮

(a)

(b)

FIGURE 11.8 Illustration of the cooperative MAC protocol.

and requests help from another user only if relaying can speed up the transmission. In other words, user 1 selects user i to be the potential helper, where

$$i = \arg\min_{r} \left(\frac{N}{R_{sr}} + \frac{N}{R_{rd}} \right),$$

and requests cooperation only if

$$\left(\frac{N}{R_{si}} + \frac{N}{R_{id}} \right) > \frac{N}{R_{sd}}.$$

To inform the destination about the availability of the relay node and reserve the transmission time for the relay node, an additional HTS control message is exchanged among the source, relay, and destination. Specifically, when the source node decides to transmit cooperatively, it will include in the RTS packet the ID of the potential helper and the corresponding rate indicated in the helper's table. When the helper receives the message, it will check whether the rate indicated by the source is achievable. Inconsistencies between the helper table and the actual achievable rate occur when the environment varies and the entries in the table become out of date. If the helper is able to cooperate, it will respond with an HTS message to notify the source and the destination that cooperation is available and reserve the channel from its neighboring users. On the other hand, if the relay node is busy (due to its own transmission or because of the transmission of a neighbor), or it cannot achieve the rate requested by the source node, it will simply remain silent, and after a certain time-out period expires, the destination node will proceed and reply with a CTS message to reserve the channel for direct transmission.

As shown above, the concept of cooperation can be easily incorporated into the IEEE 802.11 MAC with little changes to the original design. Several enhancement methods are also given in [15] to reduce the transmission overhead and increase the system throughput.

11.5   Cooperation-Enhanced Collision Resolution Methods

The essence of random access protocols is to enable users to access the channel in a distributed fashion and deal with the collision afterwards. In the cooperative MAC protocols introduced above, the advantages of relaying are exploited to combat fading in the wireless channel, allowing the users that experience bad channels to be helped by the users that experience good channels. However, these methods do not exploit the cooperative advantages in resolving the interference or collision at the destination. In this and the following sections, we show that, with cooperation among users and appropriate signal processing techniques at the destination, the mixture of signals that is received during collision can be utilized to enhance the detection performance. In particular, we describe in this section the system, proposed first by Tsatsanis et al. in [18] and then by Lin and Petropulu in [17], that utilizes the variations in channel gains of the multiple relaying paths to perform signal separation when collision occurs.

Consider a network of N nodes transmitting to an access point, similar to that shown in Figure 11.2. Conventionally, when more than one user is transmitting in the same time slot, a mixture of signals will be received at the destination, causing strong interference among each other. In this case, it is likely that no packet will be received successfully, resulting in the so-called *collision*. The mixture of signals is usually discarded at the destination, and no information is extracted from the signals, which is clearly a waste of energy and bandwidth.

Due to this reason, Tsatsanis et al. [18] proposed a network-assisted collision resolution method. In this method, the destination records the mixture of signals whenever a collision occurs and enables the same set of colliding users to continue transmitting simultaneously in a certain number of subsequent time slots. Specifically, let $\mathbf{x}_i[n] = [x_{i,1}[n], \cdots, x_{i,M}[n]]^T$ be the M-bit message transmitted by user i in the nth time slot, and let $\mathcal{I}[n] \subset \{1, 2, ..., N\}$ be the set of users that are transmitting in the nth time slot. A collision occurs when the cardinality of $\mathcal{I}[n]$ is greater than 1, i.e., $\|\mathcal{I}[n]\| > 1$. The received signal at the destination is given by

$$\mathbf{y}[n]=[y_1[n],\cdots,y_M[n]]^T = \sum_{i\in\mathcal{I}[n]} h_{i,d}[n]\mathbf{x}_i[n]+\mathbf{v}[n]$$

where $h_{i,d}[n]$ is the channel between user i and the destination in the nth time slot, and $\mathbf{v}[n] = [v_n(1), \cdots, v_n(M)]^T$ is the additive noise vector with i.i.d. elements. The channel is assumed to be static over each time slot, but varying rapidly over different time slots.

Suppose that the destination or access point is able to estimate accurately the number of users transmitting in the same time slot and, through the cyclic redundancy check (CRC), is also able to detect an error whenever it occurs. After each time slot, the destination sends a **0**, **1**, or **e** feedback, with **0** indicating that the channel was idle, **1** indicating that all packets transmitted in the time slot were correctly received, and **e** indicating that at least one packet was not correctly decoded. When an **e** feedback is received, the users that were originally transmitting in the time slot will retransmit their messages again in the next time slot. Therefore, a sequence of collisions will occur, as shown in Figure 11.9. Suppose that the retransmissions (and thus collisions) occur over $\hat{K}$ consecutive time slots. Then, with $\mathcal{I}[n] = \{i_1, \dots, i_K\}$, the destination receives the messages

$$\mathbf{Y}[n]=[\mathbf{y}[n],\mathbf{y}[n+1],\cdots,\mathbf{y}[n+K-1]]$$

$$=[\mathbf{x}_{i_1}[n],\cdots,\mathbf{x}_{i_K}[n]]\cdot\begin{bmatrix} h_{i_1,d}[n] & \cdots & h_{i_1,d}[n+k-1] \\ \vdots & \ddots & \vdots \\ h_{i_K,d}[n] & \cdots & h_{i_K,d}[n+K-1] \end{bmatrix}+[\mathbf{v}[n],\cdots,\mathbf{v}[n+K-1]] \quad (11.13)$$

$$= \mathbf{X}[n]\mathbf{H}[n]+\mathbf{V}[n].$$

In the absence of noise, the messages $\mathbf{X}[n]$ can be decoded by solving the set of linear equations $\mathbf{Y}[n] = \mathbf{X}[n]\mathbf{H}[n]$, given that the rank of $\mathbf{H}[n]$ is greater than K. Assuming that the channel coefficients for all users and all time slots are continuous random variables, e.g., Rayleigh fading, the matrix $\mathbf{H}[n]$ will have full rank with probability 1. Therefore, in the noiseless case, we need only $\hat{K} = K$ time slots in order to resolve the message of K users. Hence, there is no loss of throughput due to collision.

When noise is present, the maximum likelihood estimate of the message is given by

$$\hat{\mathbf{X}}[n]=\arg\min_{\mathbf{X}}\left\| \mathbf{Y}[n]-\mathbf{X}\cdot\mathbf{H}[n] \right\|_F^2, \quad (11.14)$$

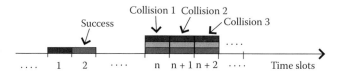

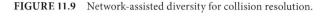

FIGURE 11.9 Network-assisted diversity for collision resolution.

where $\|\cdot\|^2$ represents the Frobenius norm. A suboptimal linear solution,

$$\hat{\mathbf{X}}[n] = \mathbf{Y}[n] \cdot \mathbf{H}^{-1}[n], \tag{11.15}$$

can also be used provided that $\mathbf{H}[n]$ is full rank.

However, when the channel varies slowly with respect to the time slot index n, one may need a large number of transmissions before the set of linear equations yield a reliable solution to the channel messages. To improve upon this, Lin and Petropulu [17] proposed the use of cooperative relaying to increase the variations of the channel gains in each transmission. Namely, upon collision, instead of asking the same users to retransmit continuously in consecutive time slots, the new strategy assigns a single node to transmit based on some predetermined order. The order can be determined, for example, by giving a unique index from 1 to N to each user upon deployment and allocating the retransmission time slot to the user that has an index equal to the time index m modulo N. The node that is assigned to transmit can be either one of the nodes participating in the collision, in which case it will simply retransmit its own message, or any other node in the network that amplifies and retransmits the mixture of signals that it overheard in the previous time slot. This period of repeated retransmissions is referred to as the *cooperative transmission epoch* (CTE). The CTE terminates when the receiver gains sufficient knowledge to recover the packets from all users. An illustration of this process is given in Figure 11.10.

Specifically, when a collision occurs in slot n, we assume that all users that are not involved in the collision will receive a mixture of signals similar to that at the destination. The signal received at user r is given by

$$\mathbf{z}_r[n] = \sum_{i \in \mathcal{I}[n]} h_{i,r}[n] \mathbf{x}_i[n] + \mathbf{v}_r[n],$$

where $h_{i,r}[n]$ is the channel coefficient between user i and user r during the nth time slot and $\mathbf{v}_r[n]$ is the noise at user r. Upon collision, the system will enter the CTE, during which $\hat{K} - 1$ users, denoted by $r_{n+1}, \cdots, r_{n+\hat{K}-1}$, will be assigned to retransmit. Suppose that the retransmitting node r_{n+k} belongs to the set $\mathcal{I}[n]$, then the user will simply transmit its own message and the signal received at the destination will be

$$\mathbf{y}[n+k] = h_{rd}[n+k] \mathbf{x}_r[n] + \mathbf{v}[n+k].$$

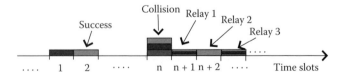

FIGURE 11.10 Cooperative diversity for collision resolution.

On the other hand, if $r_{n+k} \notin \mathcal{I}[n]$, then it will amplify and forward the signal with amplifying gain equal to $c[n + k]$, which is determined by the energy constraint at each node. The signal received at the destination is given by

$$\mathbf{y}[n+k] = h_{rd}[n+k] \cdot c[n+k] \cdot \mathbf{z}_r[n] + \mathbf{v}[n+k].$$

With amplify-and-forward relaying, the signal can also be decomposed into a signal component and a noise component, and the channel coefficient of the signal will be equal to the product of the source-to-relay and the relay-to-destination channels. Therefore, the detection methods given in (14.14) and (14.15) are equally applicable for this case.

By fully exploiting the signals that were received during collision, the system throughput may increase dramatically due to significant improvements in bandwidth inefficiency. Significant gains in energy efficiency may also be observed since all signals are combined for detection at the receiver. This concept can be exploited in ad hoc networks as described in the following section.

11.6 Asynchronous Cooperation in Multihop Ad Hoc Networks

In ad hoc wireless networks, users communicate over multihop relaying paths with no predetermined network infrastructure. In fact, multihop relaying is basically a form of cooperative communications but does not exploit diversity-combining techniques at the destinations. That is, as the messages are transmitted over a multihop route, each user in the route utilizes for detection only the signal coming from the single closest transmitter, even though the messages transmitted by all users in the same route are identical. The signals received from other users are essentially treated as interference. This can be improved upon by considering diversity combining at the destination, but must be done without the strict synchronization requirements that are prohibitive in large networks.

For network broadcasting applications, the *opportunistic large arrays* (OLA) system was proposed in [9, 19] to exploit the advantages of signal combining in cooperative multihop networks. The key idea is to treat the mixture of signals that arrive at unsynchronized time instants as an artificial multipath signal and utilize existing techniques such as the RAKE receiver or equalizers to resolve the multipath signal [20].

Consider a network broadcasting scenario where we have one source node transmitting a message to all other nodes in the network through multihop transmissions. At the beginning of the process, the source node first broadcasts its message through the wireless medium to all other users. Each user that is able to reliably decode the message will then retransmit the same message, which is again heard by all other nodes in the network. The users that have not yet transmitted will then take the mixture of signals received from their upstream nodes and decode the message using standard RAKE receivers or equalizers [9, 20]. Given that the transmission energy is sufficiently high, these operations will trigger an avalanche of signals propagating through the network. The intermediate nodes can either transmit a repetition of the decoded message or simply amplify and forward the original signal. To reduce the effect of error or noise propagation, the reliability of the signal received at each node can be defined based on different criteria, such as the received SNR or the probability of error. The mechanics of this system emulate the chanting of *Olé!* in a football stadium, where the signal first starts from a single source and propagates to the entire stadium as more and more people follow.

Suppose that there are N nodes in the network. Let us start by considering a symbol-by-symbol relaying scheme described as follows. At the beginning of the network broadcast operation, the source node first transmits an M-ary symbol represented by one of the waveforms $p_m(t)$, $m \in \{1, \ldots, M\}$, which has average energy normalized to 1,

$$\frac{1}{M} \sum_{m=1}^{M} \int |p_m(t)|^2 \, dt = 1.$$

In the case where all the intermediate nodes are eventually able to correctly decode the message, but at different instants in time, the signal received at user i can be written as

$$y_i(t) = \sum_{n=1}^{N} h_{i,n}(t) \cdot \sqrt{\varepsilon_n} \cdot p_m(t - \tau_{i,n}(t)) + n_i(t) = s_{i,m}(t) + n_i(t), \qquad (11.16)$$

where $n_i(t)$ is the additive white Gaussian noise (AWGN) at the ith node with variance N_0, $\tau_{i,n}(t)$ is the time that user n transmits plus the propagation delay between the ith node and the nth node, ε_n is the energy of the pulse transmitted by user n, and $h_{i,n}(t)$ is the channel fading coefficient between user i and user n. The relaying nodes act as active scatters that form the multipath signal $s_{i,m}(t)$. Interestingly, this signal is unique for each user i due to the different channel gains and propagation delays experienced by the incoming signals. We refer to this as the *network signature* of user i given that the mth symbol was transmitted.

Assume that $h_{i,n}(t)$ and $\tau_{i,n}(t)$ are constant over the symbol duration T_s. Notice that T_s is proportional to the maximum delay spread of the signals $s_{i,m}(t)$, for all i, which is denoted by $\Delta\tau$. Specifically, we have

$$\Delta\tau = \max_i \left\{ \sqrt{\frac{\int_{-\infty}^{\infty} (t - \bar{\tau}_i)^2 |s_{i,m}(t)|^2 \, dt}{\int_{-\infty}^{\infty} |s_{i,m}(t)|^2 \, dt}} \right\}, \qquad (11.17)$$

where

$$\bar{\tau}_i = \frac{\int_{-\infty}^{\infty} t |s_{i,m}(t)|^2 \, dt}{\int_{-\infty}^{\infty} |s_{i,m}(t)|^2 \, dt} \qquad (11.18)$$

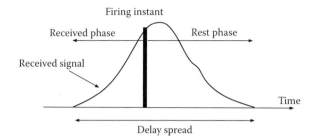

FIGURE 11.11 Signal received by active nodes.

is the average delay of the signals received at user i. With the symbol duration $T_s \approx \Delta\tau$, the rate $R_s \approx 1/\Delta\tau$ is achieved without intersymbol interference. Nevertheless, a higher rate is achievable with the use of equalizers at the destination, which is shown in [20]. An example of the signature waveform is depicted in Figure 11.11.

The signal received at each user can be divided into two phases: the receive phase and the rest phase. In the *receive phase*, the user first accumulates the signal transmitted by its upstream nodes until the signal energy is sufficient to perform reliable detection. At this instant, the symbol is detected and retransmitted to other users in the network. This instant is called the *firing instant*. Once a user has transmitted, it will then shut down until the downstream signals fade away. This period of time is called the *rest phase* and is used to prevent an infinite feedback, which may cause the system to become unstable.

The operations described above can be demonstrated with a simple four-node example, as shown in Figure 11.12. Assume node 1 is the source node and nodes 2–4 are the relays. We can see from the figure that node 2 will be the first to receive the signal from node 1. By the time user 2 accumulates sufficient signal energy for reliable detection, it will immediately retransmit the packet to nodes 3 and 4. By shutting down the receivers in the rest phase, the nodes that transmitted earlier in the symbol period will not receive signals from their downstream nodes.

Two main properties of OLA allow it to improve over the conventional multihop broadcasting scheme. The first property is its ability to combine the signals from all relays, including those that were transmitted from nodes that are located far away and were originally omitted in conventional networks. This enables a tremendous amount of energy savings, as detailed in [19]. Second, the simultaneously transmitting signals, which were originally treated as interference, are also combined for detection. Not only does this increase the energy efficiency, but it also reduces the network congestion and significantly decreases the broadcasting delay [see 9]. A specific equalizer design that performs the signal combining is proposed by Wei, Gofeckel, and Valenti in [20], where a random delay at the users is also proposed to increase the delay diversity of the system.

OLA provides a simple and efficient way to achieve broadcast communications in wireless networks. The system scales naturally to a large network since users are only required to follow a simple local rule. The integrate-and-fire mechanism at each user emulates the behavior of many biological networks, such as the flashing of fireflies, the firing of neurons, or even highway traffic patterns. These examples show that complex large-scale behaviors can result from the interaction between simple local mechanisms.

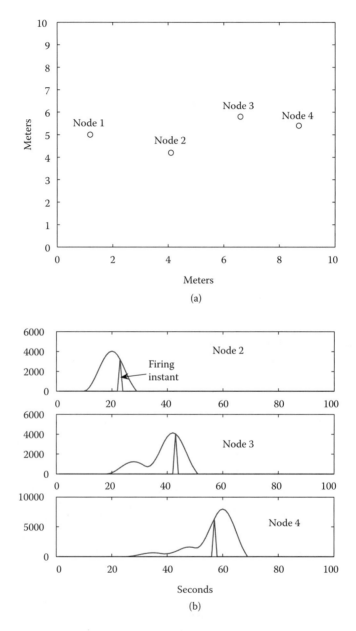

FIGURE 11.12   Demonstration of the signal at nodes 2–4. (a) Node location; (b) OLA signals received at nodes 2–4.

Biological networks are widely believed to be near optimal as a result of elimination and evolution. It is interesting to see how the efficiency of communication systems can be improved by emulating these biological networks. Discussions on these issues can be found in [36, 37] and the references therein.

11.7   Conclusion

Cooperative communications is a promising technique that has been proposed to combat fading without requiring multiple antennas on each device. Most studies in the literature focus on the signal processing aspects in the physical layer while assuming strict synchronization among the cooperating users. This strict synchronization requirement is perhaps the most difficult problem to overcome when realizing a cooperative network.

In this chapter, we studied the advantage of cooperative communications in a random access scenario where the users transmit based only on simple local rules. Some efficiency will be lost inevitably due to the lack of coordination, but significant improvements can be attained with cooperation. Two concepts have been repeatedly explored in this chapter: the user's ability to relay the information from the other users and the diversity-combining technique at the destination. With proper design of the transmission protocol, these added functionalities may significantly improve the throughput of the system. In this chapter, these advantages have been exploited in (1) the slotted ALOHA system, (2) the CSMA/CA system, (3) the collision resolution, and (4) multihop ad hoc networks. These discussions highlight the importance of cross-layered considerations when designing cooperative systems.

References

[1] J. H. Winters. 1987. On the capacity of radio-communication systems with diversity in a Rayleigh fading environment. *IEEE J. Selected Areas Commun.* 5:871–78.

[2] V. Tarokh, N. Seshadri, and A. R. Calderbank. 1998. Space-time codes for high data rate wireless communication: Performance analysis and code construction. *IEEE Trans. Inf. Theory* 44:744–65.

[3] G. J. Foschini. 1996. Layered space-time architecture for wireless communication in fading environments when using multi-element antennas. *Bell Syst. Tech. J.* 1:44–59.

[4] A. Goldsmith. 2005. *Wireless communications.* Cambridge: Cambridge University Press.

[5] L. Zheng and D. N. C. Tse. 2003. Diversity and multiplexing: A fundamental tradeoff in multiple-antenna channels. *IEEE Trans. Inf. Theory* 49:1073–96.

[6] T. Cover and A. El Gamal. 1979. Capacity theorems for the relay channel. *IEEE Trans. Inf. Theory* 25:572–84.

[7] J. N. Laneman, D. N. C. Tse, and G. W. Wornell. 2004. Cooperative diversity in wireless networks: Efficient protocols and outage behavior. *IEEE Trans. Inf. Theory* 50:3062–80.

[8] A. Sendonaris, E. Erkip, and B. Aazhang. 2003. User cooperation diversity. Part I. System description. Part II. Implementation aspects and performance analysis. *IEEE Trans. Commun.* 51(11).

[9] A. Scaglione and Y.-W. Hong. 2003. Opportunistic large arrays: Cooperative transmission in wireless multi-hop ad hoc networks for the reach back channel. *IEEE Trans. Signal Processing* 51(8).

[10] A. Nosratinia, T. E. Hunter, and A. Hedayat. 2004. Cooperative communication in wireless networks. *IEEE Commun. Mag.* 42:74–80.

[11] Y.-W. Hong, W.-J. Huang, F.-H. Chiu, and C.-C. J. Kuo. 2007. Cooperative communications in resource-constrained wireless networks. *IEEE Signal Processing Mag.* 24:47–57.

[12] K. Azarian, H. El Gamal, and P. Schniter. 2005. On the achievable diversity-multiplexing tradeoff in half-duplex cooperative channels. *IEEE Trans. Inf. Theory* 51:4152–72.

[13] Y.-W. Hong, C.-K. Lin, and S.-H. Wang. 2007. On the stability of two-user slotted ALOHA with channel-aware and cooperative users. In *Proceedings of the IEEE Workshop on Wireless Networks: Communication, Cooperation and Competition (WNC3)*.

[14] Y.-W. Hong, C.-K. Lin, and S. H. Wang. 2007. On the stability region of finite-user slotted ALOHA with cooperative relays. In *Proceedings of the International Symposium on Information Theory (ISIT)*.

[15] P. Liu, Z. Tao, and S. Panwar. 2005. A cooperative MAC protocol for wireless local area networks. In *Proceedings of the IEEE International Conference on Communications (ICC)*, vol. 5, pp. 2962–2968.

[16] P. Liu, Z. Tao, Z. Lin, E. Erkip, and S. Panwar. 2006. Cooperative wireless communications: A cross-layer approach. *IEEE Wireless Commun. Mag.* 13:84–92.

[17] R. Lin and A. P. Petropulu. 2005. A new wireless network medium access protocol based on cooperation. *IEEE Trans. Signal Processing* 53:4675–84.

[18] M. K. Tsatsanis, R. Zhang, and S. Banerjee. 2000. Network-assisted diversity for random access wireless networks. *IEEE Trans. Commun.* 48:702–11.

[19] Y.-W. Hong and A. Scaglione. 2006. Energy-efficient broadcasting with cooperative transmission in wireless ad hoc networks. *IEEE Trans. Wireless Commun.* 5(10).

[20] S. Wei, D. L. Goeckel, and M. Valenti. 2006. Asynchronous cooperative diversity. *IEEE Trans. Wireless Commun.* 5:1547–57.

[21] M. Janani, A. Hedayat, T. E. Hunter, and A. Nosratinia. 2004. Coded cooperation in wireless communications: Space-time transmission and iterative decoding. *IEEE Trans. Signal Processing* 52:362–71.

[22] G. Kramer, M. Gastpar, and P. Gupta. 2005. Cooperative strategies and capacity theorems for relay networks. *IEEE Trans. Inf. Theory* 51:3037–63.

[23] J. N. Laneman and G. W. Wornell. 2003. Distributed space-time-coded protocols for exploiting cooperative diversity in wireless networks. *IEEE Trans. Inf. Theory* 49:2415–25.

[24] Y.-W. P. Hong and A. Scaglione. 2008. Group testing for binary Markov sources: Data-driven group queries for cooperative sensor networks. *IEEE Trans. Inf. Theory*.

[25] Y.-W. Hong and A. Scaglione. 2004. On multiple access for distributed dependent sources: A content-based group testing approach. In *Proceedings of the IEEE Information Theory Workshop*, San Antonio, TX, pp. 298–303.

[26] F. H. P. Fitzek and M. D. Katz, eds. 2006. *Cooperation in wireless networks: Principles and applications: Real egoistic behavior is to cooperate*. The Netherlands: Springer.

[27] R. Rao and A. Ephremides. 1988. On a stability of interacting queues in multi-access system. *IEEE Trans. Inf. Theory* 34:918–30.

[28] W. Szpankowski. 1994. Stability conditions for some multiqueue distributed systems: Buffered random access systems. *Adv. Appl. Probability* 26:498–515.

[29] R. Loynes. 1962. The stability of a queue with non-independent inter-arrival and service times. *Proc. Cambridge Philos. Soc.* 58:497–520.

[30] D. Gross and C. M. Harris. 1998. *Fundamentals of queueing theory.* 3rd ed. New York: John Wiley & Sons.

[31] R. Knopp and P. A. Humblet. 1995. Information capacity and power control in single-cell multiuser communications. In *IEEE International Conference on Communications*, vol. 1, pp. 331–335.

[32] S. Adireddy and L. Tong. 2005. Exploiting decentralized channel state information for random access. *IEEE Trans. Inf. Theory* 51:537–61.

[33] X. Qin and R. A. Berry. 2006. Distributed approaches for exploiting multiuser diversity in wireless networks. *IEEE Trans. Inf. Theory* 52:392–413.

[34] C.-K. Lin and Y.-W. P. Hong. 2008. On the finite-user stability region of slotted ALOHA with cooperative users. In *International Conference on Communications (ICC)*.

[35] Y. E. Sagduyu and A. Ephremides. 2006. A game-theoretic look at throughput and stability in random access. In *Proceedings of the Military Communications Conference (MILCOM)*, pp. 1–7.

[36] E. Bonabeau, M. Dorigo, and G. Theraulaz. 1999. *Swarm intelligence: From natural to artificial systems.* Oxford: Oxford University Press.

[37] Y.-W. Hong and A. Scaglione. 2003. Swarming activities to distribute information in large sensor networks. In *Proceedings of the Military Communications Conference (MILCOM)*, Boston, pp. 682–687.

12

Cooperative Diversity: *Capacity Bounds and Code Designs*

V. Stanković
University of Strathclyde

Anders Høst-Madsen
University of Hawaii

Zixiang Xiong
Texas A&M University

12.1 Introduction

In this chapter we discuss methods for conveying information in a wireless network based on cooperation among network nodes. The methods are particulary suitable for decentralized wireless ad hoc networks where information is transmitted in multihops.

A wireless ad hoc network is a network without infrastructure and centralized control. It consists of a large number of possibly mobile nodes that communicate with each other over wireless links. Low cost and simple reconfiguration make ad hoc networks attractive in both commercial and military applications such as wireless local area network (LAN) (e.g., IEEE 802.11x) and metropolitan area network (MAN) (e.g., DARPA's GLOMO), home networks (e.g., HomeRF), device networks (e.g., based on Bluetooth or ZigBee), and sensor networks (e.g., SmartDust, WINS).

In contrast to a traditional infrastructure wireless network (e.g., a cellular network), where information is transmitted from one user to another via a control base station, an ad hoc network allows peer-to-peer communication from a sending node to a destination node. That is, the information can travel directly from a sending to receiving node in a single hop.

However, since wireless channels are often poor, single-hop routing requires either high transmission power, and consequently causes increased interference, or complex multiple access schemes. To achieve significant power savings and keep complexity low, information should be conveyed to a destination through multiple intermediate nodes. Whereas transmission over a single-hop channel has already been intensively studied and is well understood, cooperative communication in multinode networks is still an open research problem, which recently has received considerable attention, inspired by the papers [1, 2].

Communication over a wireless channel is limited by interference, fading, multipath, path loss, and shadowing. The main design challenge in ad hoc networks lies in devising communication methodologies in a decentralized manner, based on the knowledge of local conditions only, to overcome these limitations. An additional design issue has to do with the high dynamics of an ad hoc network, where nodes frequently join and leave the network.

One way of achieving high performance is to employ multiple transmitter and receiver antennas at nodes. This multiple antenna system increases the capacity and improves robustness to fading and interference by means of *spatial diversity* and *data rate multiplexing* [3, 4]. However, building multiple antennas at each node can be expensive, impractical, and often infeasible, especially for small and simple nodes such as those used in sensor networks.

Another recently proposed solution for achieving spatial diversity without requiring multiple antennas at any node is *cooperative diversity* [1, 2]. It is based on grouping several nodes (each with only one antenna) together into a cluster to form a large transmit or receive antenna array. Collaborative clusters are formed in an ad hoc fashion by negotiations among neighboring nodes without centralized control (see Figure 12.1). Cooperative diversity naturally arises in ad hoc networks as it enables great power savings with cheap, simple, and mobile nodes, while supporting decentralized routing and control algorithms. However, it is not limited to ad hoc networks, as it can be useful in infrastructure networks as well.

The simplest nontrivial setup is when the nodes form pairs, i.e., clusters of two. In a two-transmitter two-receiver cooperative channel, the two single-antenna transmitters want to communicate messages to the two remote single-antenna receivers over the same wireless radio channel. In *transmitter cooperation*, the two transmitters first exchange their messages, and then start to act as a single two-antenna broadcast transmitter. On the other hand, in *receiver cooperation*, the two receivers exchange their received signals and act as a single two-antenna multiple access receiver. In general, the two transmitters as well as the two receivers can collaborate among each other to form a virtual multiple-input multiple-output (MIMO) channel with two transmitter and two receiver antennas. The main goal of node cooperation is to achieve spatial diversity and

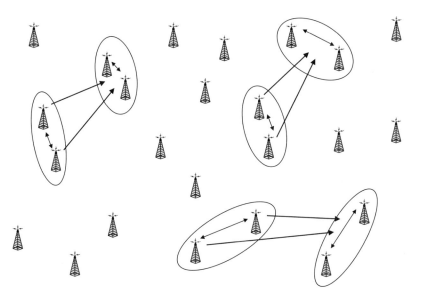

FIGURE 12.1 A wireless network with cooperative diversity. Closely located transmitters form a transmitter cluster, whereas closely located receivers form a receiver cluster.

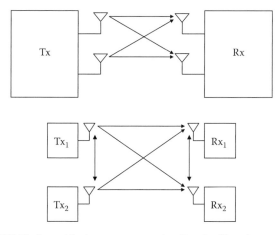

FIGURE 12.2 MIMO channel (up) versus cooperative diversity (down).

rate multiplexing (that is, to mimic a MIMO transceiver) without increasing the number of antennas at a single node.

The difference between a MIMO channel and a cooperative diversity channel is highlighted in Figure 12.2. In a two-antenna MIMO channel, a transmitter/receiver is equipped with two transmitting antennas. In a two-transmitter two-receiver cooperative diversity channel, two antennas at the transmitter/receiver side are physically separated; hence, a somewhat lower performance is expected. However, if the two transmitters/ receivers are closely located, then the performance of a well-designed cooperative diversity system should be close to that of a MIMO system.

Note that the concept of cooperative diversity can be extended to the case where multiple-antenna transceivers cooperate to build a virtual transceiver with a higher number of antennas. However, in this chapter we focus on the simplest scenario where each node is equipped with a single transceiver/receiver antenna.

This chapter is organized as follows. In the next section we give an information-theoretical model of two-transceiver two-receiver cooperative channels. In section 12.3 we present the capacity bounds of the classic three-node relay channel [5], which is a building block of cooperative diversity, and cooperative channels. The next section focuses on recent advances made in practical code designs for the relay, transmitter, and receiver cooperative channels. The final section points to open challenges and opportunities for future research.

12.2   Information-Theoretical Model

Consider the channel model depicted in Figure 12.3 (upper left), where the transmitter in node 1 wants to send message $\omega_1 \in \{1, ..., M_1\}$ to the receiver in node 3; likewise, the transmitter in node 2 intends to send message $\omega_2 \in \{1, ..., M_2\}$ to the receiver in node 4.

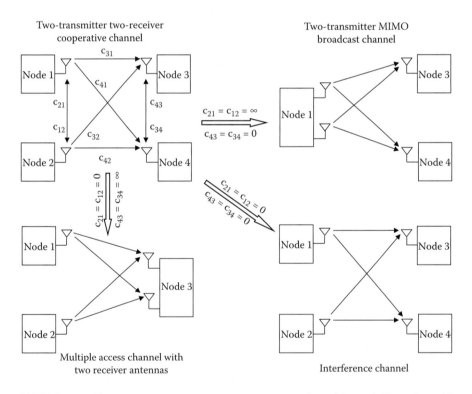

FIGURE 12.3   The two-transmitter two-receiver cooperative channel (upper left) together with its three special cases: MIMO broadcast channel (upper right), multiple access channel (lower left), and interference channel (lower right).

Specifically, node i ($i = 1, 2, 3, 4$) transmits a block $x_i[n]$ of N symbols at a time with $n = 1, \ldots, N$, while being subject to an average power constraint

$$\frac{1}{N} \sum_{n=1}^{N} |x_i[n]|^2 \le P_i.$$

The rate of the transmission from node i is then

$$R_i = \frac{\log M_i}{N}.$$

We assume that the channel between node i and node j is a Rayleigh flat-fading channel [6] with channel coefficient c_{ij}, which is an independent identically distributed (i.i.d.), complex, zero-mean, Gaussian random variable. The nodes can either be able to transmit and receive simultaneously in the same time slot or at the same frequency (full-duplex) or not, in which case the nodes transmit during one time slot, or at one frequency, and receive during the next time slot, or at another frequency (half-duplex).

At the symbol level, the received signals at nodes 1 and 2 are given by

$$y_1[n] = c_{12} x_2[n] + z_1[n] \tag{12.1}$$

and

$$y_2[n] = c_{21} x_1[n] + z_2[n], \tag{12.2}$$

respectively, and the signals received by nodes 3 and 4:

$$y_3[n] = c_{31} x_1[n] + c_{32} x_2[n] + c_{34} x_4[n] + z_3[n] \tag{12.3}$$

and

$$y_4[n] = c_{41} x_1[n] + c_{42} x_2[n] + c_{43} x_3[n] + z_4[n], \tag{12.4}$$

respectively, where z_i, $i = 1, \ldots, 4$, are i.i.d., circular, complex, zero-mean, additive, Gaussian noises. Without loss of generality, we assume that the noises are of unit power and $c_{31} = c_{42} = 1$. In the transmitter cooperative channel, we set $c_{43} = c_{34} = 0$, and for receiver cooperation, $c_{21} = c_{12} = 0$.

If cooperation is perfect, then transmitter cooperation leads to a two-antenna MIMO broadcast channel [7] ($c_{21}, c_{12} \to \infty$ in the sense that the channel between the two transmitters becomes an ideal noiseless channel with zero delay), receiver cooperation reduces to a two-user multiple access channel (MAC) [8] with two receiver antennas ($c_{43}, c_{34} \to \infty$), and the general setup with both transmitter and receiver cooperation becomes a single MIMO channel [3] with two transmitter and two receiver antennas ($c_{21}, c_{12}, c_{43}, c_{34} \to \infty$).

On the other hand, when cooperation is not allowed, i.e., $c_{21} = c_{12} = c_{43} = c_{34} = 0$, the channel degenerates to the interference channel [9]. Figure 12.3 depicts three of these four simplifications.

When we restrict the channels to be quasi-static, then all channel coefficients are constant during transmission of each block of N symbols. In the *synchronous model* of (12.1)–(12.4), we assume that the nodes are perfectly synchronized and have full channel state information (CSI), i.e., each node knows instantaneous values of all channel coefficients and their statistics. While it is relatively simple to achieve symbol/time synchronization between nodes, carrier synchronization, which requires phase-locking separated microwave oscillators, is challenging in practice. Therefore, we also consider the *asynchronous model*, where random phase offsets due to oscillator fluctuations are added to the transmitted signals. We include these random phases in the channel coefficients, so that the model stays the same as (12.1)–(12.4). Under the asynchronous model for receiver cooperation, the transmitters do not have any CSI, whereas the receivers need to know only the magnitudes of all channel coefficients, not their phases. Thus, receiver cooperation is suitable in the systems with simple transmitters. On the other hand, under the asynchronous model for transmitter cooperation, the transmitters must know the magnitudes of all channel coefficients.

We discuss the diversity and data rate gains achievable by node cooperation, while focusing on the high-signal-to-noise ratio (SNR) regime, where the data rates are mainly limited by interference. (In the low-SNR regime, the influence of channel noise prevails, and hence the gain from cooperation is reduced.) The *diversity gain* [10], defined as

$$d = -\lim_{\text{SNR}\to\infty} \frac{\log P_e(\text{SNR})}{\log \text{SNR}},$$

shows how fast the probability of decoding error P_e decays with SNR. A higher d means lower P_e at the same SNR, and thus a more reliable system. The data rate gain is usually decoupled into a *multiplexing gain* and an *additive gain*. The multiplexing gain (or degree of freedom) [10] shows how fast the rate increases with SNR and is given by

$$r = \lim_{\text{SNR}\to\infty} \frac{R(\text{SNR})}{\log \text{SNR}},$$

where $R(\text{SNR})$ denotes the sum of data rates of transmitting nodes for a given SNR. The additive gain (or the high-SNR power offset) [11, 12] is a shift of the $R(\text{SNR})$ function from the origin at high SNRs, i.e.,

$$a = \lim_{\text{SNR}\to\infty} R(\text{SNR}) - r\log(\text{SNR}).$$

If all the limits exist, then $R(\text{SNR})$ in the high-SNR regime can be approximated by a line of slope r and SNR offset a, i.e.,

$$R(\text{SNR}) \approx r\log(\text{SNR}) + a.$$

It is well known that perfect cooperation (a two-antenna MIMO transceiver) achieves a diversity gain of $d = 2$ and a multiplexing gain of $r = 2$ [3]. On the other hand, the interference channel (without node cooperation) in Figure 12.3 provides no diversity or multiplexing gain (i.e., $d = 1$ and $r = 1$).

12.3   Capacity Bounds

In this section, we first present capacity bounds for cooperative diversity, indicating a multiplexing gain of only 1 at high SNRs, which is a somewhat negative result. However, the main message is that node cooperation *can* provide a large additive gain and a diversity gain of 2.

While the capacities of most point-to-point channels are known, this is not the case for wireless multinode channels. Indeed, we only know the capacities of the Gaussian MAC and the broadcast channel. For all other multinode channels, e.g., the relay and interference channels, capacities are known only in special cases. However, it is possible to obtain upper and lower bounds on the capacity, which are often very close, thereby practically indicating the capacity. A lower bound is the rate that can be attained by some coding scheme and is therefore an achievable rate. All rates higher than the upper bound cannot be achieved. If the lower and upper bounds overlap entirely, the complete rate region is known. However, if the two bounds do not overlap, the gap between them characterizes the unknown region.

There are two main ideas in obtaining achievable rates for cooperative channels. The first idea is based on nodes decoding messages from other nodes and re-encoding them. The second lies in exploiting the joint statistics between the data at cooperating nodes by means of *coding with side information*, i.e., Wyner-Ziv coding [13] or dirty-paper coding [14]. Specifically, it turns out that Wyner-Ziv coding achieves the capacity of receiver cooperation (asymptotically as the interference and SNR approach infinity), while dirty-paper coding plays a major role in transmitter cooperation. Below we give a brief summary of coding with side information.

12.3.1   Coding with Side Information

Distributed source coding addresses separate compression and joint decompression of correlated sources [8]. Its foundation was laid by Slepian and Wolf [15], who defined the rate region for lossless compression of two correlated discrete sources showing a surprising result: separate encoding and joint decoding suffer no rate loss compared to the case when the sources are compressed jointly. The framework was extended and generalized in [16], where the problem of lossy compression under distortion constraints, called multiterminal source coding, was posed and the bounds given.

A special case of multiterminal source coding is source coding with side information at the decoder, or Wyner-Ziv coding (WZC). The WZC problem considers lossy compression of source X under the distortion constraint when a correlated source S—called side information—is available at the decoder but not at the encoder (see Figure 12.4). This rate-distortion problem was first considered by Wyner and Ziv in [13], where the

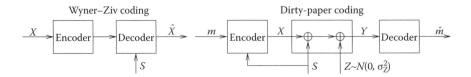

FIGURE 12.4 Coding with side information. WZC refers to lossy source coding of X with decoder side information S, whereas DPC considers channel encoding of message m with encoder side information S over an AWGN channel.

minimum rate for compressing X was derived. In general, WZC incurs a rate loss when compared to the case with S, also available at the encoder. However, if the correlation between X and S is modeled as $X = S + Z$, with Z being an i.i.d., memoryless Gaussian random variable, independent of S, then there is no rate loss with WZC under the mean-squared error (MSE) distortion measure.

The information-theoretical dual [17] of WZC is channel coding with side information at the encoder, or Gelfand-Pinsker coding [18], where the encoder has perfect (noncausal) knowledge of the side information or CSI. The limits on the rate at which messages can be transmitted to a receiver are given in [18]. In general, there is a rate loss compared to the case when the receiver also knows noncausally the CSI, i.e., the encoder side information. However, when the channel is additive white Gaussian noise (AWGN), Gelfand-Pinsker coding does not suffer any rate loss. In this case we have the celebrated dirty-paper coding (DPC) problem [14], shown in Figure 12.4 (right), where the decoder can completely cancel out the effect of the interference caused by the side information.

Practical WZC and DPC both involve source-channel coding. WZC can be implemented by first quantizing the source X, followed by Slepian-Wolf coding of the quantized X with side information S at the decoder [19]. Using syndrome-based channel coding for compression, Slepian-Wolf coding here plays the role of conditional entropy coding. For DPC, source coding is needed to quantize the side information to satisfy the power constraint. In the meantime, the quantizer induces a constrained channel, for which practical channel codes can be designed to approach its capacity. Indeed, limit-approaching code designs [20–22] have appeared for both WZC and DPC recently.

12.3.2 The Relay Channel

Since cooperative diversity is largely based on relaying messages, its information-theoretical foundation is built upon the landmark 1979 paper of Cover and El Gamal [23] on capacity bounds for relay channels. We thus start with the relay channel, give the theoretical bounds on its capacity, and describe proposed coding strategies in the Gaussian and Rayleigh flat-fading environments. Then we proceed with extensions to two-transmitter two-receiver cooperative channels.

The relay channel, introduced by van der Meulen in [5], is a three-node channel where the source communicates to the destination with the help of an intermediate relay node. It is shown in Figure 12.5. The source broadcasts encoded messages to the relay and destination. The relay processes the received information and forwards the resulting signal to the destination. The destination collects signals from both the source and relay

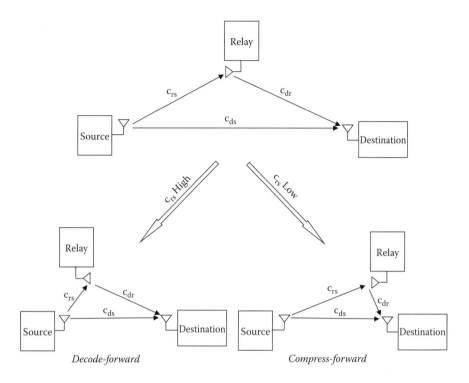

FIGURE 12.5 The wireless relay channel. Decode-forward works better when the relay is close to the source, but compress-forward is preferred when the relay is close to the destination.

node before attempting to recover the information. The relay's task is thus to facilitate decoding at the destination by means of spatial/temporal diversity; that is, it enables the message to be sent via different paths and in different time slots, so that the destination can improve decoding by exploiting two different independently corrupted looks of the same message.

The capacity for the general relay channel is still unknown, and we can only speak of the upper bound, above which reliable communication is not possible, and the lower bound, which can asymptotically be achieved with developed coding strategies; it is still unknown whether there exists a coding scheme that can operate in the uncertainty region between the two bounds.

Cover and El Gamal [23] derived upper and lower bounds on the capacity of the general relay channel using random coding and converse arguments. These two bounds coincide only in a few special cases [24, 25] (e.g., the degraded Gaussian case [23]).

The wireless relay channel is shown in Figure 12.5, where c_{rs}, c_{ds}, and c_{dr} denote channel coefficients. There are two setups in relaying: full-duplex and half-duplex. In the full-duplex setup, the relay is able to transmit and receive simultaneously on the same frequency. The capacity bounds are given by Cover and El Gamal [23], and efficient practical designs performed by Zhang and Duman [26]. Implementing full-duplex relaying, however, is a microwave design challenge (e.g., due to the large difference in the transmitting and receiving signal power levels). A simpler setup is half-duplex relaying, in which the relay does not simultaneously receive and transmit. Half-duplex relaying [24, 27–30] can be implemented with lower complexity by using either time division, frequency division, or code division, which are equivalent from an information theory point of view [8]. In the following, we will focus on time-division half-duplex relaying and discuss both capacity bounds and practical designs.

In time-division relaying, a frame of length n is divided into two parts: a relay-receive period of length $n\alpha$, $0 \le \alpha \le 1$, and a relay-transmit period of length $n(1 - \alpha)$. In the relay-receive period, the source transmits a codeword x_{s1}. The relay overhears this transmission, processes its received signal y_r in some way, and transmits a codeword $x_r = f_r(y_r)$ in the relay-transmit period. While the relay transmits, the source simultaneously transmits another codeword x_{s2}. The codeword x_{s2} is not heard by the relay, as it is in transmit mode, and is therefore transmitted directly to the destination. One way to accomplish this is to split the message $m \in \{1, ..., M\}$ into two parts, m_1 and m_2, at the source. Then, m_1 is encoded into the $n\alpha$-length codeword $x_{s1}(m_1)$, and the remaining m_2 is encoded into an $n(1 - \alpha)$-length codeword $x_{s2}(m_2)$. At the symbol level, the received signals at the relay and destination during the relay-receive period are

$$y_r[n] = c_{rs} x_{s1}(m_1)[n] + z_{rs}[n] \tag{12.5}$$

and

$$y_{d1}[n] = c_{ds} x_{s1}(m_1)[n] + z_{ds}[n], \tag{12.6}$$

respectively, where z_{rs} and z_{ds} are independent white Gaussian noises with unit power. During the relay-transmit period in the asynchronous case, the relay sends an $n(1 - \alpha)$-length codeword $x_r(m_1)$ to the destination, which receives

$$y_{d2}[n] = c_{ds} x_{s2}(m_2)[n] + c_{dr} x_r(m_1)[n] + z[n], \tag{12.7}$$

where z is again a white Gaussian noise with unit power.

In the synchronous case, the system can additionally use the antennas at the source and the relay as a two-antenna transmit array. Suppose that the source is able to completely predict what the relay will send in the relay-transmit period; then the source can transmit the same signal with a phase shift calibrated so that the two signals add up coherently at the destination. The received signal is then

$$y_{d2}[n] = c_{ds} x_{s2}(m_2)[n] + (c_{dr} + c_{ds}A) x_r(m_1)[n] + z[n], \tag{12.8}$$

where A is a complex constant subject to a power constraint and with such a phase that $|c_{dr} + c_{ds}A|$ is maximized. If the source can only partially predict what the relay will transmit, it is still possible to take advantage of this partial coherency.

The optimum operation at the relay is not known, but several coding schemes have been proposed [2, 23, 25, 27] to obtain achievable bounds on the rate region. These schemes can be classified into *decode-forward* and *observe-forward* [23], although hybrid schemes are also possible [23].

The main operation of decode-forward (DF) is full decoding at the relay node. Upon receiving y_r, the relay node first decodes m_1 and then re-encodes it before forwarding the resulting codeword $x_r(m_1)$ to the destination during the relay-transmit period. It should be emphasized that the relay might use a different codebook than the source. In any case, the source can completely predict what the relay will transmit, and full coherency is therefore possible. The destination attempts to reconstruct message m by combining the signals received during the relay-receive and relay-transmit periods using either successive list decoding [8, 23], backward decoding [31], or decoding based on parallel Gaussian channel arguments [28], which all result in the same achievable rate region. Although DF can be very efficient in some scenarios [24, 25], since the relay must perfectly decode the source message, the achievable rates are bounded by the capacity of the channel between the source and the relay. That is, the capacity of the DF relay channel cannot be higher than that of the source-relay channel; hence, if the channel between the source and relay is poor, the relay is useless, and direct transmission from the source to the destination is a better option.

To alleviate this problem of DF, a class of observe-forward schemes has been proposed, where the relay does not attempt to decode the signal from the source; it merely forwards a processed version of its received signal to the destination.

The simplest observe-forward scheme is *amplify-forward* (AF) [32], in which the relay, sticking to its rudimentary role, just amplifies the received signal before forwarding. A more sophisticated scheme is *compress-forward* (CF), which is rooted in the original work of Cover and El Gamal [23], where the relay compresses the signal it has received from the source within certain distortion. Since y_r (received by the relay node) and y_{d1} (received by the destination node) are independently corrupted versions of the same encoded message $x_{s1}(m_1)$, they are correlated. Thus, the relay node can employ WZC [13] when compressing y_r by treating y_{d1} as the decoder side information. The Wyner-Ziv compressed signal is then channel encoded to $x_r(m_1)$ before being forwarded to the destination, which recovers m_2 and m_1 using successive cancellation decoding that involves several steps. First, $\hat{x}_r(m_1)$ is reconstructed by assuming $x_{s2}(m_2)$ as the noise, and then it is subtracted from y_{d2} before m_2 is decoded. Second, to reconstruct m_1, y_r is estimated from $\hat{x}_r(m_1)$ using Wyner-Ziv decoding with y_{d1} as the decoder side information; maximum ratio combining on y_{d1} and the obtained estimate $\hat{y}_r$ is then invoked to recover m_1. CF based on WZC has higher computational complexity than DF, but it gives many rate points that are not achievable with any other coding strategies. It provides the best solution [23–25] when the relay is close to the destination node.

12.3.3 Capacity Bounds of the Gaussian Half-Duplex Relay Channel

In the flat-fading environment, channel coefficients vary in time. The upper and lower bounds on capacity can be computed by averaging over all channel realizations (with optimally allocated power). This average capacity is called *ergodic capacity*. The bounds on ergodic capacity for the full-duplex and half-duplex flat-fading relay channel are given in [24]. However, in practice, it is more convenient to use *outage capacity* [33, 34]. Since in the fading channel case, the source does not know the exact value of channel coefficients ($\mathbf{c} = [\mathbf{c}_{sr}, \mathbf{c}_{sd}, \mathbf{c}_{rd}]$) for a particular channel realization, it cannot determine the maximum achievable rate; consequently, the sending rate R can be higher than the capacity of the instantaneous channel realization $C(\mathbf{c})$, in which case the destination cannot decode. The probability of this event, $P(R > C(\mathbf{c}))$, is called the outage probability. The outage capacity is then the maximum achievable rate with the outage probability less than a certain level p; it can be computed as the $(1 - p)$ percentile of the rate for the specific value of $\mathbf{c}$.

In the following, we summarize the outage capacity bounds of the Gaussian half-duplex relay channel, which can also be used for computing the outage capacity of a wireless quasi-static flat-fading channel (as done in [24]). The bounds for the full-duplex Gaussian channel can be found in [8, 23, 24].

Under the assumption that the nodes are synchronized and have perfect CSI, i.e., each node knows instantaneous values of all channel coefficients and their statistics, an upper bound on the capacity of the Gaussian half-duplex relay channel (although channel coefficients are in general assumed to be complex, in this case they are positive real constants) is derived in [24, 27] and given by

$$C_{ub} = \max_{0 \le \rho \le 1, 0 \le \alpha \le 1, 0 \le k \le 1} \min\{C_{ub1}, C_{ub2}\}, \tag{12.9}$$

where

$$C_{ub1} = \frac{\alpha}{2} \log\left(1 + \left(|c_{rs}|^2 + |c_{ds}|^2\right) \frac{kP_s}{\alpha}\right) + \frac{1-\alpha}{2} \log\left(1 + (1-\rho^2)|c_{ds}|^2 \frac{(1-k)P_s}{1-\alpha}\right),$$

$$C_{ub2} = \frac{\alpha}{2} \log\left(1 + |c_{ds}|^2 \frac{kP_s}{\alpha}\right)$$

$$+ \frac{1-\alpha}{2} \log\left(1 + |c_{ds}|^2 \frac{(1-k)P_s}{1-\alpha} + |c_{dr}|^2 P_r + \frac{2\sqrt{\rho^2 |c_{ds}|^2 |c_{dr}|^2 (1-k)P_s P_r}}{1-\alpha}\right),$$

and P_s and P_r are the average source and relay power constraints, respectively. The parameter ρ reflects the correlation between the source and relay signals, and it can be written in closed form [24, 27]. It is clear from the bound above that the highest multiplexing gain r is 1. However, the full-diversity gain of 2 can be achieved with a simple AF scheme [2, 35].

The rate bound of DF is [24, 27]:

$$R_{DF} \le \max_{0 \le \rho \le 1, 0 \le \alpha \le 1, 0 \le k \le 1} \min\{R_{DF1}, R_{DF2}\}, \tag{12.10}$$

where

$$R_{DF1} = \frac{\alpha}{2} \log \left(1 + |c_{ds}|^2 \frac{kP_s}{\alpha} \right) + \frac{1-\alpha}{2} \log \left(1 + (1-\rho^2)|c_{ds}|^2 \frac{(1-k)P_s}{1-\alpha} \right)$$

and $R_{DF2} = C_{ub2}$. The achievable rate with CF is [24]

$$R_{CF} \leq \max_{0 \leq \alpha \leq 1, 0 \leq k \leq 1} \{ R_{CF1}(\alpha, k) + R_{CF2}(\alpha, k) \},\qquad(12.11)$$

where

$$R_{CF1}(\alpha, k) = \alpha \frac{1}{2} \log \left(1 + |c_{ds}|^2 \frac{kP_s}{\alpha} + \frac{|c_{rs}|^2 kP_s}{\alpha(1+\sigma_\omega^2)} \right)$$

and

$$R_{CF2}(\alpha, k) = (1-\alpha)\frac{1}{2} \log \left(1 + |c_{ds}|^2 \frac{(1-k)P_s}{1-\alpha} \right),$$

with σ_ω^2 being the WZC noise [13] given by

$$\sigma_\omega^2 = \frac{\alpha + \left(|c_{rs}|^2 + |c_{ds}|^2 \right) kP_s}{\left\{ \left[1 + |c_{dr}|^2 P_r \Big/ \left(1-\alpha+|c_{ds}|^2 (1-k)P_s \right) \right]^{\frac{1-\alpha}{\alpha}} - 1 \right\} \left(\alpha + |c_{ds}|^2 kP_s \right)}.$$

DF and CF give the best-known results on the achievable rates for the half-duplex relay channel (however, a hybrid approach may give a higher rate). Depending on the parameters, either DF or CF can be superior. Indeed, DF outperforms CF when the link between the source and relay is better than that between the relay and destination (e.g., when the relay is close to the source). On the other hand, CF provides higher rates when the link between the relay and destination is clean (e.g., when the relay is close to the destination). See Figure 12.5. We show in Figure 12.6 for one setup the rate bounds in (12.10) and (12.11), achievable with DF and CF, respectively, together with the upper bound given by (12.9) and the rate bound with multihop transmission, which is given by the minimum between the capacity at the source-relay link, $\frac{1}{2}\log(1 + |c_{rs}|^2 P_s)$, and the capacity at the relay-destination link, $\frac{1}{2}\log(1 + |c_{dr}|^2 P_r)$. We plot the rate gain relative to direct transmission (i.e., no relaying) as a function of $|c_{rs}|^2$. The increase in $|c_{rs}|^2$ can be construed as the result of decreased distance between the source and the relay. It is seen from Figure 12.6 that CF outperforms DF for low $|c_{rs}|^2$. When $|c_{rs}|^2 < |c_{ds}|^2 = 0$ dB, DF is worse than direct transmission. On the other hand, CF always outperforms direct

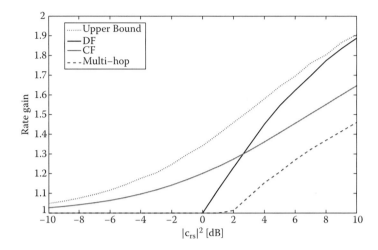

FIGURE 12.6 The multihop bound and the upper bound on the capacity together with the achievable bounds of DF and CF for the Gaussian half-duplex relay channel, assuming $|c_{ds}|^2 = 0$ dB, $|c_{dr}|^2 = 10$ dB, and $P_s = P_r = 5$ dB. The rate gain over direct transmission is shown as a function of $|c_{rs}|^2$.

transmission. Thus, even if the link between the source and relay is poor, the relay can still help somewhat by using CF.

It is instructive to compare relay channel signaling with a traditional multihop ad hoc network, where physical layer communication and networking are typically separated. Such a comparison will show how cooperative diversity can help increase the performance over traditional networking. In a traditional multihop network, the source transmits a packet either directly to the destination or to the relay, which would decode it, re-encode it, and transmit it to the destination. Relay channel signaling improves upon this in several ways:

1. The destination uses the signals from both the source and relay for decoding, as opposed to only one of them.
2. The relay uses a different codebook for encoding in DF than the source, which is similar to using error-correcting codes rather than repetition coding.
3. The relay can use soft information, as in CF, which resembles using soft decisions in decoding error-correcting codes rather than hard decisions.
4. The source is allowed to transmit new information simultaneously with the relay's transmission, which at high SNR brings a large increase in rate.
5. In the synchronous case, the relay can use coherency to combine signals constructively, achieving a gain similar to that in MIMO systems.

12.3.4 Receiver Cooperation

In receiver cooperation, two (closely located) single-antenna receivers cooperate to facilitate decoding messages from two remote single-antenna transmitters.

The channel model is shown in Figure 12.3 (upper left) with $c_{21} = c_{12} = 0$, where node 1 (node 2) wants to communicate to node 3 (node 4). We consider only the full-duplex case, i.e., nodes 3 and 4 can simultaneously transmit and receive. Since the cooperative channel can be viewed as a combination of the interference channel and the relay channel, its best achievable rate regions are obtained by combining DF or CF coding techniques for the relay channel with coding for the interference channel [9]. In receiver cooperation, a receiver node processes the received information and forwards the result to the other receiver node to help decoding. Because the distance between the two receivers is expected to be much smaller than that between a transmitter and a receiver, CF with WZC provides the highest achievable rates. Indeed, from (12.2), the received signals in nodes 3 and 4 at time instants i and $i + 1$ are

$$y_3[i] = x_1[i] + c_{32}x_2[i] + c_{34}x_4[i] + z_3[i],$$

$$y_3[i+1] = x_1[i+1] + c_{32}x_2[i+1] + c_{34}x_4[i+1] + z_3[i+1], \quad (12.12)$$

$$y_4[i] = c_{41}x_1[i] + x_2[i] + c_{43}x_3[i] + z_4[i],$$

$$y_4[i+1] = c_{41}x_1[i+1] + x_2[i+1] + c_{43}x_3[i+1] + z_4[i+1].$$

In CF [36], the receiver in node 3 (or node 4) employs WZC to compress the signal $y_3[i]$ (or $y_4[i]$) it has received, while assuming $y_4[i]$ (or $y_3[i]$) as the decoder side information, before passing the resulting codeword $x_3[i + 1]$ (or $x_4[i + 1]$) to the collaborating receiver in node 4 (or node 3). Node 3 starts by decoding $x_4[i + 1]$ from $y_3[i + 1]$ while treating $x_1[i + 1] + c_{32}x_2[i + 1]$ as part of the Gaussian noise. In *forward decoding*, $x_4[i]$, recovered in the previous time instant, is Wyner–Ziv decoded using $y_3[i]$ as the decoder side information, resulting in an estimate of $y_4[i]$. Next, the joint or individual decoding technique [9] proposed for the interference channel is employed to reconstruct $x_1[i]$ (and $x_2[i]$) from the obtained estimates

$$y_3[i] - c_{34}\hat{x}_4[i] = x_1[i] + c_{32}x_2[i] + z_3[i]$$

and

$$\hat{y}_4[i] - c_{43}x_3[i] = c_{41}x_1[i] + x_2[i] + z_4[i].$$

A similar procedure can be performed at node 4. Besides forward decoding, it is also feasible to employ *backward decoding*, where the decoder starts by decoding the previously received block of symbols and proceeding backwards. In [36], forward decoding is combined with either joint or individual decoding [9], and backward decoding is used with joint decoding, giving three different decoding choices. Since nodes 3 and 4 can use three different decoding methods each, there are nine possibilities, each providing a different rate bound. To obtain the best achievable CF rate bound, the maximum of all nine rate bounds should be taken.

In a similar manner, DF can be extended to receiver cooperation. The DF strategy for receiver cooperation can be found in [36].

12.3.5　Capacity Bounds of the Gaussian Full-Duplex Receiver Cooperative Channel

In this section we summarize the results of [36], where the upper and lower bounds on capacity of receiver cooperation are derived for both CF and DF. The derived upper bounds are tighter than the standard max-flow-min-cut bound. It is given by:

$$R_1 + R_2 \leq \log\left(1 + |c_{41}|^2 P_1 + P_2 + |c_{43}|^2 P_3\right) + \log\frac{1 + \left(1 + |c_{41}|^2\right)P_1}{1 + |c_{41}|^2 P_1}$$

in the asynchronous case, and

$$R_1 + R_2 \leq \log\left(1 + |c_{41}|^2 P_1 + P_2 + |c_{43}|^2 P_3 + 2\sqrt{|c_{43}|^2 P_2 P_3 + |c_{43}|^2 |c_{41}|^2 P_1 P_3}\right) + \log\frac{1 + \left(1 + |c_{41}|^2\right)P_1}{1 + |c_{41}|^2 P_1}$$

in the synchronous case. Note that we get a symmetric set of rate bounds if nodes 1 and 2 are exchanged with nodes 3 and 4. Achievable (lower) rate bounds for CF and DF can be found in [36]. Figure 12.7 shows the sum-rate $R_1 + R_2$ as a function of the received SNR on the direct link between nodes 1 and 3. The received SNR at the link between nodes 3 and 4 is 30 dB higher than that from the direct link—an indication that the cooperating nodes are close together. The average powers of all four nodes are the same. All channels are independent Rayleigh flat fading, meaning that each c_{ji} is an i.i.d. Gaussian random variable, and the results are averaged over simulated ensembles of channel realizations. For comparison purposes, the rate limits of the two-user MAC with two antennas at the receiver (with perfect cooperation) and the interference channel (without cooperation) are included.

It is seen from Figure 12.7 that receiver cooperation with CF gives an additive gain that can be up to 20 dB higher than no cooperation, CF always performs close to the upper bound, and there is no gain from synchronization. Interestingly, receiver cooperation performs close to using two receiver antennas at low and medium SNRs, thus providing a multiplexing gain of 2. However, at the high SNRs, the multiplexing gain drops to 1, and the rate gain over the noncooperative case boils down to a high additive gain.

We can see that, in the high-SNR regime, receiver cooperation gives a multiplexing gain of only $r = 1$, as opposed to the two-user MAC with two receiver antennas, which results in $r = 2$. However, the additive gain with receiver cooperation, which is upper bounded by

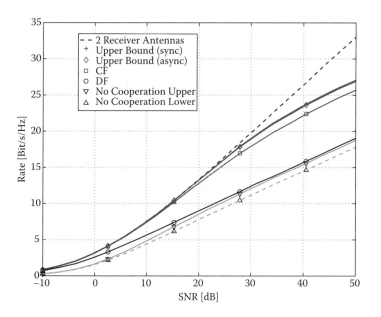

FIGURE 12.7 Bounds on the sum-rate $R_1 + R_2$ as a function of the received SNR from the direct link between nodes 1 and 3 for receiver cooperation. The received SNR at the link between nodes 3 and 4 is 30 dB higher than that at the direct link. The average powers of all four nodes are the same.

$$a \leq \min \left\{ \log \left(\left| c_{41} \right|^2 P_1 + P_2 + \left| c_{43} \right|^2 P_3 \right) + \log \left(1 + \left| c_{41} \right|^{-2} \right), \right.$$
$$\left. \log \left(P_1 + \left| c_{32} \right|^2 P_2 + \left| c_{34} \right|^2 P_4 \right) + \log \left(1 + \left| c_{32} \right|^{-2} \right) \right\},$$

(12.13)

can be very high. On the other hand, for $\left| c_{41} \right|, \left| c_{32} \right| > 1$ it is shown in [36] that CF with forward joint decoding gives an additive gain of

$$a = \min \left\{ \log \left(\left| c_{41} \right|^2 P_1 + P_2 + \left| c_{43} \right|^2 P_3 \right), \log \left(P_1 + \left| c_{32} \right|^2 P_2 + \left| c_{34} \right|^2 P_4 \right) \right\}.$$

(12.14)

Note that the gain in (12.14) is identical to that in (12.13) except for the $\log(1 + \left| c_{41} \right|^{-2})$ and $\log(1 + \left| c_{32} \right|^{-2})$ terms, which are small for large $\left| c_{41} \right|$ and $\left| c_{32} \right|$. Thus, CF with WZC achieves capacity asymptotically as $\left| c_{41} \right|$, $\left| c_{32} \right|$, and the SNRs go to infinity. All other cooperative strategies (including DF) give no additional gain over that of no cooperation, whose additive gain is

$$a = \min \left\{ \log \left(\left| c_{41} \right|^2 P_1 + P_2 \right), \log \left(P_1 + \left| c_{32} \right|^2 P_2 \right) \right\}.$$

(12.15)

However, these strategies are useful in the medium- or low-SNR regimes. From (12.14) and (12.15) we see that the gain of receiver cooperation comes from the terms $|c_{43}|^2 P_3$ and $|c_{34}|^2 P_4$, which depend on the channel between the two receivers. Since this channel is expected to be good (a node should cooperate with its "best neighbor"), this gain can be very high. An interesting conclusion from [36] is that the gain from exploiting full synchronization in receiver cooperation is very limited. Thus, in practice it is enough to resort to the asynchronous cooperation, which significantly saves the hardware cost. However, as pointed out in [37], optimal power allocation is essential in achieving the full additive gain.

12.3.6 Transmitter Cooperation

In transmitter cooperation, two (close) single-antenna transmitters collaborate in communicating to two (remote) single-antenna receivers. The channel model is depicted in Figure 12.3 (upper left) with $c_{43} = c_{34} = 0$. As in receiver cooperation, we restrict to the full-duplex case, where nodes 1 and 2 can simultaneously transmit and receive.

It is shown in [36, 37] that, in contrast to receiver cooperation, synchronization helps a lot when the transmitters cooperate. That is, if the two transmitters are synchronized, they can completely cancel out the interference using DPC.

The DPC technique was exploited in [7, 38] to find the capacity of the Gaussian MIMO broadcast channel. For the two-antenna broadcast channel with two receivers [7], the main idea is to decompose the MIMO channel into two interference channels and perform successive encoding, in which the message for the second receiver is dirty-paper encoded while assuming the previously encoded message for the first receiver is known interference (the side information). In this way, the second receiver can completely cancel out the interference from the signal for the first receiver. However, to achieve full capacity, the transmitter has to perform optimal channel decomposition using precoding with the output vector $\mathbf{x} = \mathbf{B}[u_1 \ u_2]^T$, where $\mathbf{B}$ is a 2×2 precoding matrix that has to satisfy the power constraint, and u_1 and u_2 are the encoded codewords (with unit power) intended for the first and second receivers, respectively, and obtained via successive dirty-paper encoding. Assuming a 2×2 channel matrix $\mathbf{H}$ and unit-power Gaussian noise, the achievable rates for the first and second receivers are

$$R_1 = \log\left(1 + \frac{|s_{11}|^2}{1 + |s_{12}|^2}\right)$$

and $R_2 = \log(1 + |s_{22}|^2)$, respectively, where s_{ij} are the entries of matrix $\mathbf{S} = \mathbf{HB}$.

The coding strategy of [7] is extended to transmitter cooperation in [36, 39]. In [39], it is assumed that the channel between the two transmitters is orthogonal to the channels between the transmitters and receivers (which can be achieved by means of multiple access techniques). Thus, collaboration between the transmitters does not cause interference at the receivers, which simplifies the code design. This orthogonality assumption is removed in [36], where a coding scheme that in each time instant exploits *three* DPCs is proposed.

We outline here a simplified (hence more practical) solution proposed in [40] based on *one* DPC (in conjunction with backward decoding). During the ith time instant, the transmitter in node j, $j = 1, 2$, sends

$$x_j[i] = A_j U_j\left(\omega_j[i]\right) + t_{j1} U_1^0\left(\omega_1[i-1], U_2^0\left(\omega_2[i-1]\right)\right) + t_{j2} U_2^0\left(\omega_2[i-1]\right).$$

Here U_2^0, U_1, and U_2 are Gaussian codebooks (e.g., standard channel codes in practice) of unit power that encode $\omega_2[i-1]$, $\omega_1[i]$, and $\omega_2[i]$, respectively. U_1 and U_2 are used for exchanging messages between the transmitters and appear as part of the background noise at the receivers. Assuming correct decoding of $U_1(\omega_1[i-1])$ and $U_2(\omega_2[i-1])$ in time instant $i - 1$, the two transmitters can now act as a single two-antenna broadcast transmitter, and the coding strategy of [7] described above can be applied. Thus, the unit-power codebook U_1^0 can encode $\omega_1[i-1]$ using DPC with $U_2^0(\omega_2[i-1])$ as the side information. The scaling factors A_i and t_{ij} are selected to maximize the rate while satisfying the input power constraints.

In the asynchronous case, DPC cannot be exploited, and the resulting known achievable rates are strictly below those in the synchronous case. However, so far there exist no upper bounds that actually prove that the gains cannot be obtained without synchronization. Although it is possible to use CF based on WZC in transmitter cooperation, since the two transmitters are closely located, DPC always dominates. This is why WZC is not considered in this setup.

12.3.7 Capacity Bounds of the Gaussian Full-Duplex Transmitter Cooperative Channel

In this section we summarize the capacity bounds of the Gaussian full-duplex transmitter cooperative channel. As in the case of receiver cooperation, we give only final results without derivations. All proofs can be found in [36].

Based on the argument exploited in [7] that the capacity region depends only on the marginal distribution of the noises at the receivers and not on their correlation, the following upper bounds on capacity are derived in [36]. For $|c_{41}| < 1$, the upper bound is

$$R_2 \leq \log \frac{1 + |c_{41}|^2 P_1 + P_2}{|c_{41}|^2 2^{R_1} \dfrac{1 + P_1}{1 + \left(1 + |c_{21}|^2\right)P_1} + 1 - |c_{41}|^2},$$

in the asynchronous case, and

$$R_2 \leq \log \frac{1+\left(|c_{41}|\sqrt{P_1}+\sqrt{P_2}\right)^2}{|c_{41}|^2 2^{R_1} \dfrac{1+P_1}{1+\left(1+|c_{21}|^2\right)P_1}+1-|c_{41}|^2}$$

in the synchronous case, and the sum-rate bound $R_1 + R_2$ is an increasing function of R_1. If $|c_{41}| > 1$, the upper bound is

$$R_1 + R_2 \leq \log\left(1+|c_{41}|^2 P_1 + P_2\right) + \log \frac{1+\left(|c_{41}|^2+|c_{21}|^2\right)P_1}{1+|c_{41}|^2 P_1}$$

in the asynchronous case, and

$$R_1 + R_2 \leq \log\left(1+\left(|c_{41}|\sqrt{P_1}+\sqrt{P_2}\right)^2\right) + \log \frac{1+\left(|c_{41}|^2+|c_{21}|^2\right)P_1}{1+|c_{41}|^2 P_1} \qquad (12.16)$$

in the synchronous case. There is also a symmetric set of rate bounds by exchanging nodes 1 and 2 with nodes 3 and 4.

Achievable bounds in asynchronous (without DPC) and synchronous systems (with three DPCs) can be found in [36]. The achievable bound for the scheme that used one DPC is given in [40]. Figure 12.8 shows the sum-rate bounds $R_1 + R_2$ as functions of the received SNR on the direct link between nodes 1 and 3. The simulation setup is similar to that for receiver cooperation with the received SNR at the cooperative link (between nodes 1 and 2) being 30 dB higher than that at the direct link, again indicating that the cooperating transmitters are close together.

The achievable bounds of the synchronous system with DPC are usually close to the upper bound, although noticeable gaps exist in certain SNR ranges. There is only a small performance loss if only one DPC is used instead of three. The additive gain compared to the noncooperative case is up to 15 dB in the high-SNR regime. Transmitter cooperation with DPC performs close to using two transmitter antennas at low and medium SNRs, giving a multiplexing gain of 2. However, at high SNRs, the multiplexing gain is only 1.

Similar to receiver cooperation, in the high-SNR regime, transmitter cooperation only gives a multiplexing gain of $r = 1$ (in contrast to the two-antenna broadcast channel, which results in $r = 2$). The additive gain can be high. For example, when $|c_{41}| < 1$, in the synchronous case it is bounded by

$$a \leq \log\left(\left(|c_{41}|\sqrt{P_1}+\sqrt{P_2}\right)^2\right) + \log\left(\frac{1+|c_{21}|^2}{|c_{41}|^2}\right). \qquad (12.17)$$

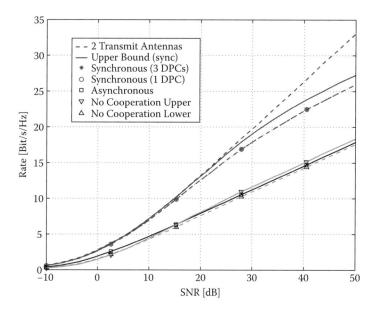

FIGURE 12.8 The bounds on the sum-rate $R_1 + R_2$ as a function of the received SNR at the direct link between nodes 1 and 3 for transmitter cooperation. The received SNR at the link between nodes 1 and 2 is 30 dB higher than that at the direct link. The average powers of all four nodes are the same. The upper bound is for the synchronous system.

By exchanging nodes 1 and 2 with nodes 3 and 4, we get another symmetric rate bound. Besides the multiplexing gain of $r = 1$, DPC achieves a high-SNR additive gain of

$$a = \log\left(\left(|c_{41}||t_{12}| + |t_{22}|\right)^2\right) + \log\frac{|c_{21}|^2}{|c_{41}|^2}. \tag{12.18}$$

Comparing (12.17) and (12.18), we see that the additive gain of using DPC is approximately equal to that from the upper bound when $|t_{12}|^2 \approx P_1$ and $|t_{22}|^2 \approx P_2$, which corresponds to the scenario with weak interference, i.e., $|c_{41}|, |c_{32}| \ll 1$. In this case, the gain compared to no cooperation in (12.15) is $\min\{\log(|c_{12}|), \log(|c_{21}|)\}$, which can be significant, because the channel between the two transmitters is expected to be good. This is illustrated in Figure 12.9, which shows the high-SNR additive gain for a symmetric cooperative channel ($|c_{41}| = |c_{32}|$, $|c_{21}| = |c_{12}|$). Note that under strong interference, i.e., $|c_{32}|, |c_{41}| \gg 1$, there is no gain from cooperation, which might seem somewhat surprising and is in contrast to receiver cooperation. For weak interference, on the other hand, there is a high gain from cooperation. This is true even when the link between the transmitters is weak ($|c_{21}|^2 = -6$ dB in Figure 12.9), and it can be explained by the fact that there is no known signaling for the interference channel with weak interference, while cooperation can help in this scenario.

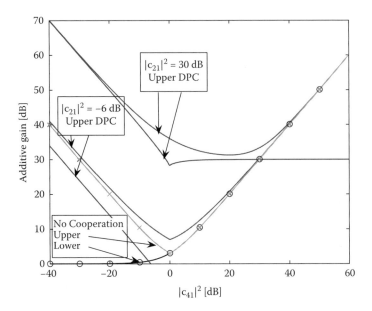

FIGURE 12.9 The bounds on the additive gain for a symmetric transmitter cooperative channel ($|c_{41}| = |c_{32}|, |c_{21}| = |c_{12}|$), averaged over the relative phases of c_{41} and c_{32}. The upper bound is for a synchronous system.

12.4   Practical Designs

In the previous section, we described coding methods that could lead to the highest rates while assuming ideal coding, signaling, and infinite block length. In this section we describe practical systems based on AF, DF, and CF for the relay and cooperative channels, with the focus on capacity-approaching code designs.

Efficient practical coding protocols for wireless relay channels appeared in [41], where, for AF and DF schemes, receivers based on maximum-likelihood and maximum SNR criteria were developed. The work was extended to wireless multirelay channels in [32], where space-time codes [3] with AF and DF are designed to enable simultaneous transmission from all relays on the same channel without receive collision. It is further shown that the proposed schemes achieve the full-diversity gain. An AF space-time code for a single relay is proposed in [35]. Note that the works of [32, 35] only outline space-time code designs without practical implementation of channel codes.

Practical DF schemes for a half-duplex flat-fading relay channel based on distributed convolutional and turbo coding are proposed in [28]. The best scheme of [28] exploits a recursive systematic convolutional code at both the source and the relay. It results in a powerful distributed turbo code, which besides a spatial diversity gain of DF, achieves extra coding gain due to interleaving.

Extending their work on practical full-duplex relaying [26], Zhang and Duman [29] recently provided a DF design for half-duplex fading relay channels, where in a given time slot, the source and relay both transmit over the same channel, resulting in a high rate gain at the price of receive collision. The design in [29] exploits turbo coding with binary phase shift keying (BPSK) modulation [6]. It is consistent with the optimal DF scheme for the half-duplex relay channel described in section 12.3.2, with the simplification that coefficient A in (12.8) is always set to zero. Decoding is based on parallel Gaussian channel arguments [28]. Similar to the MAC setting, the destination exploits a MAP detector to extract information from the received mixture signal. Recent work [42] on low-density parity-check (LDPC) code [43] design for the half-duplex relay channel based on DF also reported a similar loss of 1.2 dB to the theoretical limit.

The systems in [28, 29, 32, 35] demonstrate the great advantage of relaying as compared to direct and multihop transmission. However, because these systems exploit AF or DF, they can approach the lower bound of DF at best, which is far away from the CF limit in many cases. Indeed, as shown in section 12.3.2, when the relay is close to the destination, CF gives rate points that are not achievable with any other coding strategies. Moreover, it was shown in [44] that CF achieves optimal diversity-multiplexing trade-off in half-duplex relay systems.

Practical CF code designs for the half-duplex relay channel recently appeared in [45–47]. The design of [45] is a quantize-forward scheme that does not exploit WZC at the relay. The design of [46] is based on WZC at the relay and uses scalar quantization and convolutional codes, but it does not exploit the limit-approaching CF scheme of [24], and no theoretical bounds or performance comparisons (to the bounds) were given.

A practical CF code design based on WZC for the half-duplex Gaussian relay channel, which closely follows the CF scheme outlined in section 12.3.2, is proposed in [47]. The scheme of [47] exploits BPSK modulation; hence, the signal to be compressed by WZC at the relay and the side information at the destination are not jointly Gaussian as assumed in section 12.3.2 and [36]. Instead, the source and the side information are Gaussian mixture generated from the BPSK modulation. Although the theoretical achievable rate of WZC for this model is yet unknown, a lower bound and an upper bound are derived in [47], which in the case when the relay is close to the destination are close to each other, indicating practically system capacity. The code design relies on practical WZC based on nested lattice quantization [48] followed by Slepian-Wolf coding [15] of the nested quantization index as a second stage of binning for further compression [49]. Thus, Slepian-Wolf coding here plays the role of conditional entropy coding of the nested quantization indices (given the decoder side information).

Practical Slepian-Wolf coding is implemented via channel coding (see [19] for a review of channel code designs for Slepian-Wolf coding). Since the Slepian-Wolf compressed bitstream is to be transmitted over a noisy channel from the relay to the destination, channel coding is needed to protect them. This calls for distributed joint source-channel coding (DJSCC) [50], i.e., joint Slepian-Wolf compression and channel protection. In the practical implementation of [47], irregular repeat-accumulate (IRA) codes [51] were used by designing one multilayer code to take care of two channels: one is the physical noisy channel between the relay and the destination, and another is the "virtual" correlation channel [19], which characterizes the correlation between the quantized source at the relay and the decoder side information at the destination.

In particular, the message m is split at the source into two parts, m_1 and m_2, which are protected independently by two different LDPC codes [43] and BPSK modulated before being transmitted in two separate fractions of a time slot. The relay compresses its received signal using WZC and adds error protection against the noise (and interference) in the link between the relay and the destination. WZC and error protection are performed jointly using DJSCC; the resulting codeword $x_r(m_1)$ is sent during the relay-transmit period. The destination starts by recovering m_2 using successive cancellation decoding: $x_r(m_1)$ is first reconstructed using distributed joint source-channel decoding, and is then subtracted from y_{d2} (interference cancellation) so that m_2 can be recovered with the first LDPC decoder having z as the only noise in the channel. The main idea behind distributed joint source-channel decoding is to view the system as transmitting the symbols over two channels—the first being the actual transmission channel (i.e., the MAC) with noise $z + x_{s2}(m_2)$, which describes the distortion experienced by the parity-check symbols of the IRA code, and the second being the virtual correlation channel [19] between y_r and the side information y_{d1}. The destination estimates $\hat{y}_r$ by employing a conventional IRA decoder for these two parallel channels. Finally, m_1 is recovered using maximum ratio combining (of $\hat{y}_r$ and $y_{d1} = c_{ds}x_{s1}(m_1) + z_{ds}$) and a second LDPC decoder for the direct transmission channel.

Figure 12.10 compares this CF design with the best practical DF design of [29] for the half-duplex Gaussian relay channel. Different theoretical bounds are also depicted (deri-

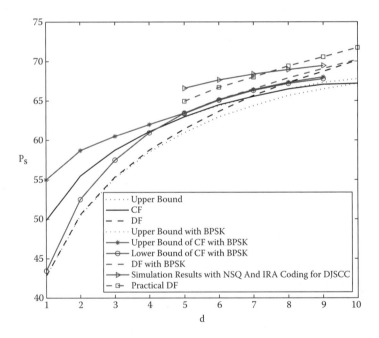

FIGURE 12.10 Half-duplex Gaussian relay channel. The average transmitting power from the source P_s is shown as a function of the distance d from the source to relay. The transmitting rate is 0.5 and the average transmitting power from the relay is $P_r = 70$ dB. The relay is moving along the line from the source to destination. BPSK signaling is assumed.

vations of the bounds in the BPSK case are not included here due to space limitations and interested readers are referred to [47] for details). The transmission rate is fixed at 0.5 bit per channel use, and the average relay power is $P_r = 70$ dB.

It is assumed that the relay is located along a straight line from the source to the destination, which are 10 m apart. The channel coefficient of the link from sender i to receiver j (sender i could be the source or relay, and receiver j could be the relay or destination) is $c_{ij}^2 = K_o d_{ij}^{-n}$ [28], where d_{ij} is the distance from sender i to receiver j, n is the path loss coefficient, $K_o = (c/4\pi d_o f_c)^2$, c is the light speed, d_o is the free-space reference distance, and f_c is the transmission frequency. The experimental setup is fixed with $f_c = 2.4$ GHz carrier frequency, path loss coefficient $n = 3$, and free-space reference distance $d_o = 1$ m.

Figure 12.10 shows the average source power P_s as a function of the distance d between the source and relay. The additive white Gaussian noises over the transmitting channels are all set to be with unit variance. It is seen that, when $d > 8$ m, CF outperforms DF theoretically, and the practical CF code of [47] is preferable to the practical DF code [29]. Note that when 7.5 m $< d <$ 8 m, where DF is superior in theory, the CF scheme still performs better than the practical DF code.

The above practical WZC-based design can be extended to the case of the two-receiver cooperative channel by following the CF coding strategy with forward individual decoding described in section 12.3.4. Each transmitter is equipped with one LDPC channel encoder, and each receiver performs one distributed joint source-channel encoding step and two channel decoding steps.

In contrast to the scarcity of WZC-based CF designs for receiver cooperation, there have been more code designs for transmitter cooperation. For example, since the publication of the work on user cooperation [1], several research groups [32, 52–55] have developed practical designs based on AF and DF for the wireless two-transmitter cooperative channel. A common characteristic of these designs is the avoidance of receive collision by transmitting signals over orthogonal channels, which simplifies the code design. Specifically, the two transmitters (or cooperative partners) send encoded messages over orthogonal channels during the first fraction of a time slot. Each transmitter decodes the signal it has received from its partner and, in the case of successful decoding, either re-encodes the recovered message using the partner's codebook (repetition-based DF) or generates additional parity symbols out of a rate-compatible code (DF based on incremental redundancy). The resulting codewords are then forwarded over orthogonal channels to the receiver during the second fraction of a time slot. If a transmitter cannot successfully decode its partner's message, it switches to either the noncooperative mode or AF. Orthogonal signaling is achieved by using time-, frequency-, or code-division or space-time coding.

A DF design based on incremental redundancy for a flat-fading transmitter cooperative channel, dubbed coded cooperation, is proposed in [52, 53]. Two users, partners, first exchange their messages; then each user sends only additional redundancy bits for its partner message. That is, after successful decoding, the user transmits only additional protection bits of a rate-compatible code for the partner's message, so that its overall code rate is decreased. In addition, whenever the partner's message cannot be decoded,

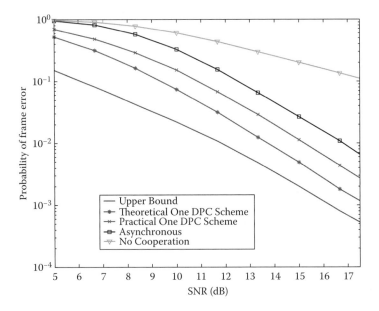

FIGURE 12.11   Simulation results obtained with the DPC-based scheme of [40] for transmitter cooperation, together with various theoretical bounds. The probability of frame error is shown versus the received SNR at the direct link. The rate constraints are fixed at $R_1^o = R_2^o = 1$ bit/sample, and the received SNR at the link between nodes 1 and 2 is 30 dB higher than that at the direct link between nodes 1 and 3.

the user switches to a noncooperative mode. A low-complexity implementation with rate-compatible punctured convolutional (RCPC) codes and a more complex design with space-time turbo coding (efficient on both slow and fast flat-fading channels) are given in [52] and [53], respectively.

A similar DF scheme, based on incremental redundancy, reported in [54] exploits convolutional codes optimized for two-transmitter cooperation in a Rayleigh flat-fading environment. The scheme is shown to be able to achieve the full-diversity gain.

Although the above DF schemes can provide the full-diversity gain, they do so at the expense of decreased rate gain. On the other hand, we know from section 12.3.6 that DPC provides the highest achievable rate over a transmitter cooperative channel. Practical DPC is exploited in [40] to design codes for a wireless two-transmitter cooperative channel. The scheme follows closely coding strategy outlined in section 12.3.6 based on one DPC only. The DPC scheme of [22] is used at the transmitters. It is based on trellis-coded quantization and turbo trellis-coded modulation. Punctured turbo trellis-coded modulation is used to facilitate message exchanges between the two transmitters. Practical design results are shown in Figure 12.11, where the frame error rate against the SNR at the direct link between nodes 1 and 3 is shown. The figure indicates a loss of 1.5 dB from the achievable bound at 2% frame error rate.

12.5   Conclusions

Due to its low complexity and decentralized nature, cooperative diversity arises as a strong candidate for conveying information in emerging wireless ad hoc networks. This motivates research in determining its ultimate performance limit. For a two-transmitter two-receiver cooperative channel, the theory shows that, at least in the high-SNR regime, in contrast to a two-antenna MIMO system, cooperative diversity cannot achieve the full multiplexing gain of 2. Thus, there is a cost paid for having only one antenna at each node. This result indicates that in achieving the full multiplexing gain of 2, tight coordination among the transmit/receive antennas is necessary, which is possible only if they are placed together. However, cooperative diversity does offer high additive rate gains when compared to the noncooperation case, and the key in achieving these gains lies in coding with side information (e.g., DPC and WZC). In transmitter cooperation, synchronization between the two transmitters is essential in obtaining high data rates; DF with DPC is the best coding strategy, coming very close to the upper limits if the interference is weak. Unfortunately, when the interference is strong, transmitter cooperation does not help (in the high-SNR regime). In contrast, receiver cooperation is beneficial in both weak and strong interference scenarios, and CF with WZC is the dominant coding strategy that asymptotically achieves the capacity as the interference and SNR approach infinity. More importantly, full synchronization between the nodes is not necessary for receiver cooperation. Interestingly, as pointed out in [37], optimal power allocation is crucial to realize the full performance gain of receiver cooperation, whereas it provides only a marginal gain in transmitter cooperation.

Owing to its promising application in wireless ad hoc networks, cooperative diversity has been studied intensively recently. Many problems are still open. For example, posed more than 30 years ago, the capacity of the simplest Gaussian relay channel—a building block of cooperative diversity—is still unknown. Recent achievements [38] in providing the full-capacity region for a Gaussian MIMO broadcast channel using DPC might inspire new ideas for solving this problem. The theoretical bounds reported so far are mainly for the full-duplex setups with up to four nodes, where either two transmitters or two receivers cooperate. Providing results for half-duplex cooperative channels should be a research priority. Treating an ad hoc network where the two transmitters and two receivers simultaneously cooperate is another possible research direction. Combining DPC and WZC could lead to the largest achievable rate region, but such a theoretical treatment is not straightforward. In addition, extensions to larger networks with more than four nodes that require cross-layer designs are very challenging because of the additional problem associated with selecting the best partner for cooperation. Also, considering cases where the relay has side information in the form of correlated source [56] or with multiantenna relaying [57] is of interest. Finally, exploiting network coding [58] for user cooperation [59] is a promising approach.

The reported practical designs still suffer performance loss compared to the theoretical limits. Closing this gap with better code designs while staying at acceptable complexity is an urgent research task. The practical designs proposed so far are only for wireless relay and two-transmitter two-receiver cooperative channels. Substantial research efforts are needed to construct practical systems based on cooperative diversity for larger ad hoc networks.

References

[1] A. Sendonaris, E. Erkip, and B. Aazhang. 2003. User cooperation diversity part I and part II. *IEEE Trans. Commun.* 51:1927–48.

[2] J. Laneman, D. Tse, and G. Wornell. 2004. Cooperative diversity in wireless networks: Efficient protocols and outage behavior. *IEEE Trans. Inform. Theory* 50:3062–80.

[3] D. Tse and P. Viswanath. 2005. *Fundamentals of wireless communications*. Cambridge: Cambridge University Press.

[4] A. Goldsmith. 2005. *Wireless communications*. Cambridge: Cambridge University Press.

[5] E. van der Meulen. 1971. Three-terminal communication channels. *Adv. Appl. Probability* 3:120–54.

[6] J. Proakis. 2000. *Digital communications*. New York: McGraw-Hill.

[7] G. Caire and S. Shamai (Shitz). 2003. On the achievable throughput of a multi-antenna Gaussian broadcast channel. *IEEE Trans. Inform. Theory* 49:1691–706.

[8] T. Cover and J. Thomas. 1991. *Elements of information theory*. New York: Wiley.

[9] A. Carleial. 1978. Interference channels. *IEEE Trans. Inform. Theory* 24:60–70.

[10] L. Zheng and D. Tse. 2003. Diversity and multiplexing: A fundamental tradeoff in multiple antenna channels. *IEEE Trans. Inform. Theory* 49:1073–96.

[11] S. Shamai (Shitz) and S. Verdu. 2001. The impact of frequency-flat fading on the spectral efficiency of CDMA. *IEEE Trans. Inform. Theory* 47:1302–27.

[12] A. Lozano, A. Tulino, and S. Verdu. 2005. High-SNR power offset in multi-antenna communication. *IEEE Trans. Inform. Theory* 51:4134–51.

[13] A. Wyner and J. Ziv. 1976. The rate-distortion function for source coding with side information at the decoder. *IEEE Trans. Inform. Theory* 22:1–10.

[14] M. Costa. 1983. Writing on dirty paper. *IEEE Trans. Inform. Theory* 29:439–41.

[15] D. Slepian and J. Wolf. 1973. Noiseless coding of correlated information sources. *IEEE Trans. Inform. Theory* 19:471–80.

[16] T. Berger. 1977. Multiterminal source coding. In *The information theory approach to communications*, ed. G. Longo. New York: Springer-Verlag, pp. 173–231.

[17] S. Pradhan, J. Chou, and K. Ramchandran. 2003. Duality between source coding and channel coding and its extension to the side information case. *IEEE Trans. Inform. Theory* 49:1181–203.

[18] S. Gelfand and M. Pinsker. 1980. Coding for channel with random parameters. *Probl. Contr. Inform. Theory* 9:19–31.

[19] Z. Xiong, A. Liveris, and S. Cheng. 2004. Distributed source coding for sensor networks. *IEEE Signal Processing Mag.* 21:80–94.

[20] U. Erez and S. ten Brink. 2005. A close to capacity dirty paper coding scheme. *IEEE Trans. Inform. Theory* 51:3417–32.

[21] A. Bennatan, D. Burshtein, G. Caire, and S. Shamai. 2006. Superposition coding for side-information channels. *IEEE Trans. Inform. Theory* 52:1872–89.

[22] Y. Sun, A. Liveris, V. Stanković, and Z. Xiong. 2005. Near-capacity dirty-paper code designs based on TCQ and IRA codes. In *Proceedings of the ISIT-2005 IEEE International Symposium on Information Theory*, Adelaide, Australia, pp. 184–188.

[23] T. Cover and A. El Gamal. 1979. Capacity theorems for the relay channel. *IEEE Trans. Inform. Theory* 25:572–84.

[24] A. Høst-Madsen and J. Zhang. 2005. Capacity bounds and power allocation for the wireless relay channel. *IEEE Trans. Inform. Theory* 51:2020–40.

[25] G. Kramer, M. Gastpar, and P. Gupta. 2005. Cooperative strategies and capacity theorems for relay networks. *IEEE Trans. Inform. Theory* 51:3037–63.

[26] Z. Zhang and T. Duman. 2005. Capacity approaching turbo coding and iterative decoding for relay channels. *IEEE Trans. Commun.* 53:1895–905.

[27] M. Khojastepour, A. Sabharwal, and B. Aazhang. 2003. On capacity of Gaussian "cheap" relay channel. In *Proceedings of IEEE Globecom-2003*, San Francisco, vol. 3, pp. 1776–1780.

[28] M. Valenti and B. Zhao. 2003. Distributed turbo codes: Towards the capacity of the relay channel. In *Proceedings of the VTC-2003 IEEE Vehicular Technology Conference*, Orlando, FL, pp. 322–326.

[29] Z. Zhang and T. Duman. 2005. Capacity approaching turbo coding for half-duplex relaying. In *Proceedings of the ISIT-2005 IEEE International Symposium on Information Theory*, Adelaide, Australia, pp. 1888–1892.

[30] Y. Liang and V. Veeravalli. 2005. Gaussian orthogonal relay channel: Optimal resource allocation and capacity. *IEEE Trans. Inform. Theory* 51:3284–89.

[31] C. Zeng, F. Kuhlmann, and A. Buzo. 1989. Achievability proof of some multiuser channel coding theorems using backward decoding. *IEEE Trans. Inform. Theory* 35:1160–65.

[32] J. Laneman and G. Wornell. 2003. Distributed space-time coding protocols for exploiting cooperative diversity in wireless networks. *IEEE Trans. Inform. Theory* 49:2415–25.

[33] L. Ozarow, S. Shamai, and A. Wyner. 1994. Information theoretic consideration for cellular mobile radio. *IEEE Trans. Veh. Technol.* 43:359–77.

[34] I. Telater. 1999. Capacity of multi-antenna Gaussian channels. *Eur. Trans. Telecommun.* 10:585–95.

[35] R. Nabar, H. Bölcskei, and F. Kneubühler. 2004. Fading relay channels: Performance limits and space-time signal design. *IEEE J. Select. Areas Commun.* 22:1099–109.

[36] A. Høst-Madsen. 2006. Capacity bounds for cooperative diversity. *IEEE Trans. Inform. Theory* 52:1522–44.

[37] C. Ng and A. Goldsmith. 2005. Capacity gains from transmitter and receiver cooperation. In *Proceedings of the ISIT-2005 IEEE International Symposium on Information Theory*, Adelaide, Australia, pp. 397–401.

[38] H. Weingarten, Y. Steinberg, and S. Shamai. 2006. The capacity region of the Gaussian multiple-input multiple-output broadcast channel. *IEEE Trans. Inform. Theory* 52:3936–64.

[39] N. Jindal, U. Mitra, and A. Goldsmith. 2004. Capacity of ad-hoc networks with node cooperation. In *Proceedings of the ISIT-2004 IEEE International Symposium on Information Theory*, Chicago, pp. 271–275.

[40] M. Uppal, V. Stanković, A. Høst-Madsen, and Z. Xiong. 2006. Code design for transmitter cooperation. In *Proceedings of the MSRI Workshop on Mathematics of Relaying and Cooperation in Communication Networks*, Berkeley, CA.

[41] J. Laneman and G. W. Wornell. 2000. Energy-efficient antenna sharing and relaying for wireless networks. In *Proceedings of the WCNC-2000 IEEE Wireless Communications and Networking Conference*, vol. 1, pp. 7–12.

[42] A. Chakrabarti, A. de Baynast, A. Sabharwal, and B. Aazhang. 2007. Low density parity check codes for the relay channel. *IEEE J. Selected Areas Commun.* 25:280–91.

[43] R. G. Gallager. 1963. *Low density parity check codes*. Cambridge, MA: MIT Press.

[44] M. Yuksel and E. Erkip. 2007. Diversity-multiplexing tradeoff in half-duplex relay systems. In *Proceedings of the ICC-2007 IEEE International Conference on Communications*, Glasgow, UK, pp. 689–694.

[45] A. Chakrabarti, A. de Baynast, A. Sabharwal, and B. Aazhang. 2006. Half-duplex estimate-and-forward relaying: Bounds and code design. In *Proceedings of the ISIT-2006 IEEE International Symposium on Information Theory*, Seattle, WA, pp. 1239–1243.

[46] R. Hu and J. Li. 2006. Practical compress-forward in user cooperation: Wyner-Ziv cooperation. In *Proceedings of the ISIT-2006 IEEE International Symposium on Information Theory*, Seattle, WA, pp. 489–493.

[47] Z. Liu. 2007. Slepian-Wolf coded nested lattice quantization for Wyner-Ziv coding: High-rate performance analysis, code designs, and application to cooperative networks. PhD thesis, Texas A&M University.

[48] R. Zamir, S. Shamai, and U. Erez. 2002. Nested linear/lattice codes for structured multiterminal binning. *IEEE Trans. Inform. Theory* 48:1250–76.

[49] Z. Liu, S. Cheng, A. Liveris, and Z. Xiong. 2006. Slepian-Wolf coded nested quantization for Wyner-Ziv coding: High-rate performance analysis and code design. *IEEE Trans. Inform. Theory* 52:4358–4379.

[50] Q. Xu, V. Stanković, and Z. Xiong. 2006. Layered Wyner-Ziv video coding for transmission over unreliable channels. *Signal Processing* 86:3112–25.

[51] H. Jin, A. Khandekar, and R. McEliece. 2000. Irregular repeat-accumulate codes. In *Proceedings of the 2nd International Symposium on Turbo Codes and Related Topics*, 1–8.

[52] T. Hunter and A. Nosratinia. 2006. Diversity through coded cooperation. *IEEE Trans. Wireless Commun.* 5:283–89.

[53] M. Janani, A. Hedayat, T. Hunter, and A. Nosratinia. 2004. Coded cooperative in wireless communications: Space-time transmission and iterative decoding. *IEEE Trans. Signal Processing* 52:362–71.

[54] A. Stefanov and E. Erkip. 2004. Cooperative coding for wireless networks. *IEEE Trans. Commun.* 52:1603–12.

[55] M. Souryal and B. Vojcic. 2004. Cooperative turbo coding with time-varying Rayleigh channels. In *Proceedings of the ICC-2004 IEEE International Conference on Communications*, Paris, pp. 356–60.

[56] D. Gunduz, C. T. K. Ng, E. Erkip, and A. J. Goldsmith. 2007. Source transmission over relay channels with correlated side information. In *Proceedings of the ISIT-2007 IEEE International Symposium on Information Theory*, Nice, France.

[57] Y. Fan, H. V. Poor, and J. Thompson. 2007. Cooperative multiplexing in a half-duplex relay network: Performance and constraints. In *Proceedings of the 45th Annual Allerton Conference on Communications, Control and Computing*, Monticello, IL.

[58] R. Ahlswede, N. Cai, S.-Y.R. Li, and R. W. Yeung. 2000. Network information flow. *IEEE Transactions on Information Theory* 46:1204–16.

[59] M. Yu, J. Li, and R. Blum. 2007. User cooperation through network coding. In *Proceedings of the ICC-2007 IEEE International Conference on Communications*, Glasgow, UK, pp. 4064–4069.

13

Time Synchronization for Wireless Sensor Networks

Kyoung-Lae Noh
Texas A&M University

Yik-Chung Wu
The University of Hong Kong

Khalid Qaraqe
Texas A&M University

Erchin Serpedin
Texas A&M University

13.1 Introduction

With the help of recent technological advances in micro-electro-mechanical systems (MEMS) and wireless communications, low-cost, low-power, and multifunctional wireless sensing devices have been developed. When these devices are deployed over a wide geographical region, they can collect information about the environment and efficiently collaborate to process such information by forming a distributed communication network, called the *wireless sensor network* (WSN). WSN is a special case of wireless ad hoc network, and assumes a multihop communication framework with no common infrastructure, where the sensors spontaneously cooperate to deliver information by forwarding packets from a source to a destination. The feasibility of WSNs keeps growing rapidly, and WSNs have been regarded as fundamental infrastructures for future ubiquitous communications due to a variety of promising potential applications: monitoring the health status of humans, animals, plants, and environment; control and instrumentation of industrial machines and home appliances; homeland security; detection of chemical and biological threats and leaks; etc. [1, 2].

Time synchronization is a procedure for providing a common notion of time across a distributed system. It is crucial for WSNs in performing a number of fundamental operations, such as:

- Data fusion: Data fusion is a main operation in all distributed networks for processing and integrating in a meaningful way the collected data, and it requires some or all nodes in the network to share a common timescale.
- Power management: Energy efficiency is a key designing factor for WSNs since sensors are usually left unattended without any maintenance and battery replacement for their lifetimes after deployment. Most energy-saving operations strongly depend on time synchronization. For instance, the duty cycling (sleep and wake-up modes control) helps the nodes to save huge energy resources by spending minimal power during the sleep mode. Thus, network-wide synchronization is essential for efficient duty cycling, and its performance is proportional to the synchronization accuracy.
- Transmission scheduling: Many scheduling protocols require time synchronization. For example, the Time Division Multiple Access (TDMA) scheme, one of the most popular communications schemes for distributed networks, is only applicable to a synchronized network.

Moreover, many localization, security, and tracking protocols also demand the nodes to time stamp their messages and sensing events. Therefore, time synchronization appears as one of the most important research challenges in the design of energy-efficient WSNs.

In general, synchronization is considered a critical problem for distributed wireless ad hoc networks due to its decentralized nature and the timing uncertainties introduced by the imperfections in hardware oscillators and message delays in physical and Medium Access Control (MAC) layers. All these uncertainties cause the local clocks of different nodes to drift away from each other over the course of a time interval. In the context of the Internet (a kind of distributed network), time synchronization has been thoroughly studied and investigated. In the Internet, the Network Time Protocol (NTP) [3] is employed ubiquitously due to its diverse advantages, such as scalability, robustness, and self-configurability. Besides, NTP does not rely on GPS and is a software-based protocol. However, NTP presents a number of challenges when applied to WSNs due to the unique nature of sensor networks: limited power resources, wireless channel conditions, and dynamic topology caused by mobility and failure. Therefore, different types of synchronization schemes have to be explicitly designed for WSN applications to cope with these challenges (see also the surveys in [4–9] for additional motivations in this direction).

Research works on time synchronization in the context of WSN roughly began to appear in 2002, where [5], for the first time, pointed out NTP cannot be directly applied to WSN and described some important characteristics and design principles of time synchronization in WSNs. In the same year and the next, two important time synchronization protocols for WSNs were reported: Timing-sync Protocol for Sensor Networks (TPSN) [17] and Reference Broadcasting Synchronization (RBS) [18]. These two protocols set the stage for two fundamental approaches of time synchronization in WSNs. After that, many extensions and generalizations of TPSN and RBS, and many different synchronization schemes based on other ideas, have been proposed in the literature. Notice that symbol timing synchronization in the physical layer, which is critical for accurate symbol detection at the receiver, is a different problem and is out of the scope of this chapter.

The purpose of this chapter is threefold. First, this chapter summarizes the fundamental features and theoretical results encountered in time synchronization of WSNs. Second, it represents the survey of existing time synchronization protocols for WSNs, focusing mainly on the signal processing aspects, the most recent developments in this field. Finally, this chapter discusses the need for adaptive time synchronization schemes for WSNs, analyzes the features of the recently reported adaptive time synchronization protocols, and proposes several research directions for improving their performance.

The rest of this chapter is organized as follows. In section 13.2, the general clock model for time synchronization is first introduced and analyzed. Some important features that have to be considered when designing time synchronization protocols for WSNs are presented. Additionally, various delay components in timing message delivery are categorized. Section 13.3 presents three general and fundamentally different time synchronization approaches: sender-receiver, receiver-receiver, and receiver-only synchronization. These basic approaches are analyzed and compared to illustrate the common and different characteristics in clock synchronization of WSNs. Section 13.4 categorizes and surveys the existing synchronization protocols and relates them to the results presented in section 13.3. In section 13.5, results concerning the importance and effectiveness of adaptive time synchronization schemes are presented, and the most important adaptive synchronization protocols are introduced as well. Finally, section 13.6 summarizes and concludes this chapter.

13.2 Signal Models for Time Synchronization

13.2.1 Definition of Clock

Every individual sensor in a network has its own clock. The counter in a sensor is increased in accordance with the zero-crossings or the edges of the periodic output signal of the local oscillator. When the counter reaches a certain threshold value, an interrupt is created and delivered to the memory. The frequency of the oscillator and the threshold value determine the resolution of the clock. Ideally, the clock of a sensor node should be configured such that $C(t) = t$, where t stands for the ideal or reference time. However, due to the imperfections of the clock oscillator, the clock function of the ith node is modeled as

$$C_i(t) = \theta_o + \theta_s \cdot t + e, \tag{13.1}$$

where the parameters θ_o and θ_s are called clock offset (phase difference) and clock skew (frequency difference), respectively, and e stands for random noise.

Assuming the effect of random noise e is negligible, from (13.1), the clock relationship between two nodes, say node 1 and node 2, can be represented by

$$C_1(t) = \theta_o^{(12)} + \theta_s^{(12)} \cdot C_2(t),$$

where $\theta_o^{(12)}$ and $\theta_s^{(12)}$ are the relative clock offset and skew between node 1 and node 2, respectively. Thus, $\theta_o^{(12)} = 0$ and $\theta_s^{(12)} = 1$ when the two clocks are perfectly synchronized. Suppose there are L nodes in the network, then the global network-wide synchronization is achieved when $C_i(t) = C_j(t)$ for all $i, j = 1, \cdots, L$.

Time synchronization in wireless sensor networks is a complicated problem due to the following reasons. First, every single oscillator has unique clock parameters regardless of its type. For instance, according to the data sheet of a typical crystal-quartz oscillator commonly used in sensor networks, the frequency of a clock varies up to 40 ppm, which means clocks of different nodes can lose as much as 40 ms in a second. In other words, every single oscillator might assume a different skew parameter ranging from –20 to 20 ppm.

Notice that in general, the clock skew θ_s is a time-dependent random variable (RV) and there are two concepts used often in clock terminology regarding the nature of time-dependent randomness present in clock parameters. These concepts are referred to as *short-term* and *long-term* stabilities, respectively. For the oscillators currently used in sensor networks, all these parameters are almost constant for short-term time intervals [10]. Besides, the total power of the noise process is too small to be effective in short time spans [11]. Therefore, the parameters of a clock are assumed to be constants for the time period of interest.

As far as the long-term stability is concerned, the clock parameters are subject to change due to environmental or other external effects such as temperature, atmospheric pressure, voltage changes, and hardware aging [10]. Hence, in general, the relative clock offset keeps changing with time, which means that the network has to perform periodic time resynchronization to adjust the clock parameters.

13.2.2 Design Considerations

Time synchronization for conventional wired networks has been thoroughly studied and a plethora of synchronization protocols have been developed as surveyed in [1]. For wireless sensor networks, there are a number of unique and important factors to be considered when designing time synchronization protocols as listed below.

- *Energy consumption*: Energy consumption is momentous in wireless sensor networks due to their limited and generally nonrechargeable power resources. Hence, the design of wireless sensor networks should be subjected to maintaining minimal energy expenditure in each sensor node. Various types of power control procedures, such as sleep/wake-up modes and dynamic routing controls, are commonly considered in this regard. Time synchronization is one of the critical components contributing to energy consumption due to the highly energy consuming radio transmissions for achieving clock synchronization. Indeed, the energy consumption required for time synchronization of a node is approximately 17% of the total energy spent by a node [12]. Pottie and Kaiser showed in [13] that the radio frequency (RF) energy required to transmit 1 bit over 100 m (i.e., 3 J) is equivalent to the energy required to execute 3 million instructions. Therefore, developing efficient synchronization algorithms represents an ideal mechanism for trading computational energy for reduced (RF) communication energy. In the sequel, energy efficiency is the main concern in designing time synchronization protocols.

- *Latency*: Latency in message delivery is a fundamental factor when designing communications networks. For networks based on multihop transmissions like wireless sensor networks, this is even more critical because the uncertainty in message delivery significantly increases as the number of hops increases. Besides, the effects of channel variations, mobility, and the ad hoc nature of wireless sensor networks make this problem more complex. Efficient localization and time synchronization protocols are necessary for reducing the latency error and jitter.

- *Security and reliability*: Network security has gained huge attention in recent years as the networks become more accessible and vulnerable due to the development of sophisticated spying techniques and devices. Besides, unlike wired networks, far more frequent message losses occur in wireless networks because of the time-varying nature of wireless channels. Therefore, a mechanism to cope with message losses and malicious attacks in time synchronization will be necessary for wireless sensor networks.

- *Network topology changes*: The performance of a time synchronization protocol is closely related to the network topology, i.e., it varies with the density and distribution of sensors in the network. Therefore, any shift in the location or scale of sensors incurs a network topology change, which requires at its turn a new self-configuration. Mobility of the sensors and battery timeouts are the main reasons for this change. Hence, for dynamic sensor networks, time synchronization protocols should be able to adapt well to frequent network topology changes.

- *Scalability*: Scalability is another important factor in the design of synchronization protocols. The computational complexity of synchronization algorithms becomes a critical problem as the number of sensors becomes very large. Besides, many other crucial MAC operations, such as multihop routing and network configuration, highly depend on the network scalability as well.

13.2.3 Delay Components in Timing Message Delivery

The main role of time synchronization in a distributed network is to ensure a common timescale for all the network nodes, and to provide the right temporal coordination among all the nodes engaged in a collaborative and distributed interaction with the physical environment. Timing mismatch arises mainly from different setup times of nodes and time variations introduced by local oscillators running at different frequencies. Environmental variations, such as temperature and aging, also drive local clock oscillators to run unpredictably. All these uncertainties cause the local clocks of different nodes to drift away from each other over the course of a time interval.

Assume two nodes need to be synchronized. One of the nodes sends its current time to a neighboring node; if there is absolutely no delay in the message delivery, that neighboring node can immediately know the difference between its clock and its neighbor's clock. Unfortunately, in a real wireless network, various delays affect the message delivery, making time synchronization much more difficult than it seems to be. In general, a series of timing message transmissions is required to estimate the relative time offsets among nodes. In some sense, time synchronization in wireless sensor networks can be regarded as a process of removing the nondeterministic delays during timing message transmission over wireless channels.

There are a number of nondeterministic delays while transferring messages between nodes. Kopetz and Ochsenreiter for the first time analyzed the structure of message delays and characterized the delay components according to the process of message delivery [14]. The delay components in message delivery can be categorized as follows:

1. Send time: The time spent in building the message at the application layer, including other delays introduced by the operating system when processing the send request. The send time is nondeterministic and can be up to hundreds of milliseconds depending on the workload of the system.
2. Access time: The time waiting for accessing the channel after reaching the MAC layer. This is the most significant factor and highly variable according to the specific MAC protocol. The access time is nondeterministic and varies from milliseconds up to seconds depending on the current network traffic.
3. Transmission time: The time for transmitting a message at the physical layer. This delay can be estimated by the length of a message and the speed of radio in the medium and is in the order of tens of milliseconds.
4. Propagation time: The actual time for a message to transmit from the sender to the receiver in a wireless channel. The propagation time is deterministic and less than 1 μs, which is almost negligible compared with the other delay components.
5. Reception time: The time required for receiving a message at the physical layer, which is the same as the transmission time. In some cases, this delay has been categorized as a part of the receive time, to be presented next.
6. Receive time: Time to construct and send the received message to the application layer at the receiver. It is a corresponding component of the send time on the transmitter side and can be varied due to the variable delays introduced by the operating system.

Note that the time delay in message transmission is also dependent on other factors, such as hardware platform, error correction code, and modulation scheme. The estimated time delay discussed above in each component is based on the Mica2 platform [15]. More detailed analysis can be found in [16].

13.3 Fundamental Approaches to Time Synchronization

Time synchronization in wireless sensor networks can be achieved by transferring a group of timing messages to the target sensors. The timing messages contain the information about the time stamps measured by the transmitting sensors. There exist two well-known approaches for time synchronization in wireless sensor networks, which are categorized as sender-receiver synchronization (SRS) and receiver-receiver synchronization (RRS). SRS is based on the traditional model of two-way message exchanges between a pair of nodes. For RRS, the nodes to be synchronized first receive a beacon packet from a common sender, then compare their receiving time readings of the beacon packet to compute the relative clock offset. Most of the existing time synchronization protocols rely on one of these two approaches. For instance, the Network Time Protocol (NTP) [3] and the Timing-sync Protocol for Sensor Networks (TPSN) [17] adopt SRS since they depend on a series of pairwise synchronizations that assume two-way timing message exchanges. Notice also that the Reference Broadcast Synchronization (RBS) protocol [18] relies on RRS since it requires pairs of message exchanges among children nodes (except the reference) to compensate their relative clock offsets.

Recently, a new approach for time synchronization, called receiver-only synchronization (ROS), was proposed. It aims at minimizing the number of required timing messages and energy consumption during synchronization while preserving a high level of accuracy [19]. This approach can be used to achieve network-wide synchronization with much less timing messages than other well-known existing protocols such as TPSN and RBS.

Next we will present and analyze each of these synchronization approaches and illustrate how the general design considerations can be resolved in them. For all these approaches, we only present the underlying signaling mechanisms for performing pairwise synchronization, i.e., synchronizing a pair of nodes, since network-wide synchronization can be simply achieved by performing a group of pairwise synchronizations.

13.3.1 Sender-Receiver Synchronization

This approach is based on the classical two-way timing message exchange mechanism between two adjacent nodes. Consider a parent node P and one of its children nodes, node A, in Figure 13.1. The clock model for the two-way message exchange is depicted in Figure 13.2, where $\theta_o^{(AP)}$ denotes the clock offset between node A and node P and timing messages are assumed to be exchanged multiple (N) times [4, 17]. Here, the time stamps made during the ith message exchange $T_{1,i}^{(A)}$ and $T_{4,i}^{(A)}$ are measured by the local clock of node A, and $T_{2,i}^{(P)}$ and $T_{3,i}^{(P)}$ are measured by the local clock of node P, respectively. Node A transmits a synchronization packet, containing the value of time stamp $T_{1,i}^{(A)}$ to node P. Node P receives it at time $T_{2,i}^{(P)}$ and transmits an acknowledgment packet to node A at $T_{3,i}^{(P)}$. This packet contains the value of time stamps $T_{1,i}^{(A)}$, $T_{2,i}^{(P)}$, and $T_{3,i}^{(P)}$. Then, node A finally receives the packet at $T_{4,i}^{(A)}$.

As discussed before, packet delays can be characterized into several distinct components: send, access, transmission, propagation, and receive times. These delay components

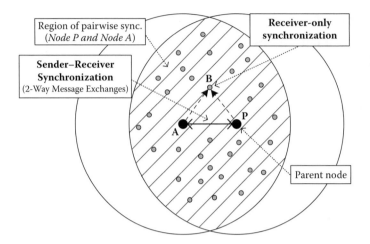

FIGURE 13.1 Sender-receiver synchronization and receiver-only synchronization.

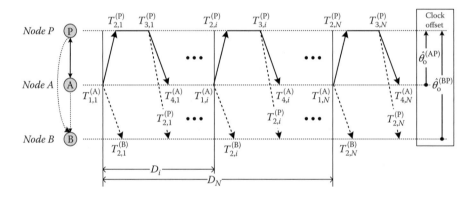

FIGURE 13.2 Clock synchronization model of SRS (node P and node A) and ROS (node B).

are divided into two parts: the fixed portion d and the variable portion X_i. The variable portion of delays depends on various network parameters (e.g., network status, traffic, etc.) and setup, and therefore no single delay model can be found to fit for every case. Thus far, several probability density function (PDF) models have been proposed for modeling random delays, the most widely deployed ones being Gaussian, Gamma, exponential, and Weibull PDFs [20, 21]. The Gaussian delay model is appropriate if the delays are thought to be the addition of numerous independent random processes. In [18], the chi-squared test showed that the variable portion of delays can be modeled as Gaussian distributed random variables (RVs) with 99.8% confidence. On the other hand, a single-server M/M/1 queue can fittingly represent the cumulative link delay for point-to-point hypothetical reference connection, where the random delays are independently modeled as exponential RVs [22]. Thus, we assume the random portions of delays are either normal or exponentially distributed RVs.

13.3.1.1 Clock Offset Estimation

Suppose that the clock frequencies of two nodes remain equal during the synchronization period, and both $X_i^{(AP)}$ and $X_i^{(PA)}$ are normal distributed RVs with mean μ and variance $\sigma^2/2$. From Figure 13.2, $T_{2,i}^{(P)}$ and $T_{4,i}^{(A)}$ can be expressed as

$$T_{2,i}^{(P)} = T_{1,i}^{(A)} + \theta_o^{(AP)} + d^{(AP)} + X_i^{(AP)}, \tag{13.2}$$

$$T_{4,i}^{(A)} = T_{3,i}^{(P)} + \theta_o^{(PA)} + d^{(PA)} + X_i^{(PA)}, \tag{13.3}$$

where $\theta_o^{(PA)} = -\theta_o^{(AP)}$, $d^{(AP)}$, and $X_i^{(AP)}$ denote the fixed and random portions of timing delays in the message transmissions from node A to node P, respectively. By defining the delays in uplink $U_i \triangleq T_{2,i}^{(P)} - T_{1,i}^{(A)}$ and downlink $V_i \triangleq T_{4,i}^{(A)} - T_{3,i}^{(P)}$, the ith delay observations corresponding to the ith timing message exchange are given by $U_i = \theta_o^{(AP)} + d^{(AP)} + X_i^{(AP)}$ and $V_i = \theta_o^{(PA)} + d^{(PA)} + X_i^{(PA)}$, respectively. Then, the likelihood function based on the observations $\{U_i\}_{i=1}^N$ and $\{V_i\}_{i=1}^N$ is given by

$$L\left(\theta_o^{(AP)}, \mu, \sigma^2\right) = (\pi\sigma^2)^{-\frac{N}{2}} e^{-\frac{1}{\sigma^2}\left[\sum_{i=1}^N (U_i - d^{(AP)} - \theta_o^{(AP)} - \mu)^2 + \sum_{i=1}^N (V_i - d^{(PA)} + \theta_o^{(AP)} - \mu)^2\right]},$$

where N is the number of message exchanges. Differentiating the log-likelihood function leads to

$$\frac{\partial \ln L\left(\theta_o^{(AP)}\right)}{\partial \theta_o^{(AP)}} = -\frac{2}{\sigma^2} \sum_{i=1}^N \left[\theta_o^{(AP)} + d^{(AP)} - d^{(PA)} - (U_i - V_i)\right].$$

The fixed portions of delays are mainly determined by the propagation delays, and both up- and downlink channels have the same distance. Thus, the fixed portions of delays $d^{(AP)}$ and $d^{(PA)}$ are assumed to be equal, and are denoted by d for the rest of this chapter. Indeed, the propagation delay is less than 1 μs for ranges under 300 m, hence almost negligible when compared to other dominant delay components whose ranges are about hundreds of milliseconds [16]. The maximum likelihood estimate (MLE) of clock offset is given by [25]

$$\hat{\theta}_o^{(AP)} = \arg\max_{\theta_o^{(AP)}} \left[\ln L\left(\theta_o^{(AP)}\right)\right] = \frac{\bar{U} - \bar{V}}{2}. \tag{13.4}$$

Thus, node A can be synchronized to the parent node P by simply taking the difference of the average delay observations $\bar{U}$ and $\bar{V}$.

For exponential random delays $X_i^{(PA)}$ and $X_i^{(AP)}$ with the same mean λ, the likelihood function based on the observations $\{U_i\}_{i=1}^N$ and $\{V_i\}_{i=1}^N$ becomes

$$L(\theta_o^{(AP)},\lambda)=\lambda^{-2N}e^{-\frac{1}{\lambda}\sum_{i=1}^{N}[U_i+V_i-2d]} \cdot \prod_{i=1}^{N} I\left[U_i-\theta_o^{(AP)}-d\geq 0, V_i+\theta_o^{(AP)}-d\geq 0\right],$$

where $I(\cdot)$ stands for the indicator function (i.e., $I(\cdot)$ is 1 whenever its inner condition holds, otherwise it is equal to 0). In [23], Jeske proved that the maximum likelihood estimator of $\theta_o^{(AP)}$ exists when d is unknown and exhibits the same form as the estimator proposed in [24], namely,

$$\hat{\theta}_o^{(AP)} = \frac{\min_{1\leq i\leq N} U_i - \min_{1\leq i\leq N} V_i}{2}. \tag{13.5}$$

Notice from (13.4) and (13.5), it is clear that if only one round of message exchange is performed ($N = 1$), the MLE of clock offset for both exponential and Gaussian delay models becomes $\hat{\theta}_o^{(AP)} = (U - V)/2$, which is exactly the same clock offset estimator adopted in [17].

13.3.1.2   Joint Clock Offset and Skew Estimation

The clock offset between two nodes generally keeps increasing due to the difference of clock parameters of each oscillator. Thus, a model with the same clock frequency is not sufficient for long-term synchronization. Indeed, applying the clock skew correction mechanism increases the synchronization accuracy and guarantees the long-term reliability of synchronization.

Figure 13.3 shows the effect of clock offset (θ_o) and skew (θ_s) on timing message exchanges between two nodes. Without loss of generality, the reference time $T_{1,1}^{(A)}$ is set to be zero. Here, the time stamp at node P in the ith uplink message $T_{2,i}^{(B)}$ is given by

$$T_{2,i}^{(P)} = T_{1,i}^{(A)}+\theta_o^{(AP)}+\theta_s^{(AP)}(T_{1,i}^{(A)}+d+X_i^{(AP)})+d+X_i^{(AP)}$$
$$= (1+\theta_s^{(AP)})(T_{1,i}^{(A)}+d+X_i^{(AP)})+\theta_o^{(AP)}, \tag{13.6}$$

where the term $\theta_s^{(AP)}(T_{1,i}^{(A)} + d + X_i^{(AP)})$ is due to the effect of clock skew. Similarly, the time stamp at node P in the ith downlink message $T_{3,i}^{(P)}$ takes the equations

$$T_{3,i}^{(P)} = T_{4,i}^{(A)}+\theta_o^{(AP)}+\theta_s^{(AP)}(T_{4,i}^{(A)}-d-X_i^{(PA)})-d-X_i^{(PA)}$$
$$= (1+\theta_s^{(AP)})(T_{4,i}^{(A)}-d-X_i^{(PA)})+\theta_o^{(AP)}, \tag{13.7}$$

where the term $\theta_s^{(AP)}(T_{4,i}^{(A)} - d - X_i^{(PA)})$ is again due to the effect of clock skew. For an easier illustration, we introduce the simplified notations $\theta_s \triangleq \theta_s^{(AP)}$, $\theta_o \triangleq \theta_o^{(AP)}$, $T_{1,i} \triangleq T_{1,i}^{(A)}$, $T_{4,i} \triangleq T_{4,i}^{(A)}$, $T_{2,i} \triangleq T_{2,i}^{(P)}$, $T_{3,i} \triangleq T_{3,i}^{(P)}$, $X_i \triangleq X_i^{(AP)}$, and $Y_i \triangleq X_i^{(PA)}$ in this section, respectively.

Assuming $\{X_i\}_{i=1}^{N}$ and $\{Y_i\}_{i=1}^{N}$ are zero-mean independent Gaussian distributed RVs with variance $\sigma^2/2$, then the joint PDF of $\mathbf{X} \triangleq \{X_i\}_{i=1}^{N}$ and $\mathbf{Y} \triangleq \{Y_i\}_{i=1}^{N}$ is given by

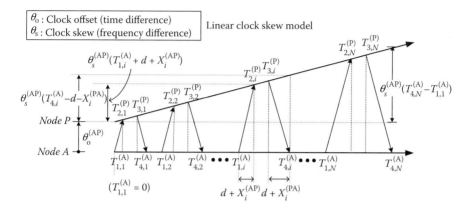

FIGURE 13.3 Two-way timing message exchange model that assumes clock offset and skew.

$$f_{X,Y}(x,y)=\left(\pi\sigma^2\right)^{-\frac{N}{2}}e^{-\frac{1}{\sigma^2}\sum_{i=1}^{N}\left[\left(\frac{\theta_o-T_{2,i}+(T_{1,i}+d)(1+\theta_s)}{1+\theta_s}\right)^2+\left(\frac{\theta_o-T_{3,i}+(T_{4,i}-d)(1+\theta_s)}{1+\theta_s}\right)^2\right]}.$$

Further assuming that the fixed portion of delay d is known and $\theta'_s \triangleq 1/(1+\theta_s)$, then the log-likelihood function (ignoring irrelevant additive and multiplicative constants) for (θ_o,θ'_s), based on observations $\{T_{1,i}\}_{i=1}^{N}$, $\{T_{2,i}\}_{i=1}^{N}$, $\{T_{3,i}\}_{i=1}^{N}$, and $\{T_{4,i}\}_{i=1}^{N}$, is given by

$$\ln L(\theta_o,\theta'_s)=-\sum_{i=1}^{N}\left\{\left[\theta'_s(\theta_o-T_{2,i})+(T_{1,i}+d)\right]^2+\left[\theta'_s(\theta_o-T_{3,i})+(T_{4,i}-d)\right]^2\right\}. \quad (13.8)$$

It has been shown in [25] that the values of θ_o and θ_s that maximize the above log-likelihood function are given, respectively, by

$$\hat{\theta}_o^{ML}=\frac{\sum_{i=1}^{N}(T_{1,i}+T_{4,i})\sum_{i=1}^{N}(T_{2,i}^2+T_{3,i}^2)-\sum_{i=1}^{N}(T_{2,i}+T_{3,i})Q}{\sum_{i=1}^{N}(T_{2,i}+T_{3,i})\sum_{i=1}^{N}(T_{1,i}+T_{4,i})-2NQ}, \quad (13.9)$$

$$\hat{\theta}_s^{ML}=\frac{-2N\left[\sum_{i=1}^{N}(T_{1,i}+T_{4,i})\sum_{i=1}^{N}(T_{2,i}^2+T_{3,i}^2)-Q\sum_{i=1}^{N}(T_{2,i}+T_{3,i})\right]}{\sum_{i=1}^{N}(T_{1,i}+T_{4,i})\left[\sum_{i=1}^{N}(T_{2,i}+T_{3,i})\sum_{i=1}^{N}(T_{1,i}+T_{4,i})-2NQ\right]}+\frac{\sum_{i=1}^{N}(T_{2,i}+T_{3,i})}{\sum_{i=1}^{N}(T_{1,i}+T_{4,i})}-1, \quad (13.10)$$

where

$$Q \triangleq \sum_{i=1}^{N} \left(T_{1,i}T_{2,i} + T_{3,i}T_{4,i} + (T_{2,i} - T_{3,i})d \right).$$

Note that the joint MLE depends on the value of the fixed portion of delays d, which is assumed to be known in this section. Although estimating d is an achievable task, we do not consider d another unknown (nuisance) parameter due to the inherent highly nonlinear and complex operations required for estimating d.

The Cramer-Rao lower bound (CRB) for the vector parameter $\theta = [\theta_o, \theta_s]^T$ can be derived from the 2×2 Fisher information matrix $\mathbf{I}(\theta)$ by taking its inverse. From (13.6), the second-order derivatives of the log-likelihood function with respect to θ_o and θ'_s are found as

$$\frac{\partial^2 \ln L\left(\theta_o, \theta'_s, \sigma^2\right)}{\partial \theta_o^2} = -\frac{4N\theta'^2_s}{\sigma^2},$$

$$\frac{\partial^2 \ln L\left(\theta_o, \theta'_s, \sigma^2\right)}{\partial \theta'^2_s} = -\frac{2}{\sigma^2} \sum_{i=1}^{N} \left[(T_{2,i} - \theta_o)^2 + (T_{3,i} - \theta_o)^2 \right],$$

$$\frac{\partial^2 \ln L\left(\theta_o, \theta'_s, \sigma^2\right)}{\partial \theta_o \theta'_s} = -\frac{2}{\sigma^2} \sum_{i=1}^{N} \left(2\theta'_s\theta_o - \theta'_sT_{2,i} + T_{1,i} - \theta'_sT_{3,i} - T_{4,i} \right).$$

Taking the negative expectations yields

$$-E\left[\frac{\partial^2 \ln L\left(\theta_o, \theta'_s, \sigma^2\right)}{\partial \theta_o^2} \right] = \frac{4N\theta'^2_s}{\sigma^2},$$

$$-E\left[\frac{\partial^2 \ln L\left(\theta_o, \theta'_s, \sigma^2\right)}{\partial \theta'^2_s} \right] = \frac{2}{\sigma^2} \sum_{i=1}^{N} E_{X_i, Y_i}\left[\frac{(X_i + T_{1,i} + d)^2 + (Y_i - T_{4,i} + d)^2}{\theta'^2_s} \right]$$

$$\overset{(a)}{=} \frac{2\sum_{i=1}^{N}\left((T_{1,i} + d)^2 + (T_{4,i} - d)^2 + 2\sigma^2 \right)}{\sigma^2 \theta'^2_s},$$

$$-E\left[\frac{\partial^2 \ln L\left(\theta_o, \theta'_s, \sigma^2\right)}{\partial \theta_o \theta'_s} \right] = -\frac{2}{\sigma^2} \sum_{i=1}^{N} E_{X_i, Y_i}\left[2\theta'_s(2\theta_o - T_{2,i} - T_{3,i}) + T_{1,i} + T_{4,i} \right]$$

$$\overset{(b)}{=} \frac{2N}{\sigma^2}\left(\overline{T_1} + \overline{T_4} \right),$$

where (*a*) and (*b*) are due to $X_i = \theta'_s(T_{2,i} - \theta_o) - (T_{1,i} + d)$ and $Y_i = \theta'_s(\theta_o - T_{3,i}) + (T_{4,i} - d)$, and

$$\bar{T}_1 = \sum_{i=1}^{N} T_{1,i}/N$$

and

$$\bar{T}_4 = \sum_{i=1}^{N} T_{4,i}/N$$

stand for the average of time observations. Therefore,

$$\mathbf{I}(\theta) = \begin{bmatrix} -E\left[\dfrac{\partial^2 \ln L\left(\theta_o,\theta'_s,\sigma^2\right)}{\partial\theta_o^2}\right] & -E\left[\dfrac{\partial^2 \ln L\left(\theta_o,\theta'_s,\sigma^2\right)}{\partial\theta_o\theta'_s}\right] \\ -E\left[\dfrac{\partial^2 \ln L\left(\theta_o,\theta'_s,\sigma^2\right)}{\partial\theta'_s\theta_o}\right] & -E\left[\dfrac{\partial^2 \ln L\left(\theta_o,\theta'_s,\sigma^2\right)}{\partial\theta'^2_s}\right] \end{bmatrix},$$ (13.11)

$$= \frac{2}{\sigma^2}\begin{bmatrix} 2N\theta'^2_s & N\left(\bar{T}_1+\bar{T}_4\right) \\ N\left(\bar{T}_1+\bar{T}_4\right) & \dfrac{1}{\theta'^2_s}\sum_{i=1}^{N}\left[\left(T_{1,i}+d\right)^2+\left(T_{4,i}-d\right)^2+\sigma^2\right] \end{bmatrix}.$$

The CRB can be obtained by taking the [i,i]th element of the inverse of the Fisher information matrix (i.e., var $(\hat{\theta}_i) \ge [\mathbf{I}^{-1}(\theta)]_{ii}$), and the inverse $\mathbf{I}^{-1}(\theta)$ is given by

$$\mathbf{I}^{-1}(\theta) = \frac{\sigma^2}{2}\begin{bmatrix} \dfrac{P}{\theta'^2_s N\left[2P-N\left(\bar{T}_1+\bar{T}_4\right)^2\right]} & \dfrac{-\left(\bar{T}_1+\bar{T}_4\right)}{2P-N\left(\bar{T}_1+\bar{T}_4\right)^2} \\ \dfrac{-\left(\bar{T}_1+\bar{T}_4\right)}{2P-N\left(\bar{T}_1+\bar{T}_4\right)^2} & \dfrac{2\theta'^2_s}{2P-N\left(\bar{T}_1+\bar{T}_4\right)^2} \end{bmatrix},$$ (13.12)

where

$$P \triangleq \sum_{i=1}^{N}\left[\left(T_{1,i}+d\right)^2+\left(T_{4,i}-d\right)^2+2\sigma^2\right].$$

Consequently, the CRBs of the joint clock offset and skew estimator are given by

$$\text{var}(\hat{\theta}_o^{ML}) \ge \frac{\sigma^2(1+\theta_s)^2 P}{2N\left[2P-N\left(\bar{T}_1+\bar{T}_4\right)^2\right]},$$ (13.13)

$$\text{var}(\hat{\theta}_s^{ML}) \geq \left(\frac{\partial \theta_s}{\partial \theta_s'}\right)^2 \cdot \frac{\sigma^2 \theta_s'^2}{2P - N\left(\bar{T_1} + \bar{T_4}\right)^2} = \frac{\sigma^2\left(1 + \theta_s\right)^2}{2P - N\left(\bar{T_1} + \bar{T_4}\right)^2}. \tag{13.14}$$

In fact, finding the joint MLE of clock skew requires quite a number of computations as in (13.10), and the fixed portion of delays d must be known (or estimated), which might not be applicable for wireless sensor networks consisting of low-end terminals. In practice, it requires an additional estimation procedure, which might deteriorate the robustness of the joint MLE. To overcome this limitation, a family of robust and simple clock offset and skew estimators that do not require prior knowledge of d have been proposed in [25].

13.3.2 Receiver-Only Synchronization

Due to the power constraint, the communication range of a sensor is strictly limited to a (radio-geometrical) circle whose radius depends on the transmission power (see Figure 13.1). In this figure, every node within the checked area (e.g., node B) can receive messages from both node P and node A. Suppose that node P is a parent (or reference) node, and node P and node A perform a pairwise synchronization using two-way timing message exchanges [17]. Then, all the nodes in the common coverage region of node P and node A (checked region) can receive a series of synchronization messages containing the information about the time stamps of the pairwise synchronization. Using this information, node B can also be synchronized to the parent node, node P, with no extra timing message transmissions. This approach is called receiver-only synchronization (ROS). In general, all the sensor nodes lying within the checked area can be synchronized by only receiving timing messages using ROS. Here, node P and node A can be regarded as super nodes since they provide synchronization beacons for all the nodes located in their vicinity.

In Figure 13.1, consider an arbitrary node, say node B, in the checked region. While node P and node A exchange time messages, node B can overhear these time messages. Hence, node B is capable of observing a set of time readings ($\{T_{2,i}^{(B)}\}_{i=1}^N$) at its local clock when it receives packets from node A, as depicted in Figure 13.2. Besides, node B can also receive the information about a set of time stamps $\{T_{2,i}^{(P)}\}_{i=1}^N$ obtained by receiving the packets transmitted by node P. Considering the effects of both clock offset and skew, the reception time at node P in the ith uplink message $T_{2,i}^{(P)}$ is given by

$$T_{2,i}^{(P)} = T_{1,i}^{(A)} + \theta_o^{(AP)} + \theta_s^{(AP)} \cdot (T_{1,i}^{(A)} - T_{1,1}^{(A)}) + d^{(AP)} + X_i^{(AP)}, \tag{13.15}$$

where $\theta_s^{(AP)}$ stands for the relative clock skew between node A and node P. Likewise, the reception time at node B in the ith uplink message $T_{2,i}^{(B)}$ can be represented by

$$T_{2,i}^{(B)} = T_{1,i}^{(A)} + \theta_o^{(AB)} + \theta_s^{(AB)} \cdot (T_{1,i}^{(A)} - T_{1,1}^{(A)}) + d^{(AB)} + X_i^{(AB)}, \tag{13.16}$$

where $\theta_o^{(AB)}$ and $\theta_s^{(AB)}$ stand for the relative clock offset and skew between node A and node B, and $d^{(AB)}$ and $X_i^{(AB)}$ denote the fixed and random portions of timing delays in the message transmission from node A to node B, respectively. Here, $X_i^{(AB)}$ is assumed to be a normal distributed RV with mean μ and variance $\sigma^2/2$.

The linear regression technique can be applied to synchronize node B and compensate the effects of the relative clock skew between node P and node B. Subtracting (13.16) from (13.15) gives

$$T_{2,i}^{(P)} - T_{2,i}^{(B)} = \theta_o^{(BP)} + \theta_s^{(BP)} \cdot (T_{1,i}^{(A)} - T_{1,1}^{(A)}) + d^{(AP)} - d^{(AB)} + X_i^{(AP)} - X_i^{(AB)}. \quad (13.17)$$

Since $d^{(AB)}$ and $d^{(AP)}$ are fixed values and $X_i^{(AB)}$ and $X_i^{(AP)}$ are normal distributed RVs, the noise component can be defined by $z[i] \triangleq \mu' + X_i^{(AP)} - X_i^{(AB)}$, where $\mu' \triangleq d^{(AP)} - d^{(AB)}$ and $z[i] \sim \mathcal{N}(\mu',\sigma^2)$. Let $x[i] \triangleq T_{2,i}^{(P)} - T_{2,i}^{(B)} - \mu'$ and $w[i] \triangleq z[i] - \mu'$, then the set of observed data can be written in matrix notation as follows:

$$\mathbf{x} = \mathbf{H}\theta + \mathbf{w},$$

where $\mathbf{x} = [x[1]\ x[2] \cdots x[N]]^T$, $\mathbf{w} = [w[1]\ w[2] \cdots w[N]]^T$, $\theta = [\theta_o^{(BP)}\ \theta_s^{(BP)}]^T$, and

$$\mathbf{H} = \begin{bmatrix} 1 & 1 & \cdots & 1 \\ 0 & T_{1,2}^{(A)} - T_{1,1}^{(A)} & \cdots & T_{1,N}^{(A)} - T_{1,1}^{(A)} \end{bmatrix}^T.$$

Note that the noise vector $\mathbf{w} \sim \mathcal{N}(0,\sigma^2 \mathbf{I})$ and the matrix $\mathbf{H}$ is the observation matrix whose dimension is $N \times 2$. From [26, theorem 3.2, p. 44], the minimum variance unbiased (MVU) estimator for the relative clock offset and skew is given by $\hat{\theta} = \mathbf{g}(\mathbf{x})$, where $\mathbf{g}(\mathbf{x})$ satisfies

$$\frac{\partial \ln p(\mathbf{x};\theta)}{\partial \theta} = \mathbf{I}(\theta)(\mathbf{g}(\mathbf{x}) - \theta). \quad (13.18)$$

Since the noise vector $\mathbf{w}$ is zero mean and Gaussian distributed, from the results in [26, p. 85], the derivative of the log-likelihood function can be written as

$$\frac{\partial \ln p(\mathbf{x};\theta)}{\partial \theta} = \frac{\mathbf{H}^T \mathbf{H}}{\sigma^2} [(\mathbf{H}^T \mathbf{H})^{-1} \mathbf{H}^T \mathbf{x} - \theta], \quad (13.19)$$

where $\mathbf{H}^T \mathbf{H}$ is assumed to be invertible. Therefore, comparing (13.18) with (13.9) yields

$$\hat{\theta} = (\mathbf{H}^T \mathbf{H})^{-1} \mathbf{H}^T \mathbf{x}, \quad (13.20)$$

$$\mathbf{I}(\theta) = \frac{\mathbf{H}^T \mathbf{H}}{\sigma^2}, \quad (13.21)$$

where $\mathbf{I}(\theta)$ is the Fisher information matrix. After some mathematical manipulations, the joint clock offset and skew estimator can be expressed as [19]

$$\begin{bmatrix} \hat{\theta}_o^{(BP)} \\ \hat{\theta}_s^{(BP)} \end{bmatrix} = \frac{1}{N\sum_{i=1}^{N}D_i^2 - \left[\sum_{i=1}^{N}D_i\right]^2} \begin{bmatrix} \sum_{i=1}^{N}D_i^2\sum_{i=1}^{N}x[i] - \sum_{i=1}^{N}D_i\sum_{i=1}^{N}\left[D_i\cdot x[i]\right] \\ N\sum_{i=1}^{N}\left[D_i\cdot x[i]\right] - \sum_{i=1}^{N}D_i\sum_{i=1}^{N}x[i] \end{bmatrix}, \quad (13.22)$$

where $D_i \triangleq T_{1,i}^{(A)} - T_{1,1}^{(A)}$. The Cramer-Rao lower bound (CRB) can be obtained by inverting the Fisher information matrix $\mathbf{I}(\theta)$. From (13.21), the Fisher information matrix is given by

$$\mathbf{I}(\theta) = \frac{1}{\sigma^2} \begin{bmatrix} N & \sum_{i=1}^{N}D_i \\ \sum_{i=1}^{N}D_i & \sum_{i=1}^{N}D_i^2 \end{bmatrix}.$$

Then, inverting $\mathbf{I}(\theta)$ yields

$$\mathbf{I}^{-1}(\theta) = \frac{\sigma^2}{N\sum_{i=1}^{N}D_i^2 - \left[\sum_{i=1}^{N}D_i\right]^2} \begin{bmatrix} \sum_{i=1}^{N}D_i^2 & -\sum_{i=1}^{N}D_i \\ -\sum_{i=1}^{N}D_i & N \end{bmatrix}. \quad (13.23)$$

Hence, from (13.23), the CRBs for the relative clock offset and skew become

$$\mathrm{var}(\hat{\theta}_o^{(BP)}) \geq \frac{\sigma^2\sum_{i=1}^{N}D_i^2}{N\sum_{i=1}^{N}D_i^2 - \left[\sum_{i=1}^{N}D_i\right]^2} \quad (13.24)$$

and

$$\mathrm{var}(\hat{\theta}_s^{(BP)}) \geq \frac{\sigma^2 N}{N\sum_{i=1}^{N}D_i^2 - \left[\sum_{i=1}^{N}D_i\right]^2}. \quad (13.25)$$

Notice further that the regularity conditions for the CRBs hold:

$$
E\left[\frac{\partial \ln p(\mathbf{x};\theta)}{\partial \theta}\right] = \begin{bmatrix} E\left[\frac{\partial \ln p(\mathbf{x};\theta)}{\partial \theta_o^{(BP)}}\right] \\ E\left[\frac{\partial \ln p(\mathbf{x};\theta)}{\partial \theta_s^{(BP)}}\right] \end{bmatrix}
$$

$$
= \begin{bmatrix} E\left[\frac{1}{\sigma^2}\sum_{i=1}^{N}\left[x[n]-\theta_o^{(BP)}-\theta_s^{(BP)}\cdot D_i\right]\right] \\ E\left[\frac{1}{\sigma^2}\sum_{i=1}^{N}\left\{\left[x[n]-\theta_o^{(BP)}-\theta_s^{(BP)}\cdot D_i\right]\cdot D_i\right\}\right] \end{bmatrix} = 0.
$$

Consequently, using the results in (13.22), node B can be synchronized to node P. Likewise, all the other nodes in the checked region in Figure 13.1 can be simultaneously synchronized to the parent node, node P, without any additional timing message transmissions, thus saving a significant amount of energy. Besides, there is no loss of synchronization accuracy when compared with other approaches [19].

13.3.3 Receiver-Receiver Synchronization

Receiver-receiver synchronization is an approach to synchronize a set of children nodes who receive the beacon messages from a common sender (a reference or parent node). Consider a parent (reference) node P and arbitrary nodes A and B, which locate within the communication range of the parent node in Figure 13.4. Suppose, in Figure 13.5 both node A and node B receive the ith beacon from node P at time instants $T_{2,i}^{(A)}$ and $T_{2,i}^{(B)}$ of their local clocks, respectively. Nodes A and B record the arrival time of the broadcast packet according to their own timescales and then exchange their time stamps. Suppose

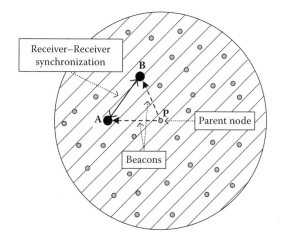

FIGURE 13.4 Receiver-receiver synchronization.

Receiver–receiver synchronization (*Node A* and *Node B*)

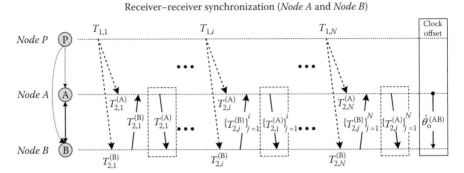

FIGURE 13.5   Clock synchronization model of RRS.

$X_i^{(PA)}$ denotes the nondeterministic delay components (random portion of delays) and $d^{(PA)}$ denotes the deterministic delay component (propagation delay) from node P to node A; then $T_{2,i}^{(A)}$ can be written as

$$T_{2,i}^{(A)} = T_{1,i} + d^{(PA)} + X_i^{(PA)} + \theta_o^{(PA)} + \theta_s^{(PA)} \cdot (T_{1,i} - T_{1,1}), \qquad (13.26)$$

where $T_{1,i}$ is the transmission time at the reference node, and $\theta_o^{(PA)}$ and $\theta_s^{(PA)}$ are the clock offset and skew of node A with respect to the reference node, respectively. Similarly, we can decompose the arrival time at node B as

$$T_{2,i}^{(B)} = T_{1,i} + d^{(PB)} + X_i^{(PB)} + \theta_o^{(PB)} + \theta_s^{(PB)} \cdot (T_{1,i} - T_{1,1}), \qquad (13.27)$$

where $d^{(PB)}$, $X_i^{(PB)}$, $\theta_o^{(PB)}$, and $\theta_s^{(PB)}$ stand for the propagation (fixed) delay, random portion of delays, clock offset, and skew of node B with respect to the reference node, respectively.

Subtracting (13.27) from (13.26), we obtain

$$T_{2,i}^{(A)} - T_{2,i}^{(B)} = \theta_o^{(BA)} + \theta_s^{(BA)} \cdot (T_{1,i} - T_{1,1}) + d^{(PA)} - d^{(PB)} + X_i^{(PA)} - X_i^{(PB)}, \qquad (13.28)$$

where $\theta_o^{(BA)} \triangleq \theta_o^{(PA)} - \theta_o^{(PB)}$ and $\theta_s^{(BA)} \triangleq \theta_s^{(PA)} - \theta_s^{(PB)}$ are the relative clock offset and skew between node A and node B at the time they receive the ith broadcast packet from the reference node, respectively. Here, we assume these random portions of delays $X_i^{(PA)}$ and $X_i^{(PB)}$ are normal distributed RVs with mean μ and variance $\sigma^2/2$. Indeed, (13.28) assumes exactly the same form as (13.17). Hence, the same steps can be applied to derive the joint clock offset and skew estimator for ROS. More specifically, let the noise component $z[i] \triangleq \mu' + X_i^{(BA)}$, where $\mu' \triangleq d^{(PA)} - d^{(PB)}$ and $z[i] \sim \mathcal{N}(\mu', \sigma^2)$. Let us also define $x[i] \triangleq T_{2,i}^{(A)} - T_{2,i}^{(B)} - \mu'$ and $w[i] \triangleq z[i] - \mu'$. Using similar steps as in ROS, it is straightforward to show that the same form of the joint clock offset and skew estimator (13.22) can also be applied to RRS. Consequently, there is no difference between ROS and RRS with regard to the accuracy of synchronization since the effects of random delays are the same. Likewise, the CRB for RRS can also be obtained using a similar procedure as in ROS. When there is no relative clock skew ($\theta_s^{(BA)} = 0$), it is straightforward to show that the maximum likelihood estimator of the relative clock offset $\hat{\theta}_o^{(BA)}$ becomes

$$\hat{\theta}_o^{(BA)} = \frac{1}{N} \sum_{i=1}^{N} \left[T_{2,i}^{(A)} - T_{2,i}^{(B)} \right], \tag{13.29}$$

which is the equivalent to the estimator presented in [18].

The main benefit of this approach is that all nondeterministic delay components on the transmitter side (send time and access time) are eliminated. Thus, a high degree of synchronization accuracy can be achieved using this approach.

13.4 Existing Time Synchronization Protocols

Thus far, a number of protocols have been suggested to solve the problem of time synchronization in distributed networks. For general computer networks, NTP has been adopted as the standard time synchronization scheme of the Internet [3]. Although NTP was shown to perform well in computer networks, it is not directly applicable to wireless sensor networks due to the unique challenges sensor networks face: limited power resources, wireless channel conditions, dynamic topology changes, etc. (recall also the design considerations presented in section 13.2.2). NTP enjoys unlimited (or rechargeable) energy resources and a relatively static topology in computer networks. However, these are not available in sensor networks. Therefore, different types of time synchronization protocols have been proposed to meet the design requirements of wireless sensor networks [2].

Ideally, a time synchronization protocol should be able to work optimally in terms of all design requirements of time synchronization, which are energy efficiency, scalability, precision, security, reliability, and robustness to network dynamics. However, the complex nature of wireless sensor networks makes it very difficult to optimize the protocol with respect to all these requirements simultaneously. Due to the trade-offs in satisfying these requirements, each protocol is designed to put distinct emphases on different requirements.

Assuming various criteria, time synchronization protocols can be categorized into different classes:

- **Master-slave versus peer-to-peer**
  - *Master-slave*: Where first a tree-like network hierarchy is arranged, and upon the completion of this arrangement only the connected nodes in the hierarchy synchronize with each other.
  - *Peer-to-peer*: Where any pair of nodes in the network can synchronize with each other.

- **Clock correcting versus untethered clock**
  - *Clock correcting*: Where the clock function in memory is modified after each run of the time synchronization process.
  - *Untethered clock*: Where every node maintains its own clock as it is, and keeps a time-translation table relating its clock to other nodes' clocks. Thus, instead of updating its clock constantly, each node translates the time information in the data packets coming from other nodes to its own clock by using the time-translation table.
- **Synchronization approach**
  - *Sender-receiver*: Where one of two nodes, which are synchronizing with each other, sends a time-stamp message while the other one receives it.
  - *Receiver-receiver*: Where a reference node transmits synchronization signals and two synchronizing nodes receive these signals and record the time of receptions (time stamps).
  - *Receiver-only*: Where a group of nodes can be simultaneously synchronized by only listening to the message exchanges of a pair of nodes.
- **Pairwise synchronization versus network-wide synchronization**
  - *Pairwise synchronization*: Where the protocols are primarily designed to synchronize two nodes, although they usually can be extended to handle synchronization of a group of nodes.
  - *Network-wide synchronization*: Where the protocols are primarily designed to synchronize a large number of nodes in the network.

Additional classifications can be found in [4]. In the following, we will summarize the existing time synchronization protocols based on the last category.

13.4.1   Pairwise Synchronization

13.4.1.1   Timing-Sync Protocol for Sensor Networks (TPSN)

TPSN [17] uses the two-way message exchange mechanism, as discussed in the sender-receiver synchronization approach described in section 13.3.1, to achieve the synchronization between two nodes. With only one round of message exchanges, and without any statistical model on the variable delay components $X_i^{(AP)}$ and $X_i^{(PA)}$ in (13.2) and (13.3), a simple estimate for $\theta_0^{(AP)}$ is proposed in [17] as

$$\hat{\theta}_0^{(AP)} = \frac{U_i - V_i}{2}, \tag{13.30}$$

where $U_i \triangleq T_{2,i}^{(P)} - T_{1,i}^{(A)}$ and $V_i \triangleq T_{4,i}^{(A)} - T_{3,i}^{(P)}$. Notice that in the original form of TPSN, it does not estimate clock skew; therefore, frequent application of TPSN is needed to keep the clock offset between two nodes under a certain limit.

13.4.1.2   Maximum Likelihood Estimation for Clock Offset Based on Two-Way Message Exchanges

Assume the clock offset $\theta_0^{(AP)}$ is constant for N rounds of message exchanges. If $X_i^{(AP)}$ and $X_i^{(PA)}$ in (13.2) and (13.3) are exponentially distributed with the same unknown mean λ, and when $d \triangleq d^{(AP)} = d^{(PA)}$ is unknown, it is proved in [23] that the ML estimator of $\theta_o^{(AP)}$ is given by

$$\hat{\theta}_o^{(AP)} = \frac{\min_{1 \le i \le N} U_i - \min_{1 \le i \le N} V_i}{2}. \tag{13.31}$$

On the other hand, with $X_i^{(AP)}$ and $X_i^{(PA)}$ in (13.2) and (13.3) being modeled as independent and normally distributed RVs with the same mean μ and variance $\sigma^2/2$, the maximum likelihood (ML) estimate for $\theta_o^{(AP)}$ takes the equation (derived in section 13.3.1)

$$\hat{\theta}_o^{(AP)} = \frac{\frac{1}{N} \sum_{i=1}^{N} U_i - \frac{1}{N} \sum_{i=1}^{N} V_i}{2}. \tag{13.32}$$

Notice from (13.30)–(13.32), it is clear that if only one round of message exchange is performed, the TPSN presented in (13.30) is the ML estimator under both exponential and Gaussian delay models.

13.4.1.3 Joint Clock Offset and Skew Estimation Based on Two-Way Message Exchanges

When clock skew exists between two nodes, the clock offset between them will increase linearly, as shown in Figure 13.3. In order to establish long-term synchronization, it is more efficient to estimate jointly the clock offset and skew. In section 13.3.1, we derived the joint offset and skew ML estimators (see equations (13.9) and (13.10)), when the variable delays $X_i^{(PA)}$ and $X_i^{(AP)}$ are modeled as independent Gaussian distributed RVs. When $X_i^{(PA)}$ and $X_i^{(AP)}$ are exponentially distributed RVs, the likelihood function for joint estimation of the clock offset and skew is very complicated. However, a solution to this problem has been recently reported in [27].

Notice that the joint offset and skew ML estimators (equations (13.9) and (13.10)) under Gaussian delay assumption are quite complicated. Besides, there is no simple closed-form solution for the ML joint offset and skew estimation when the delays are exponentially distributed. For these reasons, a family of robust and simple clock offset and skew estimators, named maximum likelihood like estimators (MLLEs), has been proposed in [25].

13.4.1.4 Tiny-Sync and Mini-Sync

Tiny-sync and Mini-sync [28] are two lightweight clock synchronization protocols that also use the two-way message exchanges. Node A and node P exchange messages just like in Figure 13.3. The only difference here is that node P replies to node A immediately after receiving the message, i.e., $T_{2,i}^{(P)} = T_{3,i}^{(P)}$. Assuming the clocks between node A and node P are linearly related, from (13.6) and (13.7) we have

$$\frac{T_{2,i}^{(P)} - \theta_o^{(AP)}}{1 + \theta_s^{(AP)}} = T_{1,i}^{(A)} + d + X_i^{(AP)},$$

$$\frac{T_{2,i}^{(P)} - \theta_o^{(AP)}}{1 + \theta_s^{(AP)}} = T_{4,i}^{(A)} - d - X_i^{(PA)}.$$

Since d, $X_i^{(AP)}$, and $X_i^{(PA)}$ are all nonnegative, defining $\theta'_s \triangleq 1/(1 + \theta_s^{(AP)})$ and $\theta'_o \triangleq \theta_o^{(AP)}/(1 + \theta_s^{(AP)})$, we obtain

$$T_{1,i}^{(A)} \leq \theta'_s T_{2,i}^{(P)} + \theta'_o \leq T_{4,i}^{(A)}. \tag{13.33}$$

The 3-tuple of time stamp $(T_{1,i}^{(A)}, T_{2,i}^{(P)}, \text{and } T_{3,i}^{(A)})$ is called a data point. With N message exchanges, the goal is to find θ'_o and θ'_s such that they satisfy (13.33) for $1 \leq i \leq N$. In general, this is a linear programming problem and there are an infinite number of solutions for this problem [29]. Although more time stamps would generate tighter bounds on θ'_o and θ'_s, unfortunately, at the same time, the computational and storage requirements of the linear programming approach also increase. Thus, such an approach appears to be not suitable for implementation in wireless sensor nodes, which have strictly limited memory and computing resources.

Tiny-sync and Mini-sync tackle the problem as finding the best-fit line that lies between the bound sets defined by the data points. Based on the observation that not all data points are useful, Tiny-sync preserves only four constraints (the ones that yield the best bounds on the estimate) out of all data points. This results in a very efficient algorithm. However, it is shown by a counterexample [28] that this scheme does not always produce the optimal solution since some data points are considered useless and discarded at a certain time, a step that actually might provide a better bound if it is properly considered with another data point that is yet to come.

Mini-sync is an improved version of Tiny-sync in the sense that it finds the optimal solution with increased complexity (but still with lesser complexity than the linear programming approach). Mini-sync basically uses an additional criterion to determine whether the data point can be safely discarded.

13.4.1.5 Reference Broadcast Synchronization (RBS)

RBS [18] is based on the RRS approach discussed in section 13.3.3. Let the time stamps recorded at node A and node B for receiving the ith common packet be denoted as $T_{2,i}^{(A)}$ and $T_{2,i}^{(B)}$, respectively. The estimate of the clock offset between node A and node B is proposed in [18] as

$$\hat{\theta}_o^{(BA)} = \frac{1}{N} \sum_{i=1}^{N} \left[T_{2,i}^{(A)} - T_{2,i}^{(B)} \right], \tag{13.34}$$

where N stands for the total number of common packets received by node A and node B. We have shown in section 13.3.3 that the above estimator is actually the ML estimator for the clock offset, assuming the random portions of the delays in message deliveries are Gaussian distributed RVs, and there is no clock skew. When there is a clock skew between node A and node B, least-squares linear regression is proposed in [18] to estimate the clock skew.

The main advantage of RBS is that by comparing the time stamps of a common packet at two different nodes, it removes the largest sources of nondeterministic error (send time and access time) from the transmission path. Thus, RBS provides a high degree of synchronization accuracy. Note also that RBS can be applied to commodity hardware and existing software in sensor networks as it does not need access to the low levels of the operating system.

13.4.1.6 Clock Offset and Skew Estimation Based on Broadcast Clock

Under the setting that a sensor node observes and synchronizes to a broadcast clock, [30] derives the ML estimator for clock offset and skew with the broadcast message delay being modeled as uniformly distributed RVs. It is shown that the ML estimate in this case is generally not unique. Furthermore, the support of the likelihood function is not convex, which leaves out the possibility of taking the mean of all equally likely solutions. This motivated [30] to consider the linear estimator for the clock offset and skew. Under the same setting, [31] derives the joint ML clock offset and skew estimator with the assumption that the broadcast message delays are modeled as exponentially distributed RVs. It is shown in [31] that a unique joint ML clock offset and skew estimate exists under certain conditions, as opposed to the case of uniformly distributed delay. Furthermore, the Gibbs sampler was introduced in [31] to further enhance the performance of the joint ML estimator.

13.4.1.7 Flooding Time Synchronization Protocol (FTSP)

In [16], it is argued that if one can time-stamp the message at the MAC layer, this immediately eliminates three sources of delay uncertainties: transmit, access, and receive times. In this case, the main delivery delay comes from transmission and reception times at the radio chips (see section 13.2.3). These delays can be further decomposed into:

1. Interrupt handling time, which is the delay between the radio chip raising and the microcontroller responding to an interrupt
2. Encoding time, which is the time it takes for the radio chip to encode and transform the message into a radio wave
3. Decoding time, which is the time for the radio chip at the receiver to transform the radio wave back into binary data
4. Byte alignment time, which is the delay at the receiver to synchronize with the byte boundary at the physical layer

FTSP [16] uses a single broadcasted message to establish synchronization points between sender and receivers, while eliminating the jitter of interrupt handling and encoding/decoding times by utilizing multiple MAC layer time stamps on both the sender and receiver sides. Furthermore, the skew of the clock between sender and receiver is estimated using multiple messages and linear regression.

13.4.2   Network-Wide Synchronization

Until this point, we have only described the time synchronization between two neighboring sensor nodes. In this section, we will discuss protocols for network-wide synchronization.

13.4.2.1   Extension of TPSN

In order to establish a global timescale for all the nodes in the sensor field based on TPSN, [17] proposes to create a hierarchical structure (spanning tree) in the network (named level discovery phase) before pairwise synchronization is performed between adjacent levels (named synchronization phase). The level discovery phase consists of the following steps:

1. Select a root node using an appropriate leader election algorithm and assign a 0 level to the root node.
2. The root node broadcasts a level discovery packet (LDP) containing the identity and level of the packet.
3. Every node that receives an LDP assigns its level to a level greater (by one) than that of the received packet and sends a new level discovery packet attaching its own level (once being assigned a level, a node neglects future packets requesting level discovery to avoid flooding congestion).
4. Repeat step 3 until every node in the network successfully assigns a level.

After the spanning tree is formed, the root node initiates the synchronization phase by synchronizing all the nodes in level 1. Next, the nodes in level 1 synchronize with the nodes in level 2, and so on, until all the nodes have been synchronized. Notice that the synchronization error of a node with respect to the root node is a nondecreasing function of the hop distance because the random signal errors over each hop add up. A number of different searching algorithms can be considered in the construction of the spanning tree. For instance, Van Greunen and Rabaey suggested some preliminary ideas on constructing spanning trees with low depth in order to improve the accuracy of synchronization [12].

13.4.2.2   Lightweight Time Synchronization (LTS)

Also based on two-way message exchanges, [12] proposes two network-wide synchronization protocols. The first one is called centralized multihop LTS, which is basically the same protocol as the extension of TPSN discussed above. The other one is called distributed multihop LTS. This distributed LTS algorithm moves the resynchronization from the root node to the nodes that need resynchronization. When a node A determines that it needs to be resynchronized, it will send a resynchronization request to the root node. In order for node A to resynchronize, all nodes along the routing path from the root node to node A will be synchronized in a pairwise fashion.

13.4.2.3   Extension of RBS

The RBS protocol discussed in the above subsection can only synchronize a set of nodes that lie within a single broadcast domain. In order to synchronize a large sensor network,

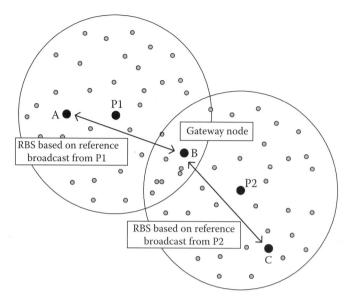

FIGURE 13.6 Extension of RBS to multihop.

[18] proposes to use *gateway* nodes for converting time stamps from one neighborhood's time base to another. The idea is illustrated in Figure 13.6. Nodes P1 and P2 send out synchronization beacons, and they create two overlapping neighborhoods, where node B lies in the overlapping area. Since node A and node B lie within the same neighborhood, their clock relationship (i.e., clock offset and skew) can be established from node P1's reference broadcast. Similarly, the clock relationship between node B and node C can be established from node P2's reference broadcast. Therefore, the clock relationship between node A and node C can be computed with node B acting as a gateway.

13.4.2.4 Extension of FTSP

FTSP can be extended to network-wide synchronization in a straightforward manner. First, a root node, to which the whole network is being synchronized, is elected by the network. Nodes that are within the broadcast radius of the root node can receive time-stamped messages from the root node. They then estimate the offset and skew of their own local clocks, thus synchronizing with the root node. The newly synchronized nodes can then broadcast synchronization messages to other nodes in the network. The advantage of this flooding process is that it begins with the root node, and there is no need to have a level hierarchy, as opposed to TPSN.

13.4.2.5 Pairwise Broadcast Synchronization

Pairwise broadcast synchronization (PBS) employs both sender-receiver and receiver-only synchronization approaches to achieve network-wide synchronization with high energy efficiency [19]. As discussed in section 13.3.2, in PBS a number of sensor nodes can be synchronized by only overhearing timing messages being exchanged between pairs of nodes, which significantly reduces the overall energy consumption by decreasing the number of required timing messages in synchronization. PBS requires a much smaller number of timing messages than other existing protocols, such as RBS, TPSN, and FTSP, and its benefits remarkably increase as the sensors are more densely deployed.

Let N_{RBS}, N_{TPSN}, N_{FTSP}, and N_{PBS} denote the numbers of required timing messages for synchronization in RBS, TPSN, FTSP, and PBS, respectively. In TPSN, since every node in the network is connected to its parent node (except the root node), there are $L - 1$ branches (edges) in a hierarchical tree, where L is the overall number of sensor nodes [17]. For TPSN, $2N$ timing messages are required in every pairwise synchronization, hence $N_{TPSN} = 2N(L - 1)$. This result can be applied to other level-based SRS protocols without loss of generality. For RBS, the reference node must broadcast the beacon packet N times. Besides, every sensor node must send time readings upon receiving the broadcast beacons of all the other nodes in the network to compensate relative clock offsets among each other [18]. Thus, $N_{RBS} = N + L(L - 1)/2$, since the number of unique pairs in the network is $L(L - 1)/2$. In FTSP, each sensor node must send its timing messages once upon receiving the timing messages from another sensor due to its flood-based communication procedure [16]. Hence, the number of required timing messages in FTSP becomes $N_{FTSP} = NL$.

It is remarkable that the required numbers of timing messages for all the above-mentioned protocols are proportional to the number of sensors in the network L or its square L^2. However, PBS needs only $2N$ timing messages in every synchronization period, i.e., $N_{PBS} = 2N$, assuming all the nodes lie within a single broadcast neighborhood. Hence, N_{PBS} does not depend on the number of sensors in the network, which incurs an enormous amount of energy savings. Moreover, this gain proportionally increases with respect to the scale of the network. Consequently, the benefit of PBS over RBS, TPSN, and FTSP is clear and huge in terms of energy savings with the cost of allocating two super nodes in the network. In case there exist other nodes located outside of the checked region in Figure 13.1, likewise RBS, the network could be divided into a number of separated groups (clusters), and they could be synchronized by additional pairwise synchronizations among the super nodes in different groups, i.e., global synchronization can be achieved by a sequence of pairwise synchronizations among the super nodes. Here, diverse grouping and pair selection algorithms can be considered according to the type of the network. For instance, assuming the level hierarchy of the network is established, there are groups of parents and children nodes, where a group consists of a parent and its children nodes. Here, every parent node can search the connectivity among its children nodes to select the best synchronization pairs that maximize the number of nodes performing ROS, i.e., minimizing the number of pairwise synchronizations. In fact, no network-wide connection search is required in this case because of its limited and known set of scanning nodes.

13.4.2.6 Time-Diffusion Synchronization Protocol (TDP)

TDP [32] is a protocol enabling the sensor network to reach an equilibrium time with the clocks of individual sensors within a small time deviation from the equilibrium time. The protocol can be understood as periodically applying three phases: (1) election of master/diffused leader nodes, (2) time-diffusion procedure, and (3) peer evaluation

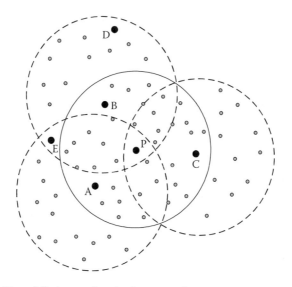

FIGURE 13.7 Time-diffusion synchronization protocol.

procedure. It is shown analytically in [32] that the TDP enables the clocks in the whole network to converge to a unique value.

In the first phase, master nodes are elected in the sensor field. The election criteria include the quality of clock and the energy resources of a particular node. Referring to Figure 13.7, assume that node P is elected to be the master node (here we illustrate the concept with one master node, while in more complicated networks, more than one master node might be possible). Node P then sends a number of time-stamped messages to its neighbors. Once the neighbors receive the messages, they self-determine if they would become diffused leader nodes, based on the results of the last round peer evaluation procedure (the third phase). In Figure 13.7, nodes A, B, and C are the elected diffused leader nodes. The elected diffused leader nodes respond to the master node, thus enabling the master node to measure the average and the standard deviation of the round-trip delay from its neighbors. At the same time, the diffused leader nodes start sending messages to their own neighbors to measure the mean and standard deviation of round-trip delay to their neighbors. The process is repeated until all the nodes have been covered.

In the second phase, the time information from the master node will be diffused (with the help of diffused leader nodes) to all the nodes in the network. The diffusion procedure takes place according to the following sequence of events. First, the master node sends a time-stamped message containing the standard deviation of the round-trip delay to its neighbors. Before transmission, the time stamp of the message is adjusted with half of the measured average round-trip delay (from the first phase) to account for the message delivery delay to its neighbors. Once the diffused leader nodes receive the time-stamped message, they set their clock according to the received time stamp and then broadcast their own time-stamped messages, containing their measured standard derivations of the round-trip times to their neighbors. Again, before transmission of the messages, the time stamps have to be adjusted with half of the measured average round-trip delay to their neighbors. For nodes that are not diffused leaders, if they only receive a message from one diffused leader node (e.g., node D in Figure 13.7), they just set their clock according to the time stamp they received. For the nodes that have received more than one time-stamped message originating from different diffused leader nodes (e.g., node E in Figure 13.7), they will use the standard deviations as weightings (the smaller the deviation, the larger the weighting) to combine the clock values and set their clocks according to the result.

The purpose of the third phase is to allow the sensor nodes to evaluate the stability of their local clock. First, the elected master nodes broadcast a number of time-stamped messages. The neighbor nodes receiving these messages calculate the two-sample Allan variance [32] of the local clock from the clock of the master nodes and send back these calculated Allan variances to the master nodes. Then the master nodes compute the average of all the Allan variances they received and send the result back to their neighbor nodes. By this procedure, all the neighbor nodes can evaluate the quality of their clocks with respect to those of their neighbors by comparing their calculated Allan variance with the average value. The above procedure is repeated, but with the elected diffused leader nodes broadcasting the time-stamped messages.

13.4.2.7   Synchronous and Asynchronous Diffusion Algorithms

In [33], two diffusion algorithms are proposed. The first one is called rate-based synchronous diffusion algorithm. The idea behind this algorithm is that in order for a network to achieve an equilibrium time, the clock at node i, denoted as c_i, should be adjusted according to the differences between its clock and its neighbors' clocks (assuming node i has exchanged clock readings with its neighbors). That is, the clock at node i should be set to

$$c_i - \sum_{j \neq i} r_{ij}(c_i - c_j),$$

where $r_{ij} > 0$ is the diffusion rate, $r_{ij} = 0$ if node i and node j cannot directly communicate, and the condition

$$\sum_{j \neq i} r_{ij} \leq 1$$

is enforced. The above algorithm can also be formulated using matrix notation. For a group of n sensor nodes, let $\mathbf{c}^t$ be the vector of length n containing the clock readings of all the sensor nodes at time t. The synchronous diffusion algorithm adjusts the clocks of different nodes using $\mathbf{c}^{t+1} = R\mathbf{c}^t$, where

$$\mathbf{R} = \begin{pmatrix} r_{11} & r_{12} & \cdots & r_{1n} \\ r_{21} & r_{22} & \cdots & r_{2n} \\ \vdots & \vdots & \ddots & \vdots \\ r_{n1} & r_{n2} & \cdots & r_{nn} \end{pmatrix} \qquad (13.35)$$

and

$$r_{ii} = 1 - \sum_{j \neq i} r_{ij}.$$

It is shown in [33] that if the second largest eigenvalue of $\mathbf{R}$ is smaller than 1, the synchronous diffusion algorithm will converge, in the sense that all the elements in $\mathbf{c}^t$ will be equal.

The synchronous diffusion algorithm requires all the nodes to operate in an ordered manner. In order to remove this constraint, [33] proposed another algorithm, named asynchronous diffusion algorithm. In this algorithm, each node asks its neighbors about their clock readings and computes the average value. Then the average value is sent back to the neighbors so they can update their clocks. This algorithm gives a very simple averaging operation of a node over its neighbors, and the averaging operations by different nodes can be carried out at different times and in any order (thus the name asynchronous). It is shown in [33] that the clocks of sensor nodes at a sensor network converge to the average value by using this asynchronous algorithm.

13.4.2.8 Protocols Based on Pulse Transmissions

Recently, synchronization schemes that operate exclusively at the physical layer by transmitting pulses instead of message packets have been proposed in [34] and [35]. In [34], inspired by the synchronously flashing fireflies, the time synchronization problem in sensor network is modeled using pulse coupled oscillators (PCOs). In this scheme, each node (say node j) in the sensor network is associated with an increasing monotonic state function $x_j(t)$ taking values from 0 to 1. If a node is isolated, the state function $x_j(t)$ increases from 0 to 1 smoothly as a function of time, and the node emits a pulse when the state function achieves the unit value ($x_j(t) = 1$). After firing a pulse, the node immediately resets its state to zero. This results in periodic emission of pulses with period T. If a node is not isolated, it can receive pulses from other nodes. When a node receives a pulse, its state variable changes as follows:

$$x_j(\tau^+) = \begin{cases} x_j(\tau) + \varepsilon, & \text{if } x_j(\tau) + \varepsilon < 1 \\ 0, & \text{otherwise} \end{cases}, \tag{13.36}$$

where τ is the time the node receives a pulse and ε is the advancement of the clock phase. This means that a node receiving a pulse either emits the pulse at the same time or shortens the waiting time for the next round of emissions. With the assumption that after a node fires a pulse, it enters a short refractory period, during which no signal can be received from other nodes (to avoid infinite feedback), it can be shown that only when the nodes emit the pulse simultaneously will they be insensitive to coupling, and therefore achieve synchronization.

In [35], a cooperative technique that constructs a sequence of pulses with equidistance zero-crossings is developed. The basic idea of this scheme is as follows. Assume there is a leader node and it emits a sequence of pulses with equidistance zero-crossings.

The surrounding nodes receive this pulse sequence, and based on the locations of the observed zero-crossings, the surrounding nodes predict when the next pulse will be transmitted. Then, these nodes emit pulses at their predicted times and an aggregate pulse sequence will be generated. It is shown in [35] that although the prediction at individual nodes may not be perfect, under certain conditions on the pulse and in asymptotically dense networks, the zero-crossings of the aggregate waveform sequence will be at the same positions as the zero-crossings of the original waveform sequence emitted by the leader node due to spatial averaging. This aggregate pulse sequence will be heard by the nodes lying further away from the leader node, and these nodes perform prediction as described before and emit their pulses at their predicted times. The procedure will be continued until all the nodes are synchronized.

Notice that the synchronization algorithms discussed in this section only provide a unified ticking rhythm across sensor nodes, but not the synchronization of clock time. A good analog is a group of people clapping together to get a rhythm. However, there exist applications in which a unified rhythm is enough, e.g., in distributed beamforming and reachback channel [36]. As another variation, a joint physical and network layer time synchronization scheme was proposed to overcome the effects of imperfect physical layer synchronization due to the nature of common wireless channels [37].

13.5 Adaptive Time Synchronization for WSNs

While all the above protocols in section 13.4 can achieve instantaneous synchronization among nodes, the timing of different nodes would drift apart as time passes; therefore, periodic resynchronization is needed to maintain long-term synchronization. Intuitively, less resynchronization needs less energy but leads to a larger synchronization error, while more frequent resynchronization leads to a smaller synchronization error but requires more energy. A natural question is what is the minimum resynchronization frequency (or equivalently, maximum resynchronization period) that can meet the desired synchronization precision. Therefore, adaptive algorithms are necessary to dynamically determine the resynchronization period, number of beacons to be used in each round of synchronization, synchronization accuracy, and so on. In this section, we will review three existing adaptive time synchronization algorithms proposed in the literature.

13.5.1 Rate-Adaptive Time Synchronization (RATS)

Consider the case where node A sends time-stamped messages to node B periodically with period τ, and node B records the receiving times of the messages. Based on a number of data points $(T_i^{(A)}, T_i^{(B)})$, where $T_i^{(A)}$ and $T_i^{(B)}$ are the time stamps made at node A and node B respectively, node B wants to determine the largest τ such that the synchronization error is smaller than a certain limit. The rate-adaptive time synchronization [38] is an algorithm that determines the optimal τ adaptively. Its idea can be summarized using the flowchart shown in Figure 13.8. First, node B calculates the optimal number of data samples for model parameters (e.g., clock offset and skew) estimation based on the current value of τ. Next, node B takes the required number of data points (stored

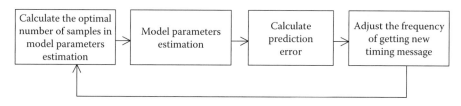

FIGURE 13.8 Flowchart of RATS run at the node receiving time-stamped messages from another node.

in memory) and estimates the model parameters. Then, node B computes the prediction error. Finally, using the calculated prediction error, node B adjusts the frequency of getting a new timing message from node A: If the prediction error is larger than the upper limit threshold E_u, then the timing message rate is not frequent enough from node A, and therefore τ should be decreased. On the other hand, if the prediction error is smaller than the lower limit threshold E_l, then there are fewer timing messages, and thus τ should be increased. Multiplicative increase and decrease strategies are used to enable fast convergence and quick response to the changing environment. After getting a new data point according to the new value of τ, the above process is repeated.

13.5.2 RBS-Based Adaptive Clock Synchronization

With the RBS setting, [39] extends the deterministic RBS protocol (discussed in section 13.4.1) to an adaptive probabilistic synchronization algorithm, allowing trade-offs between synchronization accuracy and resource requirement. It is based on the observation that if the relative clock skew error between two nodes ε, after applying RBS with one broadcast message, is a Gaussian RV with zero mean and variance σ^2, then the probability of error-free synchronization with N broadcast messages is given by

$$Pr\left(|\varepsilon|<\varepsilon_{max}\right)=2erf\left(\frac{\sqrt{N}\varepsilon_{max}}{\sigma}\right),\qquad(13.37)$$

where ε_{max} stands for the maximum specified (allowable) clock offset for communications, and

$$erf(x)\triangleq 1/2\pi\cdot\int_0^x \exp\left(-t^2/2\right)dt.$$

From the above equation, it is clear that the performance criterion is a probabilistic measure since there is always a possibility that the clock offset is greater than some limit ε_{max}. However, one can reduce this probability to an arbitrarily small value by increasing N, the number of broadcasting messages in one round of RBS.

After application of RBS, we can bound the clock skew error with certain probability. However, since clocks from different nodes would drift apart as time passes, we need to reapply RBS periodically. Reference [39] proposes a formula to determine the maximum time between resynchronization τ_{max} as

$$\tau_{max} = \frac{\gamma_{max} - \varepsilon_{max}}{\rho} - d_{max}, \tag{13.38}$$

where γ_{max} denotes the maximum allowable clock skew at any time, ρ denotes the maximum drift of clock, and d_{max} is the maximum delay of time-stamp exchanges in RBS. With different synchronization precision requirements (specified by γ_{max}), one can determine the required resynchronization period τ_{max}.

13.5.3 Adaptive Multihop Time Synchronization

Adaptive multihop time synchronization (AMTS) [40] is based on a two-way message exchange mechanism similar to that used in TPSN. AMTS adaptively optimizes some crucial network parameters, such as the synchronization mode, the resynchronization period, and the number of beacons per pairwise synchronization, with respect to the current network status. AMTS consists of three functional phases:

- *Level discovery phase*: The same as that in TPSN, and used for generating a hierarchical structure in the network.
- *Synchronization phase*: Similar to the corresponding synchronization phase in TPSN. However, as opposed to TPSN, AMTS adjusts not only the current clock offset but also the clock skew to guarantee the long-term synchronization, while TPSN only estimates the clock offset. Hence, AMTS requires far less frequent resynchronization.
- *Network evaluation phase*: The reference node investigates the current status of network traffic in order to select the synchronization mode between *always on* (**AO**) (always maintain network-wide synchronization) and *sensor initiated* (**SI**) (synchronize only when it needs to). Besides, it optimizes the resynchronization period and the number of beacons per each pairwise synchronization.

The second and third phases (i.e., synchronization and the network evaluation phases) will be periodically repeated in order to minimize overall energy consumption with respect to the current network status. Since the level discovery procedure (section 13.4.2) and pairwise synchronization based on two-way message exchanges (section 13.3.1) have been detailed previously, we only focus on the network evaluation phase below.

13.5.3.1 Synchronization Mode Selection

The idea of selecting the synchronization mode between *always on* (**AO**) and *sensor initiated* (**SI**) is based on the observation that when network traffic occurs rarely and synchronization delay is not a critical problem, keeping all the sensor nodes synchronized all the time (**AO** mode) is not a good strategy since synchronization consumes a lot of energy. In addition, for some applications, the sensor clocks might be allowed to go out of synchronization unless sensing events happen. In this case, the **SI** mode, where only nodes participating in a particular multihop data transmission synchronize with each other, is a better choice.

In order to determine whether **AO** or **SI** mode is to be used, the following network parameters are introduced in AMTS:

- B: Number of branches (edges) in a spanning tree of the network. It can be obtained after the level discovery phase.
- τ: Resynchronization period, i.e., the time between resynchronizations.
- $\bar{h}$: Average of hops per unit time. In every sensing event, the destination node accumulates the number of hops that have occurred in that particular transmission to its storage. During the synchronization phase, the reference node collects the information about the total number of hops occurred in the last synchronization period and determines the average number of hops per unit time ($\bar{h}$) in the network. This information indicates how busy the network traffic is and can be included in timing messages with a small overhead.
- δ: Latency factor ($0 \leq \delta \leq 1$) reflecting the amount of allowed delay in data transmission. Higher latency factor means less concern for network delays. For example, δ is set to be 0 for sensor networks requiring network synchronization all the time. On the other extreme, for delay-independent networks, δ should be close to 1.
- N: Number of timing message exchanges per pairwise synchronization.

In the **AO** mode, the number of timing messages per unit time is given by $\bar{M} = 2BN/\tau$, while in the **SI** mode, $\bar{M} = 2\bar{h}N$. To minimize the number of timing messages per unit time $\bar{M}$, the synchronization mode should be selected as follows:

$$\frac{2BN\delta}{\tau} \underset{\text{SI}}{\overset{\text{AO}}{\lessgtr}} 2\bar{h}N, \tag{13.39}$$

or equivalently,

$$\tau \underset{\text{SI}}{\overset{\text{AO}}{\lessgtr}} \frac{B\delta}{\bar{h}}. \tag{13.40}$$

From (13.40), the synchronization mode changes from **AO** to **SI** when τ is smaller than $B\delta/\bar{h}$ and vice versa. In the **SI** mode, the reference node periodically inquires about the number of hops that occurred during the past time interval, and then it makes a decision whether or not to switch to the **AO** mode. Notice that AMTS is an iterative process, and the resynchronization period parameter in the mode selection will be from the last iteration of AMTS.

13.5.3.2 Determination of Resynchronization Period

As the resynchronization period τ increases, the network becomes more power efficient. Thus, τ should be chosen as large as possible. However, a too large value of τ induces a critical synchronization problem since the clock difference (offset) between nodes keeps generally increasing with time. Hence, there exists a maximum resynchronization period (τ_{max}) that is determined by the oscillator specifications and the accuracy of estimators.

Suppose that the clock timing mismatch ε between the two nodes is modeled as follows: $\varepsilon = \varepsilon_o + \varepsilon_s t$, where t denotes the reference time and ε_o and ε_s stand for the clock offset and skew errors, respectively. In general, it is difficult to determine any specific mathematical model for either clock offset or skew errors. Herein, we model both clock offset and skew errors by normal distributions based on the experimental results reported in [17] and [18]:

$$\varepsilon_{o,i} \sim \mathcal{N}(0, \sigma^2_{\varepsilon_{o,i}}) \quad 1 \le i \le N,$$

$$\varepsilon_{s,i} \sim \mathcal{N}(0, \sigma^2_{\varepsilon_{s,i}}) \quad 1 \le i \le N,$$

where $\varepsilon_{o,i}$ and $\varepsilon_{s,i}$ denote the clock skew and offset estimation errors after the ith message exchanges, respectively. Note that clock skew estimation is only available when there are multiple message exchanges. Hence, $\varepsilon_{s,1}$ stands for the clock skew error when no skew estimation occurred. Here, the maximum clock mismatch can be modeled as another normal distribution $\varepsilon \sim \mathcal{N}(0, \sigma^2_\varepsilon)$, where $\sigma^2_\varepsilon = \sigma^2_{\varepsilon_{o,N}} + \sigma^2_{\varepsilon_{s,N}} \tau^2_{max}$, ($t = \tau_{max}$). Imposing the upper limit ε_{max} for the clock error via the probabilistic measure,

$$P_s = Pr\left(|\varepsilon| \ge \varepsilon_{max}\right) = erfc\left(\frac{\varepsilon_{max}}{\sqrt{2}\sigma_\varepsilon}\right),$$

where

$$erfc(x) \triangleq 2/\sqrt{\pi} \cdot \int_x^\infty \exp(-t^2)dt$$

and P_s denotes the synchronization error probability for pairwise synchronization. Thus, σ_ε can be determined when ε_{max} and the maximum allowable P_s are fixed. For instance, when P_s is limited to 0.1% and ε_{max} is 10 ms, then the standard deviation of clock mismatch (σ_ε) has to be smaller than 3.04 ms.

The maximum resynchronization period with N beacons can be written as

$$\tau^{(N)}_{max} = \sqrt{\frac{\sigma^2_\varepsilon - \sigma^2_{\varepsilon_{o,N}}}{\sigma^2_{\varepsilon_{s,N}}}}. \tag{13.41}$$

Based on the lower bounds and asymptotic performance of the estimators, one can easily infer closed-form expressions of the variances $\varepsilon_{o,N}$ and $\varepsilon_{s,N}$ in terms of the variances $\varepsilon_{o,1}$ and $\varepsilon_{s,2}$, respectively. From the lower bounds derived in [40], $\sigma^2_{\varepsilon_{o,N}}$ can be written with respect to N and $\sigma^2_{\varepsilon_{o,1}}$ as

$$\sigma^2_{\varepsilon_{o,N}} = \frac{\sigma^2_{\varepsilon_{o,1}}}{N}.$$

Similarly, since the time differences between beacons are proportional to N and by far greater than the variance of delays, the following relationship can be obtained from the lower bound for the clock skew estimator derived in [25]:

$$\sigma^2_{\varepsilon_{s,N}} = \frac{\sigma^2_{\varepsilon_{s,2}}}{(N-1)^2}, \quad N \geq 2.$$

Therefore, for $N \geq 2$, $\tau^{(N)}_{\max}$ can be rewritten as

$$\tau^{(N)}_{\max} = (N-1)\sqrt{\frac{\sigma^2_\varepsilon - \frac{\sigma^2_{\varepsilon_{o,1}}}{N}}{\sigma^2_{\varepsilon_{s,2}}}}, \quad N \geq 2. \tag{13.42}$$

Note that $\varepsilon_{s,1}$ can be obtained by the specifications of the crystal oscillator, and $\varepsilon_{o,1}$ and $\varepsilon_{s,2}$ can be determined by simple experimental tests. Therefore, the maximum resynchronization period is proportional to the number of beacons, and performing clock skew estimation will significantly increase $\tau^{(N)}_{\max}$ since $\sigma_{\varepsilon_{s,1}} \gg \sigma_{\varepsilon_{s,2}}$.

13.5.3.3 Number of Beacons Required for Each Pairwise Synchronization

The goal of AMTS is to minimize the average number of message exchanges $(\bar{M})$. Hence, from (13.41), finding the optimal number of beacons (N) resumes to solving the following optimization problem:

$$\hat{N} = \arg\min_N \bar{M}, \tag{13.43}$$

with

$$\bar{M} = \frac{2BN}{\tau^{(N)}_{sync} + \tau^{(N)}_{\max}} = \begin{cases} \dfrac{2B}{\tau^{(1)}_{sync} + \sqrt{\dfrac{\sigma^2_\varepsilon - \sigma^2_{\varepsilon_{o,1}}}{\sigma^2_{\varepsilon_{s,1}}}}} & N = 1 \\[4ex] \dfrac{2B}{\tau^{(N)}_{sync} + \dfrac{N-1}{N}\sqrt{\dfrac{\sigma^2_\varepsilon - \dfrac{\sigma^2_{\varepsilon_{o,1}}}{N}}{\sigma^2_{\varepsilon_{s,2}}}}} & N \geq 2 \end{cases},$$

where $\tau^{(N)}_{sync}$ denotes the synchronization time with N beacons and will be estimated at the reference node for different N values when the network is first established. Once N is estimated from (13.43), $\tau^{(N)}_{\max}$ can be obtained from (13.42).

Simulation results in [40] show that AMTS requires far less timing messages than TPSN when there exist multiple numbers of beacon transmissions. Moreover, the gap between the average number of required timing messages between AMTS and TPSN significantly increases as N increases, and thus AMTP is by far more energy efficient than

TPSN for large N values. Moreover, the adaptive features in AMTS make it applicable to various different types of sensor network applications.

13.6   Conclusions

In recent years, huge attention has been paid to WSNs due to their capability of serving a variety of purposes. Time synchronization is a significant part in WSNs, and a number of fundamental operations, like data fusion, power management, and transmission scheduling, require accurate time synchronization. Since the conventional time synchronization protocol for the Internet cannot be directly applied to WSNs, a number of synchronization protocols have been developed to meet the unique requirements of sensor network applications.

The importance of time synchronization also comes from the evolution of WSNs, which has been driven by technological advances in diverse areas. For instance, unlike the currently deployed WSNs, next-generation sensor networks may consist of dynamic mobile sensors or a mixture of static and dynamic sensors. In this scenario, far more sophisticated time synchronization protocols that efficiently deal with mobility of sensors will be required. Indeed, as the network becomes more complicated, the role of time synchronization becomes much more important.

In this chapter, basic features and theoretical backgrounds of the time synchronization problem in WSNs have been introduced, and three different basic approaches are analyzed and compared to reveal the general ideas and characteristics of time synchronization protocols in WSNs. In addition, a survey of existing time synchronization protocols in the literature has been provided, including the most recent results. The material of this chapter is presented from a signal processing viewpoint, which makes it distinguishable from other existing surveys. Furthermore, we shed light on adaptive time synchronization schemes for WSNs because of their huge benefits and flexibility to topology changes. We believe that the analysis and summary in this chapter will assist researchers in this area to select and develop more powerful synchronization protocols tailored specifically to the needs of their applications.

References

[1] I. F. Akyildiz, W. Su, Y. Sankarasubramaniam, and E. Cayirci. 2002. Wireless sensor networks: A survey. *Comput. Networks* 38:393–422.

[2] N. Bulusu and S. Jha. 2005. *Wireless sensor networks: A systems perspective.* Norwood, MA: Artech House.

[3] D. L. Mills. 1991. Internet time synchronization: The network time protocol. *IEEE Trans. Commun.* 39:1482–93.

[4] B. Sundararaman et al. 2005. Clock synchronization for wireless sensor networks: A survey. *Ad Hoc Networks* 3:281–323.

[5] J. Elson and K. Romer. 2003. Wireless sensor networks: A new regime for time synchronization. In *ACM SIGCOMM Comput. Commun. Rev. (CCR)* 33(1):149–154.

[6] F. Sivrikaya and B. Yener. 2004. Time synchronization in sensor networks: A survey. *IEEE Networks* 18:45–50.

[7] W. Su. 2005. Overview of time synchronization issues in sensor networks. In *Embedded systems*, ed. R. Zurawski. Boca Raton, FL: CRC Press, pp. 35–35-10.

[8] S. Ganeriwal, J. Elson, and M. B. Srivastava. 2005. Time Synchronization. In *Wireless sensor networks: A systems perspective*, ed. N Bulusu and S. Jha. Norwood, MA: Artech House, pp. 59–74.

[9] B. M. Sadler and A. Swami. 2006. Synchronization in sensor networks: An overview. In *Proceedings of the Military Communications Conference 2006*, Washington, DC, pp. 1–6.

[10] J. R. Vig. 1992. *Introduction to quartz frequency standards*. Technical Report SLCET-TR-92-1. Army Research Laboratory Electronics and Power Sources Directorate.

[11] D. A. Howe, D. W. Alan, and J. A. Barnes. 1981. Properties of signal sources and measurements methods. In *The 35th Annual Symposium on Frequency Control*, Philadelphia, PA, pp. 669–716.

[12] J. Van Greunen and J. Rabaey. 2003. Lightweight time synchronization for sensor networks. In *Proceedings of the 2nd ACM International Conference on Wireless Sensor Networks and Applications (WSNA)*, pp. 11–19.

[13] G. J. Pottie and W. J. Kaiser. 2000. Wireless integrated network sensors. *Commun. ACM* 43:51–58.

[14] H. Kopetz and W. Ochsenreiter. 1987. Clock synchronization in distributed real-time systems. *IEEE Trans. Comput.* 36:933–39.

[15] Mica2 and Mica2Dot. Accessed January 2007 from http://www.xbow.com/Products/Wireless_Sensor_Networks.htm.

[16] M. Maroti, B. Kusy, G. Simon, and A. Ledeczi. 2004. The flooding time synchronization protocol. In *Proceedings of the 2nd International Conference on Embedded Networked Sensor Systems 2004*, pp. 39–49.

[17] S. Ganeriwal, R. Kumar, and M. B. Srivastava. 2003. Timing-sync protocol for sensor networks. In *SenSys 03*, pp. 138–149.

[18] J. Elson, L. Girod, and D. Estrin. 2002. Fine-grained network time synchronization using reference broadcasts. In *Proceedings of the Fifth Symposium on Operating System Design and Implementation*, Boston, MA, pp. 147–163.

[19] K.-L. Noh and E. Serpedin. 2007. Pairwise broadcast synchronization for wireless sensor networks. In *Proceedings of IEEE International Workshop: From Theory to Practice in Wireless Sensor Networks (T2PWSN 2007)*, Helsinki, pp. 1–6.

[20] A. Papoulis. 1991. *Probability, random variables and stochastic processes*. 3rd ed. New York: McGraw-Hill.

[21] A. Leon-Garcia. 1993. *Probability and random processes for electrical engineering*. 2nd ed. Reading, MA: Addison-Wesley.

[22] H. S. Abdel-Ghaffar. 2002. Analysis of synchronization algorithm with time-out control over networks with exponentially symmetric delays. *IEEE Trans. Commun.* 50:1652–61.

[23] D. R. Jeske. 2005. On the maximum likelihood estimation of clock offset. *IEEE Trans. Commun.* 53:53–54.

[24] V. Paxson. 1998. On calibrating measurements of packet transit times. In *Proceedings of the 7th ACM Sigmetrics Conference*, vol. 26, pp. 11–21.

[25] K.-L. Noh, Q. Chaudhari, E. Serpedin, and B. Suter. 2007. Novel clock phase offset and skew estimation using two-way timing message exchanges for wireless sensor networks. *IEEE Trans. Commun.* 55:766–77.

[26] S. M. Kay. 1993. *Fundamentals of statistical signal processing: Estimation theory.* Vol. I. Englewood Cliffs, NJ: Prentice Hall.

[27] Q. Chaudhari and E. Serpedin. 2007. On maximum likelihood estimation of clock offset and skew in wireless sensor networks with exponential delays. *IEEE Trans. Signal Processing*, 56(4):1685–1697.

[28] M. L. Sichitiu and C. Veerarittiphan. 2003. Simple, accurate time synchronization for wireless sensor networks. In *Proceedings of the IEEE WCNC 2003*, pp. 1266–73.

[29] M. Lemmon, J. Ganguly, and L. Xia. 2000. Model-based clock synchronization in networks with drifting clocks. In *Proceedings of the 2000 Pacific Rim International Symposium on Dependable Computing*, Los Angeles, pp. 177–85.

[30] B. M. Sadler. 2006. Local and broadcast clock synchronization in a sensor node. *IEEE Signal Proc. Lett.* 13:9–12.

[31] I. Sari, E. Serpedin, and B. Suter. 2006. On the joint synchronization of clock offset and skew in RBS-protocol. In *Proceedings of the Military Communications Conference 2006*, Washington, DC, 1–4.

[32] W. Su and I. F. Akyildiz. 2005. Time-diffusion synchronization protocol for wireless sensor networks. *IEEE/ACM Trans. Networking* 13:384–97.

[33] Q. Li and D. Rus. 2006. Global clock synchronization in sensor networks. *IEEE Trans. Comput.* 55:214–26.

[34] Y.-W. Hong and A. Scaglione. 2005. A scalable synchronization protocol for large scale sensor networks and its applications. *J. Selected Areas Commun.* 23:1085–99.

[35] A.-S. Hu and S. D. Servetto. 2006. On the scalability of cooperative time synchronization in pulse-connected networks. *IEEE Trans. Information Theory* 52:2725–48.

[36] A.-S. Hu and S. D. Servetto. 2003. Asymptotically optimal time synchronization in dense sensor networks. In *Proceedings of the 2nd ACM International Conference on Wireless Sensor Networks and Applications (WSNA)*, pp. 1–10.

[37] Z. Tian, X. Luo, and G. B. Giannakis. 2004. Cross-layer sensor network synchronization. In *Proceedings of the 38th Asilomar Conference on Signals, Systems and Computers*, vol. 1, pp. 1276–80.

[38] S. Ganeriwal, D. Ganesan, H. Shim, V. Tsiatsis, and M. B. Srivastava. 2005. Estimating clock uncertainty for efficient duty-cycling in sensor networks. In *Proceedings of SenSys 05*, San Diego, pp. 131–141.

[39] S. PalChaudhuri, A. K. Saha, and D. B. Johnson. 2004. Adaptive clock synchronization in sensor networks. In *IPSN 04*, Berkeley, CA, pp. 340–348.

[40] K.-L. Noh and E. Serpedin. 2007. Adaptive multi-hop timing synchronization for wireless sensor networks. In *Proceedings of ISSPA2007*, Sharjah, UAE, pp. 1–6.

Adaptive Interference Nulling and Direction of Arrival Estimation in GPS Dual-Polarized Antenna Receiver

Moeness G. Amin
Villanova University

14.1　Antijam GPS Receiver Arrays

14.1.1　Interference Nulling

Four fundamental approaches based on spatial discrimination are commonly adopted in antijam Global Positioning Systems (GPS). The application of these approaches requires the GPS receiver to be equipped with an antenna array. The choice between which approach to undertake depends on various parameters, including the number of available satellites in the field of view (FOV), the nature of the jamming environment, the antenna polarization characteristics, and the ability to utilize the satellite direction information. The four approaches, placed in broad categories, are:

Null-steering arrays: These antenna arrays adjust their reception pattern to have low gain in the direction of the interferences, without considering the GPS satellite direction of arrival (DOA). This is the case for the power inversion technique, also known as minimum variance [1]. The power inversion (PI) approach for jammer nulling may unintentionally allow satellite signals to be suppressed along with the jammer.

Beam-steering arrays: These arrays make full use of the GPS satellite DOA information in the adaptive algorithm. They adjust the array reception pattern to have high gain in the direction of the GPS satellites and to form nulls in the direction of the interferers. This is the case of the minimum-variance distortionless response (MVDR) beamforming approach [2]. The MVDR-based interference nulling technique may inadvertently allow jammers to be passed on, along with the GPS signals, to the correlation loops of the receiver.

GPS signal-dependent arrays: These arrays utilize the GPS coarse/acquisition (C/A) temporal and periodic structure, and maximize the cross-correlation between the received signal and its one code-length delayed version. By this action, beams are formed toward the GPS satellites in the field of view and nulls are placed toward the jammers. This approach is known as a self-coherence antijamming GPS receiver [3]. It has a shortcoming of treating period jammers, which have the same fundamental period as the GPS code period, as desired signals.

Dual-polarized arrays: These arrays replace the circularly polarized GPS antennas by dual-polarized patch antennas. This approach requires a dual-feed dual-polarized antenna array, and can implement either beam-steering or nulling arrays. Including the polarization diversity in GPS adaptive antenna arrays increases the maximum number of possible jammer cancellations and gives more flexibility in the choice of the optimality criteria that can be applied to reduce jammer contaminations [4, 5].

14.1.2 Interference Direction Estimation

Interference nulling can be pursued jointly with the equally important task of estimating the interference directions of arrival. The latter can be achieved by implementing high-resolution subspace methods operating on the data covariance matrix, as is typically the case in commercial and military radar and communication receivers. Particular to navigation GPS receivers is the capability of providing simultaneous multiple beams, each oriented toward one satellite. For this reason, the sets of adaptive steering weights corresponding to the different satellites embed the interference spatial information and, as such, can be directly operated on by the eigndecomposition methods.

This chapter deals with interference nulling and DOA estimation for GPS receivers. Both tasks are performed prior to GPS signal dispreading, which is executed via the correlation loops of the receiver, utilizing the signal, jamming, and noise relative power underlying satellite geolocation application, and incorporating the navigation receiver information on polarization and spatial signatures of the satellites. Although focused on GPS, the approaches presented are applicable to Global Navigation Satellite Systems (GNSS), including Galileo, GLONASS, and Beidou [6]. All GNSS share the same operating principle. The receiver position is computed based on the distances between the receiving antenna and a set of satellites, and the receiver determines these distances by measuring the propagation time of the signals transmitted by the satellites. This propagation time can be obtained from the delay (referred to as pseudorange or code phase) of the complex envelope and from the carrier phase.

Dual-polarized beam-steering antenna array receivers are considered. We discuss the modeling and performance of optimum and adaptive antijam receivers implementing constrained minimization techniques.

14.2   Background

The Global Positioning System (GPS) is a satellite-based system used in localization and navigation for both military and civilian applications [7, 8]. It employs direct-sequence spread spectrum (DSSS) on two carriers, L1 at 1575.42 MHz and L2 at 1227.6 MHz. Each GPS satellite broadcasts a C/A code with chip rate at 1.023 Mchips/s or a precision (P) code at 10.23 Mchips/s. New navigation signal and code structures are considered, aiming at improving receiver positioning and combating multipath [6].

Depending on the operating environment, the signal-to-noise ratio (SNR) at the GPS receivers can be as low as −30 dB. For a strong interference, the jammer-to-noise ratio (JNR) may exceed 30 dB, which will severely degrade the GPS performance and the code synchronization process. Although the spread spectrum modulation provides some antijamming protection to weak signals, a high power interference (e.g., more than 5 dB above the noise) cannot be suppressed adequately by the spreading gain, causing inaccurate tracking and positioning errors. Many antijam mitigation techniques have been proposed, which are based on temporal processing [9–11], spectral-based processing [12–14], subspace projection [15], and spatial signal processing [1, 16, 17]. Combinations of the above techniques, such as time-frequency processing [18] and space-time processing [19, 20], provide superior jammer suppression compared to single-antenna or single-domain processing.

Adaptive techniques are commonly adopted for fast antijam implementations [21–30], and shown to be effective in spatial discriminations and interference nulling. The array response is continuously adjusted so that the interferers have low receiver gains. For spatial-only processing, each antenna is allocated a complex weight, which is adaptively chosen to minimize a desired cost function subject to satisfying a single or a set of desirable constraints. The two main types of adaptive antenna arrays are the nulling arrays [1, 4] and the beam-steering arrays [2]. In both arrays, interference cancellation can be achieved, irrespective of the signal temporal properties. It is noted that the MVDR method incorporates knowledge of the directions of arrival (DOAs) of the GPS signals to form beams toward the GPS satellites, while placing nulls toward the directions of the interferers. Although this method provides significant cancellation of the interference, it has high computational complexity, and its performance relies on accurate and prior knowledge of the DOAs of the desired signals. This knowledge is obtained from the navigation data and, therefore, could be difficult to acquire at "cold" start when the receiver has not yet acquired the signal and is already subjected to jamming. The power inversion method works in a different way. Because the GPS signals are typically −20 ~ −30 dB below the noise, the interferers are the dominant data components and, as such, are removed by suppressing the high-power-signal arrivals. In essence, the array weights are organized such that deep nulls are obtained at the directions of the high power signals, irrespective of their angular positions relative to the GPS DOAs. As the GPS signal directions are not taken into account in forming the array weights, it is possible that the GPS receiver array, in the power inversion method, nonintentionally offers a deep null or low gain toward one or more satellites in the field of view.

Equally important to jammer nulling is jammer direction estimation. High-resolution direction-finding techniques are widely used for the interference DOA estimation. These techniques are signal dependent and include the MVDR method as well as eigenstructure-based methods [31–35]. The latter include the multiple signal classification (MUSIC) algorithm [31], estimation of signal parameters via rotational invariance techniques (ESPRIT) [36–38], and Pisarenko harmonic decomposition (PHD) [39]. The eigenstructure methods utilize the property that the noise subspace eigenvectors of the data covariance matrix are orthogonal to the signal vectors. The DOA estimates are extracted from a low-dimension subspace of the estimated data covariance matrix via standard eigendecomposition or singular value decomposition (SVD).

Two important features separate the multiantenna GPS receiver from other receivers used in different applications. First, the GPS signal is of extremely low power compared to noise and potential interference, which makes the desired signal negligible prior to dispreading. The latter removes the spreading PN C/A code and compresses the signal to a narrow bandwidth. Second, the GPS in utilizing the direction of the satellites in the FOV can implement simultaneous beamformers, each with an adaptive set of weights. The above two features apply independent of the navigation satellite system and transparent to the navigation data signal structure [6].

Interference nulling and interference DOA estimation are typically two separate problems; each operates on the incoming data or the estimate of its covariance matrix. The two problems, however, can be jointly addressed by applying eigendecomposition to a matrix of the adaptive weights, rather than the covariance matrix. In this case, DOA estimation follows adaptive nulling and its accuracy, therefore, depends on the interference cancellation performance of the receiver array.

We show that the interference subspace information is embedded in the weight matrix, and thereby, interference DOA can be obtained directly from the nulling weights. The columns of the adaptive weight matrix pertain to the beamformers corresponding to different satellites. Each beamformer weight vector attempts to minimize the output power subject to a unit-gain constraint, i.e., a single constraint MVDR. In essence, instead of obtaining the data spatial covariance matrix for DOA estimation, the signal and noise orthogonal subspaces can be directly provided from the eigendecomposition of the weight matrix. The beamformer weights can be obtained using the least mean square (LMS) or other adaptive gradient algorithms [40, 41, 42]. The adaptive algorithms can be applied in baseband or instantaneous frequency (IF), and if required, the weights can even be adjusted using analog adaptive loops [43]. This approach was introduced in [43] and was referred to as concurrent adaptive nulling and localization (CANAL). The work in [43], however, is very sensitive to the noise power estimate, has not been applied to GPS, and did not consider a receiver with dual-polarized antennas.

Interference nulling and DOA estimation for GPS receivers can be pursued using dual-polarized antenna arrays. This improves the receiver capability for counteracting different types of the jammers, such as vertically polarized, horizontally polarized, and left-hand circularly polarized (LHCP) interferers. Further, the dual-polarized antenna array has around twice the number of degrees of freedom as its single-polarized antenna array counterpart.

14.3   GPS Receiver Model with Dual-Polarized Antenna Array

A GPS antenna array with N dual-polarized antennas is depicted in Figure 14.1. The dual-polarized antenna can be viewed as the linear combination of the vertical and horizontal elements. At each antenna, two received signals corresponding to the vertical and horizontal polarizations are collected. Each signal is assigned a separate weight. In baseband processing, an adaptive beamformer with complex weights, operating on the in-phase and quadrature signal components, is applied to linearly combine the dual-polarized antenna outputs.

Assume an N uniform linear array (ULA) with interelement spacing d. Consider D GPS signals incident on the array from the directions $\theta_1, \theta_2, \ldots, \theta_D$, and M interferers

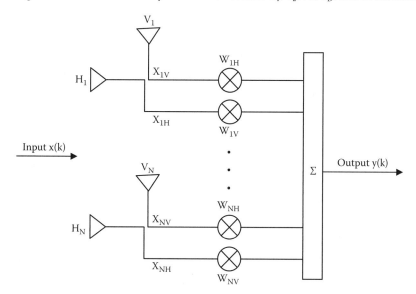

FIGURE 14.1   GPS receiver with dual-polarized antenna array.

arrive from the directions φ_1, φ_2, ..., φ_M. The channel is an additive white Gaussian noise (AWGN). We assume that the GPS signal is a direct line-of-sight signal with no reflection or diffraction components [44]. The kth data samples received at the horizontal element and the vertical element of the ith antenna are denoted as $x_{iH}(k)$ and $x_{iV}(k)$, respectively. Thus, the 2N-by-1 data vector $\mathbf{x}(k)$ is given by

$$\mathbf{x}(k)=[x_{1H}(k) \quad x_{1V}(k) \quad ... \quad x_{NH}(k) \quad x_{NV}(k)]^T, \tag{14.1}$$

where $(.)^T$ denotes transpose. Let a D-by-1 vector $\mathbf{s}_D(k)$ denote the D complex GPS signals at the kth sample,

$$\mathbf{s}_D(k)=[s_1(k), \ s_2(k), \ ..., \ s_D(k)]^T. \tag{14.2}$$

Similarly, the M-by-1 interference vector $\mathbf{i}_M(k)$ represents the M complex interferers at the kth sample,

$$\mathbf{i}_M(k)=[i_1(k), \ i_2(k), \ ..., \ i_M(k)]^T. \tag{14.3}$$

Let $\mathbf{A}_D(\theta)$ denote the GPS signal 2N-by-D steering matrix,

$$\mathbf{A}_D(\theta)=\begin{bmatrix} \mathbf{a}(\theta_1) & \mathbf{a}(\theta_2) & ... & \mathbf{a}(\theta_D) \end{bmatrix}, \tag{14.4}$$

where $\mathbf{a}(\theta_i)$ is the 2N-by-1 steering vector of the ith GPS signal incident on the antenna array from direction θ_i,

$$\mathbf{a}(\theta_i)=[a_{1H}(\theta_i) \ a_{1V}(\theta_i) \ ... \ a_{NH}(\theta_i) \ a_{NV}(\theta_i)]^H, \tag{14.5}$$

where $(.)^H$ denotes complex conjugate transpose. $\mathbf{A}_I(\varphi)$ represents the interference 2N-by-M steering matrix,

$$\mathbf{A}_I(\varphi)=\begin{bmatrix} \mathbf{a}(\varphi_1) & \mathbf{a}(\varphi_2) & ... & \mathbf{a}(\varphi_M) \end{bmatrix}. \tag{14.6}$$

In the above equation, $\mathbf{a}(\varphi_i)$ represents the 2N-by-1 steering vector of the ith interference,

$$\mathbf{a}(\varphi_i)=[a_{1H}(\varphi_i) \ a_{1V}(\varphi_i) \ ... \ a_{NH}(\varphi_i) \ a_{NV}(\varphi_i)]^H. \tag{14.7}$$

In GPS, two binary phase shift keying (BPSK) codes are applied to the GPS information symbols. The coarse acquisition (C/A) code is a repeating 1 MHz pseudo random noise (PRN) code (1,023 bits) transmitted with the carrier frequency L1 = 1,575.42 MHz by the satellite vehicles (SVs). The precise (P) code is a very long 10 MHz PRN code (7 days) transmitted with both L1 = 1,575.42 MHz and L2 = 1,227.60 MHz. Over one navigation data symbol, the received GPS signal in the absence of noise and interference is given by

$$s_d(t) = Ag(t - \gamma)\exp\{j[\omega_c t + \phi]\},\tag{14.8}$$

where A is the signal amplitude, $g(t)$ is the ± 1 valued C/A code, γ is the time delay of the original signal, ω_c is the carrier frequency, and ϕ is the carrier phase. After downconversion and phase tracking, the baseband signal is expressed as

$$s_{db}(t) = Ag(t - \gamma)\exp[j(\phi - \hat{\phi})],\tag{14.9}$$

where $\hat{\phi}$ is the estimated carrier phase. The code replica, generated by the receiver, is therefore

$$s_D(t) = g(t - \hat{\gamma}),\tag{14.10}$$

where $\hat{\gamma}$ is the estimated time delay prior to processing through the GPS correlation loops. The received data vector $\mathbf{x}(k)$ is the superposition of the GPS signals, interferers, and AWGN noise, and can be expressed as

$$\mathbf{x}(k) = \mathbf{A}_D(\theta)\mathbf{s}_D(k) + \mathbf{A}_I(\varphi)\mathbf{i}_M(k) + \mathbf{n}(k),\tag{14.11}$$

where the $2N$-by-1 vector $\mathbf{n}(k)$ represents the AWGN noise at the $2N$ elements. The SNR of the GPS is in the vicinity of -20 dB, and the JNR is typically 20 dB. It is noted that with RHCP signals, $a_{iV} = -j a_{iH}$ ($i = 1, 2, \ldots, N$). In this case, assuming all the antennas are omnidirectional, the steering vectors $\mathbf{a}(\theta_i)$ and $\mathbf{a}(\varphi_i)$ can be expressed as

$$\mathbf{a}(\theta_i) = \begin{bmatrix} 1 & -j & \cdots & e^{j2\pi\frac{d}{\lambda}(N-1)\sin\theta_i} & -je^{j2\pi\frac{d}{\lambda}(N-1)\sin\theta} \end{bmatrix}^H,\tag{14.12}$$

$$\mathbf{a}(\varphi_i) = \begin{bmatrix} 1 & -j & \cdots & e^{j2\pi\frac{d}{\lambda}(N-1)\sin\varphi_i} & -je^{j2\pi\frac{d}{\lambda}(N-1)\sin\varphi_i} \end{bmatrix}^H,\tag{14.13}$$

where λ is the wavelength. For a horizontally polarized interference, $a_{iV} = 0$ ($i = 1, 2, \ldots, N$), whereas for a vertically polarized interference, $a_{iH} = 0$ ($i = 1, 2, \ldots, N$), and for LHCP interference, $a_{iV} = j a_{1H}$ ($i = 1, 2, \ldots, N$). Let $\mathbf{R}$ represent the data spatial correlation matrix,

$$\mathbf{R} = E[\mathbf{x}(k)\mathbf{x}^H(k)],\tag{14.14}$$

where $E[.]$ denotes statistical expectation. In practice, $\mathbf{R}$ is replaced by its time-average estimates, $\hat{\mathbf{R}}$, and can be expressed as

$$\hat{\mathbf{R}} = \frac{1}{T}\sum_{k=1}^{T}\mathbf{x}(k)\mathbf{x}^{H}(k). \tag{14.15}$$

Denote the 2N-by-1 complex beamformer weight vector for the N dual-polarized antennas as

$$\mathbf{w} = [w_{1H} \ \ w_{1V} \ \ w_{2H} \ \ w_{2V} \ \ \dots \ \ w_{NH} \ \ w_{NV}]^{T}. \tag{14.16}$$

The corresponding antenna array output $y(k)$ at the dual-polarized antenna array is given by

$$y(k) = \mathbf{w}^{H}\mathbf{x}(k). \tag{14.17}$$

Equations (14.11)–(14.17) and the analyses hereafter are also valid and applicable to the European, Russian, and Chinese satellite navigation systems [6].

14.4   Single versus Multiple Antijam Beamformers

The minimum variance distortionless response (MVDR) technique is an efficient tool to mitigate interference without compromising the desired signals. The array weights are computed to form unit gains toward the DOAs of the GPS signals, and to place deep nulls toward the directions of the interferers. This is achieved by minimizing the output power subject to unit-gain constraints,

$$\min_{w} \mathbf{w}^{H}\mathbf{R}\mathbf{w} \text{ subject to } \mathbf{C}^{H}\mathbf{w} = \mathbf{f}, \tag{14.18}$$

where the constraint matrix $\mathbf{C}$ represents the GPS steering matrix $\mathbf{A}_{D}(\theta)$. The optimal weights for the above constrained minimization problem can be obtained as [40]

$$\mathbf{w}_{opt} = \mathbf{R}^{-1}\mathbf{C}(\mathbf{C}^{H}\mathbf{R}^{-1}\mathbf{C})^{-1}\mathbf{f}, \tag{14.19}$$

where $\mathbf{f}$ is a D-by-1 vector of unit values, $\mathbf{f} = [1 \ \ 1 \ \ \dots \ \ 1]^{T}$. Applying the optimal weights to the received data vector $\mathbf{x}(k)$ yields the interference beamformer output

$$y(k) = \mathbf{w}_{opt}^{H}\mathbf{x}(k). \tag{14.20}$$

It is noted that the power inversion method is a special case of equation (14.19), where the constraint matrix is a vector of zero values except the first element, which has a unit value. Also, $\mathbf{f}$ is a scalar equal to 1.

In minimizing $E\{|y(k)|^2\}$, the dual-polarized antenna array will attempt to reduce the output power of the interference, which could be horizontally polarized, vertically polarized, right-hand circularly polarized (RHCP), or LHCP. Each antenna may act alone, by setting one of the two values of its corresponding complex weights to zero, or by forcing the two polarization weight components into a complex conjugation relationship. This would be clearly the case for cancelling any large number of horizontally polarized, vertically polarized, or LHCP interferers. The antenna may also act in conjunction with other receiver antennas, in which case the interference DOA plays a role in determining the weight values.

Since GPS signals are RHCP, the number of RHCP interferers that can be suppressed with N dual-polarized antennas is limited to $N - D$. The cancellation of a large or infinite number of horizontally polarized interferers is achieved by setting the weights of the horizontally polarized elements to zero. In this case, the total number of degrees of freedom is reduced to N, out of which $N - D$ degrees can be used to cancel, through spatial nulling, $N - D$ types of jammers, such as RHCP, LHCP, or vertically polarized interferers. Similarly, an infinite number of vertically polarized interferers can be cancelled along with a maximum of $N - D$ RHCP, LHCP, or horizontally polarized interferers. Further, an infinite number of LHCP interferers can be cancelled by the RHCP antenna array, which can additionally cancel $N - D$ RHCP, horizontally polarized, or vertically polarized interferers through spatial nulling.

It is noted, however, that depending on the number of dual-polarized antennas and the number of satellites in the field of view (FOV), D, one set of array weights may not be sufficient to cancel all interferers in dense jamming environments, even with the use of dual-polarized antenna array. This problem can be mitigated by using multiple MVDR beamformers, each corresponding to one satellite. In this case, several sets of weight vectors are applied; each weight vector is associated with one satellite and designed to satisfy a single unit-gain constraint. As a result, up to $N - 1$ RHCP interferers can be cancelled by the multiple beamformer approach, compared to only $N - D$ interferers in the case of the single beamformer approach. In addition to increasing the number of degrees of freedom, another important advantage of the multiple MVDR beamformers is in achieving regular array patterns. It is noted that in order to keep unit gain at all GPS satellite directions, a single MVDR beamformer will encounter difficulty suppressing the undesired signal, if it is close in angle to any one of the GPS satellites. With close angular separation between one satellite and one jammer position, the array response will be highly irregular, giving rise to several lobes in a random manner that makes the receiver vulnerable to newly borne interferers or on-off interferers with a long duty cycle. Insufficient interference cancellation compromises the receiver delay lock loops and provides undesirable tracking and positioning errors. With multiple MVDR beamformers, however, each of the $D - 1$ beamformer responses will null all interferers impinging on the GPS receiver, subject to the available degrees of freedom. Only the one response associated with the satellite that is aligned or close to an interference DOA will fail and provide an irregular pattern. In this respect, loss of acquisition, if it occurs, will be confined to only one satellite. With typically more than four satellites in the FOV, the loss of one satellite's information is not detrimental to the receiver pseudo-range estimate calculations. The multiple MVDR beamformers approach is shown in

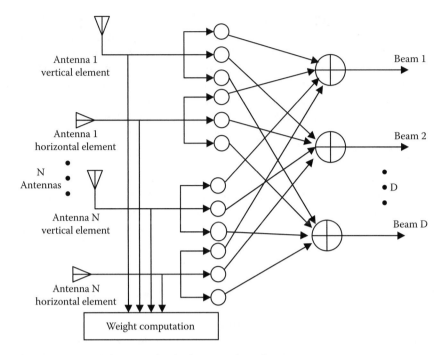

FIGURE 14.2   Block diagram of multiple MVDR beamforming.

the block diagram of Figure 14.2. The array response of each weight vector provides unit gain toward one direction of the GPS satellites and places nulls toward the directions of the interferers, irrespective of their temporal characteristics.

For the ith satellite, the 2N-by-1 optimum weight vector in the multiple MVDR beam receiver is denoted as

$$\mathbf{w}_{i\_opt} = \mathbf{R}^{-1}\mathbf{c}_i / (\mathbf{c}_i^H \mathbf{R}^{-1}\mathbf{c}_i), \tag{14.21}$$

where c_i represents the steering vector toward the direction of the ith GPS signal. The corresponding array output is

$$y_i(k) = \mathbf{w}_{i\_opt}^H \mathbf{x}(k). \tag{14.22}$$

For illustration, we consider eight dual-polarized antennas. We compare the performance of single versus multiple MVDR beamformers in various jamming environments. Figure 14.3 shows the RHCP array response and horizontally polarized array response for twelve interferers, each with JNR of 20 dB. Five RHCP interferers are incident on the array with angles [−65, −50, −30, 40, 65] degrees, along with seven horizontally polarized interferers with angles [−80, −40, −15, 0, 30, 50, 80] degrees. Four GPS signals arrive from [−5, 5, 20, 27] degrees of elevation angles with an SNR of −20 dB. Figure 14.3(a) demonstrates that the single MVDR beamformer (solid curve) fails to cancel all five

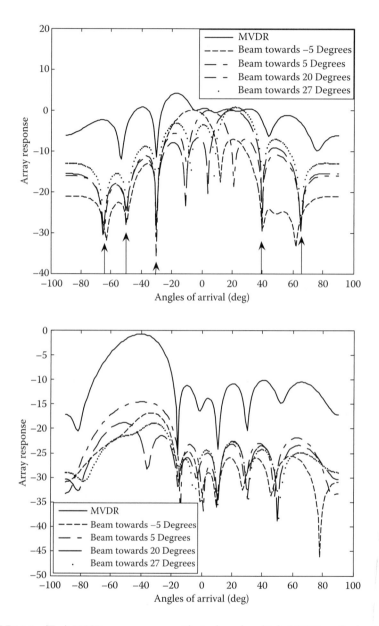

FIGURE 14.3 (Top) RHCP array response with single and multiple MVDR methods with five RHCP and seven horizontally polarized interferences. (Bottom) Horizontally polarized array response with single and multiple MVDR methods with five RHCP and seven horizontally polarized interferences.

RHCP interferers, whereas the multiple MVDR beamforming method, due to the existence of a sufficient number of degrees of freedom, places deep nulls at all interferer DOAs. Figure 14.3(b) demonstrates that both single and multiple MVDR methods null

TABLE 14.1 Weight Vector in Single and Multiple MVDR Methods

| | MVDR | Multiple Output Beamforming | |
		Beam 1	Beam 2
W_{1H}	0.0182 + 0.0395i	−0.0013 + 0.0010i	0.0011 + 0.0019i
W_{1V}	0.7563 − 0.2890i	0.0435 − 0.0011i	0.0831 − 0.0081i
W_{2H}	−0.0518 + 0.0866i	−0.0010 − 0.0009i	0.0008 − 0.0006i
W_{2V}	0.5980 − 0.0485i	0.1228 + 0.0152i	0.1466 − 0.0469i
W_{3H}	−0.0320 + 0.0181i	0.0013 − 0.0022i	−0.0016 + 0.0029i
W_{3V}	0.0921 + 0.2146i	0.1681 + 0.0541i	0.1335 − 0.0646i
W_{4H}	0.0566 − 0.0037i	0.0028 + 0.0015i	−0.0046 − 0.0041i
W_{4V}	−0.0948 + 0.1721i	0.1371 + 0.1086i	0.0906 − 0.0839i
W_{5H}	0.0964 − 0.0094i	−0.0028 + 0.0023i	0.0066 − 0.0027i
W_{5V}	−0.1696 + 0.0930i	0.0606 + 0.1661i	0.0521 − 0.1188i
W_{6H}	0.0268 + 0.0249i	−0.0015 − 0.0032i	−0.0000 + 0.0079i
W_{6V}	−0.1854 − 0.1926i	−0.0099 + 0.1735i	0.0256 − 0.1356i
W_{7H}	−0.0652 − 0.0562i	0.0030 − 0.0006i	−0.0065 − 0.0023i
W_{7V}	−0.0871 − 0.2303i	−0.0281 + 0.1067i	−0.0131 − 0.1407i
W_{8H}	−0.0167 − 0.1025i	−0.0004 + 0.0024i	0.0038 − 0.0030i
W_{8V}	0.1837 + 0.0734i	−0.0131 + 0.0330i	−0.0267 − 0.0809i

all the horizontally polarized interferers. In both cases, this is achieved by reducing the weights of the horizontally polarized antennas to approximately zero values. The weight vector values are given in Table 14.1. It is clear that while the horizontal weight at each antenna assumes a small value to suppress the horizontally polarized interferers, the vertical weights are organized as a beamformer to spatially cancel the vertical components of the RHCP interferers and to receive the GPS signals with unit gain.

In the next example, one of the interferers arrives from the same direction as one of the GPS satellites, e.g., 27 degrees of elevation angle. In this case, the optimum antenna array has to maintain a unit gain toward this direction, while trying to null a very close interference. Figure 14.4 depicts the array responses at the eight dual-polarized antenna arrays when two 20 dB JNR interferers arrive from 50 and 27 degrees. In this case, the single MVDR method will maintain a unit gain at 27 degrees, permitting the nonintentional GPS signal along with the interference to be received with equal sensitivity. Further, the null placed by the beamformer at 50 degrees appears shallow, with only −10 dB depth, and as such, the respective jammer may not be fully suppressed. On the other hand, with the multiple MVDR beamforming method, only the beamformer toward 27 degrees acts similar to the single MVDR beamformer and fails to cancel the interference. The other three beamformers show regular array responses and successfully suppress the two interferers with approximately −30 dB null.

In space-time processing, where spatial and temporal processing are applied jointly or sequentially, jammers that escape the spatial nulling can be dealt with in the temporal domain using excision filters [19, 40]. However, depending on the jammer signal characteristics, temporal and spatial filtering may not be able to provide proper jammer mitigation without compromising the desired signal.

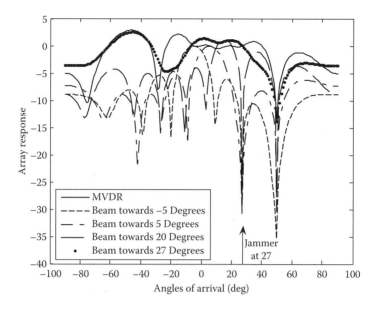

FIGURE 14.4 Array response performance at the eight RHCP antenna arrays with single and multiple MVDR methods with two RHCP interferers at 27 and 50 degrees.

14.5 Combined Nulling and DOA Estimation

Interference DOA can be obtained using subspace methods such as the MUSIC algorithm. Eigendecomposition is typically applied to the covariance matrix estimated from the data collected over an observation period,

$$\hat{\mathbf{R}} = \mathbf{U}_1\mathbf{\Sigma}_1\mathbf{U}_1^H,$$
(14.23)

where $\mathbf{\Sigma}_1$ is a $2N$-by-$2N$ matrix representing the eigenvalue diagonal matrix, and $\mathbf{U}_1$ is a unitary matrix whose columns represent the corresponding eigenvectors. The signal subspace is spanned by the eigenvectors associated with the dominant eigenvalues of $\hat{\mathbf{R}}$. The MUSIC spectrum is given by [31]

$$\mathbf{P}_{MUSIC}(\theta) = \frac{1}{\mathbf{a}^H(\theta)\mathbf{U}_n\mathbf{U}_n^H\mathbf{a}(\theta)},$$
(14.24)

where $\mathbf{U}_n$ is a matrix of the eigenvectors corresponding to the noise eigenvalues. The MUSIC algorithm as well as other eigenstructure methods require proper sampling of the data for covariance matrix estimation. These methods can be computationally demanding for GPS receivers, often call for large memory, and cannot be used effectively in simple analog adaptive loops. An alternative way is to build on the fact that multiple beamformers produce different sets of weights, and each carries interference information. One may therefore proceed to estimate the interferer DOAs by applying subspace methods, such as the MUSIC technique, to a matrix made of the weight vectors corresponding to different satellites. These weights are already generated for interference nulling purposes and can be obtained recursively through digital, or even analog, implementations. It is noted that this approach is different from beamspace MUSIC, where the covariance matrix of the beam data outputs is considered for eigendecomposition [45].

To elaborate on the combined cascade implementation of interference nulling and DOA estimation, we consider the optimum nulling weight matrix solution, given by

$$\mathbf{W}_a = \mathbf{W}_{opt}\boldsymbol{\alpha} \tag{14.25}$$

where

$$\mathbf{W}_a = \mathbf{R}^{-1}\mathbf{A}_D(\theta). \tag{14.26}$$

The 2N-by-D matrix $\mathbf{W}_{opt} = [\mathbf{w}_{1\_opt} \ \mathbf{w}_{2\_opt} \ \cdots \ \mathbf{w}_{D\_opt}]$ consists of the weight vectors in the multiple MVDR beamforming approach, and

$$\boldsymbol{\alpha} = diag\left\{\frac{1}{\alpha_1} \ \frac{1}{\alpha_2} \ \cdots \ \frac{1}{\alpha_D}\right\}$$

is a diagonal D-by-D matrix, where α_i ($i = 1, 2, \ldots, D$) is the output power at each beamformer, $\alpha_i = y_i y_i^*$.

The spatial covariance matrix of the received data can be written as

$$\mathbf{R} = \mathbf{x}\mathbf{x}^H = \mathbf{A}_D\mathbf{P}_s\mathbf{A}_D^H + \mathbf{A}_I\mathbf{P}_I\mathbf{A}_I^H + \sigma^2\mathbf{I} = \mathbf{R}_s + \mathbf{R}_I + \mathbf{R}_n, \tag{14.27}$$

where $\mathbf{P}_s = diag(P_{si})$ ($i = 1, 2, \ldots, D$) and $\mathbf{P}_I = diag(P_{Ii})$ ($i = 1, 2, \ldots, M$) represent the diagonal power matrix of the GPS signals and the interferers, respectively. Due to the low power of the GPS signals, matrix $\mathbf{R}_s$ is negligible compared to the interference and noise covariance matrix. Therefore, equation (14.27) can be simplified to

$$\mathbf{R} = \mathbf{A}_I\mathbf{P}_I\mathbf{A}_I^H + \sigma^2\mathbf{I}. \tag{14.28}$$

The eigendecomposition of (14.28) is

$$\mathbf{R} = \mathbf{E}\boldsymbol{\Lambda}\mathbf{E}^H, \tag{14.29}$$

where $\boldsymbol{\Lambda}$ is the diagonal eigenvalue matrix, $\boldsymbol{\Lambda} = diag\{\sigma_1^2 \ \cdots \ \sigma_M^2 \ \sigma_n^2 \ \cdots \ \sigma_n^2\}$, and $\mathbf{E}$ is the corresponding eigenvector matrix. In the above equation, σ_i^2 ($i = 1, 2, \ldots, M$) represents the ith significant eigenvalue, and σ_n^2 represents the noise eigenvalue. The first M columns of $\mathbf{E}$ span the interference subspace, i.e.,

$$\text{span}\begin{bmatrix} \mathbf{i}_1 & \mathbf{i}_2 & \cdots & \mathbf{i}_M \end{bmatrix} = \text{span}\begin{bmatrix} \mathbf{e}_1 & \mathbf{e}_2 & \cdots & \mathbf{e}_M \end{bmatrix}, \tag{14.30}$$

where $\mathbf{e}_i$ represents the ith column of $\mathbf{E}$. If we adopt the approach in [43] for interference DOA estimation, then we define

$$\mathbf{W}_c = \frac{1}{\sigma^2}\mathbf{A}_D(\theta) - \mathbf{W}_a, \tag{14.31}$$

where σ^2 represents the power of the AWGN. Substituting equations (14.26) and (14.29) into (14.31), the cancellation weight vector for the kth satellite can be expressed as

$$\begin{aligned}
\mathbf{w}_{ck} &= \frac{1}{\sigma^2}\mathbf{a}_k(\theta) - \mathbf{E}\mathbf{\Lambda}^{-1}\mathbf{E}^H\mathbf{a}_k(\theta) = \frac{1}{\sigma^2}\mathbf{I}\mathbf{a}_k(\theta) - \mathbf{E}\mathbf{\Lambda}^{-1}\mathbf{E}^H\mathbf{a}_k(\theta) \\
&= \left(\frac{1}{\sigma^2}\mathbf{I} - \sum_{i=1}^{M}\frac{1}{\sigma_i^2}\mathbf{e}_i\mathbf{e}_i^H - \sum_{j=M+1}^{2N}\frac{1}{\sigma_n^2}\mathbf{e}_j\mathbf{e}_j^H\right)\mathbf{a}_k(\theta),
\end{aligned} \tag{14.32}$$

where $\mathbf{I}$ is the identity matrix. Typically,

$$\frac{1}{\sigma_i^2} \ll \frac{1}{\sigma^2}, \sigma^2 = \sigma_n^2,$$

and equation (14.32) can be simplified to

$$\mathbf{w}_{ck} = \left(\frac{1}{\sigma^2}\mathbf{I} - \sum_{j=M+1}^{2N}\frac{1}{\sigma^2}\mathbf{e}_j\mathbf{e}_j^H\right)\mathbf{a}_k(\theta) = \left(\sum_{i=1}^{M}\frac{1}{\sigma^2}\mathbf{e}_i\mathbf{e}_i^H\right)\mathbf{a}_k(\theta) = \sum_{i=1}^{M}\frac{1}{\sigma^2}\left(\mathbf{e}_i^H\mathbf{a}_k(\theta)\right)\mathbf{e}_i. \tag{14.33}$$

The above vector lies in the interference subspace. If $D > M$, then the columns of the cancellation weight matrix span the M interference subspace, i.e.,

$$\text{span}\begin{bmatrix} \mathbf{i}_1 & \mathbf{i}_2 & \cdots & \mathbf{i}_M \end{bmatrix} = \text{span}\begin{bmatrix} \mathbf{w}_{c1} & \mathbf{w}_{c2} & \cdots & \mathbf{w}_{cD} \end{bmatrix}. \tag{14.34}$$

It is noted that if the interferers are LHCP, the eigenvectors of the covariance matrix in the presence of weak GPS signals are presented as LHCP. Accordingly, the coefficients $\mathbf{e}_i^H\mathbf{a}_k(\theta)$ in (14.33) become zero, and the corresponding weight vector $\mathbf{w}_{ck}$ becomes zero as well. As a result, the DOAs of the interferers cannot be estimated from $\mathbf{W}_a$ or $\mathbf{W}_c$. This can be viewed as a drawback of using the adaptive weights for DOA estimation. If the interferers are RHCP, horizontally polarized, or vertically polarized, the above coefficients are nonzero, and equation (14.34) is valid.

If the exact order statistics are available, then one would generate the nulling weights, proceed with eigendecomposition of $\mathbf{W}_c$, and obtain a corresponding MUSIC spectrum, followed by peak finding and non-LHCP interference DOA estimation. This technique is known as concurrent adaptive nulling and localization (CANAL) [43] (although it performs nulling and localization in a sequential manner). In practice, however, only a short observation interval of the data is available, and as such, the noise perturbation and estimation inaccuracies will violate the interference subspace in equation (14.33), rendering equation (14.34) nonapplicable. For illustration, define

$$\lambda_i = \frac{1}{\sigma^2} - \frac{1}{\sigma_i^2}$$

for $i = 1, 2, \ldots, M$, and

$$\lambda_j = \frac{1}{\sigma^2} - \frac{1}{\sigma_{nj}^2}$$

for $j = M + 1, \ldots, 2N$, where σ_{nj}^2 represents the jth noise eigenvalue obtained from the estimated data covariance matrix $\hat{\mathbf{R}}$, and it is a perturbed version of σ^2. If $\sigma^2 \neq \sigma_{nj}^2$, then equation (14.32) can be rewritten as

$$\mathbf{w}_{ck} = \left(\frac{1}{\sigma^2} \mathbf{E}\mathbf{E}^H - \sum_{i=1}^{M} \frac{1}{\sigma_i^2} \mathbf{e}_i \mathbf{e}_i^H - \sum_{j=M+1}^{2N} \frac{1}{\sigma_{nj}^2} \mathbf{e}_j \mathbf{e}_j^H \right) \mathbf{a}_k(\theta)$$

$$= \sum_{i=1}^{M} \lambda_i \left(\mathbf{e}_i^H \mathbf{a}_k(\theta) \right) \mathbf{e}_i + \sum_{j=M+1}^{2N} \lambda_j \left(\mathbf{e}_j^H \mathbf{a}_k(\theta) \right) \mathbf{e}_j. \tag{14.35}$$

It is clear that the cancellation weight vector is a linear combination of all interference and the noise subspace vectors $\mathbf{e}_i$ ($i = 1, 2, \ldots, 2N$) pertaining to the spatial covariance matrix. Proceeding with the singular value decomposition (SVD) for $\mathbf{W}_c$ yields

$$\mathbf{W}_c = \mathbf{U}\mathbf{\Sigma}\mathbf{V}, \tag{14.36}$$

where $\mathbf{U}$ is a 2N-by-2N unitary matrix, $\mathbf{V}$ is a D-by-D unitary matrix, and $\mathbf{\Sigma}$ is a 2N-by-D diagonal matrix with singular values, γ_i, $i = 1, \ldots, D$,

$$\mathbf{\Sigma} = \begin{bmatrix} \gamma_1 & 0 & 0 \\ 0 & \cdots & 0 \\ 0 & 0 & \gamma_D \\ 0 & \cdots & 0 \\ 0 & \cdots & 0 \\ 0 & \cdots & 0 \end{bmatrix}. \tag{14.37}$$

If $\mathbf{W}_c$ is optimally calculated, and providing $D > M$, then its rank is M. However, with adaptive processing, the matrix rank is no longer M. In this way, the first D columns of $\mathbf{U}$ are the column subspace, containing the interference information ($M < D$), and the last $2N - D$ columns of $\mathbf{U}$ are the noise subspace. That is,

$$a_k^H(\theta)u_i = 0, \ (i = D + 1, \ ..., \ 2N) \tag{14.38}$$

for $k = 1, \ ..., \ M$. The DOAs of the interferers can be obtained by searching the peaks of the CANAL spectrum

$$P(\theta) = \frac{1}{\displaystyle\sum_{i=D+1}^{2N} \left\| a^H(\theta)u_i \right\|^2}. \tag{14.39}$$

It is noteworthy that the rank of the cancellation weight matrix $\mathbf{W}_c$ is $min(D, N)$. That is, only up to $D - 1$ RHCP interferers can be considered, compared to $N - 1$ by the MUSIC algorithm when operating on the data covariance matrix. By increasing D ($D < N$), the interference information will be further represented by a larger column subspace, producing enhanced interference DOA estimation performance. Accordingly, it is desirable to add $N - D$ beamformers to achieve a total number of N beamformers. For simplicity, we define the number of beamformers as L. If $D < N$, then the L beamformers consist of D beamformers toward the GPS directions and the $L - D$ beamformers toward arbitrary directions. In this case, equation (14.26) can be expressed as

$$\mathbf{W}_a = \mathbf{R}^{-1}\mathbf{T}(\theta), \tag{14.40}$$

and equation (14.31) becomes

$$\mathbf{W}_c = \frac{1}{\sigma^2}\mathbf{T}(\theta) - \mathbf{W}_a. \tag{14.41}$$

Therefore, equation (14.39) is rewritten as

$$P(\theta) = \frac{1}{\displaystyle\sum_{i=L+1}^{2N} \left\| a^H(\theta)u_i \right\|^2}. \tag{14.42}$$

The optimal performance will occur when $L = N$, and a maximum number of $N - 1$ interference DOAs can be estimated.

The multiple MVDR beamforming method can null a large number of different types of the interferers as presented in section 14.4. However, not all interferers can be considered with the CANAL method. For instance, if the interferers are LHCP, the inner product $\mathbf{e}_i^H \, \mathbf{a}_k(\theta)$ ($i = 1, 2, \ldots, M$) corresponding to the ith signal subspace becomes zero. This is because $\mathbf{e}_i$ is LHCP and $\mathbf{a}_k(\theta)$ is RHCP. The eigenvectors $\mathbf{e}_j$ corresponding to the noise eigenvalues do not assume specific polarization. As a result, the cancellation weight vector $\mathbf{w}_{ck}$ no longer carries the interference DOA information, and is expressed as the linear combination of the noise subspace. Another important case occurs when the number of vertically polarized or horizontally polarized jammers exceeds the number of dual-polarized antennas N. In this case, the vertical elements or horizontal elements of the nulling weight matrix, respectively, will become zero, as this is the only option available to the array to suppress all jammers. The interferers' DOAs cannot be estimated based on the zero weight values. In other words, for the interference DOA to be estimated from the cancellation weight matrix, the interferers must have been already suppressed by spatial nulling.

Several simulation results of the interference localization in GPS are presented. We examine the interference localization performance as a function of the number of look directions L. Assume eight dual-polarized antennas, uniform linear antenna array with the interelement spacing of a half wavelength, are employed. Four RHCP GPS signals arrive at –20, 0, 15, and 45 degrees of elevation angles with SNRs of –20 dB. The number of snapshots is set to 1,000. Eight look directions are considered. Eight beamformers are formed, four of those toward the four GPS directions and the rest to –74, –55, 60, and 85 degrees of elevation angle. The latter were arbitrarily chosen. Seven ($M = 7$) 20 dB RHCP interferers incident on the GPS receiver from –40, –30, –10, 10, 35, 40, and 70 degrees were generated. In Figure 14.5, the MUSIC spectrum and CANAL spectrum are presented and compared. The dashed line represents the spectrum with the CANAL approach applied to the cancellation weight matrix obtained in equation (14.41), and the solid line represents the spectrum with the MUSIC technique applied to the estimated data covariance matrix, presented by equation (14.24). Both techniques show seven clear peaks at the interferers' directions.

If the number of beamformers L reduces to six, toward –74, –20, 0, 15, 45, and 60 degrees, the rank of the cancellation matrix $\mathbf{W}_c$ reduces to six, and subsequently, the maximum number of RHCP interferers that can be cancelled is six. The performance is shown in Figure 14.6. It is clear that in this case, the covariance matrix-based MUSIC algorithm can estimate the DOAs of the interferers, while the CANAL method fails, since $M > L$.

In order to examine the sensitivity of the choice of the noise subspace dimension on the CANAL performance, we assume two interferers arrive from –40 and 35 degrees. We use the same eight look directions as in the previous example. We choose the dominant singular vectors $\mathbf{u}_1$ and $\mathbf{u}_2$ to represent the interference subspace, instead of $\mathbf{u}_1, \mathbf{u}_2, \ldots, \mathbf{u}_8$. The spectrum is shown in the solid curve in Figure 14.7. This figure demonstrates that the interference subspace is not properly captured in $\mathbf{u}_1$ and $\mathbf{u}_2$, but rather has a nonzero projection on all singular vectors $\mathbf{u}_1, \mathbf{u}_2, \ldots, \mathbf{u}_8$. The example below further underscores the above point.

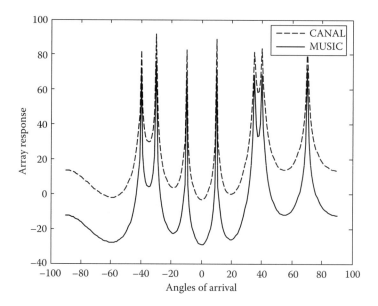

FIGURE 14.5 CANAL and MUSIC spectra with eight beamformers toward −74, −55, 60, 80, −20, 0, 15, and 45 degrees, where seven jammers arrive from −40, −30, −10, 10, 35, 40, and 70 degrees.

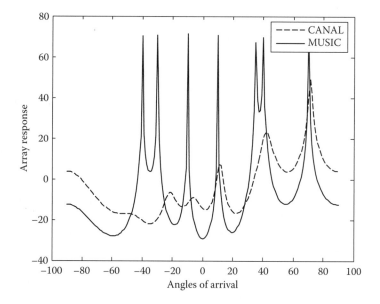

FIGURE 14.6 CANAL and MUSIC spectra with six beamformers toward −74, 60, −20, 0, 15, and 45 degrees, where seven jammers arrive from −40, −30, −10, 10, 35, 40, and 70 degrees.

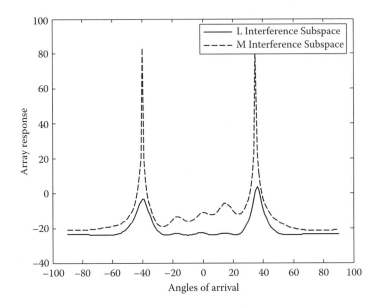

FIGURE 14.7   CANAL spectra where the noise subspace is of dimension $2N - M$ and $2N - L$ with eight dual-polarized antennas. Two interferers are located at −40 and 40 degrees.

We examine, in Figure 14.8, the normalized eigenvalues of the estimated data covariance matrix in the MUSIC technique as well as the normalized eigenvalues of the cancellation weight covariance matrix in the CANAL technique. The latter is obtained by the product $\mathbf{W}_c \mathbf{W}_c^H$. The solid line represents the eigenvalues in the MUSIC technique, where the first two eigenvalues are the interference eigenvalues, and the rest are small values representing the noise eigenvalues. The dotted line represents the eigenvalues in the CANAL technique. It is clear that the zero eigenvalues start from the eighth, and not from the third, entry, as expected in the ideal case.

In Figure 14.9, we compare the CANAL spectra under different numbers of MVDR beamformers, L. Two RHCP interferers arrive from −40 and 35 degrees of elevation angle with JNR of 20 dB. The number of beamformers is set to 5, 8, and 11. The number of dual-polarized antennas, N, is eight. The look directions for the three cases are the four GPS directions and other arbitrary directions, namely, [−20, 0, 15, 45, −55], [−20, 0, 15, 45, −55, 60, −74, 85], and [−20, 0, 15, 45, −55, 60, −74, 85, −10, 20, 70] degrees, respectively. It is clear that when $L < N$, the CANAL method estimates the interferer DOAs with poor resolution. On the other hand, when $L > N$, the rank of the cancellation matrix is N, and the CANAL spectrum performance will not be improved significantly.

In the next example, nine interferers arrive from [−40, 35, 70, −30, 40, −10, 10, 50, −60] degrees with JNR of 20 dB, and eight beamformers are formed toward the directions of 85, −74, 60, −55, 45, −20, 0, and 15 degrees. The first interferer from −40 degrees is RHCP, and all the others are horizontally polarized. With eight antennas, the multiple MVDR method will suppress the eight horizontally polarized interferers by placing zero weight values at the horizontal elements of the nulling weight matrix, and cancel the RHCP

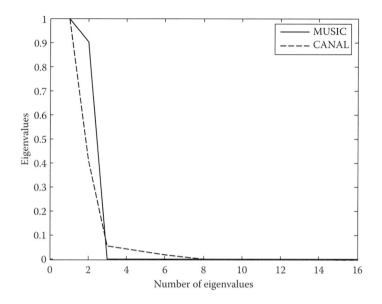

FIGURE 14.8 Eigenvalues of data covariance matrix and the cancellation weight covariance matrix in the MUSIC and CANAL techniques, respectively.

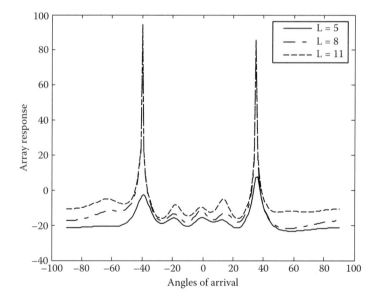

FIGURE 14.9 CANAL spectra with five, eight, and eleven beamformers, where two interferers are from −40 and 35 degrees.

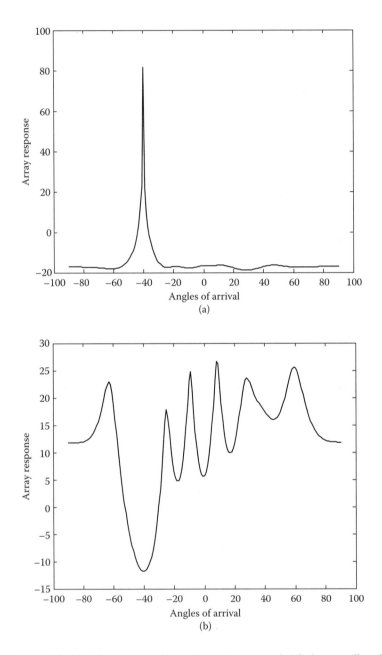

FIGURE 14.10 (a) CANAL spectrum with one RHCP jammer and eight horizontally polarized jammers. (b) CANAL spectrum with one RHCP jammer and eight horizontally polarized jammers.

interferer spatially using the MVDR method. The CANAL spectrum of the RHCP interferer and the horizontally polarized interferers are presented in Figure 14.10(a) and (b), respectively. The RHCP interferer is clearly located at −40 degrees. The CANAL spectrum for the horizontally polarized interferers shows random peaks. This demonstrates that it fails to perform direction finding of the interferers that are suppressed by the array orthogonal dual-polarization property.

14.6 Constrained Adaptive Algorithm

In this section, we recall a simple adaptive algorithm (constrained LMS algorithm) and use it to provide the cancellation weight matrix without the need to estimate the data covariance matrix. The constraint LMS algorithm is based on the gradient-descent algorithm. It iteratively adapts the weights of an antenna array to minimize the noise power at the output while retaining a unit response at the look directions. The multiple MVDR beamforming method solves for $\mathbf{w}$ according to

$$\min_{w} \mathbf{w}^H \mathbf{R} \mathbf{w} \text{ subject to } \mathbf{c}^H \mathbf{w} = 1. \tag{14.43}$$

Denote $\mathbf{F} = \mathbf{c}(\mathbf{c}^H\mathbf{c})^{-1}$ and $\mathbf{P} = \mathbf{I} - \mathbf{c}(\mathbf{c}^H\mathbf{c})^{-1}\mathbf{c}^H$. The weight vector can be recursively updated as [40]

$$\mathbf{w}(k+1) = \mathbf{P}\mathbf{w}(k) - \mu\mathbf{P}\mathbf{R}\mathbf{w}(k) + \mathbf{F} = \mathbf{P}(\mathbf{w}(k) - \mu\mathbf{R}\mathbf{w}(k)) + \mathbf{F}$$
$$= \mathbf{P}(\mathbf{w}(k) - \mu y(k)\mathbf{x}(k)) + \mathbf{F}. \tag{14.44}$$

The constraint LMS algorithm at the kth time sample can be expressed as

$$\mathbf{w}(0) = \mathbf{F}$$
$$\mathbf{w}(k+1) = \mathbf{P}(\mathbf{w}(k) - \mu y(k)\mathbf{x}(k)) + \mathbf{F}, \tag{14.45}$$

where μ is the adaptation step size. The output power of the antenna array is given by

$$P\_out = yy^H = \mathbf{w}^H \mathbf{x}\mathbf{x}^H \mathbf{w} = \mathbf{w}^H \mathbf{R}\mathbf{w}$$

$$= \left(\frac{\mathbf{R}^{-1}\mathbf{c}}{\mathbf{c}^H\mathbf{R}^{-1}\mathbf{c}}\right)^H \mathbf{R} \frac{\mathbf{R}^{-1}\mathbf{c}}{\mathbf{c}^H\mathbf{R}^{-1}\mathbf{c}} \tag{14.46}$$

$$= \frac{\mathbf{c}^H\mathbf{R}^{-1}}{\mathbf{c}^H\mathbf{R}^{-1}\mathbf{c}} \mathbf{R} \frac{\mathbf{R}^{-1}\mathbf{c}}{\mathbf{c}^H\mathbf{R}^{-1}\mathbf{c}} = \frac{\mathbf{c}^H\mathbf{R}^{-1}\mathbf{c}}{(\mathbf{c}^H\mathbf{R}^{-1}\mathbf{c})(\mathbf{c}^H\mathbf{R}^{-1}\mathbf{c})} = \frac{1}{\mathbf{c}^H\mathbf{R}^{-1}\mathbf{c}}.$$

Details of the derivations of equations (14.44)–(14.46) can be found in [40]. The optimum nulling weight vector in equation (14.25) then becomes

$$\mathbf{w}_{ai} = \mathbf{w}_i / P\_out\_i. \tag{14.47}$$

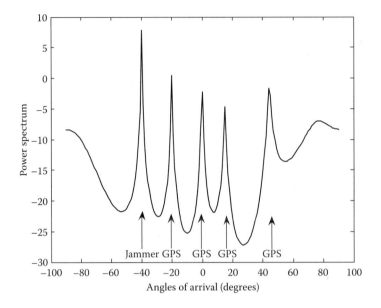

FIGURE 14.11 CANAL spectrum by LMS algorithm with eight dual-polarized antenna arrays, when one RHCP interferer arrives at -40 degrees.

In the following examples, the interference localization performance based on adaptive cancellation weight vector using the constraint LMS algorithm is examined. We consider eight dual-polarized ULAs with interelement spacing of a half wavelength. The four RHCP GPS signals arrive from –20, 0, 15, and 45 degrees with SNRs of –20 dB. Eight look directions include the four GPS directions and the additional directions of –74, –55, 60, and 85 degrees. The number of snapshots is set to 4,000, and the step size is 0.000005. In Figure 14.11, the CANAL spectrum is presented when one RHCP interferer arrives from –40 degrees with 20 dB JNR. It is evident that the directions of both the interferer and the GPS signals are presented by clear peaks, when applying the constraint LMS algorithm. This is the result of equation (14.35) and the mismatching of the true noise power value and the one offered implicitly by the converged weights. With this mismatch, the subtraction gives rise to the GPS signal subspace, which can no longer be ignored in the equation.

Figure 14.12 shows the output powers at the GPS receiver for each iteration with step sizes of 0.00005 and 0.000005. The output power is calculated by $\mathbf{w}_1^H(k)\mathbf{Rw}_1(k)$, where $\mathbf{w}_1(k)$ represents the nulling weights of the first beamformer at the kth iteration. The dashed line represents the output power with step size of 0.00005, whereas the solid line represents the output power with step size of 0.000005. It is obvious that the LMS converges faster when increasing the step size. In Figure 14.13, the first weight at the first beamformer is presented. The dashed line represents the optimum weight value obtained by the covariance matrix, whereas the solid line represents the adaptive weight. It is clear that the adaptive weight converges to the optimum value.

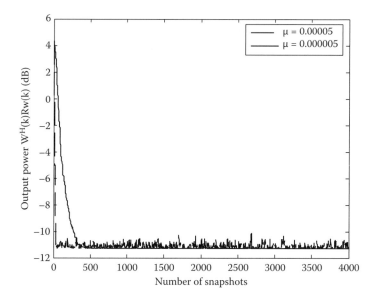

FIGURE 14.12   Output power at the GPS receiver for each snapshot with different step sizes.

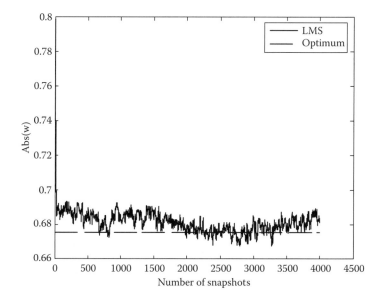

FIGURE 14.13   The adaptively obtained and optimum weight at the first beamformer for each iteration.

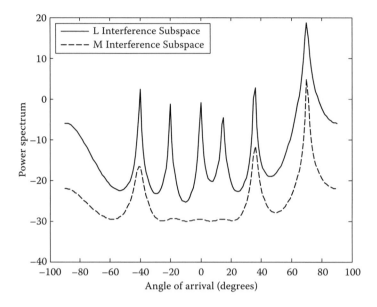

FIGURE 14.14 CANAL spectra by LMS algorithm where the noise subspace is of dimension $2N - M$ and $2N - L$ with eight dual-polarized antennas. Three RHCP interferers arrive at -40, 35, and 70 degrees.

In the next example, we consider three RHCP interferers arriving from -40, 35, and 70 degrees of elevation angle. We choose the dominant singular vectors $\mathbf{u}_1$, $\mathbf{u}_2$, and $\mathbf{u}_3$ to represent the interference subspace, instead of $\mathbf{u}_1$, $\mathbf{u}_2$, ..., $\mathbf{u}_8$. The CANAL spectrum is shown in Figure 14.14. When confining the subspace to dimension three, only the interference DOAs become visible.

In Figure 14.15, one horizontally polarized interferer arrives at -40 degrees with 20 dB JNR. We use the same GPS directions, look directions, and step size as in the previous example. It is evident that the horizontally polarized interferer has a clear peak at -40 degrees.

14.7   An Alternative Cascade Processing Approach

It is evident from the examples provided in this section that the CANAL method is sensitive to subspace dimension and the cancellation weight perturbation. Instead of applying the cancellation weight matrix, the nulling weight matrix itself can be used to obtain the DOAs of the interference. The nulling weight matrix is expressed in equation (14.40) as

$$\mathbf{W}_a = \mathbf{R}^{-1}\mathbf{T}(\theta). \tag{14.48}$$

The nulling weight vector for the kth look direction is

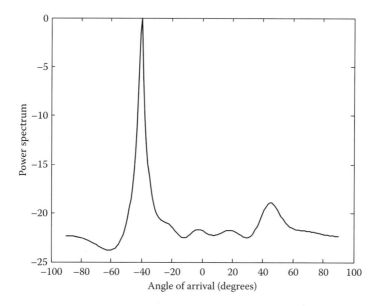

FIGURE 14.15 CANAL spectrum, by LMS algorithm for one horizontally polarized interferer from –40 degrees.

$$\mathbf{w}_{ak} = \mathbf{E}\mathbf{\Lambda}^{-1}\mathbf{E}^{H}\mathbf{a}_{k}(\theta) = \left(\sum_{i=1}^{M} \frac{1}{\sigma_{i}^{2}} \mathbf{e}_{i}\mathbf{e}_{i}^{H} + \sum_{j=M+1}^{2N} \frac{1}{\sigma_{n}^{2}} \mathbf{e}_{j}\mathbf{e}_{j}^{H} \right) \mathbf{a}_{k}(\theta). \tag{14.49}$$

For large JNR, $\sigma_{i}^{2} \gg \sigma_{n}^{2}$ and equation (14.49) can be simplified as

$$\mathbf{w}_{ak} = \sum_{j=M+1}^{2N} \frac{1}{\sigma_{n}^{2}} \mathbf{e}_{j}\mathbf{e}_{j}^{H}\mathbf{a}_{k}(\theta) = \frac{1}{\sigma_{n}^{2}}\mathbf{E}_{n}\mathbf{E}_{n}^{H}\mathbf{a}_{k}(\theta) = \frac{1}{\sigma_{n}^{2}}(\mathbf{I} - \mathbf{E}_{j}\mathbf{E}_{j}^{H})\mathbf{a}_{k}(\theta), \tag{14.50}$$

where $\mathbf{E}_{n}$ and $\mathbf{E}_{j}$ are long matrices representing the 2N-by-(2N – M) and 2N-by-M eigenvector matrices corresponding to the noise and the jammer eigenvalues, respectively. It is noted that the nulling weight matrix is orthogonal to the jammer subspace, and the MUSIC spectrum can be directly obtained as

$$P(\theta) = \frac{1}{\sum_{i=1}^{L} \left\| \mathbf{a}^{H}(\theta)\mathbf{w}_{ai} \right\|^{2}}. \tag{14.51}$$

If $L = 1$, then the problem simplifies to looking for peaks by displaying the inverse spectrum. If $L > 1$, the common nulls are emphasized, whereas spurious system nulls are eliminated. In practice, however, the adaptive weights, at convergence, are perturbed, especially for low-power jammers and under fast convergence. The SVD of the nulling weight matrix can be expressed as

$$\mathbf{W}_a = \mathbf{U}_a \mathbf{\Sigma}_a \mathbf{V}_a, \tag{14.52}$$

where $\mathbf{U}_a$ is a 2N-by-2N unitary matrix, $\mathbf{V}_a$ is an L-by-L unitary matrix, and $\mathbf{\Sigma}_a$ is a 2N-by-L singular value matrix. The matrix $\mathbf{U}_{an}$ of the first $L - M$ singular vectors corresponding to the nonzero singular values in $\mathbf{\Sigma}_a$ is orthogonal to the jammers' subspace, and the spectrum can be expressed as

$$P(\theta) = \frac{1}{\displaystyle\sum_{i=1}^{L-M} \left\| \mathbf{a}^H(\theta)\mathbf{u}_{an\_i} \right\|^2}. \tag{14.53}$$

In the following examples, four RHCP GPS signals arrive at –20, 0, 15, and 45 degrees with SNR of –20 dB. The eight look directions considered are the four GPS directions and additional directions of –74, –55, 60, and 85 degrees. The number of snapshots is set to 4,000. The nulling weight matrix is adaptively obtained by the constraint LMS algorithm with the step size of 0.000005. In Figure 14.16, the spectrum with one RHCP jammer is examined. The jammer signal is incident on the array with –40 degrees of

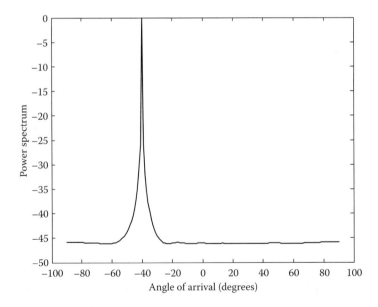

FIGURE 14.16 Spectrum based on the nulling weight matrix obtained by the LMS algorithm with eight dual-polarized antenna arrays, when one RHCP interferer arrives at –40 degrees.

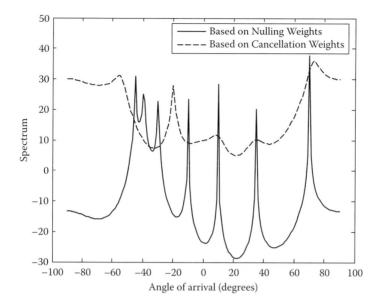

FIGURE 14.17 Spectrum based on the nulling and cancellation weight matrix obtained by the LMS algorithm with eight dual-polarized antenna arrays for seven RHCP interferers.

elevation angle and 20 dB JNR. The number of nonzero singular values is seven. The spectrum is obtained using equation (14.53).

Figure 14.17 compares the spectra based on the nulling weight matrix and the cancellation weight matrix for seven RHCP jammers, arriving from −40, 35, 70, −30, −45, −10, and 10 degrees. It is clear that the jammers can be properly localized if we apply the nulling weight matrix, whereas the cancellation weight matrix-based spectrum fails to exhibit the correct peaks at the jammer DOAs.

14.8 Summary

In this chapter, we addressed interference nulling and interference DOA estimation for GPS receivers with dual-polarized antenna arrays. It is shown that interference nulling can be achieved by invoking the orthogonal polarization properties of the array or by coherently combining the antenna outputs through optimum and adaptive weight values. For antijam GPS, the GPS receiver may be designed to employ one set of weights that forms one beamformer and places low gain toward the interference DOAs and unit gain toward satellites. It is shown that this approach has shortcomings, as it offers a limited number of degrees of freedom and provides poor nulling capabilities when the jammers and the satellites have close angular separations. Multiple weight vectors corresponding to multiple beamformers mitigate the aforementioned problems at the expense of increased receiver complexity. We have also presented a DOA estimation approach that operates on the adaptive weights rather than the estimated covariance matrix. Two different techniques under this approach are represented. The first deals with cancellation weight vectors that span the interference subspace, and the other is based on the adaptive nulling weight vectors that are orthogonal to the interference subspace. We provided simulation results comparing the two techniques under various jamming environments and using different interference signal polarizations.

Acknowledgment

The author thanks Ms. Jing Wang for providing the computer simulations.

References

[1] Zoltowski, M. D., and Gecan, A. S. 1995. Advanced adaptive null steering concepts for GPS. In *Proceedings of the 1995 Military Communications Conference (MIL-COM 1995)*, vol. 3, p. 1214.

[2] Van Veen, B., and Buckley, K. 1988. Beamforming: A versatile approach to spatial filtering. *IEEE ASSP Mag.* 5:4.

[3] Sun, W., and Amin, M. 2005. A self-coherence based anti-jam GPS receiver. *IEEE Trans. Signal Processing* 53:3910.

[4] Trinkle, M., and Cheuk, W. C. 2003. Null-steering GPS dual-polarized antenna arrays. In *6th International Symposium on Satellite Navigation Technology.*

[5] Fante, F. L., and Vaccaro, J. J. 2001. *Jammer cancellation performance of a fully-polarimetric GPS array.* MITRE technical report.

[6] GNSS. 2006. GNSS album images and spectral signatures of the new GNSS signals. In *GNSS: Engineering solutions for the global navigation satellite system*, pp. 46–56.

[7] Kaplan, E. D. 1996. *Understanding GPS, principles and application.* Norwood, MA: Artech House Publishers.

[8] Grewal, M. S., Weill, L. R., and Andrews, A. P. 2001. *Global positioning systems, inertial navigation, and integration.* New York: Wiley.

[9] Ketchum, J. W., and Proakis, J. G. 1982. Adaptive algorithms for estimating and suppressing narrow-band interference in PN spread-sprectrum systems. *IEEE Trans. Commun.* 30:913.

[10] Milstein, L. B. 1988. Interference rejection techniques in spread spectrum communications. *Proc. IEEE* 76:657.

[11] Amin, M. G., and Zhang, Y. 2002. Interference suppression in spread spectrum communication systems. In *The Wiley encyclopedia of telecommunications*, ed. J. G. Proakis. New York: John Wiley.

[12] Badke, B., and Spanias, A. 2002. Partial band interference excisions for GPS using frequency-domain exponents. In *IEEE International Conference on Acoustics, Speech, and Signal Processing (ICASSP 2002)*, vol. 4, p. 3936.

[13] Capozza, P. T., Holland, B. J., Hopkinson, T. M., and Landrau, R. L. 2000. A single-chip narrow-band frequency domain excisor for a global positioning system (GPS) receiver. *IEEE J. Solid State Circuits* 35:401.

[14] DiPietro, R. C. 1989. An FFT based technique for suppressing narrow-band interference in PN spread spectrum communications systems. In *IEEE International Conference on Acoustics, Speech, and Signal Processing (ICASSP 1989)*, vol. 2, p. 1360.

[15] Amin, M. G., Zhao, L., and Lindsey, A. R. 2004. Subspace array processing for the suppression of FM jamming in GPS receivers. *IEEE Trans. Aerosp. Electronics Syst.* 40:80.

[16] Subbaram, H., and Abend, K. 1993. Interference suppression via orthogonal projections: A performance analysis. *IEEE Trans. Antennas Propagation* 41:1187.

[17] Fante, R. L., and Vaccaro, J. J. 2000. Wideband cancellation of interference in a GPS receive array. *IEEE Trans. Aerosp. Electronic Syst.* 36:549.

[18] Zhang, Y., Amin, M. G., and Lindsey, A. R. 2001. Anti-jamming GPS receivers based on bilinear signal distributions. In *Proceedings of the 2001 Military Communications Conference (MILCOM 2001)*, vol. 2, p. 1070.

[19] Xiong, P., Medley, M. J., and Batalama, S. N. 2003. Spatial and temporal processing for global navigation satellite systems: The GPS receiver paradigm. *IEEE Trans. Aerosp. Electronics Syst.* 39:1471.

[20] Myrick, W. L., Goldstein, J. S., and Zoltowski, M. D. 2001. Low complexity anti-jam space-time processing for GPS. In *International Conference on Acoustics, Speech, and Signal Processing*, vol. 4, p. 2233.

[21] Vaccaro, F. R. 2002. Evaluation of adaptive space-time-polarization cancellation of broadband interference. In *Proceedings of the IEEE Precision Location and Navigation Symposium*, p. 1.

[22] GNSS. 2006. GNSS solutions: Adaptive antenna arrays, multi-GNSS tropospheric monitoring, and high-dynamic receivers. In *GNSS: Engineering solutions for the global navigation satellite system*.

[23] Seco, G., Fernández-Rubio, J. A., and Fernández-Prades, C. 2005. ML estimator and hybrid beamformer for multipath and interference mitigation in GNSS receivers. *IEEE Trans. Signal Processing* 53:1194–1208.

[24] Moelker, D., van der Pol, E., and Bar-Ness, Y. 1996. Adaptive antenna arrays for interference cancellation in GPS and GLONASS receivers. In *Proceedings of the IEEE Position Location Navigation Symposium*, p. 191.

[25] Hatke, G. 1998. Adaptive array processing for wideband nulling in GPS systems. In *Proceedings of the 32nd Asilomar Conference on Signals, Systems, and Computers*, vol. 2, p. 1332.

[26] Gecan, A., and Flikkema, P. 1996. Jammer cancellation with adaptive arrays for GPS signals. In *Proceedings of the IEEE Southeast Conference: Bringing Together Education, Science, and Technology*, p. 320.

[27] Mingjie, D., Xinjian, P., Fang, Y., and Jianghong, L. 2001. Research on the technology of adaptive nulling antenna used in anti-jam GPS. In *Proceedings of the CIE International Conference on Radar*, Beijing, p. 1178.

[28] Zoltowski, M. D., and Gecan, A. S. 1995. Advanced adaptive null steering concepts for GPS. In *Proceedings of the Military Communication Conference*, vol. 3, p. 1214.

[29] Moelker, D., van der Pol, E., and Bar-Ness, Y. 1996. Adaptive antenna arrays for interference cancellation in GPS and GLONASS receivers. In *Proceedings of the IEEE Position Location Navigation Symposium*, pp. 191–98.

[30] Hatke, G. 1998. Adaptive array processing for wideband nulling in GPS systems. In *Proceedings of the 32nd Asilomar Conference on Signals, Systems, and Computers*, vol. 2, p. 1332.

[31] Schmidt, R. O. 1986. Multiple emitter location and signal parameter estimation. *IEEE Trans. Antennas Propagation* 34:276.

[32] Schweppe, F. C. 1968. Sensor array data processing for multiple signal sources. *IEEE Trans. Inform. Theory* 14:294.

[33] Krim, J., and Viberg, M. 1996. Two decades of array signal processing research: The parametric approach. *IEEE Signal Processing Mag.* 13:67.

[34] Kumaresan, R., and Tufts, D. W. 1983. Estimating the angles of arrival of multiple plane waves. *IEEE Trans. Aerosp. Electronics Syst.* 19:134.

[35] Grover, R., Pados, D. A., and Medley, M. J. 2007. Subspace direction finding with an auxiliary-vector basis. *IEEE Trans. Signal Processing* 55:758.

[36] Roy, R., and Kailath, T. 1989. ESPRIT—Estimation of signal parameters via rotational invariance techniques. *IEEE Trans. Acoust. Speech Signal Processing* 37:984.

[37] Swindlehurst, A., Ottersten, B., Roy, R., and Kailath, T. 1992. Multiple invariance ESPRIT. *IEEE Trans. Signal Processing* 40:867.

[38] Yu, S. J., and Lee, J. H. 1999. Efficient eigenspace-based array signal processing using multiple shift-invariant subarrays. *IEEE Trans. Antennas Propagation* 47:186.

[39] Fuchs, J. J. 1996. Rectangular Pisarenko method applied to source localization. *IEEE Trans. Signal Processing* 44:2377.

[40] Frost, O. L., III. 1972. An algorithm for linearly constrained adaptive array processing. *Proc. IEEE* 60:926.

[41] Griffiths, L. J. 1969. A simple adaptive algorithm for real-time processing in antenna arrays. *Proc. IEEE* 57:1696.

[42] Widrow, B., Mantey, P. E., Griffiths, L. J., and Goode, B. B. 1967. Adaptive antenna systems. *Proc. IEEE* 55:2143.

[43] Amin, M. G. 1992. Concurrent nulling and locations of multiple interferences in adaptive antenna array. *IEEE Trans. Signal Processing* 40:2658.

[44] Ercek, R., De Doncker, P., and Grenez, F. 2005. Study of pseudo-range error due to non-line-of-sight-multipath in urban canyons. In *Proceedings of ION GNSS 2005*, Long Beach, CA, pp. 1083–1094.

[45] Xu, X., and Buckley, K. 1993. Analysis of beam-space source localization. *IEEE Trans. Signal Processing* 41:501–509.

15

Reconfigurable Baseband Processing for Wireless Communications

André B. J. Kokkeler
University of Twente

Gerard K. Rauwerda
University of Twente

Pascal T. Wolkotte
University of Twente

Qiwei Zhang
University of Twente

Philip K. F. Hölzenspies
University of Twente

Gerard J. M. Smit
University of Twente

15.1   Introduction

Within wireless communications, advances in digital processing are pushing system performance to higher levels, providing increasingly larger bandwidths and introducing more complex coding and modulation schemes. Increasingly larger parts of processing of both the radio access network (RAN) and user equipment (UE) are mapped onto digital computing devices. Looking at this mapping process, one can distinguish different classes of processes (or tasks) and different types of computing devices.

A classification of processing is general purpose (GP processing), domain specific (DS processing), and application specific (AS processing). GP processing is generally characterized by a nonpredictable execution flow, whereas AS processing has a predictable execution flow. A characteristic of DS processing is that several similar tasks or algorithms are to be executed on a single computing device.

Computing devices or processors are generally classified as general-purpose processors (GPPs), digital signal processors (DSPs), reconfigurable processors (RPs), and

TABLE 15.1 Overview of Types of Processing
and Computing Devices

	GPP	DSP	RP	ASIC
GP processing	*	(*)		
DS processing	*	*	*	
AS processing	*	*	*	*
Power efficiency	– –	–	+	++
Area efficiency	– –	–	+	++
Engineering efficiency	++	+	–	– –
Flexibility	++	+	–	– –

application-specific integrated circuits (ASICs). The ability to execute different types of processing for the different types of computing devices is indicated in Table 15.1.

GPPs are able to execute all types of processing. DSPs are processors that are tailored toward the signal processing domain. This is a relatively broad application domain, and DSPs therefore resemble GPPs to a large extent; both GPPs and DSPs are typically von Neumann machines [48]. Reconfigurable processors can be adapted in such a way that different types of processing within a specific application domain can be executed. ASICs generally are only suited for AS processing.

When mapping computing tasks onto computing devices, performance criteria of computing devices are evaluated. Performance criteria are presented at the lower part of Table 15.1. In general, the performance criteria are power, area, and engineering efficiency and flexibility. Because ASICs basically have no control overhead, most dissipated energy contributes to the execution of the tasks at hand. On the other hand, since GPPs are based on the von Neumann principle, substantial energy is dissipated for fetching and decoding instructions. From a power consumption perspective, this can be considered as overhead leading to low power efficiency. DSPs are also based on the von Neumann principle but are more power efficient. RPs show more ASIC-like behavior and are therefore more power efficient than DSPs. Area efficiency relates to the area required on an integrated circuit (IC) for realizing a specific task. ASICs are far more area efficient than GPPs. Engineering efficiency refers to the relative engineering effort required to have the computing device executing the demanded processing. For GPPs and DSPs, in general, compilers are available to efficiently compile high-level sequential code into machine instructions. Compilers for mapping applications onto RPs are not yet as mature as for GPPs and DSPs. For ASICs, functionality can be specified by means of schematic capture or a hardware specification language. However, the trajectory from this specification to an integrated circuit is time-consuming, risky, and expensive. Flexibility basically refers to the range of applications for which a computing device can be used. It is clear the GPP offers the most flexibility and ASIC the least.

Because of the diminishing feature size of integrated circuits, more and more functionality can be packed into a single chip. This has led to the creation of heterogeneous system-on-chips (SoCs), where multiple processing cores of different types are interconnected by means of a network-on-chip (NoC). An example can be found in [47].

The developments described above have their impact on wireless communications. If the RAN is considered, (chip) area and power consumption are not of utmost importance. Engineering efficiency is important because RAN equipment is sold in relatively low quantities. Flexibility is necessary in case new services have to be offered by the RAN. Therefore, its functionality is implemented in GPPs and DSPs as much as possible. For example, in UMTS base stations, chip-level processing is done by means of DSPs [54]. In UE, low power consumption has been the driving force, and large parts of functionality are traditionally implemented in ASICs. However, nonrecurrent engineering costs and flexibility become more and more important. This has led to a trend where, also in the UE, reconfigurable devices are becoming feasible.

In the remaining part of this chapter we will first discuss reconfigurable platforms since they are attractive for both RAN and UE. Second, the implementation of wireless communication applications on a reconfigurable platform will be described where we separately deal with licensed communication and a form of unlicensed communication called cognitive radio. Third, we zoom into the process of mapping applications onto a reconfigurable platform. Both the design time mapping and runtime mapping are discussed.

15.2 Reconfigurable Platforms

In [11], a definition of reconfigurable platforms is given: "Systems incorporating some form of hardware programmability—customizing how the hardware is used using a number of physical control points. These control points can then be changed periodically in order to execute different applications using the same hardware." Based on this definition, reconfigurable platforms can consist of fine-grained or coarse-grained functional units [45]. Fine-grained functional units implement a function on a single bit or a small number of bits, e.g., small lookup tables in a standard field programmable gate array (FPGA). Coarse-grained functional units are typically much larger and may consist of arithmetic and logic units (ALUs) and a significant amount of storage. Fine-grained and coarse-grained reconfigurable devices are described separately.

For the wireless communications domain, conventional computing architectures will be replaced by reconfigurable multiprocessor system-on-chips (MP-SoCs). On MP-SoCs, several instances of a single type of processing device can be implemented. These types of MP-SoCs are referred to as homogeneous tiled architectures. MP-SoCs with several instances of a few different types of processing devices are referred to as heterogeneous tiled architectures. We will give examples of both types of architectures. To interconnect different processors on an MP-SoC, we use a NoC. NoCs are described at the end of this section.

15.2.1 Fine-Grained Reconfigurable Devices

Recent advances in reconfigurable devices are based on the success of FPGAs. Within an FPGA, static random access memory (SRAM) bits are connected to configuration points. For example, writing to these SRAM bits, routing structures are configured and

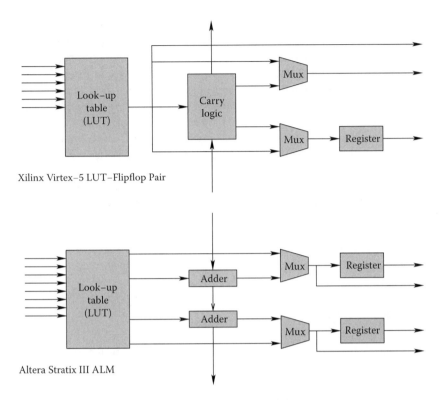

FIGURE 15.1   Basic computational elements of two state-of-the-art FPGAs.

(de)multiplexers can be set. Routing structures and (de)multiplexers are used to inter-connect the basic computational elements of an FPGA. The basic computational elements of two state-of-the-art FPGAs are presented in Figure 15.1.

The computational elements are from recent FPGA families of the two dominant manufacturers: the adaptive logic module (ALM) of the Stratix III family of Altera and the combination of a lookup table (LUT) and flipflop of the Virtex-5 family of Xilinx. The ALM has eight inputs, connected to a LUT. At most two single-bit functions can be realized by the LUT. Furthermore, two adders and two registers are available. The unit of a Virtex-5 device that is comparable to an ALM is a LUT-flipflop pair. The LUT has six inputs to which carry logic and a single register are added.

In Table 15.2 an overview of the characteristics of two devices out of the Stratix III and Virtex-5 families is given. These devices are the ones with the maximum number of LUTs within their family.

The numbers that are given for features of the devices cannot be compared directly. A slice within the Virtex-5 series consists of four six-input LUT-flipflop pairs. Further-more, the basic element within a Virtex-5 DSP slice is an 18-by-25-bit multiplier. Memory is organized in block RAMs. ALMs of the Stratix III series are comparable with a LUT-flipflop pair of the Virtex-5 series. An ALM has eight inputs. The multipliers within

TABLE 15.2   Characteristics of Two State-of-the-Art Fine-Grained Reconfigurable Devices

XilinxVirtex-5 XC5VLX330T		AlteraStratix III EP3SC340	
Slices	51,840	Adaptive logic modules	135,300
⇒ Number of LUTs	207,360	⇒ Number of LUTs	135,300
DSP slices	192	Multipliers	576
(18 × 25 bit multipliers)		(18 × 18 multipliers)	
Clock frequency	550 MHz	Clock frequency	600 MHz
Memory	11.664 Mbit	Memory	17.208 Mbit
In block RAM		In memory blocks	

the Stratix III device have two 18-bit input operands. Memory is organized in memory blocks. The frequencies given for both devices are the maximum internal clock speed.

Fine-grained reconfigurable devices are in principle only used in the RAN, for example, in base stations. Fine-grained devices offer a large flexibility at the cost of a reduced (chip) area efficiency and power efficiency. Area and power are of less importance for the RAN but are of utmost importance in the UE. An example of a HiperLAN/2 base station implemented on an FPGA is given in [37]. Furthermore, there are numerous examples of the implementation of parts of UMTS transmitters and receivers on an FPGA. Recent examples can be found in [5], [7], [34], and [43]. FPGA-based platforms are also used for the design and analysis of communication systems. An example is the BEE2 platform [8]. This is a modular platform, consisting of multiple FPGAs used to speed up the emulation and design of wireless communications systems.

15.2.2   Coarse-Grained Reconfigurable Devices

Within wireless communication systems, the computational data paths have path widths greater than a single bit, and for that reason, fine-grained architectures are much less efficient because of the routing area overhead. Coarse-grained architectures have word-level data paths and are therefore more efficient. While the basic elements within FPGAs are slices or LUT-flipflop pairs, the basic elements within a coarse-grained architecture are reconfigurable data path units (DPUs). Different coarse-grained reconfigurable devices distinguish themselves through the functionality provided by the DPUs and the structure of interconnections between different DPUs. A DPU can be as simple as a single arithmetic and logic unit (ALU) or relatively obese, containing multiple ALUs, multiple memories, and control/configuration logic. In case relatively small DPUs are used, FPGA-like structures interconnect the different DPUs. In case of multiple larger DPUs, usually a NoC interconnects different processors.

15.2.2.1   Coarse-Grained Reconfigurable Architectures

Recently, several word-level reconfigurable architectures have been proposed, very often in the context of a tiled architecture template. The Pleiades architecture [3] is a template to create an instance of a domain-specific processor. In the template, a control processor is surrounded by an array of heterogeneous autonomous special-purpose satellite processors. All processors are interconnected via a reconfigurable communication network. The adaptive computing machine by QuickSilver [25], is a heterogeneous system-on-chip. Nodes with different flexibility (adaptive nodes, domain-specific [ASIC-like] nodes, and programmable nodes) can be interconnected by means of a mesh interconnect. The XPP extreme processor platform [4] consists of clusters of computing elements and a packet-oriented communication network. A cluster consists of a set of parameterizable tiles. There are two types of tiles: memory tiles and ALU tiles. A memory tile has a capacity of 256 or 512 words. The word size can be either 16, 24, or 32 bits. The architecture uses a large number of processors, which makes the chip suitable for high-end applications. Silicon Hive provides reconfigurable accelerators designed according to a hierarchical approach. At the lowest level, the basic component is a VLIW-like processing and storage element (PSE). At the next level, multiple PSEs form a cell. A cell is a processor, capable of executing complete algorithms. One level higher, multiple cells can be combined. The accelerators are to be integrated onto a SoC. The ADRES architecture template [6] consists of a tightly coupled VLIW processor and a coarse-grained reconfigurable array. The reconfigurable array is intended to process computationally intensive kernels of applications. The VLIW host processor uses parts of the reconfigurable array to execute its instructions. The host processor and reconfigurable array are therefore tightly coupled and share resources. This architecture is dedicated to applications that require tight control of data flow operations. A similar approach is used in the Chimaera architecture [24]. In this architecture, the reconfigurable part has direct access to the host processor's register file. The Kilocore KC256 chip is a commercial version of the PipeRench chip [21]. The chips are characterized by a multicore computing kernel where cores can be cascaded to constitute multiple processing pipelines. Besides the configurable ALU, a processing core only contains a register file and no memory. Configuration and information data are stored in separate SRAMs. Because PipeRench is specially designed for pipelined applications, best performance is achieved when pipeline stages are identical or perfectly balanced. A specific coarse-grained reconfigurable architecture, developed at the University of Twente, is the Montium processor. The Montium will be described in more detail below and will be used throughout this chapter in the examples of coarse-grained reconfigurable processors.

15.2.2.2   The Montium

The Montium is described in detail in [27], and in this section the general structure is discussed. A single Montium processing tile is depicted in Figure 15.2.

The lower part of Figure 15.2 shows the communication and configuration unit (CCU), which deals with the off-tile communication and configuration of the upper part, the reconfigurable tile processor (TP). The TP is the computing part that can be dynamically reconfigured to implement a particular algorithm. At first glance the TP has a VLIW structure. However, the control structure of the Montium is very different. For (energy) efficiency it is imperative to minimize the control overhead. This is, for example, accomplished by scheduling instructions statically at compile time. A relatively simple sequencer controls the entire tile processor. The sequencer selects configurable tile instructions that are stored in the instruction decoding block (see Figure 15.2).

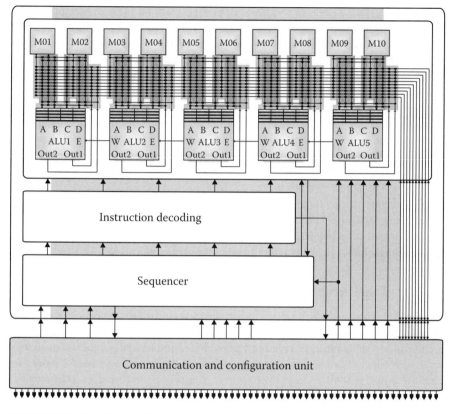

FIGURE 15.2 The Montium tile processor and network interface.

Furthermore, there are multiple ALUs (ALU1 ... ALU5) and multiple memories (M01 ... M10). A single ALU has four inputs (A, B, C, D). Each input has a private input register file that can store up to four operands. The input register file cannot be bypassed, i.e., an operand is always read from an input register. Input registers can be written by various sources via a flexible interconnect. An ALU has two outputs (OUT1, OUT2), which are connected to the interconnect. The ALU is entirely combinational, and consequently, there are no pipeline registers within the ALU. Neighboring ALUs can also communicate directly: the west output (W) of an ALU connects to the east input (E) of the ALU neighboring on the left.

The ALUs support both signed integer and signed fixed-point arithmetic. The five identical ALUs in a tile can exploit spatial concurrency to enhance performance. This parallelism demands a very high memory bandwidth, which is obtained by having ten local memories in parallel.

An address generation unit (AGU; not shown in Figure 15.2) accompanies each memory. The AGU can generate the typical memory access patterns found in common DSP algorithms, e.g., incremental, decremental, and bit-reversal addressing. It is also possible to use the memory as a lookup table for complicated functions that cannot be calculated using an ALU such as sine or division with a single constant value. A memory can be used for both integer and fixed-point lookups.

The reconfigurable elements within the Montium are the sequencer, the instruction decoding block, and the AGUs. Their functionality can be changed at runtime. The Montium is programmed in two steps. In the first step, the AGUs are configured and a limited set of instructions is defined by configuring the instruction decoding block. The sequencer is then instructed to sequentially select the required instructions. A predefined instruction set is available using an assembly type of mnemonics (Montium assembly). A compiler has been constructed to convert a Montium assembly program into configuration data for both the instruction decoding block and the sequencer.

15.2.2.2.1 Energy Consumption

Using power estimation tooling, the dynamic power consumption of a typical multiply-accumulate (MAC) operation in the Montium is estimated to be about 0.5 mW/MHz, realized in 130 nm complementary metal oxide semiconductor (CMOS) technology. The area of a single Montium TP is about 2 mm^2 in this technology [26].

15.2.3 Tiled Architecture

Tiled architectures are where relatively complex elements (tiles) are replicated on a single integrated circuit (IC). The tiles are interconnected via an on-chip network. Tiled architectures are becoming increasingly popular because a tile has to be designed only once, after which it can be copied onto a single IC multiple times. By adding more tiles onto the chip, it is relatively easy to profit from diminishing feature sizes. The computation model, programming model, interconnection structure, and memory organization can stay the same. Below, the following prominent tiled architectures are discussed: the RAW processor [49], the cell processor [29], the Polaris processor [47], and the Tile64 processor (see www.tilera.com). Furthermore, the Chameleon heterogeneous tiled architecture is introduced.

15.2.3.1 The RAW Processor

The RAW processor is one of the earliest tiled architectures. A RAW processor consists of a set of relatively simple tiles interconnected by a set of switches. Each tile contains instruction memory, data memories, an ALU, registers, configurable logic, and a programmable switch with an associated instruction memory. The general idea is that the internal hardware structure of both the tile and the switches is exposed to the compiler. This way there are two sets of control logic: operation control for the processor and sequencing routing instructions for the switches. A consequence is that the burden on the compiler is high, which leads to relatively long compile times. The configurable logic in each tile supports a few wide-word or many narrow-word operations and is coarser than FPGA-based processors.

15.2.3.2 The Cell Processor

The cell processor consists of eight replicas of a synergistic processor element (SPE) and a (single) power processor element (PPE) with a power core. An SPE consists of a division multiple access (DMA) unit, a local store memory (256 KB), execution units, and memory and bus interface controllers. The PPE is accompanied by first- and second-level caches. The nine cores are interconnected by a coherent on-chip bus. The SPEs and PPEs are SIMD processors where the element width of the SPE can range from 2×64 bits to 128×1 bits.

15.2.3.3 Polaris

Intel has recently released information on an 80-tile, 1.28 TFLOPS processor in 65 nm CMOS. The tiles are arranged as a 10-by-8 mesh network. Each tile consist of a processing engine connected to a five-port router that forwards packets between tiles. The PE contains two independent fully pipelined single-precision floating-point multiply-accumulate units, 3 KB single-cycle instruction memory, and 2 KB data memory. The NoC connects the PEs using packet-based communication, and the router in each PE can send and receive packets from any of the other tiles on the network.

15.2.3.4 Tile64

Tile64 is a processor based on the mesh architecture, originally developed for the RAW machine. The chip consists of a grid of processor tiles arranged in a network, where each tile consists of a GPP, cache, and a nonblocking router that the tile uses to communicate with the other tiles on the chip. Each processor has a register file and three functional units: two integer ALUs and a load-store unit. A processor also has a split L1 cache and an L2 cache. When there is a miss in the L2 cache of a specific processor, the L2 caches of the other processors are searched for the data before external memory is consulted. This way, a large L3 cache is emulated. The delay that is caused by the on-chip mesh network is exposed to the programmer, and during compilation, the compiler takes the delays into account when scheduling different tasks on different tiles.

15.2.3.5 The Chameleon Tiled Architecture Template

Trends in wireless communications show that many standards with the same purpose coexist. On one hand, new standards are being developed, but on the other hand, revisions of standards evolve quickly. Different contradicting requirements are identified for the platform that implements wireless communications systems:

- High performance: Due to the increasing complexity of signal processing in digital receivers, high-performance signal processors are needed.
- Flexibility: Radio receivers need to switch quickly between different standards, as they have to support a wide variety of standards. Within standards, fast switching between different configurations is required.
- Low power: Portable receivers need to be very energy efficient because they are battery powered.

To meet contradicting requirements for high performance, flexibility, and energy efficiency, a heterogeneous SoC template is designed. A heterogeneous SoC contains different types of processing parts, like general-purpose, fine-grained reconfigurable, coarse-grained reconfigurable, and dedicated hardware parts. Multiple processing parts are interconnected by means of the NoC. The idea behind a heterogeneous SoC is that

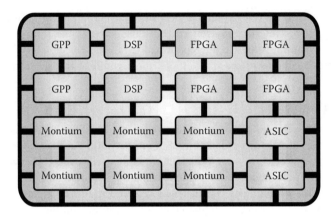

FIGURE 15.3   The Chameleon SoC template.

all parts of an algorithm can run on a processing part on which it can be implemented most efficiently. The ideas above translated into the definition of the Chameleon SoC template. This template is illustrated in Figure 15.3.

A Chameleon SoC that contains sixteen processing tiles is depicted. The black grid in the figure represents the NoC. The Chameleon SoC attempts to combine the best of all worlds: the performance and energy efficiency of ASICs, the bit-level flexibility of FPGAs, the domain-specific flexibility of the Montium and DSPs, and the general-purpose applicability of GPPs. Information exchange between the different tiles is realized by means of the NoC.

15.2.4   Network-on-Chip

A key element within the Chameleon SoC template is the NoC. In a NoC, a processing tile is connected to a router. Routers of different processing tiles are interconnected. Communication between two processing tiles involves at least the two routers of the corresponding processing tiles, but other routers might be involved as well. A NoC that routes data items in a SoC has a higher bandwidth than an on-chip bus, as it supports multiple concurrent communications. The well-controlled electrical parameters of an on-chip interconnection network enable the use of high-performance circuits that result in significantly lower power dissipation, shorter propagation time, and higher bandwidth than is possible with a bus (see also [52]). To describe the network traffic in a system, multiple types are identified [22]. According to the type of service required, the following types of traffic can be distinguished in the network:

- Guaranteed throughput (GT) is the part of the traffic for which the network has to give real-time guarantees (i.e., guaranteed bandwidth, bounded latency).
- Best effort (BE) is the part of the traffic for which the network guarantees only fairness but does not give any bandwidth and timing guarantees.

Within wireless communications, most traffic is in the GT category. Besides the mainstream of GT communication, a minor part (assumed to be less than 5%) of BE communication is foreseen, e.g., control, interrupts, and configuration data.

Several NoC solutions have been proposed that use different techniques to provide guarantees. The Æthereal NoC [23] combines a global time division multiplexing (TDM) schedule to provide contention free (i.e., guaranteed throughput) routes to network streams. The best-effort streams are handled in the slack time of the schedule via wormhole routing. The Nostrum NoC [38] uses a TDM-related technique called temporally disjoint networks. Containers are used to route the GT traffic. A third packet-switched solution uses virtual channels and wormhole routing to provide guarantees [33]. Using deterministic local arbitration mechanisms (e.g., round-robin) results in a guaranteed throughput per virtual channel. Assigning a single stream of traffic per virtual channel will give network-wide guarantees.

The performance of this virtual channel network is illustrated by a HiperLAN/2 case study. For a 6 × 6 network running at 333 MHz, the processes of multiple HiperLAN/2 receivers are placed on the processor tiles of the platform. Each process in the HiperLAN/2 process graph sends GT traffic at a rate of 256 bytes (i.e., one OFDM symbol) per 4 µs. Extra BE messages are offered to the network to control the receivers and emulate other applications running on the system concurrently. Figure 15.4 shows how the latency of the GT and BE messages depends on the offered BE load.

For the GT traffic, the mean and maximal latency of packets are given. When the offered BE load is low, the latency of the GT packets is lower than the guaranteed (or

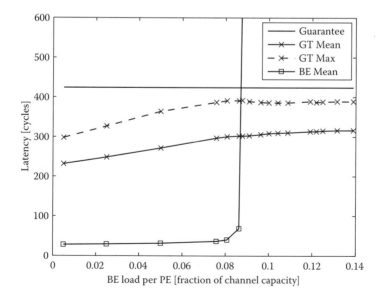

FIGURE 15.4 Message latency of the GT and BE traffic versus BE load for a 6-by-6 network (queue size 2 flits).

allowed) latency. The reason is that the GT traffic utilizes the bandwidth unused by the BE traffic. The latency of the GT packets is higher than the latency of the BE traffic because the GT packets are larger (256 bytes compared with 10 bytes for BE packets). With the increase of the BE load, the latency of the GT traffic increases too until the maximum delay reaches the limit. Further increase of the BE load increases the GT mean latency, but the GT maximum latency never exceeds the guaranteed latency.

Combining guaranteed traffic with best-effort traffic is hard [42]. When using dedicated techniques for both types of traffic, it is possible to reduce the total area and power consumption. The reasons for reconsidering circuit switching are that the flexibility of packet switching is not needed because a connection between two tiles will remain open for a long period (e.g., seconds or longer). Furthermore, large amounts of the traffic between tiles will need guaranteed throughput, which is easier to satisfy in a circuit-switched connection. Circuit switching also eases the implementation of asynchronous communication techniques, because data and control can be separated and circuit switching has a minimal amount of control in the data path (e.g., no arbitration). This increases the energy efficiency per transported bit and the maximum throughput.

Further, scheduling communication streams over non-time-multiplexed channels is easier because, by definition, a stream will not have collisions with other communication streams. In contrast with this, the Æthereal [23] routers are using time-multiplexed channels, and therefore have large interaction between data streams and have to guarantee contention free paths. Determining the static time slots table for these systems requires considerable effort. Because data streams are physically separated in a circuit-switched NoC, collisions do not occur. Therefore, no buffering and arbitration is required in the individual router. An example of a circuit-switched NoC is described in [51]. The number of parallel physical channels between routers is increased to increase the amount of simultaneously active circuitries. This solution utilizes the huge amount of wire resources provided by current and future silicon technologies.

15.3   Applications

In this section, we will discuss the mapping onto a heterogeneous SoC of different licensed communication standards and the mapping of the emerging cognitive radio application. Concerning licensed communication standards, first the realization of a UMTS receiver is given. Next, realizations of an OFDM receiver are presented. Finally, several digital broadcasting systems are discussed.

15.3.1   Licensed Communications

15.3.1.1   UMTS Receiver on Reconfigurable Hardware

The Universal Mobile Telecommunications System (UMTS) standard, defined by European Telecommunication Standards Institute (ETSI) [2], is an example of a third-generation (3G) mobile communication system. The communication system has an air interface that is based on direct-sequence code division multiple access (DS-CDMA)

[30]. The bit rate of UMTS at the physical level depends on the modulation type and the spreading factor (SF). Table 15.3 depicts the general characteristics of the UMTS communication system. Figure 15.5 depicts the basic operation principle of CDMA.

TABLE 15.3   Downlink UMTS Properties in the FDD Mode

Chip rate	3.84 Mcps
Chips per frame	38,400
Chips per slot	2,560
Slots per frame	15
Frame period	10 ms
Slot period	666.67 μs
Scrambling code length	38,400 chips
Scrambling code period	10 ms
SF	4–512
Symbol rate	7.5–960 ksps
Modulation	QPSK, QAM-16

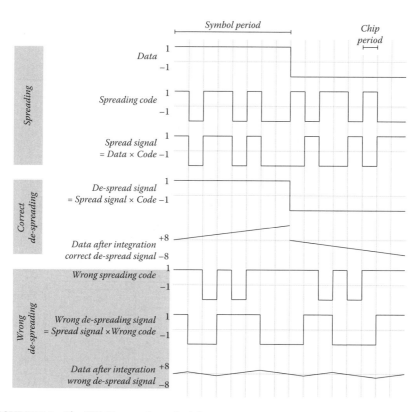

FIGURE 15.5   The CDMA operation principle.

We investigate the possibilities for implementing the digital signal processing functionality of a UMTS receiver on the Chameleon SoC template (see Figure 15.3). We only focus on the downlink of the UMTS receiver at the mobile terminal in the FDD mode.

15.3.1.2   Rake Receiver Implementation

Figure 15.6 shows the baseband processing, performed in the WCDMA receiver using a Rake receiver with four so-called Rake fingers. This WCDMA receiver has been implemented on an implemented platform, based on the Chameleon SoC template. This platform consists of four Montiums, two GPPs, and an FPGA part. Most of the computationally intensive baseband functionality has been implemented in coarse-grained reconfigurable hardware (Montiums), whereas fine-grained reconfigurable hardware (FPGA part) and GPPs are supposed to be used for additional control functionality.

Figure 15.7 shows the functional blocks of the WCDMA receiver that are intended to be implemented in the tiles of the heterogeneous reconfigurable SoC.

The first receive filter is commonly implemented as a pulse-shape filter. This pulse-shape filter is implemented on Montium tile 1. The output streams of the pulse-shape filter are

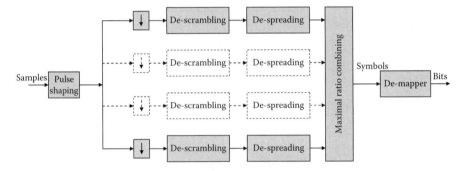

FIGURE 15.6   Baseband processing in the Rake-based WCDMA receiver.

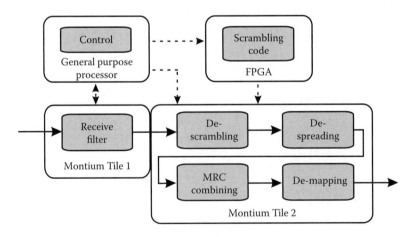

FIGURE 15.7   WCDMA receiver mapped to the heterogeneous reconfigurable SoC.

the input for the Rake receiver, which has been implemented on Montium tile 2. Generated scrambling codes can be mapped efficiently on an FPGA. The three blocks, Montium tile 1, Montium tile 2, and FPGA, are controlled by a general-purpose processor.

15.3.1.2.1   Block versus Streaming Communication

Two different mechanisms to exchange data with a Montium can be distinguished: block mode and streaming mode. Some applications require all the input data to be stored in local memories before execution can be started. This operation mode is called block mode. The most important feature of this communication mode is that, during the data transfers, the Montium is halted to make sure the execution is not started until it is sure that the data are valid. In streaming mode, the Montium processes data while simultaneously performing data transfers.

The timing properties of the UMTS communication system concern both the data processing and the control part of the receiver. It seems natural that data processing is performed according to the streaming communication principle, while control-oriented functions are partly done according to the block communication. The implemented WCDMA receiver has been realized according to the streaming communication principle because for block communication too many resources (memory for storing the scrambling code) are required during the baseband processing. According to Table 15.3 the scrambling code sequence consists of 38,400 samples. Hence, if the WCDMA receiver operates in block mode, 38,400 samples have to be stored in the local memory of the Montium. Even when data would be processed on a slot basis instead of a frame basis, still 2,560 data samples would have to be stored in local memory. Therefore, we may conclude that the block communication principle in the WCDMA receiver is not efficient, since blocks are too large.

15.3.1.2.2   Communication Requirements

The implemented WCDMA receiver thus operates according to the streaming communication principle. The implemented receiver can process four individual paths of the received signal (see also Figure 15.6). Consequently, the receiver requires four complex-number data streams for the four implemented fingers. All implemented fingers require the same scrambling code. The scrambling code can be generated using, e.g., an FPGA tile. The implemented receiver takes the complex-number scrambling code stream as an input. Before de-spreading starts, the appropriate spreading code is stored in the local memory of the Montium. The spreading code is stored in local memory because the code has a maximum length of 512 samples. Thus, a relatively small amount of data has to be stored in contrast to the scrambling code. Furthermore, the spreading code is assigned to a particular user in the UMTS communication system, and therefore, the spreading code will not change frequently. After de-spreading and before de-mapping, the received symbols of the individual signal paths are combined. During this combining phase, each symbol is scaled according to a complex-number coefficient before summation. These complex-number coefficients are provided by the channel estimator. The receiver outputs a bit stream with the received data. The characteristics of all the input and output streams are given in Table 15.4.

TABLE 15.4 Signal Stream Characteristics of the Implemented WCDMA Receiver

Signal Stream	Direction	Data Rate (Msps)
Data finger 1	Complex input	3.84
Data finger 2	Complex input	3.84
Data finger 3	Complex input	3.84
Data finger 4	Complex input	3.84
Scrambling code	Complex input	3.84
MRC coefficient finger 1	Complex input	3.84/SF
MRC coefficient finger 2	Complex input	3.84/SF
MRC coefficient finger 3	Complex input	3.84/SF
MRC coefficient finger 4	Complex input	3.84/SF
De-mapped bits	Real output	3.84/SF

15.3.1.2.3 Dynamic Reconfiguration

The complete Rake receiver is implemented in the Montium. This receiver includes the de-scrambling and de-spreading of four individual fingers, combining the results of the four fingers, and de-mapping (according to Figure 15.6). The configuration size of the complete Rake receiver in the Montium is only 858 bytes. Since two bytes per clock cycle can be loaded in the configuration memory, one Montium can be configured for Rake receiving in 429 clock cycles. With a configuration clock frequency of 100 MHz, a Rake receiver with four fingers can be configured in 4.29 μs.

In case the spreading code changes, and so the SF, only the new spreading code has to be loaded in the local memory of the Montium. Loading a particular spreading code into the local memory costs SF clock cycles; furthermore, one constant in the sequencer program (i.e., the loop counter) has to be changed.

The signal streams for the different fingers are buffered in local memories inside the Montium. When the delay of one of the paths changes, the buffering strategy of the local memories has to be changed (i.e., the AGU instructions are reconfigured). The buffering strategy of the memories is configured with 24 bytes. These 24 bytes can be reconfigured in twelve clock cycles. Consequently, the Rake receiver can update its complete path delay profile in 120 ns.

In total, for reconfiguring the number of fingers from, e.g., four to two, only 24 bytes have to be reconfigured in the configuration memory of the Montium. So, the Rake receiver can be reconfigured in 120 ns, which corresponds to twelve clock cycles. During this reconfiguration process from four to two fingers, the Montium sequencer program is changed as well as the ALU instructions.

15.3.1.2.4 Dynamic Power Consumption

Voltage and frequency scaling are important measures to control the dynamic power consumption of embedded systems. Because of the modular, regular structure of the Rake receiver, frequency scaling can easily be applied. The clock frequency of the Montium during Rake processing of four fingers is 20 MHz. Moreover, when the Rake

TABLE 15.5 Average Dynamic Power Consumption
of the Montium Rake Receiver

	Clock Frequency Montium (MHz)	Dynamic Power Consumption Montium (mW)
2 fingers	10	5
4 fingers	20	10

receiver is reconfigured to process two fingers, the clock frequency of the Montium can be reduced to 10 MHz.

Table 15.5 summarizes the average power consumption figures for the implemented Rake receiver.

15.3.1.2.5   UMTS Performance Verification

The functional behavior of the implemented Rake receiver in the Montium has been verified by means of hardware/software co-simulations. Using a co-simulation environment, we can simulate the baseband functionality of the UMTS communication system in software (i.e., using Matlab and SaSUMTSSim [41]) and partly in hardware (i.e., using a VHDL simulator). SaSUMTSSim has been developed to evaluate the performance of the de-spreading and demodulation functions of the UMTS downlink [41].

The performance and functional behavior of the Montium-based Rake receiver have been evaluated under different propagation conditions. The propagation conditions for performance measurements in a multipath fading environment are defined in [1] for the UMTS communication system and are called cases. One of these cases, case 4, has been used to evaluate the Montium-based Rake receiver with four fingers.

Figure 15.8 shows the BER versus SNR performance for the Rake receiver with four fingers under case 4 conditions. The reception of one UMTS frame is simulated. The individual results of five simulation runs are given in Figure 15.8 for a reference Rake receiver (the SaSUMTSSim receiver) and for the Montium-based Rake receiver, which are labeled "Reference" and "textscMontium," respectively. For both receivers, the average of the five simulation runs is presented as well. The simulation results show that the performance of the Montium-based Rake receiver and the SaSUMTSSim Rake receiver are almost equal. Only for bad channel conditions (i.e., low E_c/N_0) does a small performance gap arise. This gap is caused by saturation in the ALUs. Scaling the input data of the Montium reduces the effect of saturation. The effect of proper input scaling is depicted in Figure 15.9 by the additional simulation results (labeled "Additional input scaling, Montium"). In Figure 15.9, only the averaged simulation results are shown, not the five individual results as in Figure 15.8. The additional simulation results are obtained by more rigorously scaling the input data of the Montium in bad channel conditions.

15.3.1.3   OFDM Receiver on Reconfigurable Hardware

The concept of orthogonal frequency division multiplexing (OFDM) has been known since 1966 [9]; however, due to implementational complexity, it has only been adopted since the 1990s. Currently OFDM is widely used as a modulation technique in wireless communication systems. OFDM is a special multicarrier modulation technique that

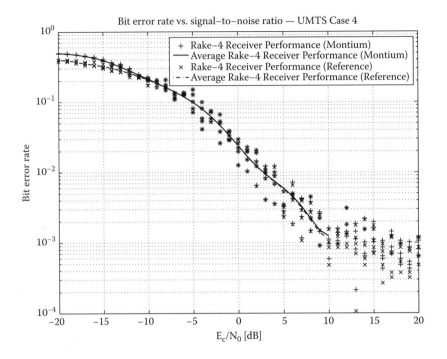

FIGURE 15.8 The BER before error correction of the Rake-4 receiver under case 4 propagation conditions with ideal channel estimation (+ and × indicate individual simulation points).

utilizes multiple subcarriers within a single channel. The modulation technique divides the high-data-rate information into several parallel bit streams, and each of these bit streams modulates a separate subcarrier.

OFDM-based communication systems are all designed according to a generic framework. Figure 15.10 shows the generic OFDM framework of an OFDM receiver. In this framework the characteristic properties are based on specific OFDM standards. This means that, for example, the number of subcarriers and the length of the guard interval may differ for different OFDM standards. The characteristics of an OFDM receiver for a single standard can even differ in the case where it has different modes defined. Characteristics of some OFDM-based standards, HiperLAN/2, DAB, and DRM are summarized in Table 15.6 [13–15].

15.3.1.4   OFDM Case Study: HiperLAN/2 Receiver Implementation

Parts of the baseband processing of a HiperLAN/2 receiver are implemented in the heterogeneous reconfigurable SoC of Figure 15.3. The coarse-grained Montium architecture [26, 28] is used as the target architecture for mapping the baseband DSP algorithms. The physical layer of the HiperLAN/2 receiver [13] is implemented on Montiums in combination with a GPP. Figure 15.11 shows the baseband processing blocks of the receiver that are implemented in the SoC. The solid arrows in Figure 15.11 indicate the data, which is processed in consecutive processing tiles. The input consists of data samples,

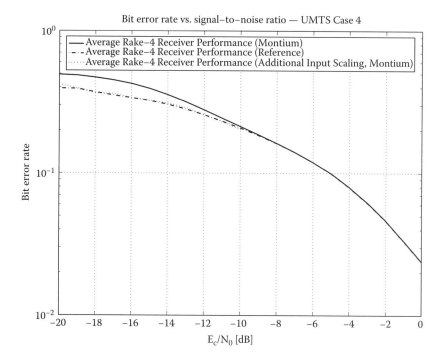

FIGURE 15.9 The influence of additional input scaling on the average BER before error correction under case 4 propagation conditions (additional results to Figure 15.8).

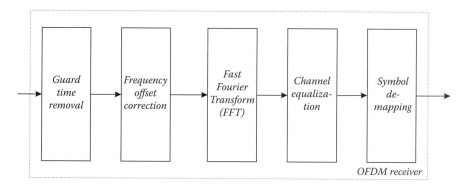

FIGURE 15.10 Generic OFDM receiver framework.

which form OFDM symbols. The output of the baseband processing part consists of encoded channel bits. The baseband processing functions are supported by a GPP, which is used for control purposes. The control streams between the processing tiles are shown as dashed arrows in Figure 15.11. The data streams between the processing tiles are mapped on channels of the NoC. The NoC provides the on-chip communication network in the SoC.

TABLE 15.6 Characteristics of Different OFDM Standards from [13–15]

		Hiper LAN/2	DAB				DRM			
			I	II	III	IV	A	B	C	D
Bandwidth	(MHz)	16.25	1.536	1.536	1.536	1.536	0.01	0.01	0.01	0.01
No. of carriers (FFT size), N		64	2,048	512	256	1 024	288	256	176	112
No. of modulated carriers, K		52	1,536	384	192	768	225	205	137	87
Symbol time, T_{OFDM}	[µs]	4	1,246	312	156	623	26,667	26.667	20,000	16,667
Guard time, T_g	[µs]	0.8	246	62	31	123	2,667	5,333	5.333	7.333
Useful time, T	[µs]	3.2	1,000	250	125	500	24,000	21,333	14 667	9,333
Subcarrier spacing, Δf	(kHz)	312.5	1.0	4.0	8.0	2.0	0.0417	0.0469	0.0682	0.1071

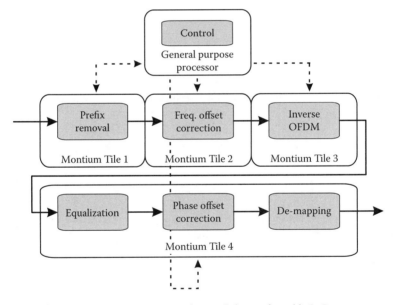

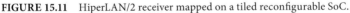

FIGURE 15.11 HiperLAN/2 receiver mapped on a tiled reconfigurable SoC.

Irregular tasks, which are outside the algorithm domain of the Montium, are performed in software on the GPP. The irregular processes in the HiperLAN/2 receiver are the channel estimation functions like frequency offset estimation and computation of equalization coefficients. Because of its relatively large coherence time, the channel estimates have to be updated only once per MAC frame, i.e., once per 2 ms. Table 15.7 shows the results of partioning the receiver's functionality over the Montium and the GPP.

The characteristics of the DSP kernels in the HiperLAN/2 receiver that are mapped on the tiled reconfigurable SoC (according to Figure 15.11) are summarized in Table 15.8.

TABLE 15.7 Reconfigurable Hardware/Software Partitioning of the HiperLAN/2 Functionality

	Implemented in	Multiplies per MAC Frame	Additions per MAC Frame
Determine frequency offset	Software	64	64
Determine equalizer coefficients	Software	0	0
Frequency offset correction	Montium	127,744	95,309
Inverse OFDM	Montium	383,232	574,848
Equalizer, phase offset, de-mapper	Montium	203,184	104,082

TABLE 15.8 Properties of the HiperLAN/2 Receiver Implementation

		Frequency Offset Correction	Inverse OFDM	Equalizer, Phase Offset, De-Mapper
Execution time	[Cycles]	67	204	110
Block mode				
Communication time				
(input + output)	[Cycles]	128	116	<100
Minimum Montium + NoC clock streaming communication	[MHz]	17	51	28
Minimum Montium + NoC clock block communication	[MHz]	49	80	53
Minimum Montium clock with block communication (NoC @ 100 MHz)	[MHz]	25	72	37
Configuration size	[Bytes]	274	946	576
Configuration time	[Cycles]	137	473	288

The table shows the impact of the communication overhead in the NoC, which can be performed by streaming or block mode communication. Furthermore, the configuration overhead of the DSP kernels implemented in the coarse-grained Montium is given.

15.3.1.4.1 Configuration

The configuration sizes of the Montium are small for the different functions (Table 15.8). Montium tile 3 (see Figure 15.11), on which the inverse OFDM is performed, requires the largest configuration. The configuration of tile 3 contains less than 1 kB of configuration data. The configuration data are written into the configuration memory of the Montium in 473 clock cycles, as every clock cycle 16 bits are written. So, tile 3 can be configured in 4.73 μs, which is dominating the time to switch from receive to transmit mode. Notice that the maximum radio turnaround time* of the HiperLAN/2 communication system

---

* The radio turnaround time indicates the time to switch from transmit to receive mode in a communication transceiver, and vice versa.

is 6 µs [12], so the Montium HiperLAN/2 implementation can be considered a real-time dynamically reconfigurable HiperLAN/2 system.

15.3.1.4.2   *Frequency Scaling*

Since the DSP kernels in the HiperLAN/2 receiver are mapped on different Montiums in the SoC architecture, frequency scaling can easily be applied. Frequency scaling in combination with voltage scaling is an important means to control the power consumption of embedded systems. The idea of dynamic voltage scaling is to keep the supply voltage as low as possible. The maximum operating frequency is tightly coupled to the supply voltage level. This means that by downscaling the clock frequency of hardware, the supply voltage can be lowered as well, resulting in a quadratic decrease of the power consumption.

All DSP operations in the physical layer of the HiperLAN/2 communication system are performed on OFDM symbols. Every OFDM symbol has a time period of 4 µs. So, it should be ensured that every 4 µs a new OFDM symbol can be processed by the receiver.

Typically, the clock frequency of the NoC is fixed and the clock frequency of the tiles can be varied. In case of block mode communication and under the assumption that the clock frequency of the NoC is fixed at 100 MHz, the clock frequency of the Montium tile for frequency offset correction has to be at least 25 MHz.

In case of block mode communication, the clock frequency of the NoC is equal to the clock frequency of the reconfigurable processor tile and the clock frequency can be adapted, and the clock frequency of the entire SoC (i.e., processor tiles and NoC) can be scaled to its minimum value. The minimum clock frequency of the system would in this case be reduced to 49 MHz for frequency offset correction. However, this situation is fairly unlikely to happen because managing the adaptable clock frequency of the common NoC is rather complex.

Introducing the streaming communication mode variant of the DSP kernels in the HiperLAN/2 receiver provides a situation where the data processing and input and output communication are performed in parallel. In this case the data words are processed immediately as they become available from the on-chip network, while previously processed data are sent to the next stage of processing. Consequently, the communication time is not a bottleneck. For example, the data processing for frequency offset correction is performed during sixty-seven clock cycles, and because input and output are done simultaneously, communication is reduced to sixty-four clock cycles. Processing needs to complete in 4 µs. Hence, the minimum clock frequency of the Montium tile is 17 MHz for frequency correction. In case of streaming communication, the clock frequency of the NoC should be at least 17 MHz. Typically, the clock speed of the NoC is fixed to the maximum operating frequency of the processing tiles.

15.3.1.5   Digital Broadcasting Systems

Many digital replacements of current analog broadcasting systems are available nowadays. It is expected that within a few years all analog broadcasting services will be switched off and replaced by digital standards like Digital Audio Broadcasting, Digital Video Broadcasting, and Digital Radio Mondiale [14–16]. For example, in the United States, television broadcasters have to switch from analog to digital before 2009. In the Netherlands, almost 98% coverage of DVB-T has been achieved and analog television broadcast has been completely turned off. Additionally, 88% of the United Kingdom is already covered with DAB.

One common characteristic of the digital radio systems is that they all employ orthogonal frequency division multiplexing (OFDM). Although digital radio standards are all OFDM based, many differences exist. Characteristic OFDM properties vary between the different standards, but the characteristics of the OFDM receiver can even change within one standard, because different modes are defined (see Table 15.6). Moreover, for all digital radio standards different source coding techniques are applied, i.e., MPEG-1 Layer 2 Audio Coding (MP2) or Advanced Audio Coding (AAC).

The identified DSP kernels in digital broadcasting receivers are based on the generic OFDM receiver framework (Figure 15.10). Mapping digital broadcasting receivers on the heterogeneous reconfigurable SoC can be done in a manner similar to the Hiper-LAN/2 implementation approach. The DSP kernels in, e.g., the DAB, DRM, and DVB receiver are similar to those in the HiperLAN/2 receiver and can be mapped to the SoC in the same manner as in Figure 15.11.

15.3.1.5.1 Digital Audio Broadcasting

The Digital Audio Broadcasting standard has been in existence since the early 1980s. The standard was initiated to replace the analog FM radio services and has already been adapted by many countries all over the world. DAB is capable of delivering data and audio services to end users. Data services are performed via packet transfers.

Figure 15.12 depicts the basic DAB receiver structure. After frequency and gain correction (AFC/AGC) of the DAB signal, the received signal is OFDM demodulated by means of a fast Fourier transform (FFT) (referred to as inverse OFDM) and differential quadrature phase shift keying (QPSK) demodulation. Then, the FEC decoding is applied by means of deinterleaving and Viterbi decoding. In the last stage of the receiver, the user data are demultiplexed and source decoded using the MP2 codec to audio.

15.3.1.5.2 Digital Radio Mondiale

Digital Radio Mondiale (DRM), approved as an ETSI standard in 2001, is proposed to replace the analog AM radio service for frequencies below 30 MHz. The DRM system delivers near-FM quality sound over short-wave, medium-wave, and long-wave radio channels.

Figure 15.13 shows the basic structure of the DRM receiver. In the first stage of the receiver, the FFT is applied to the received OFDM signal. All separate OFDM carriers are demodulated using the appropriate hierarchical constellation (BPSK, QPSK,

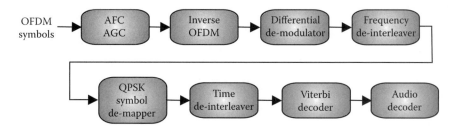

FIGURE 15.12 DAB receiver structure.

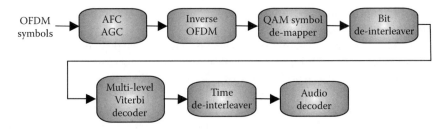

FIGURE 15.13 DRM receiver structure.

16-QAM [quadrature amplitude modulation], or 64-QAM). The multilevel coded data are decoded using an iterative Viterbi decoder approach. The last stage of the receiver comprises demultiplexing of the user data and decoding using the AAC codec to audio.

15.3.1.5.3 Digital Video Broadcasting

Digital Video Broadcasting–Handheld (DVB-H) is a new digital broadcast standard for the transmission of audio and video content to handheld terminals. DVB-H has been based on the standard for digital terrestrial television, Digital Video Broadcasting–Terrestrial (DVB-T). The DVB-H standard is similar to DVB-T, but it considers properties specific for mobile terminals that are portable, small, and battery powered.

In contrast to the digital audio counterparts, the DVB standard defines more robust forward error correction (FEC). DVB has been implemented with cascaded FEC; the convolutional inner coder is cascaded with an outer coder. The FEC outer coder is implemented using Reed-Solomon coding. Reed-Solomon proved to be a very powerful algorithm to solve (burst) errors. The inner decoder step is usually implemented by the Viterbi decoder. The outer decoding step, implemented by the Reed-Solomon decoder, should result in a quasi-error-free output, containing less than one uncorrected erroneous event per hour. The user data have to be decoded by the MPEG-2 codec to video (see Figure 15.14).

In future mobile platforms, one or more of these standards have to be supported. Therefore, reconfigurable multicore SoC architectures are needed to support all these and future OFDM standards.

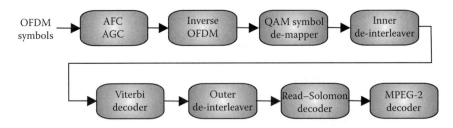

FIGURE 15.14 DVB receiver structure.

15.3.2   Cognitive Radio

In November 2002, the Federal Communications Commission (FCC) in the United States released a report [18] aimed at improving the management of spectrum resources in the United States. The report concluded that the current spectrum scarcity problem is largely due to the strict regulation on spectrum access. Spectrum measurements conducted by the FCC indicated that only small portions of the spectrum are heavily used, while other frequency bands are either partially used or unoccupied most of the time. So, spectrum utilization can be improved by making it possible for an unlicensed user (secondary user) to access the spectrum that is not occupied by the licensed user (primary user). The unlicensed user has the awareness of the spectrum and adapts its transmission accordingly on a noninterference basis. This spectrum access and awareness scheme is referred to as cognitive radio by the FCC.

The idea of cognitive radio was first presented by Joseph Mitola III in his paper [40], where he proposed that cognitive radio can enhance the personal wireless service by a radio knowledge representation language (RKRL). This language represents knowledge of radio at all aspects, from transmission to application scenarios, in such a way that automated reasoning about the needs of the user is supported. Cognitive radio thus is able to autonomously observe and learn the radio environment, generate plans, and even correct mistakes. A comprehensive conceptual architecture of cognitive radio was later presented in [31], where cognitive radio was thought to be a final point of the software-defined radio platform evolution: a fully reconfigurable radio that changes its communication functions depending on network and user demands. Recently, cognitive radio has become a very hot topic due to its impact on future spectrum policy, which could fundamentally change the current status of radio communication. At the Berkeley Wireless Research Center (BWRC), a dedicated Cognitive Radio Research (CRR) project is in progress. Their motivation is to improve the spectrum utilization by opportunistic use of the spectrum, which is the same as the FCC's initiative. Spectrum pooling is investigated in [50]. The basic idea is that a secondary user can dynamically access the licensed band by switching on and off OFDM subcarriers to avoid interference to the licensed user (primary user). The cognitive radio project at Virginia Tech does not specifically aim to improve spectrum utilization. This project is based on the observation that cognitive radio distinguishes itself by awareness and learning. In [10], a genetic algorithm-based cognitive engine is proposed to learn its environment and respond with an optimal adaptation. The European Union 6th Framework End-to-End Reconfigurability (E2R) project studies reconfigurability, software-defined radio, and cognitive radio. The key objective of the E2R project is to devise, develop, and test the architectural design of reconfigurable devices and supporting system functions for users, application and service providers, operators, and regulators in the context of heterogeneous systems. Although the project does not specifically address cognitive radio, dynamic spectrum allocation and evolution from software-defined radio to cognitive radio have been envisioned. In parallel with the ongoing research projects around the world, international standardization organizations also have proposals to improve the spectrum utilization. An example is IEEE 802.22, which is a new standard for a cognitive point-to-point (P2P) air interface for spectrum sharing with television bands.

The AAF project [53] focuses on spectrum awareness and access. The objective is to demonstrate a cognitive radio system to improve current emergency networks, which work in limited frequency bands. Each radio node works in an ad-hoc-based network. The node adopts the AAF protocol stack, which is consistent with the five-layer protocol reference model (physical layer, data link layer, network layer, transport layer, and application layer). In the physical layer, free spectrum is discovered. Whenever free spectrum is found, the AAF system will create an infrastructure using this spectrum. In cognitive radio, the first step is to focus on identification of free resources in the frequency domain by spectrum sensing. An OFDM-based system can approach the Shannon capacity in a segmented spectrum; the capacity to nullify individual carriers poses interesting opportunities for cognition, as was also observed in [50]. Cognitive radio has to operate in multiple bands and under different channel conditions and supports various multimedia services. A heterogeneous reconfigurable platform supports the reconfigurability of the physical layer of cognitive radio.

15.3.2.1 Spectrum Sensing

In order to identify the licensed user and locate unused spectrum, the system has to sense the spectrum. Spectrum sensing is not a new topic since a lot of research has been done in the area of signal detection and estimation. Three signal processing techniques are commonly used for signal detection: matched filtering, energy detection, and cyclostationary feature detection.

Matched filtering is an optimal way for signal detection in communication systems. It correlates the received signal with a known signal pattern that maximizes the received signal-to-noise ratio. However, cognitive radio may not have prior knowledge of the licensed user signal, and strict timing and frequency synchronizations are required for coherent detection. Therefore, matched filtering is not an option for spectrum sensing.

In situations where not much knowledge concerning the signal is available, energy detection [46] is often used to determine the presence of the signal. It measures the signal power within a certain time interval and frequency band. The detection decision is based on a noise threshold. However, limitations for the energy detection are: (1) the decision threshold is subject to changing signal-to-noise ratios; (2) it cannot distinguish interference from signals; and (3) it is not effective for spectrum spreading signals whose power has been spread. Therefore, energy detection is not always adequate.

Cyclostationary feature detection [19] is used to extract signal features in the background of noise. Since the modulated signal can be modeled as a cyclostationary process in which a signal varies in time with certain periodicities, it contains spectral redundancy information that can be exploited by analysis of the cyclic spectrum. The advantages of cyclostationary feature detection over energy detection have been recognized in [20]. Within the AAF project, the focus is on energy detection and cyclostationary feature detection. A system-level architecture for spectrum sensing is presented in Figure 15.15. The detailed theory can be found in [46] and [19]. At system level, energy detection can be implemented as an FFT algorithm, which has a computational complexity of $O(\frac{N}{2}\log_2 N)$, where N is the size of FFT. Cyclostationary feature detection is a combination of an FFT and spectral correlation, which has a computational complexity of $O(N^2 + \frac{N}{2}\log_2 N)$. When a large N is used, the processing of cyclostationary feature

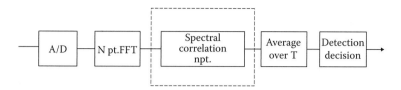

FIGURE 15.15 System-level architecture of spectrum sensing with the cyclostationary feature detection option.

detection can be prohibitive in terms of performance and computational power. It means that building a dedicated cyclostationary feature detector is simply too expensive. Therefore, cyclostationary feature detection is a complementary option when energy detection fails. So, energy detection can be switched to cyclostationary feature detection by turning on the spectral correlation functional module (see the dashed box in Figure 15.15). This option can be supported by a reconfigurable platform where the processing elements for spectral correlation can be switched on and off.

Based on the spectrum sensing result, an individual node needs to make a local decision on whether or not the band under consideration is empty. However, this local decision may not be reliable due to fading and shadowing. Therefore, collaborative sensing is proposed [39] to improve the quality of the licensed user detection.

15.3.2.2 OFDM-Based Baseband

The benefit of OFDM is that the high data rate of the whole system is transformed into relatively low-data-rate streams on each subcarrier that is more robust to intersymbol interference (ISI) caused by multipath delay spread. In the hardware design, an OFDM radio is also easy to integrate with spectrum sensing because both use FFT cores. The system robustness and hardware resource sharing make OFDM a good choice for cognitive radio baseband systems. More importantly, an OFDM system can optimally approach the Shannon capacity in the segmented spectrum by adaptive resource allocation on each subcarrier, which includes adaptive bit loading and adaptive power loading [53]. In an OFDM-based cognitive radio system, information bits are loaded as different modulation types onto each available subcarrier depending on the subcarrier's signal-to-noise ratio, while the subcarriers currently not available to cognitive radio are switched off. Two optimization methods for the adaptive resource allocation can be used for cognitive radio: using the power constraint or using the data-rate constraint.

We could maximize the data rate of the system under a certain power constraint. It is formulated as follows:

$$Max \ R = \sum_{k=1}^{K} \frac{F_k}{K} \log_2 \left(1 + \frac{h_k^2 p_k}{N_0 \frac{B}{K}} \right) \qquad (15.1)$$

$$\text{Subject to: } \sum_{k=1}^{K} p_k \leq P_{total}$$

$$F_k \in \{0,1\} \text{ for all } k$$

$$p_k = 0 \text{ for all } k \text{ which satisfy } F_k = 0,$$

where R is the data rate, K is the number of the subcarriers, N_0 is the noise power density, B is the band of interest for cognitive radio, h_k is the subcarrier gain, and p_k is the power allocated to the corresponding subcarrier. F_k is the factor indicating the availability of subcarrier k to cognitive radio, where $F_k = 1$ means the kth carrier can be used by cognitive radio.

The system power minimization can also be applied under the constraint of a constant data rate. We formulate it as follows:

$$Min \sum_{k=1}^{K} p_k = P_{total}$$

$$\text{Subject to: } R = \sum_{k=1}^{K} \frac{F_k}{K} \log_2 \left(1 + \frac{h_k^2 p_k}{N_0 \frac{B}{K}} \right) \tag{15.2}$$

$$F_k \in \{0,1\} \text{ for all } k$$

$$p_k = 0 \text{ for all } k \text{ which satisfy } F_k = 0.$$

A functional diagram of the system is presented in Figure 15.16. A bit allocation vector indicates how many bits are loaded on each subcarrier. The number of bits corresponds to the different modulation types used for each subcarrier. The bit allocation vector is determined by the spectrum occupancy information from spectrum sensing and the SNR of subchannels. The bit allocation vector is disseminated via a signaling channel so that both transmitter and receiver have the same information. The bit allocation vector does not change frequently, for instance, only after several frames. The basic idea is to load more bits on good subcarriers and zeros on carriers that cause interference to the licensed user or lead to poor transmissions.

15.3.2.3 Reconfigurable Physical Layer

As already foreseen by Mitola [31], a cognitive radio is the final point of software-defined radio platform evolution: a fully reconfigurable radio that changes its communication functions depending on network and user demands. His definition on reconfigurability is very broad, but we focus on the physical layer reconfigurability. We will discuss some possibilities for reconfiguration.

The proposed architecture for spectrum sensing is shown in Figure 15.15, where cyclostationary feature detection is used as a complementary option. This option has to be supported by a reconfigurable platform that can efficiently perform the spectrum correlation. In [35], a two-step methodology to analyze the mapping of cyclostationary feature detection (CFD) onto a Montium-based multicore processing platform is proposed. In the first step, the tasks to be executed by each core are determined in a

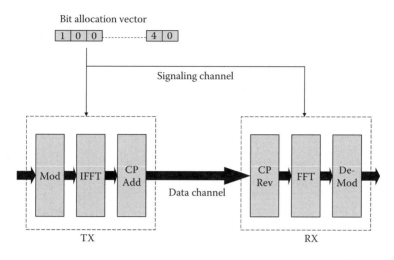

Bit allocation vector

FIGURE 15.16 OFDM for cognitive radio.

structured way using techniques known from the design of array processors. In the second step, the implementation of the tasks onto the Montium is analyzed. It is shown that calculating a 127×127 discrete spectral correlation function requires approximately 140 μs on a system-on-chip (SoC) with 4 Montium cores. For energy detection, spectrum sensing can be done in different frequency resolutions with which the trade-offs between performance and computational power can be made. This can be achieved by a size-reconfigurable FFT on the Montium.

OFDM baseband processing for cognitive radio is a parameterizable OFDM processing chain that is configured by a configuration manager (see Figure 15.17). By applying different parameter settings in each task, the OFDM system is adaptive to various channel conditions and provides various data rates.

Relevant parameters are shown in Table 15.9, but are not limited by this table and can be extended to add more flexibility to the system. The number of OFDM symbols per frame (N_{sym}) is limited by the channel coherence time, the time during which the channel characteristics are constant. The number of guard samples (N_g) is chosen to deal with different channel delay spreads. Generally not all data are used to carry user information. A part of the data (e.g., pilots) is used to guarantee reliable transmissions. Different pilots are used for different purposes, such as channel estimation or phase offset estimation. They can be placed in the preamble section prior to each frame or embedded in the

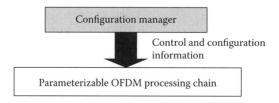

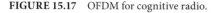

FIGURE 15.17 OFDM for cognitive radio.

TABLE 15.9   The Parameter Set for the
Parameterizable OFDM

B	Bandwidth of OFDM system
N_{sym}	Number of OFDM symbols per frame
$N_{preamble}$	Preamble length per frame
N	Number of OFDM samples per symbol
N_g	Number of guard samples per symbol
N_{data}	Number of useful data per symbol
mod	Modulation modes
tab_{pilot}	Table for pilot information
tab_{format}	Table for format information

OFDM symbol. Modulation modes (mod) indicate the modulation type for the OFDM samples that carry useful information. The modulation mode in one OFDM symbol can differ on a subcarrier basis. In the cognitive radio case, the modulation mode can be set to zero to nullify carriers.

15.4   Mapping Applications to Multicore Architectures

In the most general terms, spatial mapping is the allocation of spatial resources to applications. As such, spatial mapping is to space what scheduling is to time. In the context of a multiprocessor system-on-chip (MP-SoC), the spatial resources to allocate are tiles and interconnections between tiles. These tiles may be processors or memory (controllers). Although the allocation of columns of fabric inside an FPGA could also be described as a spatial mapping problem, the system-on-chip context is considerably more coarse grained.

Baseband processing applications in wireless communications are streaming applications. Such streaming applications can be decomposed into multiple communicating computational kernels. Most streaming applications have intuitive decompositions into chains of such kernels, which is illustrated by the fact that most standards contain block diagrams of the components that make up the entire application [e.g., 17]. This decomposition can be as general as a Kahn process network (KPN) [32], or if more detailed information is available, some specialized form of a data flow graph (DFG).

A commonly used specialized DFG is the synchronous data flow graph (SDFG) [36]. The nodes of an SDFG are generally referred to as actors. The edges between actors are annotated with tokens. Tokens are an abstract notion of data that is communicated between two actors.

Actors are annotated for every incoming and outgoing edge. On incoming edges, the annotation is the number of tokens required on that specific input to fire the actor once, i.e., to execute the corresponding computational kernel. On outgoing edges, the annotation is the number of tokens the actor will produce after firing. Actors in an SDFG are self-timed, i.e., there is no global arbitration of what actor fires when, but rather a simple local firing rule: an actor may fire when all incoming edges have at least as many tokens on them as annotated.

Decomposing an application into actors and their (data) interdependence allows for more fine-grained resource management. Now, actors are the atomic units for which tiles need to be allocated. Whenever two different tiles are allocated to two interdependent actors, the interactor communication requires services from the interconnect between the tiles of the MP-SoC. In other words, communication needs to be routed through the network-on-chip (NoC).

If a tile supports multitasking, it can be allocated to multiple actors. In order to guarantee nonfunctional properties of the application (e.g., throughput and latency), all multitasking tiles must have deterministic schedulers. More formally, a tile's scheduler must be a latency-rate server [44]. This guarantees an upper bound on the latency between the first request for service and the first service response and a lower bound on the throughput (rate) from that point onward, provided it is kept busy. This restriction on multitasking tiles makes it possible to make allocations to actors independent of each other (composability), i.e., when a tile is already allocated to an actor, allocating it to another actor does not influence the nonfunctional properties of the first (if the sum of the actors' rate requirements do not exceed the rate of the scheduler).

To be able to perform the mapping of an application to tiles, a spatial mapping algorithm needs:

- A model of the application
- A model of the (MP-SoC) platform
- The constraints on the nonfunctional properties of the application
- Data on the resource requirements of process implementations (e.g., time, memory, and energy)

The constraints on the nonfunctional properties of the application can only be checked after it has been mapped. Thus, when a spatial mapping has been determined and latencies and throughputs of processes running on tiles are known, the constraints can be checked. A spatial mapping that lets the application meet its constraints on its nonfunctional properties is considered feasible.

15.4.1   Design Time Mapping

The spatial mapping of applications can be performed at design time, if all applications are known and the availability of the MP-SoC is guaranteed (what happens when this is not the case is described in section 15.4.2). This is a (quite often manual) optimization problem, in which the best allocation of tiles to actors should be found. A valid solution must work for all possible user scenarios. When one application is already running on an MP-SoC and a user starts another, then—assuming the combination of applications is allowed—the new application should be runnable with the remaining resources.

For sufficiently small applications and MP-SoCs, exhaustive search techniques can be used to find optimal allocations for all possible scenarios. The mapping of slightly larger applications is generally tackled by heuristic search methods that try to direct the search for a solution, possibly missing the optimum, but yielding an acceptable result.

Very large applications or failed heuristic searches are often tackled with manual help from a designer.

There is no general method for spatial mapping to all kinds of MP-SoC, because what constitutes a good mapping depends heavily on the available resources of the MP-SoC and the resource demands of the application. For example, when the number of processors per NoC node (multiple processors can be connected to a single router) is relatively high, communication is likely to become a problem in the mapping. In such a case, searches generally start off with determining the minimum bandwidth clustering of the process graph into clusters that fit on a tile.

When using SDFGs, a great advantage of performing a spatial mapping at design time is that a so-called static-order schedule can be calculated for actors mapped onto the same tile. This means that the order in which the actors fire is known and the actors are ordered such that they fire in sequence. If a static-order schedule is found, the schedule can be implemented by grouping the actors together and compiling them into a single piece of binary code for the tile, thereby eliminating the need for scheduling at runtime and the overhead it would induce.

It is worth noting that tiles of architectures that do not have support for multitasking can still be allocated to multiple actors, if such a static-order schedule can be found. This follows directly from the fact that actors in a static-order schedule can be compiled into a single binary executable.

15.4.2   Runtime Mapping

Performing the spatial mapping at runtime is necessary, whenever the application set is not known completely at design time. This happens for a wide variety of reasons; for example, when the platform allows the user to use software from any vendor, developed for that platform. When different software vendors produce software for the same platform independently, no one knows the complete application set. Also, when the resource availability of the MP-SoC is not known beforehand, the spatial mapping must be found at runtime. This can happen when tiles can be broken (by wear or by faults in production), or when applications are developed, not for a specific MP-SoC, but for a class of MP-SoCs that have the same types of tiles and the same type of interconnect, but in a different configuration. Runtime spatial mapping in this sense is inherently ad hoc, i.e., only when an application is started will the system know what resources are available.

When mapping to a heterogeneous MP-SoC, having multiple implementations of computational kernels for different types of tiles increases the options for the spatial map. Performance figures—(expected) worst-case execution time, energy consumption, etc.—of these implementations can be determined at design time. However, some figures of the applications simply cannot be determined at design time. Interprocess communication parameters (e.g., estimated latency), for example, need to be determined at runtime, as these are dependent on the specific mapping of the communicating processes. Likewise, it is only known at runtime on which tile a process will be executed and which processes are already running on this tile, so the actual execution time (often referred to as response time in the synchronous data flow world) of a process is only known at runtime.

15.5   Conclusion

Future wireless communication devices will contain integrated circuits where multiple processing cores are implemented on one chip. These multiprocessor systems-on-chip (MP-SoC) will be heterogeneous; they will consist of multiple types of processing cores. Bit-level functions will be executed by FPGA-type processors, word-level functions by coarse-grained reconfigurable processors, and control-oriented functions by GPPs. In wireless communication systems, processing is dominated by word-level signal processing that is efficiently implemented on coarse-grained reconfigurable processors. The Montium coarse-grained reconfigurable processor is an example of such a processor, enabling energy-efficient implementation of wireless communication receivers. This is illustrated for three classes of receiver baseband processing: UMTS Rake receivers, OFDM baseband processing (HiperLAN/2), and digital broadcasting receivers (DAB, DRM, and DVB). Furthermore, an MP-SoC, containing multiple Montiums, is a suitable platform for the implementation of cognitive radio (CR). The flexibility offered by coarse-grained reconfigurable processors supports the efficient implementation of new communication paradigms.

Because of the large number of transistors envisaged on future integrated circuits, a processing platform will not be fixed during its lifetime due to faults in production or wear. Furthermore, the applications are not fixed. Quality-of-service requirements and environmental conditions change, and cognitive radio has changing requirements by definition. Consequently, applications need to be mapped onto the processing platform at runtime.

Realizing that wireless communication applications are streaming applications explains the use of data flow graphs to model these applications. Data flow graphs naturally fit platforms with multiple processing cores where nodes of the graph are mapped onto the processing cores and edges are mapped onto the interconnecting network-on-chip. The use of data flow graphs supports the runtime mapping.

References

[1] 3rd Generation Partnership Project. 2004. Base station (BS) radio transmission and reception (FDD). 3GPP TS 25.104 v6.4.0.

[2] 3rd Generation Partnership Project. 2004. Physical layer—General description. 3GPP TS 25.201 v6.0.

[3] A. Abnous, H. Zhang, M. Wan, G. Varghese, V. Prabhu, and J. Rabaey. 2002. The Pleiades architecture. In *The application of programmable DSPs in mobile communications*, ed. A. Gatherer and A. Auslander, 327–60. New York: Wiley.

[4] V. Baumgarte, G. Ehlers, F. May, A. Nuckel, M. Vorbach, and M. Weinhardt. 2003. Pact xpp—A self-reconfigurable data processing architecture. *J. Supercomputing* 260:167–84.

[5] L. Bissi, P. Placidi, G. Baruffa, and A. Scorzoni. 2006. A multi-standard reconfigurable Viterbi decoder using embedded FPGA blocks. In *Proceedings of the 9th Euromicro Conference on Digital Systems Design: Architectures Methods and Tools* (DSD2006), pp. 146–54.

[6] B. Mei, S. Vernalde, D. Verkest, H. D. Man, and R. Lauwereins. 2003. Adres: An architecture with tightly coupled VLIW processor and coarse-grained reconfigurable matrix. In *Field-Programmable Logic and Applications*, pp. 61–70.

[7] J. Chandran, R. Kaluri, J. Singh, V. Owall, and R. Veljanovski. 2004. Xilinx Virtex ii pro implementation of a reconfigurable UMTS digital channel filter. In *Proceedings of the Second IEEE International Workshop on Electronic Design, Test and Applications*, pp. 77–82.

[8] C. Chang, J. Wawrzynek, and R. W. Brodersen. 2005. Bee2: A high-end reconfigurable computing system. *IEEE Des. Test. Comput.* 220:114–25.

[9] R. W. Chang. 1966. Synthesis of band-limited orthogonal signals for multi-channel data transmission. *Bell Syst. Tech. J.* 46:1775–96.

[10] C. J. Rieser. 2004. *Biologically inspired cognitive radio engine model utilizing distributed genetic algorithms for secure and robust wireless communications and networking.* PhD thesis, Virginia Tech.

[11] K. Compton and S. Hauck. 2002. Reconfigurable computing: A survey of systems and software. *ACM Comput. Surv.* 340:171–210.

[12] European Telecommunication Standards Institute (ETSI). 2000. Broadband radio access networks (BRAN); HiperLAN type 2; Data link control (DLC) layer. Part 1. Basic data transport functions. ETSI TS 101 761-1 v1.1.1.

[13] European Telecommunication Standards Institute (ETSI). 2001. Broadband radio access networks (BRAN); HiperLAN type 2; Physical (PHY) layer. ETSI TS 101 475 v1.2.2.

[14] European Telecommunication Standards Institute (ETSI). 2001. Digital Radio Mondiale (DRM); System specification. ETSI TS 101 980 v1.1.1.

[15] European Telecommunication Standards Institute (ETSI). 2001. Radio broadcasting systems; Digital Audio Broadcasting (DAB) to mobile, portable and fixed receivers. ETSI EN 300 401 v1.3.3.

[16] European Telecommunication Standards Institute (ETSI). 2004. Digital Video Broadcasting (DVB); Framing structure, channel coding and modulation for digital terrestrial television. ETSI EN 300 744 v1.5.1.

[17] European Telecommunication Standards Institute (ETSI). 2003. Digital Radio Mondial (DRM); System specification. ETSI ES 201 980 v1.2.2.

[18] Federal Communication Commision. 2002. Spectrum policy task force. Technical report 02-135.

[19] W. Gardner. 1988. Signal interception: A unifying theorectical framework for feature detection. *IEEE Trans. Commun.* 36:897–906.

[20] W. Gardner and C. M. Spooner. 1992. Signal interception: Performance advantages of cyclic-feature detectors. *IEEE Trans. Commun.* 40:149-159.

[21] S. C. Goldstein, H. Schmit, M. Budiu, S. Cadambi, M. Moe, and R. R. Taylor. 2000. Piperench: A reconfigurable architecture and compiler. *IEEE Comput.* 33:70–77.

[22] K. Goossens, J. van Meerbergen, A. Peeters, and P. Wielage. 2002. Networks on silicon: Combining best-effort and guaranteed services. In *Proceedings of the Design, Automation and Test in Europe Conference and Exhibition (DATE)*, pp. 423–25.

[23] K. Goossens, J. Dielissen, and A. Rădulescu. 2005. The Æthereal network on chip: Concepts, architectures, and implementations. *IEEE Des. Test Comput.* 220:414–21.

[24] S. Hauck, T. W. Fry, M. M. Hosler, and J. P. Kao. 2004. The Chimaera reconfigurable functional unit. *IEEE Trans. VLSI Syst.* 120:206–17.

[25] G. Heidari and K. Lane. 2001. Introducing a paradigm shift in the design and implementation of wireless devices. In *Proceedings of Wireless Personal Multimedia Communications*, Aalborg, Denmark, pp. 225–30.

[26] P. M. Heysters. 2004. *Coarse-grained reconfigurable processors—Flexibility meets efficiency.* PhD thesis, University of Twente, Enschede, The Netherlands.

[27] P. M. Heysters, G. K. Rauwerda, and G. J. M. Smit. 2004. Implementation of a HiperLAN/2 receiver on the reconfigurable Montium architecture. In *Proceedings of the 11th Reconfigurable Architectures Workshop* (RAW 2004), Santa Fé, NM, p. 1476.

[28] P. M. Heysters, G. J. M. Smit, and E. Molenkamp. 2003. A flexible and energy-efficient coarse-grained reconfigurable architecture for mobile systems. *J. Supercomputing* 260:283–308.

[29] H. Peter Hofstee. Power efficient processor architecture and the cell processor. In *Proceedings of the 11th International Symposium on High-Performance Computer Architecture*, pp. 258–262.

[30] H. Holma and A. Toskala. 2001. *WCDMA for UMTS: Radio access for third generation mobile communications.* New York: John Wiley & Sons.

[31] J. Mitola III. 2000. *Cognitive radio: An integrated agent architecture for software defined radio.* PhD thesis, Royal Institute of Technology, Sweden.

[32] G. Kahn. 1974. The semantics of simple language for parallel programming. In *IFIP Congress*, pp. 471–75.

[33] N. K. Kavaldjiev. 2007. *A run-time reconfigurable network-on-chip for streaming DSP applications.* PhD thesis, University of Twente, Enschede, The Netherlands.

[34] I. O. Kennedy and F. J. Mullany. 2004. Design of a reconfigurable UMTS channel processing engine. In *Proceedings of the 59th IEEE Vehicular Technology Conference (VTC 2004)*, vol. 3, pp. 1300–4.

[35] A. B. J. Kokkeler, G. J. M. Smit, and T. Krol. 2007. Cyclostationary feature detection on a tiled-SoC. In *DATE2007 Proceedings*, pp. 171–176.

[36] E. A. Lee and D. G. Messerschmitt. 1987. Synchronous data flow: Describing signal processing algorithm for parallel computation. In *COMPCON*, pp. 310–15.

[37] K. Masselos and N. S. Voros. 2007. Implementation of wireless communications systems on FPGA-based platforms. *EURASIP J. Embedded Syst.*, article ID 12192, doi:10.1155/2007/12192.

[38] M. Millberg, E. Nilsson, R. Thid, and A. Jantsch. 2004. Guaranteed bandwidth using looped containers in temporally disjoint networks within the Nostrum network on chip. In *Proceedings of Design Automation and Test Europe Conference*, p. 20890.

[39] S. M. Mishra, A. Sahai, and R. Brodersen. 2006. Cooperative sensing among cognitive radios. In *Proceedings of IEEE ICC*, pp. 1658–1663.

[40] J. Mitola. 1999. Cognitive radio: Making software radios more personal. *IEEE Pers. Commun.* 6:13–18.

[41] J. Potman, F. Hoeksema, and K. Slump. 2003. Tradeoffs between spreading factor, symbol constellation size and Rake fingers in UMTS. In *Proceedings of ProRISC 2003*, Veldhoven, The Netherlands, pp. 543–48.

[42] J. Rexford and K. G. Shin. 1994. Support for multiple classes of traffic in multi-computer routers. In *Proceedings of the First International Workshop on Parallel Computer Routing and Communication*, pp. 116–30.

[43] A. La Rosa, L. Lavagno, and C. Passerone. 2005. Implementation of a UMTS turbo decoder on a dynamically reconfigurable platform. *IEEE Trans. Computer-Aided Des.* 240:100–6.

[44] D. Stiliadis and A. Varma. 1998. Latency-rate servers: A general model for analysis of traffic scheduling algorithms. *IEEE/ACM Trans. Netw.* 60:611–24.

[45] T. J. Todman, G. A. Constantinides, S. J. E. Wilton, O. Mencer, W. Luk, and P. Y. K. Cheung. 2005. Reconfigurable computing: Architectures and design methods. *IEE Proc. Comput. Digital Tech.* 1520:193–207.

[46] H. Urkowitz. 1967. Energy detection of unknown deterministic signals. *Proc. IEEE*, 55(4):523–531.

[47] S. Vangal, J. Howard, G. Ruhl, S. Dighe, H. Wilson, J. Tschanz, D. Finan, P. Iyer, A. Sing, T. Jacob, S. Jain, S. Venkataraman, Y. Hoskote, and N. Borkar. 2007. An 80-tile 1.28 tflops network-on-chip in 65 nm cmos. In *Proceedings of the ISSCC*, pp. 98–99.

[48] J. von Neumann. 1993. First draft of a report on the edvac. *IEEE Ann. Hist. Comput.* 15:11–21.

[49] E. Waingold, M. Taylor, V. Sarkar, W. Lee, V. Lee, J. Kim, M. Frank, P. Finch, S. Devabhaktuni, R. Barua, J. Babb, S. Amarasinghe, and A. Agarwal. 1997. Baring it all to software: Raw machines. *Computer.* 30:86–93.

[50] T. A. Weiss and F. K. Jondral. 2004. Spectrum pooling: An innovative strategy for the enhancement of spectrum efficiency. *IEEE Commun. Mag.* 24:S8–14.

[51] P. T. Wolkotte, G. J. M. Smit, G. K. Rauwerda, and L. T. Smit. 2005. An energy-efficient reconfigurable circuit switched network-on-chip. In *Proceedings of the 19th IEEE International Parallel and Distributed Processing Symposium (IPDPS'05)—12th Reconfigurable Architecture Workshop (RAW 2005)*, Los Alamitos, CA, p. 155.

[52] H. Zhang, M. Wan, V. George, and J. Rabaey. 1999. Interconnect architecture exploration for low-energy reconfigurable single-chip dsps. In *WVLSI'99: Proceedings of the IEEE Computer Society Workshop on VLSI'99*, Washington, DC, pp. 2–8.

[53] Q. Zhang, A. B. J. Kokkeler, and G. J. M. Smit. 2006. A reconfigurable radio architecture for cognitive radio in emergency networks. In *Proceedings of European Conference on Wireless Technology*, pp. 35–38.

[54] L. Zhigang and G. Luoning. 2004. Design and implementation of umts-fdd/hsdpa baseband. In *Proceedings ICSP '04*, vol. 3, pp. 1857–60.

Index